Progress in Nonlinear Differential Equations and Their Applications
Volume 32

Geometrical Optics and Related Topics

Ferruccio Colombini
Nicolas Lerner
Editors

Springer Science+Business Media, LLC

Ferruccio Colombini
Dipatimento di Matematica
Università di Pisa
Pisa, Italy

Nicolas Lerner
IRMAR
Université de Rennes 1
35042 Rennes, France

Library of Congress Cataloging-in-Publication Data

Geometrical optics and related topics / Ferruccio Colombini , Nicolas Lerner, editors.
p. cm. -- (Progress in nonlinear differential equations and their applications ; v. 32)
Includes bibliographical references and index.
ISBN 978-0-8176-3958-7 ISBN 978-1-4612-2014-5 (eBook)
DOI 10.1007/978-1-4612-2014-5

1. Geometrical optics--Mathematics. 2. Nonlinear wave equations.
I. Colombini , F. (Ferruccio) II. Lerner , Nicolas , 1953 -
III. Series .
QC383 . G46 1997
535--dc21 97-1244
CIP

AMS Classifications: 35L67, 35L70, 35L70, 35P25, 35L65, 35L10, 46E35, 35S05, 35L65, 35L10, 35S05, 35L65, 35L10, 35S05, 35A07, 35A22, 78A45, 35L70, 78A05, 47A10, 35H05, 35B65, 58G20

Printed on acid-free paper

ISBN 978-0-8176-3958-7

Reformatted from contributors' disks by Texniques, Inc., Boston, MA.

9 8 7 6 5 4 3 2 1

Contents

Foreword

This book contains fourteen research papers which are expanded versions of conferences given at a meeting held in September 1996 in Cortona, Italy. The topics include blowup questions for quasilinear equations in two dimensions, time decay of waves in L^p, uniqueness results for systems of conservation laws in one dimension, concentration effects for critical nonlinear wave equations, diffraction of nonlinear waves, propagation of singularities in scattering theory, caustics for semi-linear oscillations. Other topics linked to microlocal analysis are Sobolev embedding theorems in Weyl-Hörmander calculus, local solvability for pseudodifferential equations, hypoellipticity for highly degenerate operators. The book also contains a result on uniqueness for the Cauchy problem under partial analyticity assumptions and an article on the regularity of solutions for characteristic initial-boundary value problems.

On each topic listed above, one will find new results as well as a description of the state of the art. Various methods related to nonlinear geometrical optics are a transversal theme of several articles. Pseudodifferential techniques are used to tackle classical PDE problems like Cauchy uniqueness.

We are pleased to thank the speakers for their contributions to the meeting: Serge Alinhac, Mike Beals, Alberto Bressan, Jean-Yves Chemin, Christophe Cheverry, Daniele Del Santo, Nils Dencker, Patrick Gérard, Lars Hörmander, John Hunter, Richard Melrose, Guy Métivier, Yoshinori Morimoto, and Tatsuo Nishitani. The meeting was made possible in part by the financial support of a European commission program, "Human capital and mobility CHRX-CT94-044". The organizers also wish to thank the Scuola Normale Superiore for its hospitality in the magnificent Palazzone.

October 31, 1996

Ferruccio Colombini
Nicolas Lerner

Blowup of small data solutions for a class of quasilinear wave equations in two dimensions: an outline of the proof

Serge Alinhac

0. Introduction

We consider here quasilinear wave equations in $\mathbf{R}^{2+1}$

$$\partial_t^2 u - \Delta_x u + \sum_{0\leq i,j\leq 2} g_{ij}(\nabla u)\, \partial_{ij}^2 u = 0, \tag{0.1}$$

where

$$x_0 = t,\ x = (x_1, x_2),\ g_{ij} = g_{ji},\ g_{ij}(0) = 0.$$

We assume that the Cauchy data are C^∞ and small

$$u(x,0) = \epsilon u_1^0(x) + \epsilon^2 u_2^0(x) + \ldots,\ \partial_t u(x,0) = \epsilon u_1^1(x) + \epsilon^2 u_2^1(x) + \ldots,$$

and supported in a fix ball of radius M. Later on, we will need the function

$$g(\omega) = \sum g_{ij}^k \omega_i \omega_j \omega_k,$$

where

$$r = \sqrt{x_1^2 + x_2^2},\ x_1 = r\cos\omega,\ x_2 = r\sin\omega$$

are the usual polar coordinates in space, and

$$\omega_0 = -1,\ \omega_1 = \cos\omega,\ \omega_2 = \sin\omega,\ g_{ij}(\eta) = \sum g_{ij}^k \eta_k + O(\eta^2).$$

Our aim is to study the existence of smooth solutions to this problem, more precisely the lifespan $\tilde{T}_\epsilon$ of these solutions and, if any, the breakdown mechanism when these solutions stop being smooth.

In space dimensions two or three, this problem was introduced and extensively studied by John (see his survey paper [10] and the references therein), then by Klainerman [12], [13], Hörmander [8], [9] and many authors. Using some crude approximation by solutions of Burgers' equation, Hörmander [8] has obtained in dimensions two and three explicit lower bounds for the lifespan. The result in dimension two is

$$\liminf \epsilon \tilde{T}_\epsilon^{1/2} \geq (\max g(\omega)\partial_\sigma^2 R^{(1)}(\sigma,\omega))^{-1}. \tag{0.2}$$

Here, the "first profile" $R^{(1)}$ is defined as

$$R^{(1)}(\sigma,\omega) = \frac{1}{2\sqrt{2\pi}} \int_{s\geq\sigma} \frac{1}{\sqrt{s-\sigma}} \{R(s,\omega,u_1^1) - \partial_s R(s,\omega,u_1^0)\} ds, \tag{0.3}$$

where $R(s,\omega,v)$ denotes the Radon transform of the function v

$$R(s,\omega,v) = \int_{x\omega=s} v(x)dx.$$

Very soberly, Hörmander writes in his 1986 Lectures on nonlinear hyperbolic equations [9]: "Even if it is hard to doubt that (0.2) always gives the precise asymptotic lifespan of the solutions there is no proof except that of John [11] for the rotationally symmetric three dimensional case."

As far as we know, the situation today is still the same.[1] Our aim in this conference is to present a subclass of (0.1) for which we can prove Hörmander's conjecture that (0.2) gives indeed the correct asymptotic of the lifespan. We will only outline the rather lengthy proof, explaining the various steps.

1. Results and method of proof

1.1 We are considering here equation (0.1) with small data. The first free profile, defined by (0.3), gives a first order approximation to the true solution close to the boundary of the light cone. More precisely, let u_1 be the solution of the linearized problem at 0

$$\partial_t^2 u_1 - \Delta u_1 = 0,\ u_1(x,0) = u_1^0(x),\ \partial_t u_1(x,0) = u_1^1(x). \tag{1.1.1}$$

[1] Added in proof: the general case has been proven in [5].

We have, for $r \to \infty,\ r - t \geq -C_0$,

$$u_1 \sim \frac{R^{(1)}(r-t,\omega)}{r^{1/2}}. \tag{1.1.2}$$

Let us now define u_2 by

$$\begin{aligned} &\partial_t^2 u_2 - \Delta u_2 + \sum g_{ij}^k \partial_k u_1 \partial_{ij}^2 u_1 = 0, \\ &u_2(x,0) = u_2^0(x),\ \partial_t u_2(x,0) = u_2^1(x). \end{aligned} \tag{1.1.3}$$

We prove in [1] that, also for $r \to \infty,\ r - t \geq -C_0$,

$$u_2 - \frac{g(\omega)}{2}(\partial_\sigma R^{(1)}) \sim \frac{R^{(2)}(r-t,\omega)}{r^{1/2}}, \tag{1.1.4}$$

for a certain smooth $R^{(2)}$ that we call the "second free profile". Throughout this paper, we make the nondegeneracy assumption

(ND) The function $-g(\omega)\partial_\sigma^2 R^{(1)}(\sigma,\omega)$ has a unique negative quadratic minimum at a point (σ_0, ω_0).

We then set

$$\bar{\tau}_0 = (g(\omega_0)\partial_\sigma^2 R^{(1)}(\sigma_0,\omega_0))^{-1}, \tag{1.1.5}$$

$$\bar{\tau}_1 = -\bar{\tau}_0^2 g(\omega_0)\partial_\sigma^2 R^{(2)}(\sigma_0,\omega_0). \tag{1.1.6}$$

We recall the general result of [2] which is only of asymptotic nature.

Asymptotic theorem. *Under the nondegeneracy assumption* **(ND)**, *there exists a function $\bar{T}_\epsilon^a$ with the following properties:*
(i) For all N, $\bar{T}_\epsilon \geq \bar{T}_\epsilon^a - \epsilon^N$, for $0 < \epsilon \leq \epsilon_N$,
(ii) For some $C > 0$ and $(C\epsilon^2)^{-1} \leq t \leq \bar{T}_\epsilon^a - \epsilon^N$,

$$\frac{1}{C}\frac{1}{\bar{T}_\epsilon^a - t} \leq |\nabla^2 u(.,t)|_{L^\infty} \leq C\frac{1}{\bar{T}_\epsilon^a - t}.$$

The function $\bar{T}_\epsilon^a$ is of the form $\bar{T}_\epsilon^a = \epsilon^{-2}(\bar{\tau}_\epsilon^a)^2(\epsilon, \epsilon^2 \ln \epsilon)$, where $\bar{\tau}_\epsilon^a$ is a smooth function satisfying $\bar{\tau}_\epsilon^a = \bar{\tau}_0 + \epsilon\bar{\tau}_1 + O(\epsilon^2 \ln \epsilon)$.

Thus, for numerical purposes, the *asymptotic lifespan* $\bar{T}^a_\epsilon$ looks like the true lifespan $\bar{T}_\epsilon$: this feature would certainly make numerical experiments, designed to test whether or not the solution actually blows up at time $\bar{T}^a_\epsilon$, very hard to realize. We turn now to the special class of equations (0.2) of the form

$$\partial_t^2 u - \Delta u + L_1 u L_2 L_1 u = 0, \tag{1.1.7}$$

where L_1 and L_2 are two constant coefficients timelike vector fields

$$L_i = l_i^0 \partial_t + l_i^1 \partial_1 + l_i^2 \partial_2.$$

Remark that this class contains classical cases like the rotationally invariant wave equation

$$\partial_t^2 u - c^2(\partial_t u)\Delta u = 0, \ c'(0) \neq 0.$$

Of course, in this last case, we consider *non*rotationally invariant data. For this class we can prove the following theorem.

Blowup Theorem 1.1. *The lifespan $\bar{T}_\epsilon$ of the small data solution u of (1.1.7) satisfies*

$$\bar{\tau}_\epsilon \equiv \epsilon(\bar{T}_\epsilon)^{1/2} = \bar{\tau}_0 + \epsilon\bar{\tau}_1 + O(\epsilon^2 log\epsilon). \tag{1.1.8}$$

Moreover, there is a strip

$$-C_0 \leq r - t \leq M, \tau_0 \leq \epsilon t^{1/2} \leq \bar{\tau}_\epsilon, 0 < \tau_0 < \bar{\tau}_0 \tag{1.1.9}$$

and a point $\tilde{M}_\epsilon = (\tilde{m}_\epsilon, \bar{T}_\epsilon)$ in this strip, closed to a fixed point $\tilde{M}_0$, such that, in this strip,
(i) the solution u is of class C^1,
(ii) the solution u is of class C^2 away from $\tilde{M}_\epsilon$, and satisfies

$$|\nabla^2 u(.,t)|_{L^\infty} \leq \frac{C}{\bar{T}_\epsilon - t}, \tag{1.1.10}$$

$$|\partial_t L_1 u(.,t)|_{L^\infty} \geq \frac{1}{C}\frac{1}{\bar{T}_\epsilon - t}. \tag{1.1.11}$$

We give here only the approximation (1.1.8) for simplicity. In fact, it can be easily seen that the lifespan $\bar{T}_\epsilon$ and the location of the point $\tilde{M}_\epsilon$ can be computed to any order (for small enough ϵ) by the implicit function arguments of [2]. In particular, $\bar{T}_\epsilon \sim \bar{T}^a_\epsilon$ in the sense of asymptotic series. The inequalities (1.1.10), (1.1.11) give a rough idea of how the second order derivatives of the solution blow up. In fact, a much better description of the solution close to point $\tilde{M}_\epsilon$ can be obtained from the following theorem.

Geometric blowup Theorem 1.2. *Let $0 < \tau_0 < \bar{\tau}_0$. There exists a domain*

$$D = \{(s, \omega, \tau), -A_0 \leq s, \omega \in S^1, \tau_0 \leq \tau \leq \bar{\tau}_\epsilon\},$$

a point $M_\epsilon = (m_\epsilon, \bar{\tau}_\epsilon)$ and functions ϕ, $v \in C^3(D)$ with the following properties:
(i) ϕ satisfies

$$\textbf{(H)} \qquad \partial_s \phi \geq 0, \partial_s \phi(s, \omega, \tau) = 0 \Leftrightarrow (s, \omega, \tau) = M_\epsilon,$$

and

$$\partial_{s\tau}\phi(M_\epsilon) < 0, \nabla_{s,\omega}(\partial_s\phi)(M_\epsilon) = 0, \nabla^2_{s,\omega}(\partial_s\phi)(M_\epsilon) >> 0.$$

(ii) $\phi(s, \omega, \tau_0) = s, s \geq M \Rightarrow \phi \equiv s$.
(iii) $\partial_s v(M_\epsilon) \neq 0, s \geq M \Rightarrow v \equiv 0$. Introducing the mapping

$$\Phi(s, \omega, \tau) = (\sigma = \phi(s, \omega, \tau), \omega, \tau),$$

one sees that properties (i) and (ii) of ϕ allow one to define on

$$\Phi(D) = \{(\sigma, \omega, \tau), \phi(-A_0, \omega, \tau) \leq \sigma, \omega \in S^1, \tau_0 \leq \tau \leq \bar{\tau}_\epsilon\}$$

a function F by

$$F(\Phi) = v. \tag{1.1.12}$$

Then the solution u, supported in $\{r - t \leq M\}$, is determined in the strip (1.1.9) by

$$L_1 u = \frac{\epsilon}{r^{1/2}} F(r - t, \omega, \epsilon t^{1/2}). \tag{1.1.13}$$

Finally, the functions ϕ and v are of class C^k for $\epsilon \leq \epsilon_k$.

In this theorem, we see that the singularities of u in the strip (1.1.9) come only from the singularities of F; these in turn arise from the fact that the mapping Φ is not invertible at the point M_ϵ. More precisely, condition (**H**) implies that the singularity of Φ is a cusp singularity. Thus, describing the behavior of the derivatives of u near M_ϵ is just a local geometric problem. This is the reason for calling this behavior of u "geometric blowup" (see [3] or [6] for details).

Note that here we describe the behavior of u only in the strip (1.1.9). This is enough to prove an upper bound of the lifespan. We conjecture that in fact $\tilde{M}_\epsilon$ is the only blowup point of u for $t \leq \bar{T}_\epsilon$. We explain now the four main steps of the proof, each subsequent section giving some more detailed explanations on each step.

Step 1: Asymptotic analysis, normalization of variables and reduction to a local problem. We fix $\tau_0 < \bar{\tau}_0$, and discuss the behavior of u *close to the light cone* for $\epsilon t^{1/2}$ close to τ_0. Thus, we are still far away from a possible blowup at this stage. It turns out that the solution can be described in terms of the new variables

$$\sigma = r - t,\ \omega,\ \tau = \epsilon t^{1/2}.$$

The problem is now of solving from τ_0 to (unknown) $\bar{\tau}_\epsilon$, a local problem.

Step 2: Blowup of the problem. In this step, we introduce a singular change of variables Φ, which takes the singular solution F to a smooth solution v according to (1.1.12). Of course, this change of variables will be determined together with the solution, thus becoming an additional unknown. This step is essential, because it turns the search of the *singular* solution of the problem of Step 1 into the search of a *smooth* solution of a new (local) problem.

Step 3: Existence and tame estimates for a linear Goursat problem. The local problem of Step 2 can be reduced to a problem in a fixed domain. Linearization of this problem leads to a third order Goursat problem for which we have to prove some stability. Energy estimates are obtained using a multiplier method.

Step 4: Back to the solution u. In the previous steps, we obtained the functions Φ and v. We can reconstruct from them a piece of blowup solution $\tilde{u}$ in the strip (1.1.9) close to the light cone. It remains to prove that this piece coincides with the actual solution u.

2. Step 1: Asymptotic analysis, normalization of variables and reduction to a local problem

The asymptotic analysis of problem (0.1) has being carried out in [1], in the framework of what is called "nonlinear geometrical optics" (see also [6]). The idea is that the formal approximation

$$u \sim \epsilon u_1 + \epsilon^2 u_2 + \dots,$$

where u_1 and u_2 are defined in (1.1.1) and (1.1.3) (the other terms being defined the obvious way), can be written down in the form

$$u \sim \frac{\epsilon}{r^{1/2}} K(\sigma, \omega, \frac{1}{r}, \tau),$$

where K is a smooth function of $\sigma = r - t, \omega, \frac{1}{r}$ and the "slow time" $\tau = \epsilon t^{1/2}$. This statement is in fact not correct, a careful formal analysis yielding a second slow time $\zeta = \epsilon^2 \ln t$ and an expression of the form

$$u \sim \frac{\epsilon}{r^{1/2}} H(\sigma, \omega, \frac{1}{r}, \tau, \zeta).$$

However, for $\tau \geq \tau_0 > 0$, a smooth function H of $\epsilon, \sigma, \omega, 1/r, \tau, \zeta$ reduces to a smooth function G of $\epsilon, \epsilon^2 \ln \epsilon, \sigma, \omega, \tau$ only. This accounts for the appearence of the factor $\epsilon^2 \ln \epsilon$ in the asymptotic theorem: it will always be understood, without writing it, that G depends smoothly on ϵ and $\epsilon^2 \ln \epsilon$.

Using energy inequalites (as in [8] for example), one can compare approximate solutions (obtained by truncating the formal power series approximation) with the true solution. For instance, we obtain the following result: for each k, there is a function G_k such that, for $r - t \geq -C_0$ and $0 < \tau_1 \leq \tau \leq \tau_2 < \bar{\tau}_0$, we have for small ϵ and all α

$$|\partial^\alpha (u - \frac{\epsilon}{r^{1/2}} G_k(r - t, \omega, \tau))| \leq \epsilon^k.$$

We will use this in the following form: if we set

$$u = \frac{\epsilon}{r^{1/2}} G(r - t, \omega, \tau), \tag{2.1}$$

the function G will be bounded in C^k (independently of ϵ) if $\epsilon \leq \epsilon_k$ (ϵ_k depends of course on C_0 and τ_i). For $\epsilon = 0$, G reduces to the function, abusively denoted by $R^{(1)}(\sigma, \omega, \tau)$, which is the solution of the Cauchy problem

(2.2) $$\partial_\tau G - \frac{g}{2}(\partial_\sigma G)^2 = 0, G(\sigma, \omega, 0) = R^{(1)}(\sigma, \omega).$$

For technical reasons, we introduce

(2.3) $$w = L_1 u = \frac{\epsilon}{r^{1/2}} F(r - t, \omega, \tau),$$

which satisfies the nonlinear wave equation

(2.4) $$P(w) \equiv \partial_t^2 w - \Delta w + \frac{1}{2} L_1 L_2(w^2) = 0.$$

Thus we have to solve in an appropriate (still unknown) domain

$$-A_0 \leq \sigma \leq M, \omega \in S^1, \tau_0 \leq \tau \leq \bar{\tau}_\epsilon$$

the equation $\tilde{P}(F) = 0$ corresponding to $P(w) = 0$ ($\bar{\tau}_\epsilon$ being such that ∇F is unbounded for $\tau < \bar{\tau}_\epsilon$). To fix the ideas, without going into any details, we give here the result of the elementary computation:

(2.5)
$$\tilde{P}(F) = qFF_{\sigma\sigma} + \frac{-R^{1/2} + \epsilon^2 F(\beta' - \alpha_0)}{\tau} F_{\sigma\tau} + \frac{\epsilon^2(R^{1/2} + \epsilon^2 \alpha_0 F)}{4\tau^2} F_{\tau\tau} +$$

$$+ \frac{2\epsilon^2(B - \beta'')}{R} FF_{\sigma\omega} + \frac{\epsilon^2(-R^{1/2} + C\epsilon^2 F)}{R^2} F_{\omega\omega} + \frac{\epsilon^4 \beta''}{R\tau} FF_{\tau\omega} +$$

$$+q(\cos\omega F_\sigma - \frac{\epsilon^2 \sin\omega}{R} F_\omega, \sin\omega F_\sigma + \frac{\epsilon^2 \cos\omega}{R} F_\omega, -F_\sigma + \frac{\epsilon^2}{2\tau} F_\tau) +$$

$$+\epsilon^2 h(\omega, R, \tau, F) \nabla F + \epsilon^2 h_0(\omega, R, \tau, F).$$

Here, $q(\xi_1, \xi_2, \tau) = l_1 l_2$ is the symbol of $L_1 L_2$ and q stands for $q(\cos\omega, \sin\omega, -1)$, which is nonvanishing because the fields L_i are timelike. The coefficients A, B, C, β', β'' are various functions of ω,

$R = \tau^2 + \epsilon^2\sigma$ and h, h_0 are smooth functions. In addition to this equation, F has to fulfill the trace conditions on $\{\tau = \tau_0\}$ corresponding to the traces of u, and $F(M,.,.) = 0$. This is the local problem that we are going to blow up.

3. Step 2: Blowup of the problem and reduction to a Goursat problem on a fixed domain

1. To blow up the equation on F, we just set

$$F(\phi(s,\omega,\tau),\omega,\tau) = v(s,\omega,\tau),$$

with the idea that $\partial_s\phi$ will have to vanish somewhere at time $\bar{\tau}_\epsilon$. We find

$$\tilde{P}(F) = \frac{1}{\phi_s^3}\phi_{ss}v_s\mathcal{T}_0 - \frac{1}{\phi_s^2}\partial_s(v_s\mathcal{T}_0) + \frac{1}{\phi_s}\mathcal{T}_2 + \mathcal{T}_3, \tag{3.1.1}$$

where

$$\mathcal{T}_0 = q_1 v - \frac{R^{1/2}}{\tau}\phi_\tau - \epsilon^2\frac{R^{1/2}}{4\tau^2}\phi_\tau^2 + \frac{\epsilon^2}{R^{3/2}}\phi_\omega^2,\ q_1 \neq 0,\ R = \tau^2 + \epsilon^2\phi,$$
$$\mathcal{T}_2 = Z\partial_s v - \epsilon^2\partial_s v N\phi + \epsilon^2 v_s h_2(\omega,\tau,v,v_\omega,v_\tau,\phi,\phi_\omega,\phi_\tau),$$
$$\mathcal{T}_3 = \epsilon^2 Nv + \epsilon^2 h_3(\omega,\tau,v,v_\omega,v_\tau,\phi).$$

Here $Z = \delta_1\partial_\tau + \epsilon^2\delta_2\partial_\omega$ is a field which plays a crucial role, and N is a second order operator

$$N = N^{(1)}\partial_\tau^2 + 2\epsilon^2 N^{(2)}\partial_{\tau\omega}^2 + N^{(3)}\partial_\omega^2.$$

For the stability Step 3, it is essential to note that

$$N^{(1)} > 0,\ N^{(3)} < 0. \tag{3.1.2}$$

Looking at the form of the $\mathcal{T}_i$, we see that, to obtain $\tilde{P}(F) = 0$, it is enough to impose the "eikonal equation" $\mathcal{T}_0 = 0$ and the additional equation $\mathcal{T}_2 + \phi_s\mathcal{T}_3 = 0$. A similar computation can be found in [2]. The key fact is that we can use $\mathcal{T}_0 = 0$ to eliminate $v = V(\omega,\tau,\phi,\phi_\omega,\phi_\tau)$ from the equations (because q_1 is not zero), and thus reduce the whole problem to a third order local problem on ϕ. In other words, we have reduced the blowup system (see [4], [6] for a general definition or [2] for a special case) for the original quasilinear wave equation on u to a *scalar* equation on ϕ; this "miracle" is a consequence of the structure condition we imposed on the equation. As a matter of fact, the blowup system has multiple characteristics: if we knew how to symmetrize it, we could obtain energy inequalities the standard way. It is because we do not know how to handle the blowup system in general that we need to reduce it to a scalar equation, for which a multiplier method is always available.

Note that in the slow time variable τ, the characteristics of the wave equation corresponding roughly to $\partial_t - \partial_r$ are now nearly horizontal and known. We can straighten them out to obtain a Goursat problem for ϕ with characteristic boundaries. More precisely, by an appropriate known change of the form

$$X = s,\ Y = \omega,\ T = T(s,\omega,\tau),$$

which is only a τ_0-translation outside a small neighbourhood of $\{\tau = \tau_0\}$, we change the domain to

$$\tilde{D} = \{(X,Y,T), -A_0 \leq X \leq M, Y \in S^1, 0 \leq T \leq \bar{T} = \bar{\tau}_\epsilon - \tau_0\}$$

for which the lower boundary $\{T = 0\}$ is characteristic (ϕ satisfying the appropriate trace conditions there). The equation on ϕ has the form

$$\text{(3.1.3)}\quad \mathcal{L}(\phi) \equiv \tilde{Z}S\tilde{V} + \epsilon^2(S\phi)\tilde{N}\tilde{V} - \epsilon^2 S\tilde{V}\tilde{N}\phi + \epsilon^2 S\tilde{V}\tilde{h}_2 + \epsilon^2 S\phi\tilde{h}_3 = 0,$$

where the tildes indicate that the quantities from (3.1.1) have been transformed by the change of variables, and $S = \tilde{\partial}_s$. The boundary conditions are ϕ and $\partial_T\phi$ given on $\{T = 0\}$, and $\phi - X$ flat on $\{X = M\}$.

For $\epsilon = 0$, note that $\mathcal{T}_0 = 0$ and $\mathcal{T}_2 + \phi_s\mathcal{T}_3 = 0$ reduce to $-qv = \phi_\tau$ and $\partial^2_{\tau s}v = 0$, hence the ϕ equation is just $\partial^3_{\tau\tau s}\phi = 0$.

2. We have to consider more closely the vanishing of ϕ_X. The key point is the following: on the one hand, we want ϕ_X to vanish somewhere (to obtain a blowup solution F through (1.1.12)). On the other hand, it seems that for negative ϕ_X the Goursat problem on ϕ develops an instability. Hence we are forced to look for ϕ with ϕ_X vanishing only on the upper boundary of the domain. From the study of the case $\epsilon = 0$, it seems reasonable, taking into account the nondegeneracy condition **(ND)**, to expect the following situation: for some

point $M = (m, \bar{T})$,

$$\phi_X \geq 0,\ \phi_X = 0 \Leftrightarrow (X, Y, T) = M,$$
$$\phi_{XT}(M) < 0, \nabla_{X,Y}(\phi_X)(M) = 0, \nabla^2_{X,Y}(\phi_X)(M) >> 0.$$

We call this condition "condition **(H)**". The difficulty is now the following: in the approximation process used to solve the nonlinear equation (3.3) on ϕ, even if we start with a $\phi^{(0)}$ satisfying **(H)** for, say, $\bar{T} = T_0$, the next approximation will not satisfy **(H)** again. We have to adjust at each step the upper boundary of the domain to make it coincide with the first time of vanishing of ϕ_X.

To achieve this, we introduce a real parameter λ and perform the change of variables (the small (x, y, t) should not, of course, be confused with the original variables!)

$$X = x,\ Y = y,\ T \equiv T(t, \lambda) = T_0(t + \lambda(1 - \chi_1(t))),$$

where χ is 1 near 0 and 0 near 1. We now work in a fixed domain

$$D_0 = \{(x, y, t), -A_0 \leq x \leq M, y \in S^1, 0 \leq t \leq 1\}$$

and the ϕ equation becomes $\mathcal{L}(\lambda, \phi) = 0$. The linearized equation

$$\mathcal{L}'(\lambda, \phi)(\dot{\lambda}, \dot{\phi}) = \partial_\lambda \mathcal{L}(\lambda, \phi)\dot{\lambda} + \partial_\phi \mathcal{L}(\lambda, \phi)\dot{\phi}$$

has a special structure because it is coming from $\mathcal{L}(\lambda, \phi) = 0$ through the above change of variables. More precisely, if $\mathcal{L}(\lambda, \phi) = f$, we have

$$\mathcal{L}'(\lambda, \phi)(\dot{\lambda}, \dot{\phi}) = \dot{f} - \dot{\lambda}\partial_t f \frac{\partial_\lambda T}{\partial_t T}$$

provided

$$\partial_\phi \mathcal{L}(\lambda, \phi)\dot{\Psi} = \dot{f} \text{ and } \dot{\Psi} = \dot{\phi} - \dot{\lambda}\partial_t \phi \frac{\partial_\lambda T}{\partial_t T}.$$

In the process of solving $\mathcal{L}(\lambda, \phi) = 0$, f is small, hence the term $\dot{\lambda}\partial_t f$ is quadratically small, that is, negligible. The term $\dot{\Psi}$ being known, we see that we now have an additional degree of freedom which allows us to pick $\dot{\lambda}$ to make sure that the new $\phi + \dot{\phi}$ satisfies **(H)** again. We skip here the precise statement ensuring the possibility of such a choice (see for instance the "Lemme fondamental" of [4] for a similar procedure).

Thus, if we can solve (with good estimates) the linearized equation $\partial_\phi \mathcal{L}$ in the fixed domain D_0, we will solve the full nonlinear problem $\mathcal{L}(\lambda, \phi) = 0$ using, for instance, a standard Nash-Moser scheme. We refer the reader to [7] for details.

4. Step 3: Existence and tame estimates for the linearized Goursat problem. In Section **3**, we saw how solving our nonlinear blowup problem in an appropriate unknown domain can be reduced to solving a linear Goursat problem in a fixed domain

$$D_0 = \{(x,y,t), -A_0 \leq x \leq M, y \in S^1, 0 \leq t \leq 1\}.$$

We now give some details about this Goursat problem, and beginning with the structure of the operator. The linearized operator $\mathcal{L}_\phi$ has the form (up to a nonzero factor)

$$\mathcal{L}_\phi = ZSZ + \epsilon^2(S\phi)NZ + \epsilon^2 B\partial_y^2 + \epsilon^2 \ell + b_0 SZ.$$

One can also see [2] for a similar (and explicit) computation. We explain what the different objects are:

(i) $Z = \partial_t + \epsilon^2 z_0 \partial_y$ is a real field, $z_0 = z_0(x,y,t,\lambda,\phi,\phi_y,\phi_t)$,
(ii) $S = \partial_x + \epsilon^2 s_0 \partial_t$ is a real field, $s_0 = s_0(x,y,t,\lambda)$, $s_0 = 0$ close to $\{t = 1\}$,
(iii) $N = N_1 Z^2 + 2\epsilon^2 N_2 Z\partial_y + N_3 \partial_y^2$,
$$N_i = N_i(x,y,t,\lambda,\phi,\phi_y,\phi_t),\ N_1 < 0, N_2 > 0,$$
(iv) $B = -(ZS\phi)N_3$,
(v) ℓ is a linear combination of $\dot\phi$, $S\dot\phi$, $Z\dot\phi$, $\partial_y\dot\phi$, $SZ\dot\phi$, $Z^2\dot\phi$, $Z\partial_y\dot\phi$, $\partial_y^2\dot\phi$ with no $\partial_y^2\dot\phi$ term for $\epsilon = 0$.

We recall that the boundary conditions are two trace conditions on $\{t = 0\}$ (characteristic boundary) and ϕ supported for $\{x \leq M\}$.

The basic energy inequality is the following:

Energy inequality. *Let ϕ satisfy* **(H)** *and be close enough to $\phi^{(0)}$ in $C^4(\bar D_0)$. Fix $\mu > 1$. For smooth u with $\mathcal{L}_\phi u = f, f(x,y,0) = 0$, we have*

$$h|pSZ^2u|_0^2 + h|pZ^3u|_0^2 + \epsilon^2 h|pZ^2\partial_y u|_0^2 +$$

$$+\epsilon^4|pZ\partial_y^2 u|_0^2 + \epsilon^4 h \int p^2(S\phi)|\partial_y^2 Zu|^2 \le C|pZf|_0^2.$$

Here, we have set

$$\delta = 1 - t, \; p = \delta^{\mu/2} \exp \frac{h}{2}(x - t),$$

where h is big enough and $|.|_0$ denotes the norm in $L^2(D_0)$.

Without entering the tedious computations, let us only emphasize that the more difficult terms to control are the "angle" derivatives $\partial_y^2 u$: they are multiplied by an ϵ^4 coefficient, and contain h only through $h(S\phi)$ (recall that $S\phi$ vanishes at the point of interest). The value of B which is specified in iv) is important in the computation. This energy inequality is obtained by multiplying the equation by an expression of the form

$$Mu = aSZu + \epsilon^2 c\partial_y^2 u + dZ^2 u,$$

where the coefficients have to be chosen appropriately. It turns out that a good choice is

$$a = (S\phi)^{-1}\delta^{\mu} g, \; c = c'\delta^{\mu} g,$$
$$d = -d'\delta^{\mu} g, \; g = \exp h(x - t), \; 0 < c' < N_3, d' > 0.$$

A simpler but analogous computation is carried out in [2]. We do not know if similar or more general results on Goursat problems are available in the literature. Obtaining higher order inequalities goes without difficulties. In fact, in the process of solving the nonlinear equation $\mathcal{L}(\lambda, \phi) = 0$, we can arrange to solve the linear equation

$$\mathcal{L}_\phi u = f$$

only in smooth functions flat on $\{t = 0\}$ and $\{x = M\}$. This simplifies the computations and estimations of the various traces of the derivatives of u (see for instance [2]), and reduces the problem to a commutator analysis. We obtain tame estimates of the form

$$|u|_{H^s} \le C|f|_{H^{s+1}} + C|f|_{H^3}(1 + |\phi|_{H^{s+7}}).$$

Finally, we prove existence of flat smooth solutions by truncating the y derivatives, that is, replacing ∂_y by $\chi(\alpha D_y)\partial_y$ in $\mathcal{L}_\phi$: we can then solve as usual the remaining two dimensional Goursat problem, and pass to the limit when $\alpha \to 0$ thanks to the Sobolev estimates we already proved.

4. Step 4: Going back to the original solution u

From the function ϕ constructed in Steps 3, 4, we obtain F in the corresponding domain

$$\phi(-A_0, \omega, \tau) \leq \sigma \leq M, \tau_0 \leq \tau \leq \bar{\tau}_\epsilon,$$

by (1.1.12). It is understood that A_0 is big enough. This domain then contains a rotationally invariant domain

$$\Omega = \{-\tilde{A}_0 + C\epsilon^2 t \leq r - t, \tau_0 \leq \epsilon t^{1/2} \leq \bar{\tau}_\epsilon\}.$$

From F and L_1, we obtain a function $\tilde{u}$ in Ω. If we have taken C big enough in the definition of Ω, this domain is an influence domain for equation (1.1.7) on $\tilde{u}$. By uniqueness, $u = \tilde{u}$ in Ω. The statements (1.1.10), (1.1.11) in the blowup theorem then follow by direct computation.

Bibliography

[1] Alinhac S., Approximation près du temps d'explosion des solutions d'équations d'ondes quasilinéaires en dimension deux, *SIAM J. Math. Anal.* **26** (3) (1995), 529–565.

[2] Alinhac S., Temps de vie et comportement explosif des solutions d'équations d'ondes quasi-linéaires en dimension deux II, *Duke Math. J.* **73** (3) (1994), 543–560.

[3] Alinhac S., Explosion géométrique pour des systèmes quasi-linéaires, *Amer. J. Math.* **117** (4) (1995), 987–1017.

[4] Alinhac S., 'Explosion des solutions d'une équation d'ondes quasi-linéaire en deux dimensions d'espace, *Comm. PDE* **21** (5,6) (1996), 923–969.

[5] Alinhac S., Blowup of small data solutions for a class of quasilinear wave equations in two space dimensions I and II, Preprint, 1997, University of Paris-Sud.

[6] Alinhac S., Blowup for nonlinear hyperbolic equations, in: "Progress in Nonlinear Differential Equations and their Applications," Birkhäuser, Boston, (1995).

[7] Alinhac S. and Gérard P., *Opérateurs Pseudo-Différentiels et Théorème de Nash–Moser*, InterEditions, Paris, (1991).

[8] Hörmander L., The lifespan of classical solutions of nonlinear hyperbolic equations, *Lecture Notes Math.* **1256**, Springer Verlag, (1986), 214–280.

[9] Hörmander L., Nonlinear hyperbolic differential equations, Lectures, University of Lund, (1986–1987).

[10] John F., Nonlinear wave equations. Formation of singularities, *Leghigh University, University Lecture Series*, Amer. Math. Soc., Providence, (1990).

[11] John F., Blowup of radial solutions of $u_{tt} = c^2(u_t)\Delta u$ in three space dimensions, *Mat. Apl. Comput.* **V** (1985), 3–18.

[12] Klainerman S., Uniform decay estimates and the Lorentz invariance of the classical wave equation, *Comm. Pure Appl. Math.* **38** (1985), 321–332.

[13] Klainerman S., The null condition and global existence to nonlinear wave equations, *Lect. Appl. Math.* **23** (1986), 293–326.

Département de Mathématiques
Université de Paris-Sud
91045 Orsay Cedex, France

Concentration effects in critical nonlinear wave equation and scattering theory

Hajer Bahouri and Patrick Gérard

1. Introduction

This paper is devoted to generalized geometrical optics for the following nonlinear wave equation with critical exponent,

$$\Box u + |u|^4 u = 0, \tag{1}$$

where $\Box = \partial_t^2 - \Delta_x$, $t \in \mathbf{R}$, $x \in \mathbf{R}^3$. The global well-posedness of equation (1) in the energy space was proved rather recently by Shatah-Struwe [16]. Let us recall precisely this result: given $\varphi \in \dot{H}^1(\mathbf{R}^3)$, $\psi \in L^2(\mathbf{R}^3)$, there exists a unique solution u to (1) satisfying $u|_{t=0} = \varphi$, $\partial_t u|_{t=0} = \psi$ and $u \in L^5_{\text{loc}}(\mathbf{R}, L^{10}(\mathbf{R}^3))$. Observe that this latter property means exactly that the nonlinear term $|u|^4 u$ in (1) belongs to $L^1_{\text{loc}}(\mathbf{R}, L^2(\mathbf{R}^3))$, which allows us to consider it as a source term in the energy method. In particular, this implies $(u, \partial_t u) \in C(\mathbf{R}_t, \dot{H}^1(\mathbf{R}^3) \times L^2(\mathbf{R}^3))$ with conservation of energy,

$$\int_{\mathbf{R}^3} \left[\frac{1}{2}|\partial_t u(t,x)|^2 + \frac{1}{2}|\nabla u(t,x)|^2 + \frac{1}{6}|u(t,x)|^6\right] dx = \text{cst} := E(u). \tag{2}$$

This result is to be compared to the similar one for equations with subcritical exponents,

$$\Box u + |u|^{p-1} u = 0, \quad p < 5, \tag{3}$$

which is due to Ginibre-Velo [5]. In the latter case, every weak solution in the energy space satisfies the additional regularity property $u \in L^5_{\text{loc}}(\mathbf{R}, L^{10}(\mathbf{R}^3))$, while this fact is not known in the critical case. Similarly, in the subcritical case, the $L^5 L^{10}$ norm of the solution on $[0,T] \times \mathbf{R}^3$ can be estimated in terms of the energy only, while such an estimate is open for the solution obtained by Shatah-Struwe (except in the case of radial solutions, see [7]). It is then natural to wonder whether these technical differences are actually due to more qualitative differences between solutions to (1) and (3). This question was made more precise in [3] where the following problem of generalized geometrical optics was addressed: given a sequence (φ_n, ψ_n) of Cauchy data

Typeset by $\mathcal{A}_{\mathcal{M}}\mathcal{S}$-TeX

at $t = 0$, with bounded energy and no loss of energy at infinity in $\mathbf{R}^3$, describe how the lack of compactness of this sequence in the energy space is propagated by equations (1) and (3).

A first result, maybe not too surprising, is that subcritical solutions behave exactly as solutions to the linear wave equation $\Box v = 0$, while critical solutions may display some nonlinear behaviour, precisely when the linear solution v_n with the same Cauchy data (φ_n, ψ_n) displays a lack of compactness in $L_t^\infty L_x^6$ (see Section 3 below for a precise statement).

The purpose of this paper is to describe this nonlinear phenomenon. It turns out that the whole description can be achieved by means of wave and scattering operators associated to equation (1). As a byproduct of this complete description, we show that the $L^5 L^{10}$ norm of a critical solution can actually be estimated in terms of the energy only. However, our method does not provide an explicit estimate.

This paper is organized as follows. In Section 2, we recall two basic estimates for the analysis of nonlinear wave equations. In Section 3, we discuss briefly the cases where the critical solution u_n looks like the corresponding linear solution v_n. Section 4 is devoted to the particular case where the linear solution v_n satisfies $v_n(t_0, x) = h_n^{-1/2}\Phi(\frac{x}{h_n})$, $\partial_t v_n(t_0, x) = h_n^{-3/2}\Psi(\frac{x}{h_n})$, for some $t_0 \in \mathbf{R}$, $(\Phi, \Psi) \in \dot{H}^1 \times L^2$, and $h_n > 0$, $h_n \to 0$. In this case, after the focusing time t_0, the critical solution u_n looks like another linear solution $\tilde{v}_n$, with corresponding profiles $(\tilde{\Phi}, \tilde{\Psi})$ given by scattering theory for equation (1). Section 5 deals with the case of general Cauchy data (φ_n, ψ_n). The key lies in a refined analysis of loss of compactness for the Sobolev imbedding $\dot{H}^1(\mathbf{R}^3) \hookrightarrow L^6(\mathbf{R}^3)$. We prove that such a loss of compactness is always due to a superposition of concentrating profiles, which allows us to use the results of Section 4.

Finally, let us indicate that these results will appear in more detail in a forthcoming paper [1].

2. Two basic estimates

In this section, we recall without proof two important inequalities which are of constant use in the analysis of nonlinear wave equations. The first one is due to Ginibre-Velo [5] and is known as the "generalized Strichartz inequality",

$$(4) \quad ||v||_{L^q(0,T;L^r)} \leq C_q\Big(||\partial_t v|_{t=0}||_{L^2} + ||\nabla v|_{t=0}||_{L^2} + ||\Box v||_{L^1(0,T;L^2)}\Big)$$

where $v = v(t,x)$, $(t,x) \in \mathbf{R} \times \mathbf{R}^3$, $\frac{1}{q} + \frac{3}{r} = \frac{1}{2}$, $q > 2$, and C_q is independent of T. For a proof of (4), we refer to [6] or to [10].

The second estimate is a variant of an inequality due to Morawetz [13] and was used for the critical wave equation by Struwe [17], Grillakis [8,9] and Shatah-Struwe [15,16]. Before stating it, let us introduce some notation. Let u be a solution to (1) with $u \in L^5_{\text{loc}}(\mathbf{R}, L^{10})$. For every $t > 0$, we set

$$E(t) = \int_{|x|<t} \left[\frac{1}{2}|\partial_t u(t,x)|^2 + \frac{1}{2}|\nabla u(t,x)|^2 + \frac{1}{6}|u(t,x)|^6\right] dx. \tag{5}$$

Then we have, for every $a, b > 0$ such that $0 < a < b$,

$$\int_{|x|<b} |u(b,x)|^6\, dx \leq C_0\left[\frac{a}{b}(E(a) + E(a)^{1/3}) + E(b) - E(a) + (E(b) - E(a))^{1/3}\right]. \tag{6}$$

Inequality (6) is obtained by combining the energy identity and the identity associated to the multiplier $t\partial_t u + x \cdot \nabla u + u$.

3. Solutions without concentration effect

The following result is a slight generalization of Theorem A in [3].

Proposition 1. *Let (φ_n, ψ_n) be a bounded sequence in $\dot{H}^1(\mathbf{R}^3) \times L^2(\mathbf{R}^3)$ such that*

$$\sup_n \int_{|x|>R} (|\nabla\varphi_n(x)|^2 + |\psi_n(x)|^2)\, dx \underset{R\to+\infty}{\longrightarrow} 0. \tag{7}$$

Let u_n, v_n be such that $\Box u_n + |u_n|^4 u_n = 0$, $\Box v_n = 0$, $u_n|_{t=0} = v_n|_{t=0} = \varphi_n$, $\partial_t u_n|_{t=0} = \partial_t v_n|_{t=0} = \psi_n$. Then, for every $T > 0$, the following are equivalent:

(i) *$\{(u_n - v_n, \partial_t u_n - \partial_t v_n), n \in \mathbf{N}\}$ is contained in a compact subset of $C([0,T], \dot{H}^1 \times L^2)$.*

(ii) *$\{v_n, n \in \mathbf{N}\}$ is contained in a compact subset of $C([0,T], L^6)$.*

Moreover, if (i) *or* (ii) *holds, $\{u_n - v_n, n \in \mathbf{N}\}$ is contained in a compact subset of $L^5([0,T], L^{10})$.*

The proof of (i) $\Rightarrow$ (ii) is based on conservation of energy for equation (1) and for the linear equation. Comparison of these two conservation laws yields that the set

$$\left\{t \mapsto \|v_n(t,\cdot)\|^6_{L^6}, n \in \mathbf{N}\right\}$$

is contained in a compact subset of $\mathcal{C}([0,T])$. Then the conclusion comes from inequality (4) with $q = r = 8$, which implies, in view of (7), that $\{v_n, n \in \mathbf{N}\}$ is contained in a compact subset of $L^6([0,T] \times \mathbf{R}^3)$. The proof of (ii) $\Rightarrow$ (i) involves a systematic use of (4) and (6) on small backward cones, in the spirit of [16].

4. Solutions associated with one profile

Let $(\Phi, \Psi) \in \dot{H}^1 \times L^2$, $t_0 > 0$, and, for every $h \in]0,1]$, let v_h, u_h be defined by

$$(8)\quad \Box v_h = 0, \quad v_h(t_0, x) = \frac{1}{\sqrt{h}}\,\Phi\Big(\frac{x}{h}\Big), \quad \partial_t v_h(t_0, x) = \frac{1}{h^{3/2}}\,\Psi\Big(\frac{x}{h}\Big),$$

$$(9)\qquad \Box u_h + |u_h|^4 u_h = 0, \quad (u_h, \partial_t u_h)|_{t=0} = (v_h, \partial_t v_h)|_{t=0}$$

with $u_h \in L^5_{\text{loc}}(\mathbf{R}, L^{10}(\mathbf{R}^3))$. It is easy to check that $\{v_h, h > 0\}$ belongs to a compact subset of $\mathcal{C}([0,T], L^6)$ for every $T < t_0$, but not for $T = t_0$. Hence, by proposition 1, as $h \to 0$, u_h can be approximated by v_h up to a small energy term only for $t < t_0$. The purpose of this section is to describe u_h for $t \geq t_0$. For this we need some preliminary global properties of solutions to (1), which we list below.

a) If v is a solution of $\Box v = 0$ with finite energy, inequality (4) with large T implies that $v \in L^q(\mathbf{R}; L^r)$ for $\frac{1}{q} + \frac{3}{r} = \frac{1}{2}$, $q > 2$. It turns out that this result can be generalized to solutions of equation (1).

Proposition 2. (Bahouri-Shatah [2]). *Let u be a finite energy solution to* (1) *with $u \in L^5_{\text{loc}}(\mathbf{R}, L^{10})$. Then $\|u(t,\cdot)\|_{L^6} \to 0$ as $t \to \infty$, and $u \in L^q(\mathbf{R}, L^r)$ for any (q,r) with $q > 2$, $\frac{1}{q} + \frac{3}{r} = \frac{1}{2}$.*

Let us show how this result can be deduced from estimates recalled from Section 2. With the notation (5), the classical energy identity shows that $t \mapsto E(t)$ is a nondecreasing bounded function, hence has a limit as $t \to +\infty$. Applying inequality (6) with $a = \varepsilon T$, $b = T$ and passing to the limit as T$\to +\infty$, we get

$$\limsup_{T \to +\infty} \int_{|x|<T} |u(T,x)|^6\, dx \leq C\varepsilon$$

for every $\varepsilon > 0$, hence this limit is 0. Using translations in time, this implies, for every $R > 0$,

$$(10)\qquad \int_{|x|<T+R} |u(T,x)|^6\, dx \underset{T\to\infty}{\longrightarrow} 0.$$

Finally, by finite propagation speed, we have, for every $T > 0$,

(11)
$$\int_{|x|\geq T+R} |u(T,x)|^6\,dx \leq \int_{|x|\geq R} \Big[3|\partial_t u(0,x)|^2 + 3|\nabla u(0,x)|^2 + |u(0,x)|^6\Big]\,dx$$

which goes to 0 as $R \to +\infty$. In view of (10), we conclude that $||u(T,\cdot)||_{L^6} \to 0$ as $T \to +\infty$. Now apply (4) with $q = 4$, $r = 12$. We get, with $S > T$,

$$||u||_{L^4(T,S;L^{12})} \leq C\Big(E(u)^{1/2} + ||u||^5_{L^5(T,S;L^{10})}\Big). \tag{12}$$

By Hölder's inequality, this leads to

$$||u||_{L^4(T,S;L^{12})} \leq C\Big(E(u)^{1/2} + ||u||_{L^\infty(T,S;L^6)}||u||^4_{L^4(T,S;L^{12})}\Big). \tag{13}$$

By taking T large enough so that

$$||u||_{L^\infty(T,S;L^6)}\,E(u)^{3/2} \leq \frac{3^3}{(4C)^4},$$

we obtain

$$||u||_{L^4(T,S;L^{12})} \leq \frac{4}{3}C\,E(u)^{1/2},$$

hence $u \in L^4(0,\infty;L^{12})$. Using Hölder's inequality again, $u \in L^5(0,\infty;L^{10})$ and the case with any (q,r) follows from inequality (4) on $[0,\infty[$.

b) Now we come to scattering theory. Given a finite energy solution v of $\Box v = 0$, a standard iterative scheme allows us to construct two solutions $u_\pm$ of equation (1) such that $(u_\pm, \partial_t u_\pm) - (v, \partial_t v)$ goes to 0 in $\dot H^1 \times L^2$ as $t \to \pm\infty$. This defines the wave operators $\Omega_\pm : (v(0,\cdot), \partial_t v(0,\cdot)) \mapsto (u_\pm(0,\cdot), \partial_t u_\pm(0,\cdot))$. Moreover, Proposition 2 shows that $\Omega_\pm$ are onto, hence invertible, which allows us to define the scattering operator $S = \Omega_+^{-1}\Omega_-$ on the whole energy space generalizing a result by Pecher [14].

Now we are ready to state our first result.

Theorem 3. *Let V be defined by $\Box V = 0$, $(V, \partial_t V)|_{t=0} = (\Phi, \Psi)$. Let U_- be the solution to (1) such that $U_- \in L^5_{\mathrm{loc}}(\mathbf{R}, L^{10})$ and $(U_- \partial_t U_-) - (V, \partial_t V) \to 0$ in $\dot H^1 \times L^2$ as $t \to -\infty$. Set*

$$U_-^h(t,x) = \frac{1}{\sqrt{h}}\,U_-\Big(\frac{t-t_0}{h}, \frac{x}{h}\Big). \tag{14}$$

Then $(u_h, \partial_t u_h) - (U_-^h, \partial_t U_-^h)$ goes to 0 in $L^\infty(\mathbf{R}, \dot{H}^1 \times L^2)$ as h goes to 0, and $u_h - U_-^h$ goes to 0 in $L^5(\mathbf{R}, L^{10})$ as h goes to 0.

Observe that Theorem 3 and Proposition 2 imply that u_h is bounded in $L^5(\mathbf{R}, L^{10})$. Moreover, we are able to describe it for $t > t_0$.

Corollary 4. *Let $\tilde{v}_h$ be defined by*

$$(15) \quad \Box \tilde{v}_h = 0, \quad \tilde{v}_h(t_0, x) = \frac{1}{\sqrt{h}} \tilde{\Phi}\Big(\frac{x}{h}\Big), \quad \partial_t \tilde{v}_h(t_0, x) = \frac{1}{h^{3/2}} \tilde{\Psi}\Big(\frac{x}{h}\Big),$$

with $(\tilde{\Phi}, \tilde{\Psi}) = S(\Phi, \Psi)$. Then, for every $T > t_0$, $(u_h, \partial_t u_h) - (\tilde{v}_h, \partial_t \tilde{v}_h)$ goes to 0 in $L^\infty([T, \infty[, \dot{H}^1 \times L^2)$.

Proof of Theorem 3. After rescaling, we just need to estimate

$$(16) \qquad r_h(t,x) = \sqrt{h}\, u_h(ht + t_0, hx) - U_-(t,x).$$

We have

$$(17) \qquad \Box r_h + |U_- + r_h|^4 (U_- + r_h) - |U_-|^4 U_- = 0$$

with $(r_h, \partial_s r_h)\big|_{t=-t_0/h} = (V - U, \partial_s(V-U))\big|_{t=-t_0/h}$, which goes to 0 in $\dot{H}^1 \times L^2$ as $h \to 0$. Hence inequality (4) applied to r_h reads, for any $T > -\frac{t_0}{h}$,

$$(18) \quad ||r_h||_{L^5(-t_0/h, T; L^{10})} \le C\Big[\varepsilon_1(h) + \sum_{j=1}^{5} ||r_h^j U_-^{5-j}||_{L^1(-t_0/h, T; L^2)}\Big]$$

where $\varepsilon_1(h) \to 0$. Of course we have

$$(19) \qquad ||r_h^j U_-^{5-j}||_{L_t^1 L_x^2} \le ||r_h||_{L_t^5 L_x^{10}}^j \, ||U_-||_{L_t^5 L_x^{10}}^{5-j}$$

hence, if $T < 0$ and $|T|$ is large enough such that $C||U_-||_{L^5(-t_0/h, T; L^{10})}^4 \le \frac{1}{2}$, say, then $||r_h||_{L^5(-t_0/h, T; L^{10})} \to 0$ as $h \to 0$. Let

$$(20) \qquad T_{\max} = \sup\Big\{T \in \mathbf{R},\ ||r_h||_{L^5(-t_0/h, T; L^{10})} \underset{h \to 0}{\longrightarrow} 0\Big\}.$$

Let us show that $T_{\max} = +\infty$ by contradiction. Assuming $T_{\max} < \infty$, for any $T_1 < T_{\max}$, we have $||r_h||_{L^5(-t_0/h, T_1; L^{10})} \to 0$, hence

$$(21) \quad \begin{aligned} &||r_h U_-^4||_{L^1(-t_0/h, T_{\max}; L^{10})} \\ &\qquad \le \varepsilon_2(h) + ||r_h||_{L^5(T_1, T_{\max}; L^{10})} \times ||U_-||_{L^5(T_1, T_{\max}; L^{10})}^4, \end{aligned}$$

with $\varepsilon_2(h) \to 0$. Now apply (18) with $T = T_{\max}$, choose T_1 so close to $T_{\max}$ so that $C||U_-||^4_{L^5(T_1,T_{\max};L^{10})} \leq \frac{1}{2}$, and we conclude that $||r_h||_{L^5(-t_0/h,T_{\max};L^{10})} \to 0$. Applying the energy inequality to equation (17) on the time interval $\left[-\frac{t_0}{h}, T_{\max}\right]$, we then deduce, in view of (19),

$$\sup_{-t_0/h \leq t \leq T_{\max}} \left(||\partial_t r_h(t,\cdot)||_{L^2} + ||\nabla r_h(t,\cdot)||_{L^2}\right) \to 0. \tag{22}$$

Then apply inequality (4) to r_h on the time interval $[T_{\max}, T_{\max} + \delta]$, with δ so small that $C||U_-||^4_{L^5(T_{\max},T_{\max}+\delta,L^{10})} \leq \frac{1}{2}$, and we obtain as above that

$$||r_h||_{L^5(T_{\max},T_{\max}+\delta,L^{10})} \to 0,$$

whence the contradiction. Hence $T_{\max} = +\infty$. To finish the proof, we apply inequality (4) to r_h on the time interval $[T, +\infty[$, where T is chosen large enough so that $C||U_-||_{L^5(T,+\infty;L^{10})} \leq \frac{1}{2}$ (here Proposition 2 is crucial). Then, in view of (22) with T in place of $T_{\max}$, we obtain $r_h \to 0$ in $L^5(T, +\infty; L^{10})$ and $(r_h, \partial_t r_h) \to 0$ in $L^\infty(T, +\infty; \dot{H}^1 \times L^2)$ by the energy estimate. The estimate of r_h on $\left]-\infty, -\frac{t_0}{h}\right[$ is easier and left to the reader. ∎

5. Loss of compactness in Sobolev imbeddings and profiles

In order to generalize Theorem 3 to an arbitrary sequence of Cauchy data, we need to analyze the possible loss of compactness arising in the Sobolev imbedding $\dot{H}^1(\mathbf{R}^3) \hookrightarrow L^6(\mathbf{R}^3)$ carefully. This is the purpose of the concentration-compactness lemma (see for instance [11]) and of its microlocal refinement ([3,4]). However, the following result is more precise, since it yields the existence of concentration profiles.

First let us introduce a definition. Let us call *concentration* a triple $(\mathbf{h}, \mathbf{x}, \psi)$ where $\mathbf{h} = (h_n)$ is a sequence of positive numbers converging to 0, $\mathbf{x} = (x_n)$ is a convergent sequence of points in $\mathbf{R}^3$, $\psi \in \dot{H}^1(\mathbf{R}^3)$.

Theorem 5. *Let (u_n) be a bounded sequence of $\dot{H}^1(\mathbf{R}^3)$ satisfying*

$$u_n \rightharpoonup 0, \quad \sup_n \int_{|x|>R} |\nabla u_n(x)|^2 \, dx \underset{R\to+\infty}{\longrightarrow} 0. \tag{23}$$

Then there exists a subsequence (u'_n) and a sequence $(\mathbf{h}^{(j)}, \mathbf{x}^{(j)}, \psi_j)_{j\geq 1}$ of concentrations such that, for every $\varepsilon > 0$, we have, for some p,

(i) $u'_n(x) = \displaystyle\sum_{j=1}^{p} \frac{1}{\sqrt{h_n^{(j)}}} \psi_j\Big(\frac{x - x_n^{(j)}}{h_n^{(j)}}\Big) + r_n(x), \ \limsup_{n\to\infty} ||R_n||_{L^6} \leq \varepsilon.$

(ii) *If* $j \neq k$, $\left|\log \frac{h_n^{(j)}}{h_n^{(k)}}\right| + \left|\frac{x_n^{(j)}}{h_n^{(j)}} - \frac{x_n^{(k)}}{h_n^{(k)}}\right| \underset{n\to\infty}{\longrightarrow} \infty.$

(iii) $\lim_{n\to\infty} ||\nabla u_n'||_{L^2}^2 = \sum_{j=1}^{p} ||\nabla \psi_j||_{L^2}^2 + \lim_{n\to\infty} ||\nabla r_n||_{L^2}^2.$

In other words, any loss of compactness in the Sobolev imbedding is due to a superposition of concentration profiles. Observe that property (ii) means that these profiles are almost orthogonal, while property (iii) means that the remainder term r_n interacts very weakly with them.

Let us give an outline of the proof of Theorem 5. It is a combination of an inequality in [3], ideas from concentration-compactness and from a recent paper by Métivier-Schochet [12]. First recall the following

Lemma 6 [3]. *Let* $I = S_0 + \sum_{k=0}^{\infty} \Delta_k$ *be a dyadic decomposition of the identity. There exist* $\theta \in]0,1[$ *and* $C > 0$ *such that, for every sequence* (u_n) *satisfying the assumptions of Theorem 5, we have*

$$\limsup_{n\to\infty} ||u_n||_{L^6} \leq C \limsup_{n\to\infty} \big(\sup_{k\geq 0} ||\Delta_k u_n||_{L^6}\big)^{\theta} ||\nabla u_n||_{L^2}^{1-\theta}. \tag{24}$$

Let us write inequality (24) in a slightly different way. Because of (23), we have

$$\begin{aligned} \limsup_{n\to\infty} \sup_{k\geq 0} ||\Delta_k u_n||_{L^6} &= \sup_{\{k_n\}\to\infty} \limsup_{n\to\infty} ||\Delta_{k_n} u_n||_{L^6} \\ &\leq \sup_{\{h_n\}\to 0} \limsup_{n\to\infty} ||\varphi(h_n D) u_n||_{L^6} := \delta(\mathbf{u}) \end{aligned} \tag{25}$$

for some $\varphi \in C_0^\infty(\mathbf{R}^3 \setminus \{0\})$. Hence, if we set $M(\mathbf{u}) = \limsup_{n\to\infty} ||\nabla u_n||_{L^2}$, we have

$$\limsup_{n\to\infty} ||u_n||_{L^6} \leq C \delta(\mathbf{u})^{\theta} M(\mathbf{u})^{1-\theta}. \tag{26}$$

Given such a sequence $\mathbf{u} = (u_n)$, either $\delta(\mathbf{u}) = 0$, and the theorem is proved, or there exists – up to extracting a subsequence from $\mathbf{u}$ – a sequence $(h_n^{(1)})$ such that

$$\lim_{n\to\infty} ||\varphi(h_n^{(1)} D) u_n|| \geq \frac{1}{2} \delta(\mathbf{u}). \tag{27}$$

Using the Plancherel formula and a standard method of concentration-compactness [11], one can find – up to extracting a subsequence from $\mathbf{u}$ – a sequence $v_n^{(1)}$ satisfying (23) and

$$\limsup_{n\to\infty} \big|\big|\mathbf{1}_{h_n^{(1)}|D|<\frac{1}{R}} \nabla v_n^{(1)}\big|\big|_{L^2} + \big|\big|\mathbf{1}_{h_n^{(1)}|D|>R} \nabla v_n^{(1)}\big|\big|_{L^2} \underset{R\to\infty}{\longrightarrow} 0, \tag{28}$$

(29) $\quad \forall a > 0,\ \forall b > a, \quad \|\mathbf{1}_{a<h_n|D|<b}(\nabla u_n - \nabla v_n^{(1)})\|_{L^2} \longrightarrow 0.$

In other words, (28) means that $v_n^{(1)}$ oscillates only in frequencies of order $\frac{1}{h_n^{(1)}}$, and (29) means that $u_n - v_n^{(1)}$ does not oscillate at all in this range of frequencies. Then set

(30) $$u_n^{(1)} = u_n - v_n^{(1)}.$$

By (28), (29), we get

(31) $$M(\mathbf{u}^{(1)})^2 = M(\mathbf{u})^2 - M(\mathbf{v}^{(1)})^2.$$

Moreover, if $h_n \to 0$ and $\frac{h_n}{h_n^{(1)}} \to C \in]0, \infty[$, we have, for every $a > 0$, $b > a$,

$$\|\mathbf{1}_{a<h_n|D|<b} \nabla u_n^{(1)}\|_{L^2} \longrightarrow 0,$$

hence *a fortiori*

$$\|\mathbf{1}_{a<h_n|D|<b}\, u_n^{(1)}\|_{L^6} \longrightarrow 0,$$

which shows that $\delta(\mathbf{u}^{(1)}) \leq \delta(\mathbf{u})$, and that the only sequences (h_n) arising in $\mathbf{u}^{(1)}$ will satisfy – up to extracting –

(32) $$\left|\log \frac{h_n}{h_n^{(1)}}\right| \longrightarrow +\infty.$$

In view of (26), (27), (30), (31) and (32), we obtain, by iterating this process,

Lemma 7. *Under the assumptions of Theorem 4, there exists a sequence $(\mathbf{h}^{(j)}, \mathbf{v}^{(j)})_{j\geq 1}$, where, for every j, $\mathbf{h}^{(j)} = (h_n^{(j)})$ is a sequence of positive numbers converging to 0, $\mathbf{v}^{(j)} = (v_n^{(j)})$ is a sequence of functions satisfying assumptions of Theorem 4, such that, for every $\varepsilon > 0$, we have, for some p,*

(i) $u_n' = \sum_{j=1}^{p} v_n^{(j)} + r_n,\ \lim_{n\to\infty} \|r_n\|_{L^6} \leq \varepsilon,$

(ii) $\forall j$, $(v_n^{(j)})$ *and* $(h_n^{(j)})$ *satisfy* (28),

(iii) $j \neq k \Rightarrow \left|\log\left(\frac{h_n^{(j)}}{h_n^{(k)}}\right)\right| \longrightarrow +\infty,$

(iv) $M(\mathbf{u})^2 = \sum_{j=1}^{p} M(\mathbf{v}^{(j)})^2 + M(\mathbf{r})^2$

As a second step, we shall decompose each $\mathbf{v}^{(j)}$ as a sum of profiles. Let $\mathbf{v} = (v_n)$ be a sequence of $\dot{H}^1(\mathbf{R}^3)$ satisfying the assumptions of Theorem 5 and (28) for some scale $\mathbf{h} = (h_n)$ of positive numbers going to 0. Set

$$w_n(y) = h_n^{1/2} v_n(h_n y), \tag{33}$$

so that $||\nabla w_n||_{L^2} = ||\nabla v_n||_{L^2}$, $||w_n||_{L^6} = ||v_n||_{L^6}$ and

$$\sup_n \int_{|y|>R/h_n} |\nabla w_n(y)|^2 \, dy \underset{R\to\infty}{\longrightarrow} 0. \tag{34}$$

Moreover, by (28), we may assume, up to an error of order ε, that the Fourier transform of w_n is supported in a fixed compact subset of $\mathbf{R}^3 \setminus \{0\}$. This implies that $||w_n||_{L^2}$ and $||w_n||_{L^\infty}$ are bounded. Denote by ψ_0 the weak limit of w_n — up to extraction — and set

$$w_n^{(0)} = w_n - \psi_0 .$$

Because of spectral localization, $w_n^{(0)}$ goes to 0 on every compact subset, hence

$$\limsup ||w_n^{(0)}||_{L^\infty} = \sup_{\{y_n\}\to\infty} |w_n^{(0)}(y_n)| . \tag{35}$$

We then define $\gamma(\mathbf{w}^{(0)})$ to be the supremum of $||\psi||_{L^\infty}$, where there exists $\{y_n\} \to \infty$ in $\mathbf{R}^3$ such that $w_n^{(0)}(y + y_n)$ goes weakly to $\psi(y)$ (up to a subsequence). Then, in view of spectral localization and (35),

$$\limsup ||w_n^{(0)}||_{L^\infty} \leq C\gamma(\mathbf{w}^{(0)}) \tag{36}$$

and, by Hölder's inequality,

$$\limsup ||w_n^{(0)}||_{L^6} \leq C\gamma(\mathbf{w}^{(0)})^{2/3} M(\mathbf{w}^{(0)})^{1/3} . \tag{37}$$

We now proceed as in the first step. Either $\gamma(w^{(0)}) = 0$, and the theorem is proved in view of (37), or there exists $\{y_n^{(1)}\} \to \infty$ in $\mathbf{R}^3$ and a subsequence such that

$$w_n^{(0)}(y + y_n^{(1)}) \rightharpoonup \psi_1(y), \quad ||\psi_1||_{L^\infty} \geq \frac{1}{2}\gamma(\mathbf{w}^{(0)}). \tag{38}$$

Then we set $w_n^{(1)}(y) = w_n^{(0)}(y) - \psi_1(y - y_n^{(1)})$, and we observe

$$M(\mathbf{w}^{(1)})^2 = M(\mathbf{w}^{(0)})^2 - ||\nabla\psi_1||_{L^2}^2 \tag{39}$$

$$\gamma(\mathbf{w}^{(1)}) \leq \gamma(\mathbf{w}^{(0)}) \tag{40}$$

and the only sequences (y_n) arising in $\mathbf{w}^{(1)}$ must satisfy $|y_n - y_n^{(1)}| \to \infty$. By iterating the process in view of (37), (39), (40), we obtain

$$w_n(y) = \psi_0(y) + \sum_{j=1}^{q} \psi_j(y - y_n^{(j)}) + r_n\,, \quad \limsup ||r_n||_{L^6} \leq \varepsilon\,, \tag{41}$$

and the theorem follows after rescaling. ■

Let us now come back to wave equations. Similarly to above, we call *concentrating wave* a triple $(\mathbf{h}, (\mathbf{t}, \mathbf{x}), V)$, where $\mathbf{h} = (h_n)$ is a sequence of positive numbers going to 0, $(\mathbf{t}, \mathbf{x}) = ((t_n, x_n))$ is a convergent sequence of points in $\mathbf{R} \times \mathbf{R}^3$, and V is a finite energy solution to $\Box V = 0$ in $\mathbf{R} \times \mathbf{R}^3$. For such a solution, we set, for any t,

$$E_0(V) = \frac{1}{2}||\partial_t V(t,\cdot)||_{L^2}^2 + \frac{1}{2}||\nabla V(t,\cdot)||_{L^2}^2\,.$$

Theorem 8. *Let (v_n) be a sequence of solutions to $\Box v_n = 0$, with bounded energy, and such that*

$$\sup_n \int_{|x|>R} (|\partial_t v_n(0,x)|^2 + |\nabla v_n(0,x)|^2)\, dx \underset{R\to\infty}{\longrightarrow} 0\,. \tag{42}$$

Then there exists a subsequence (v_n'), a sequence $(\mathbf{h}^{(j)}, (\mathbf{t}^{(j)}, \mathbf{x}^{(j)}), V_j)_{j\geq 1}$ of concentrating waves, and a finite energy solution V_0 to $\Box V_0 = 0$, such that, for every $\varepsilon > 0$, we have, for some p,

(i) $v_n'(t,x) = V_0(t,x) + \sum_{j=1}^{p} \frac{1}{\sqrt{h_n^{(j)}}} V_j\left(\frac{t - t_n^{(j)}}{h_n^{(j)}}, \frac{x - x_n^{(j)}}{h_n^{(j)}}\right) + r_n(t,x),$

$\limsup_{n\to\infty} ||r_n||_{L^\infty(\mathbf{R},L^6)} \leq \varepsilon.$

(ii) *If* $j \neq k$, $\left|\log\left(\frac{h_n^{(j)}}{h_n^{(k)}}\right)\right| + \left|\frac{t_n^{(j)}}{h_n^{(j)}} - \frac{t_n^{(k)}}{h_n^{(k)}}\right| + \left|\frac{x_n^{(j)}}{h_n^{(j)}} - \frac{x_n^{(k)}}{h_n^{(k)}}\right| \underset{n\to\infty}{\longrightarrow} \infty.$

(iii) $\lim_{n\to\infty} E_0(v'_n) = \sum_{j=0}^{p} E_0(V_j) + \lim_{n\to\infty} E_0(r_n).$

Now we can combine Theorem 8 with Proposition 1 and Theorem 3. This leads to

Theorem 9. *Let (v_n) satisfy assumptions of Theorem 8, and let u'_n be the solution to (1) with $u'_n \in L^5(\mathbf{R}, L^{10})$ and $(u'_n, \partial_t u'_n)|_{t=0} = (v'_n, \partial_t v'_n)|_{t=0}$. Then, for $\varepsilon > 0$ small enough, with the notation of Theorem 8, we have*

$$(43)\quad u'_n(t,x) = U_0(t,x) + \sum_{j=1}^{p} \frac{1}{\sqrt{h_n^{(j)}}} U_j\Big(\frac{t - t_n^{(j)}}{h_n^{(j)}}, \frac{x - x_n^{(j)}}{h_n^{(j)}}\Big) + \tilde{r}_n(t,x),$$

where $\tilde{r}_n$ satisfies

$$(44)\quad \limsup_{n\to\infty} ||\tilde{r}_n||_{L^5(\mathbf{R},L^{10})} + \sup_{t\in\mathbf{R}} ||\partial_t(\tilde{r}_n - r_n)(t,\cdot)||_{L^2} + ||\nabla(\tilde{r}_n - r_n)(t,\cdot)||_{L^2} \le K\varepsilon$$

and U_j, $0 \le j \le p$, are the solutions to (1) with $U_j \in L^5(\mathbf{R}, L^{10})$ characterized by

$$(45)\quad (U_j, \partial_t U_j)|_{t=0} = (V_j, \partial_t V_j)|_{t=0} \ \text{if } j = 0 \text{ or } t_n^{(j)} \equiv 0;$$

$$(U_j, \partial_t U_j) - (V_j, \partial_t V_j) \to 0 \text{ in } \dot{H}^1 \times L^2 \text{ as } t \to \mp\infty \text{ if } \frac{t_n^{(j)}}{h_n^{(j)}} \to \infty.$$

The proof of Theorem 9 is based on two main remarks: first, property (ii) in Theorem 8 allows us to neglect interactions between two profiles, and moreover the explicit form of these profiles allows us to make the corresponding rescaling in order to simplify the estimate of the remainder term as in the proof of Theorem 3 above. The second remark is that, if a solution to (1) has a small enough energy, then, by inequality (4) and superlinear bootstrap as in the proof of Theorem 3, its $L^5 L^{10}$ norm is actually estimated linearly by the energy. We refer to [1] for details.

Finally, let us observe that Theorem 9 shows in particular that $||u'_n||_{L^5_t L^{10}_x}$ is bounded. Hence we get, using an argument by contradiction,

Corollary 10. *There exists a nondecreasing function* $C : [0, \infty[\to [0, \infty[$ *such that, for every solution* u *to (1) with* $E(u) < \infty$ *and* $u \in L^5(\mathbf{R}, L^{10})$, *we have*

$$||u||_{L^5(\mathbf{R}, L^{10})} \leq C(E(u)).$$

Proof.

Using the scaling $R \mapsto R^{1/2} u(Rt, Rx)$, it is enough to prove

$$||u||_{L^5_t L^{10}_x(|x|+|t|\leq 1)} \leq C(E(u)). \tag{46}$$

Assuming (46) fails, let (u_n) be a sequence of solutions such that $E(u_n)$ is bounded and the left hand side of (46) goes to infinity. Let $\chi \in C_0^\infty(\mathbf{R}^3)$ such that $\chi = 1$ near $|x| \leq 1$, and $\tilde{u}_n$ be the solution to (1) such that $(\tilde{u}_n, \partial_t \tilde{u}_n)|_{t=0} = (\chi u_n, \chi \partial_t u_n)|_{t=0}$. Then $E(\tilde{u}_n)$ is bounded, and, by finite propagation speed,

$$||\tilde{u}_n||_{L^5_t L^{10}_x(|x|+|t|\leq 1)} = ||u_n||_{L^5_t L^{10}_x(|x|+|t|\leq 1)}.$$

By applying Theorem 9 to $(\tilde{u}_n)$ we get a contradiction, which proves Corollary 10. ∎

References

[1] H. Bahouri, P. Gérard, High frequency approximation of solutions to critical nonlinear wave equations, preprint 97–34, Orsay, 1997.

[2] H. Bahouri, J. Shatah, Global estimate for the critical semilinear wave equation, preprint, Courant Institute, 1996, to appear in *Ann. IHP*, Analyse non linéaire.

[3] P. Gérard, Oscillations and Concentration effects in semilinear dispersive wave equations, *J. Functional Analysis* **141** (1994).

[4] ———, A microlocal version of concentration-compactness, in *Partial Differential Equations and Mathematical Physics*, 143–157, L. Hörmander and A. Melin editors, Birkhäuser, 1996.

[5] J. Ginibre, G. Velo, The global Cauchy problem for the Nonlinear Klein-Gordon equation, *Math. Z.* **189** (1985), 487–505.

[6] ———, Generalized Strichartz inequalities for the wave equation, *J. Functional Analysis* **133** (1995), 50–68.

[7] J. Ginibre, H. Soffer, G. Velo, The global Cauchy problem for the critical nonlinear wave equation, *J. Functional Analysis* **110** (1992), 96–130.

[8] M. Grillakis, Regularity and asymptotic behavior of the wave equation with a critical nonlinearity, *Ann. of Math.* **132** (1990), 485–509.
[9] ______, Regularity for the wave equation with a critical nonlinearity, *Comm. Pure and Applied Math.* **45** (1992), 749–774.
[10] H. Lindblad, C. Sogge, On existence and scattering with minimal regularity for semilinear wave equations, *J. Functional Analysis* **130** (1995), 357–426.
[11] P.-L. Lions, The concentration-compactness principle in the calculus of variations. The limit case, Part 1, *Rev. Mat. Iberoamericana* **1** (1985), 145–201.
[12] G. Métivier, S. Schochet, Interactions trilinéaires résonantes, Séminaire Equations aux Dérivées Partielles, 1995–1996, exposé n° VI, Ecole Polytechnique, Palaiseau.
[13] C. Morawetz, Time decay for the nonlinear Klein-Gordon equation, *Proc. Royal Soc. London A* **306** (1968), 291–296.
[14] H. Pecher, Nonlinear small data scattering for the wave and Klein-Gordon Equations, *Math. Z.* **185** (1984), 261–270.
[15] J. Shatah, M. Struwe, Regularity Results for Nonlinear Wave Equations, *Ann. Math.* **138** (1993), 503–518.
[16] ______, *Well-posedness in the energy space for semilinear wave equations with critical growth, IMRN* **7** (1994), 303–309.
[17] M. Struwe, Globally regular solutions to the u^5 Klein-Gordon equation, *Annali Scuola Norm. Pisa* **15** (1988), 495–513.

Hajer Bahouri
Faculté des Sciences de Tunis
Département de Mathématiques
1060 Tunis
Tunisie

Patrick Gérard
Université de Paris-sud
Département de Mathématiques
91405 Orsay cedex
France
e-mail address: patrick.GERARD@math.u-psud.fr

Lower semicontinuity of weighted path length in BV

Paolo Baiti and Alberto Bressan

Abstract. We establish some basic lower semicontinuity properties for a class of weighted metrics in **BV**. These Riemann-type metrics, uniformly equivalent to the $\mathbf{L}^1$ distance, are defined in terms of the Glimm interaction potential. They are relevant in the study of nonlinear hyperbolic systems of conservation laws, being contractive w.r.t. the corresponding flow of solutions.

Introduction

For a scalar conservation law, it is well known [7, 11] that the entropy-admissible solutions determine a semigroup which is contractive w.r.t. the $\mathbf{L}^1$ distance. This fundamental property plays a key role in the study of uniqueness, stability and perturbations of weak solutions.

On the other hand, for a nonlinear $n \times n$ hyperbolic system

$$u_t + \left[F(u)\right]_x = 0, \tag{1.1}$$

the contractivity of the flow is no longer true [14]. For this reason, when $u \in \mathbb{R}^n$, to establish the uniqueness and continuous dependence of solutions of (1.1) is a considerably more difficult problem than in the scalar case.

In a recent series of papers [1, 2, 3, 5], it was shown that, restricted to a suitable domain $\mathcal{D}$ of functions with small total variation, the system (1.1) does generate a continuous flow. More precisely, this flow is contractive w.r.t. a suitable Riemann-type metric, uniformly equivalent to the standard $\mathbf{L}^1$ distance. The construction of this weighted distance involves

- a closed set $\mathcal{D} \subset \mathbf{L}^1(\mathbb{R}; \mathbb{R}^n)$ consisting of functions with small total variation, positively invariant for the flow of (1.1);
- a dense subset $\mathcal{D}_{PL} \subset \mathcal{D}$ of piecewise Lipschitz functions;

- for each $u \in \mathcal{D}_{PL}$, a space T_u of first order generalized tangent vectors (v, ξ), with weighted norm $\|(v,\xi)\|_u$. If, say, u is piecewise Lipschitz with N jumps, then $T_u \simeq \mathbf{L}^1 \times \mathbb{R}^N$.

Given a suitably regular path $\gamma : [a,b] \mapsto \mathcal{D}_{PL}$, its weighted length is then defined as the integral of the weighted norm of its tangent vector $D\gamma$, i.e.

$$\|\gamma\|_* \doteq \int_a^b \|D\gamma(\theta)\|_{\gamma(\theta)}\, d\theta. \tag{1.2}$$

In turn, the weighted distance between two functions $u, v \in \mathcal{D}$ is defined as the infimum of the lengths of (suitably regular) paths joining u with v, i.e.

$$d_*(u,v) \doteq \inf \left\{ \|\gamma\|_* ; \quad \gamma(a) = u,\ \gamma(b) = v \right\}. \tag{1.3}$$

We remark that this weighted distance does not fit within the standard framework of Riemann or Finsler manifolds [8, p. 362], because our tangent spaces T_u are defined not for all u in some open set $\mathcal{U} \subset \mathbf{L}^1$, but only for u in a dense subset with empty interior.

The aim of this paper is to establish some basic properties of the distance d_*. We first study the behavior of the Glimm interaction functional under $\mathbf{L}^1$ convergence, and prove the semicontinuity of the coefficients in the weighted norms. Then we establish the lower semicontinuity of the weighted length (1.2) w.r.t. uniform convergence of paths in $\mathbf{L}^1$. More precisely,

$$\|\gamma\|_* \le \liminf_{\nu \to \infty} \|\gamma_\nu\|_* \tag{1.4}$$

for every sequence of paths γ_ν converging uniformly to γ. This result naturally complements and clarifies the constructions in [3, 5].

We remark that, for any Lipschitz continuous path $\gamma : [a,b] \mapsto \mathbf{L}^1$, the usual $\mathbf{L}^1$ length is defined as

$$\begin{aligned} \|\gamma\|_{\mathbf{L}^1} \doteq \sup \Bigg\{ & \sum_{i=1}^N \|\gamma(\theta_i) - \gamma(\theta_{i-1})\|_{\mathbf{L}^1} ; \\ & a = \theta_0 < \cdots < \theta_N = b,\quad N \ge 1 \Bigg\}. \end{aligned} \tag{1.5}$$

In this case, given the uniform convergence $\gamma_\nu \to \gamma$, the relation

$$\|\gamma\|_{\mathbf{L}^1} \leq \liminf_{\nu \to \infty} \|\gamma_\nu\|_{\mathbf{L}^1}$$

follows from standard convexity arguments. On the other hand, (1.4) cannot be obtained by general weak convergence methods. Indeed, the numerical value of the constants defining the weighted norms $\|(v,\xi)\|_u$ and the smallness of the total variation of functions $u \in \mathcal{D}$ both play a key role for the validity of (1.4).

All the basic definitions, including generalized tangent vectors and weighted path lengths, are collected in Section 2. Our main lower semicontinuity result is stated in Section 4. The proof relies on the construction of a family of nonlinear functionals, which approximate the weighted distance function in **BV**.

Preliminaries

Let Ω be an open and convex subset of $\mathbb{R}^n$ containing the origin, and let $F : \Omega \mapsto \mathbb{R}^n$ be three times continuously differentiable. For every $u \in \Omega$, consider the $n \times n$ Jacobian matrix $A(u) \doteq DF(u)$. We assume that the system (1.1) is strictly hyperbolic, so that each $A(u)$ has real distinct eigenvalues $\lambda_1(u) < \cdots < \lambda_n(u)$. Given $u, u' \in \Omega$, define

$$A(u,u') \doteq \int_0^1 A(\theta u + (1-\theta)u')\, d\theta. \tag{2.1}$$

Of course, $A(u,u) = A(u)$. Call $\lambda_i(u,u')$ the i-th eigenvalue of the matrix $A(u,u')$. By possibly shrinking the size of Ω, there exist n disjoint intervals $[\lambda_i^{min}, \lambda_i^{max}]$ such that

$$\lambda_i(u,u') \in [\lambda_i^{min}, \lambda_i^{max}] \qquad u, u' \in \Omega, \quad i = 1, \ldots, n.$$

We assume that each characteristic field is either genuinely nonlinear or linearly degenerate in the sense of Lax [12]. So we can choose $\mathcal{C}^2$ families of right and left eigenvectors $r_i(u,u')$, $l_i(u,u')$ of $A(u,u')$, normalized according to

$$\big|r_i(u,u')\big| \equiv 1, \tag{2.2}$$

$$l_j(u,u')\cdot r_i(u,u') = \begin{cases} 1 & \text{if} \quad i=j, \\ 0 & \text{if} \quad i\neq j. \end{cases} \tag{2.3}$$

We also write $r_i(u)$ and $l_j(u)$ respectively for $r_i(u,u)$ and $l_j(u,u)$, and choose the orientation of the r_i so that

$$r_i \bullet \lambda_i(u) \doteq \lim_{\varepsilon\to 0} \frac{\lambda_i(u+\varepsilon r_i(u)) - \lambda_i(u)}{\varepsilon} \geq 0. \tag{2.4}$$

We briefly recall the solution of the Riemann problem [12] and the interaction estimates [10, 13]. Given $u\in\Omega$, call $\sigma\mapsto S_i(\sigma)(u)$ and $\sigma\mapsto R_i(\sigma)(u)$ respectively the i-shock and i-rarefaction curve through u, parametrized by arc-length. As is customary, the orientation is chosen so that the i-th characteristic speed is non-decreasing along the curves S_i, R_i. Define the composite curves

$$\Psi_i(\sigma)(u) \doteq \begin{cases} R_i(\sigma)(u) & \text{if} \quad \sigma\geq 0, \\ S_i(\sigma)(u) & \text{if} \quad \sigma<0. \end{cases} \tag{2.5}$$

Given two states u^-, u^+ sufficiently close to the origin, by the implicit function theorem there exist unique wave sizes $\sigma_1,\dots,\sigma_n$ such that

$$u^+ = \Psi_n(\sigma_n)\circ\cdots\circ\Psi_1(\sigma_1)(u^-). \tag{2.6}$$

The solution of the Riemann problem with data (u^-,u^+) thus consists of $n+1$ constant states $\omega_0=u^-,\omega_1,\dots,\omega_n=u^+$, where each couple of states ω_{i-1},ω_i is connected by an i-wave of size σ_i (a shock or a rarefaction, depending on the sign of σ_i). We write

$$E_i(u^-,u^+) \doteq \sigma_i, \tag{2.7}$$

for the size of the i-wave determined by the Riemann data (u^-,u^+). This quantity satisfies the well known estimate

$$E_i(u^-,u^+) = l_i(u^-)\cdot(u^+-u^-) + O(|u^+-u^-|^2). \tag{2.8}$$

We also define

$$J_i(u^-,u^+) \doteq |E_i(u^-,u^+)|. \tag{2.9}$$

Let us now recall the basic interaction estimates [10, 13]. Given a left, a middle and a right state $u^l, u^m, u^r \in \Omega$, one has

$$\sum_{i=1}^{n} \left| E_i(u^l, u^r) - E_i(u^l, u^m) - E_i(u^m, u^r) \right| \qquad \qquad \leq C_3 \Delta\big(E(u^l, u^m), E(u^m, u^r)\big), \tag{2.10}$$

where

$$\Delta(\alpha, \beta) \doteq \sum_{k>j} |\alpha_k \beta_j| + \sum_{k \in \mathrm{GNL},\ \min\{\alpha_k, \beta_k\} < 0} |\alpha_k \beta_k|, \tag{2.11}$$

and GNL is the set of indices corresponding to genuinely nonlinear families. As in [4], the Glimm functional can be defined for a general **BV** function. Let $u : \mathbb{R} \mapsto \Omega$ have bounded variation. By possibly changing the values of u at countably many points, we can assume that u is right continuous. Its distributional derivative $\mu \doteq D_x u$ is then a vector measure, which can be decomposed into a continuous and an atomic part: $\mu = \mu^c + \mu^a$. For $i = 1, \ldots, n$ define the signed measure $\mu_i = \mu_i^c + \mu_i^a$ as follows. The continuous part of μ_i is the Radon measure such that

$$\int \phi \, d\mu_i^c = \int l_i(u) \cdot \phi \, d\mu^c, \tag{2.12}$$

for every scalar continuous function ϕ with compact support. The atomic part of μ_i is the measure concentrated on the countable set $\{x_\alpha;\ \alpha = 1, 2, \ldots\}$ where u has a jump, such that

$$\mu_i^a(\{x_\alpha\}) = E_i\big(u(x_\alpha^-),\ u(x_\alpha^+)\big), \tag{2.13}$$

is the size of the i-th wave in the solution of the corresponding Riemann problem at x_α. Call μ_i^+, μ_i^- the positive and negative parts of the signed measure μ_i, so that

$$\mu_i = \mu_i^+ - \mu_i^-, \qquad \qquad |\mu_i| = \mu_i^+ + \mu_i^-.$$

The *total strength of waves* in u is then defined as

$$V(u) \doteq \sum_{i=1}^{n} V_i(u), \qquad \qquad V_i(u) \doteq |\mu_i|(\mathbb{R}), \tag{2.14}$$

while the *interaction potential* of waves in u is

$$\begin{aligned} Q(u) \doteq& \sum_{i<j} \big(|\mu_j| \times |\mu_i|\big)\Big(\big\{(x,y);\ x<y\big\}\Big) \\ &+\sum_i \big(\mu_i^- \times |\mu_i|\big)\Big(\big\{(x,y);\ x\neq y\big\}\Big). \end{aligned} \tag{2.15}$$

When u is piecewise constant, one easily checks that the definitions (2.14)-(2.15) reduce to the usual ones. On the other hand, if u is Lipschitz continuous, then its derivative $u_x(x)$ exists at almost every point $x \in \mathbb{R}$. In this case, setting

$$u_x^i(x) \doteq l_i\big(u(x)\big) \cdot u_x(x), \tag{2.16}$$

one has

$$V_i(u) = \int_{-\infty}^{\infty} \big|u_x^i(x)\big|\, dx.$$

The quantities V, Q satisfy two basic properties. The first is a straightforward consequence of the definitions:

(P1) Let $\varphi : \mathbb{R} \mapsto \mathbb{R}$ be a continuous, increasing one-to-one mapping. Then, for every $u \in \mathbf{BV}$, the composed function $v(x) \doteq u\big(\varphi(x)\big)$ satisfies

$$Q(v) = Q(u), \qquad V_i(v) = V_i(u), \qquad i = 1, \ldots, n. \tag{2.17}$$

The second property follows from the Glimm interaction estimates. It can be proved first for piecewise constant functions, then extended to all $\mathbf{BV}$ functions by an approximation argument.

(P2) There exist constants $\kappa_0, \delta_0 > 0$ such that the following holds. Assume $u \in \mathbf{L}^1$, with Tot.Var.$(u) \le \delta_0$. Then, for any $a < b$ and $\bar{x} \in [a, b[$, the function

$$w(x) \doteq \begin{cases} u(x) & \text{if} \quad x \notin [a, b[, \\ u(\bar{x}) & \text{if} \quad x \in [a, b], \end{cases} \tag{2.18}$$

satisfies

$$(2.19)\qquad Q(w) \le Q(u), \qquad\qquad V(w) + \kappa_0 Q(w) \le V(u) + \kappa_0 Q(u),$$

$$(2.20)\qquad V_i(w) + \kappa_0 Q(w) \le V_i(u) + \kappa_0 Q(u), \qquad\qquad i = 1, \dots, n.$$

Observe that the function w in (2.18) is obtained from u by collapsing all wave-fronts in $[a, \bar{x}]$ onto the point a, and all wave-fronts in $]\bar{x}, b]$ onto the point b. With the same constants κ_0, δ_0 as in **(P2)**, we now define the domains

$$(2.21)\qquad \mathcal{D} \doteq \big\{ u \in \mathbf{L}^1(\mathbb{R}; \mathbb{R}^n), \quad V(u) + \kappa_0 Q(u) \le \delta_0 \big\},$$

$$(2.22)\qquad \mathcal{D}_{PL} \doteq \big\{ u \in \mathcal{D}, \quad u \text{ is piecewise Lipschitz continuous} \big\}.$$

We recall below the definition of generalized differential of a path $\gamma : [a, b] \mapsto \mathbf{L}^1$, introduced in [6]. For any $u \in \mathbf{L}^1$, on the family Σ_u of all continuous paths $\gamma : [0, \theta_0] \mapsto \mathbf{L}^1$ such that $\gamma(0) = u$, consider the equivalence relation

$$(2.23)\qquad \gamma \sim \gamma' \quad \text{iff} \quad \lim_{\theta \to 0+} \frac{1}{\theta} \big\| \gamma(\theta) - \gamma'(\theta) \big\|_{\mathbf{L}^1} = 0 \qquad (\gamma, \gamma' \in \Sigma_u).$$

Now assume that u is piecewise Lipschitz with, say, jumps at the points $x_1 < \cdots < x_N$. The space of *generalized tangent vectors at* u is then defined as $T_u \doteq \mathbf{L}^1 \times \mathbb{R}^N$. To each $(v, \xi) \in T_u$, with $\xi = (\xi_1, \dots, \xi_N)$, we associate the path $\gamma_{(v,\xi;u)} \in \Sigma_u$ defined by

$$(2.24)\qquad \begin{aligned} \gamma_{(v,\xi;u)}(\theta) \doteq u + \theta v &+ \sum_{\xi_\alpha < 0} \big[u(x_\alpha^+) - u(x_\alpha^-) \big] \chi_{[x_\alpha + \theta\xi_\alpha, x_\alpha]} \\ &- \sum_{\xi_\alpha > 0} \big[u(x_\alpha^+) - u(x_\alpha^-) \big] \chi_{[x_\alpha, x_\alpha + \theta\xi_\alpha]}. \end{aligned}$$

More generally, we say that a path $\gamma \in \Sigma_u$ *generates the generalized tangent vector* $(v, \xi) \in T_u$, if γ is equivalent to $\gamma_{(v,\xi;u)}$, under the relation (2.23).

In other words, for small values of θ, the function $u^\theta \doteq \gamma(\theta)$ can be obtained from u by adding θv and by shifting the positions of the jumps from x_α to $x_\alpha + \theta\xi_\alpha$. As $\theta \to 0+$, this procedure yields a first order approximation to u^θ, with an error $o(\theta)$ in the $\mathbf{L}^1$ norm, with $o(\theta)/\theta \to 0$ as $\theta \to 0$. In connection with the above differential structure, one can define a kind of continuous differentiability property for maps $\gamma : \theta \mapsto u^\theta \in \mathbf{L}^1$, with piecewise Lipschitz values. Following [6], we say that a map $\gamma :]a, b[\mapsto \mathbf{L}^1$ is a *regular path* if there exists an integer N such that:

(i) Every function $u^\theta \doteq \gamma(\theta)$ is piecewise Lipschitz continuous with jumps at points $x_1^\theta < \cdots < x_N^\theta$ continuously depending on θ. Outside the jumps, each u^θ is continuous with a Lipschitz constant L independent of θ. All functions u^θ coincide outside some interval $[-M, M]$.

(ii) The map $\theta \mapsto u_x^\theta$ is continuous with values in $\mathbf{L}^1$.

(iii) There exists a continuous map $\theta \mapsto (v^\theta, \xi^\theta) \in \mathbf{L}^1 \times \mathbb{R}^N$ such that for every θ

$$\lim_{\varepsilon \to 0+} \frac{1}{\varepsilon} \Big\| \gamma(\theta + \varepsilon) - \gamma_{(v^\theta, \xi^\theta; u^\theta)}(\varepsilon) \Big\|_{\mathbf{L}^1} = 0. \tag{2.25}$$

More generally, we say that a continuous map $\gamma : [a, b] \mapsto \mathbf{L}^1$ is a *piecewise regular path* if there exist points $a = \theta_0 < \theta_1 < \cdots < \theta_\nu = b$ such that the restriction of γ to each open subinterval $]\theta_{j-1}, \theta_j[$ is a regular path.

Now consider any $u \in \mathcal{D}_{PL}$, say with jumps at $x_1 < \cdots < x_N$. For every $\alpha = 1, \ldots, N$, $i = 1, \ldots, n$, define the strength of the i-th wave in the Riemann problem at x_α as

$$J_{x_\alpha}^i \doteq \Big| E_i\big(u(x_\alpha^-), u(x_\alpha^+)\big) \Big|.$$

Given a generalized tangent vector $(v, \xi) \in T_u = \mathbf{L}^1 \times \mathbb{R}^N$, recalling (2.16) we define its weighted norm as

$$\text{(2.26)}\quad \|(v,\xi)\|_u \doteq \sum_{\alpha=1}^{N}\sum_{i=1}^{n} J^i_{x_\alpha}|\xi_\alpha|W_i^u(x_\alpha)+\sum_{i=1}^{n}\int_{-\infty}^{\infty}|v_i(x)|W_i^u(x)\,dx.$$

Here $v_i(x) \doteq l_i(u(x)) \cdot v(x)$ is the i-th component of v, Q is the interaction potential (2.15) and the weights W_i^u are defined by

$$\text{(2.27)}\qquad W_i^u(x) \doteq 1 + \kappa_1 R_i^u(x) + \kappa_1\kappa_2 Q(u),$$

(2.28)

$$R_i^u(x) \doteq \left[\sum_{j\le i}\int_x^{\infty} + \sum_{j\ge i}\int_{-\infty}^{x}\right]|u_x^j(y)|\,dy + \left[\sum_{\substack{k\le i\\ x_\alpha > x}} + \sum_{\substack{k\ge i\\ x_\alpha < x}}\right] J^k_{x_\alpha}$$
$$+\begin{cases} J^i_{x_\beta} & \text{if } x = x_\beta \text{ and } E_i\big(u(x_\beta^-), u(x_\beta^+)\big) > 0,\\ 0 & \text{otherwise,}\end{cases}$$

for suitably large constants κ_1, κ_2. Intuitively, $R_i^u(x)$ can be regarded as the total strength of all waves in u which approach an infinitesimal i-shock located at x. Finally, let $\gamma : \theta \mapsto u^\theta$ be a piecewise regular path defined on $[a, b]$, and let (v^θ, ξ^θ) be its generalized tangent vector at u^θ. The *weighted length* of γ is then defined as

$$\text{(2.29)}\qquad \|\gamma\|_* \doteq \int_a^b \|(v^\theta, \xi^\theta)\|_{u^\theta}\, d\theta.$$

By the standard interaction estimates (2.10), one can choose constants $\delta_0 > 0$ small and $\kappa_0, \kappa_1, \kappa_2$ large enough in (2.21), (2.27) so that the following property holds.

(P3) Let $u \in \mathcal{D}_{PL}$. For any $a < b$ and $\bar{x} \in [a, b]$, the function w in (2.18) satisfies

$$\text{(2.30)}\qquad W_i^w(x) \le W_i^u(x) \qquad x \notin [a, b[,\quad i = 1, \dots, n.$$

From the definition of regular path, the following continuity properties can be easily derived:

Lemma 1. *Let $\gamma :]a, b[\mapsto \mathcal{D}_{PL}$ be a regular path. Let $u^\theta = \gamma(\theta)$ have jumps at the points $x_1^\theta < \cdots < x_N^\theta$. Then the following holds.*

a) *The map* $(\theta,x)\mapsto u^\theta(x)$ *is continuous outside the jump set* $\mathcal{J}\doteq\{(\theta,x_\alpha^\theta),\quad \alpha=1,\ldots,N,\quad \theta\in\,]a,b[\}$. *For every* $\bar\theta$, *at each jump point* $x_\alpha^{\bar\theta}$ *one has*

$$u^\theta\left(x_\alpha^{\theta\pm}\right)\to u^{\bar\theta}\left(x_\alpha^{\bar\theta\pm}\right)\quad as\quad \theta\to\bar\theta. \tag{2.31}$$

b) *The map* $(\theta,x)\mapsto W_i^{u^\theta}(x)$ *is continuous outside the jump set* $\mathcal{J}$. *For every* $\bar\theta$, *at each jump point* $x_\alpha^{\bar\theta}$ *one has*

$$W_i^{u^\theta}(x_\alpha^\theta)\to W_i^{u^{\bar\theta}}(x_\alpha^{\bar\theta})\quad as\quad \theta\to\bar\theta. \tag{2.32}$$

c) *The map* $\theta\mapsto\left\|(v^\theta,\xi^\theta)\right\|_{u^\theta}$ *is continuous.*

We conclude this section with a useful approximation lemma. By an *elementary path* we mean a path of the form

$$\theta\mapsto w^\theta=w_1\cdot\chi_{]-\infty,\ \alpha\theta+\beta]}+w_2\cdot\chi_{]\alpha\theta+\beta,+\infty[},\qquad \theta\in[a,b],$$

where the functions w_1,w_2 are piecewise constant. A *piecewise elementary path* is a finite concatenation of elementary paths.

Lemma 2. *Let* $\gamma:[a,b]\mapsto\mathbf{BV}$ *be a piecewise regular path. Then there exists a sequence of piecewise elementary paths* γ_ν *such that, as* $\nu\to\infty$,

$$\sup_{\theta\in[a,b]}\left\|\gamma_\nu(\theta)-\gamma(\theta)\right\|_{\mathbf{L}^1}\to 0,\qquad \left\|\gamma_\nu\right\|_*\to\|\gamma\|_*. \tag{2.33}$$

The construction of the paths γ_ν goes as follows. For simplicity, assume that γ is a regular path, with each $u^\theta=\gamma(\theta)$ having the same number of jumps, say at $x_1^\theta<\cdots<x_N^\theta$. Let all functions u^θ coincide when $x\notin[-M,M]$. Fix $\nu\geq 1$ and define

$$\theta_m\doteq a+\frac{m}{\nu}(b-a)\qquad m=1,\ldots,\nu.$$

For ν sufficiently large, for each m we can choose points p_i such that

$$-M<p_0<x_1^\theta<p_1<\cdots<x_N^\theta<p_N<M,\qquad \theta\in[\theta_{m-1},\theta_m].$$

We now approximate each u^{θ_m} with a piecewise constant function u^m. The restriction of the original path γ to the subinterval $[\theta_{m-1}, \theta_m]$ is then replaced by a new path γ' defined as follows. If $\vartheta \in [-M, p_0] \cup [p_N, M]$ we set

$$\gamma'(\vartheta) \doteq u^m \cdot \chi_{]-\infty,\vartheta]} + u^{m-1} \cdot \chi_{]\vartheta,+\infty[}. \tag{2.34}$$

The same definition (2.34) is valid if $\vartheta \in]p_{i-1}, p_i]$ and $x_i^{\theta_{m-1}} \le x_i^{\theta_m}$. On the other hand, if $\vartheta \in]p_{i-1}, p_i]$ but $x_i^{\theta_{m-1}} > x_i^{\theta_m}$, we set
(2.35)

$$\gamma'(\vartheta) \doteq u^m \cdot \chi_{]-\infty, p_{i-1}] \cup]p_{i-1}+p_i-\vartheta,\ p_i]} + u^{m-1} \cdot \chi_{]p_{i-1},\ p_{i-1}+p_i-\vartheta] \cup]p_i,+\infty[}.$$

Clearly, γ' is piecewise elementary, with $\gamma'(-M) = u^{m-1}$, $\gamma'(M) = u^m$. We now perform a suitable parameter rescaling $\theta \mapsto \vartheta(\theta)$ mapping $[\theta_{m-1}, \theta_m]$ onto $[-M, M]$ and define the path $\gamma_\nu(\theta) \doteq \gamma'(\vartheta(\theta))$. Applying the same procedure to each subinterval $[\theta_{m-1}, \theta_m]$ we obtain the piecewise elementary path $\gamma_\nu : [a, b] \mapsto \mathbf{BV}$. If the approximations u^m of u^{θ_m} are chosen in a suitably accurate way, the properties (2.33) follow.

Lower semicontinuity of the Glimm functionals

In this section we establish the lower semicontinuity of the functionals Q and $V + \kappa_0 Q$ on the domain $\mathcal{D}$ defined at (2.21).

Theorem 1. *Consider a sequence of functions $u_\nu \in \mathcal{D}$, with $u_\nu \to u$ in $\mathbf{L}^1$, as $\nu \to \infty$. Then*

$$Q(u) \le \liminf_{\nu \to \infty} Q(u_\nu), \tag{3.1}$$

$$V(u) + \kappa_0 Q(u) \le \liminf_{\nu \to \infty} \{V(u_\nu) + \kappa_0 Q(u_\nu)\}. \tag{3.2}$$

In particular, the functional $V + \kappa_0 Q$ is lower semicontinuous on $\mathcal{D}$, and $\mathcal{D}$ is closed in $\mathbf{L}^1$.

For the proof, we shall need

Lemma 3. *For some constant C_0 and every $\varepsilon > 0$ the following holds. If $u \in \mathcal{D}$ satisfies $|u(x) - \bar{u}| \le \varepsilon$ for some constant state $\bar{u}$ and all x in* *an open interval I, then*

$$\left|\int_I l_i(\bar{u})\cdot\varphi\,Du - \int_I \varphi\,d\mu_i\right| \le C_0\varepsilon \int_I |\varphi|\,|Du|, \tag{3.3}$$

for every $\varphi \in C_c(I;\mathbb{R})$, $i = 1,\ldots,n$.

Proof. By (2.8), at each point x where u has a jump there exists a vector $\tilde{l}_i(x)$ such that

$$\left|\tilde{l}_i(x) - l_i\big(u(x-)\big)\right| \le C\cdot\left|u(x+) - u(x-)\right|, \tag{3.4}$$

$$E_i\big(u(x-), u(x+)\big) = \tilde{l}_i(x)\cdot\big(u(x+) - u(x-)\big), \tag{3.5}$$

for some constant C depending only on the system (1.1). We can now write

$$\int \varphi\,d\mu_i = \int \tilde{l}_i\cdot\varphi\,Du, \tag{3.6}$$

where $\tilde{l}_i(x) = l_i\big(u(x)\big)$ at points where u is continuous, while $\tilde{l}_i(x)$ is some vector which satisfies (3.4)-(3.5) at points of jump. In all cases, the assumptions of the lemma imply an estimate of the form

$$\left|l_i(\bar{u}) - \tilde{l}_i(x)\right| \le C_0\varepsilon.$$

Hence,

$$\left|\int l_i(\bar{u})\cdot\varphi\,Du - \int \varphi\,d\mu_i\right| \le \int \left|l_i(\bar{u}) - \tilde{l}_i(x)\right|\cdot|\varphi|\,|Du| \le C_0\varepsilon \int |\varphi|\,|Du|.$$

□

We can now prove Theorem 1. Let μ_i be as in (2.12)-(2.13) and let $\mu_{\nu,i}$ be analogously defined, with u replaced by u_ν. By passing to a subsequence we can assume that $\lim_{\nu\to\infty} \big\{V(u_\nu) + \kappa_0 Q(u_\nu)\big\}$ exists, that $u_\nu(x) \to u(x)$ for all $x \in \mathbb{R}$ and that $\left|Du_\nu\right| \rightharpoonup \tilde{\mu}$ weakly in the sense of measures as $\nu \to \infty$, where $\tilde{\mu}$ is a non-negative Radon measure. Now fix $\varepsilon > 0$. Since the total mass of $\tilde{\mu}$ is finite, one can select finitely many points $y_1, \ldots, y_N$ such that

$$\tilde{\mu}(\{x\}) < \varepsilon, \qquad \forall x \notin \{y_1, \ldots y_N\}. \tag{3.7}$$

We now choose disjoint intervals $I_k \doteq (y_k - \rho, y_k + \rho)$ such that

$$\sum_{k=1}^{N} \tilde{\mu}(I_k \setminus \{y_k\}) < \frac{\varepsilon}{N}. \tag{3.8}$$

Moreover, there exists $R > 0$ such that

$$\bigcup_{k=1}^{N} I_k \subset [-R, R], \qquad \tilde{\mu}(\mathbb{R} \setminus [-R, R]) < \varepsilon. \tag{3.9}$$

Because of (3.7), we can now choose points $p_0 < -R < p_1 < \cdots < R < p_r$ which are continuity points for u and for every u_ν, and such that either

$$p_{h-1} < y_k < p_h, \qquad [p_{h-1}, p_h] \subset I_k, \tag{3.10}$$

for some k, or else

$$\tilde{\mu}([p_{h-1}, p_h]) < \varepsilon. \tag{3.11}$$

Call $J_h \doteq [p_{h-1}, p_h]$. By weak convergence, for some ν_0 sufficiently large one has

$$|Du_\nu|(J_h) < \varepsilon \qquad \text{whenever} \quad J_h \cap \{y_1, \ldots, y_N\} = \emptyset, \quad \nu \geq \nu_0. \tag{3.12}$$

Moreover, if (3.10) holds, from (3.8) it follows that

$$|Du|(J_h \setminus \{y_k\}) \leq \tilde{\mu}(J_h \setminus \{y_k\}) < \frac{\varepsilon}{N}. \tag{3.13}$$

On the other hand, if (3.12) holds, then the oscillation of u_ν on the interval J_h is very small. Indeed, for every $x, y \in J_h$ and ν sufficiently large,

$$|u_\nu(x) - u_\nu(y)| \leq C|Du_\nu|(J_h) \leq 2\tilde{\mu}(J_h) < 2\varepsilon. \tag{3.14}$$

The same is also true for u. Set $\bar{u}_h \doteq u(p_h)$. By pointwise convergence and (3.14) it follows that

$$(3.15) \qquad \big|u_\nu(x) - \bar{u}_h\big| < C\varepsilon, \qquad \big|u(x) - \bar{u}_h\big| < C\varepsilon, \qquad x \in J_h,$$

for all ν sufficiently large. Hence, by (3.3) we get

$$(3.16) \qquad \begin{aligned} \left|\int_{J_h} \varphi \, d\mu_i - \int_{J_h} \varphi \, d\mu_{\nu,i}\right| \leq C_0\varepsilon \int_{J_h} |\varphi|\big(|Du| + |Du_\nu|\big) \\ + \left|\int_{J_h} l_i(\bar{u}_h) \cdot \varphi(Du - Du_\nu)\right|, \end{aligned}$$

for all $\varphi \in \mathcal{C}(J_h; \mathbb{R})$. By weak convergence and by taking the supremum over all $|\varphi| \leq 1$, we obtain

$$(3.17) \qquad \big|\mu_i\big|(J_h) \leq \liminf_{\nu \to \infty} \big|\mu_{\nu,i}\big|(J_h) + 2C_0\varepsilon\tilde{\mu}(J_h).$$

On the other hand, inserting $\varphi \equiv 1$ in (3.16) we obtain
(3.18)

$$\big|\mu_i(J_h) - \mu_{\nu,i}(J_h)\big| \leq C_0\varepsilon \int_{J_h} \big(|Du| + |Du_\nu|\big) + \left|\int_{J_h} l_i(\bar{u}) \cdot (Du - Du_\nu)\right|.$$

Letting $\nu \to \infty$ and using (3.17) this yields

$$(3.19) \qquad \begin{aligned} \mu_i^-(J_h) &= \frac{1}{2}\Big[|\mu_i|(J_h) - \mu_i(J_h)\Big] \\ &\leq \liminf_{\nu \to \infty} \mu_{\nu,i}^-(J_h) + 2C_0\varepsilon\tilde{\mu}(J_h). \end{aligned}$$

We now take care of the intervals J_h containing a point y_k of large oscillation. For each $k = 1, \ldots, N$, let $h = h(k) \in \{1, \ldots, r\}$ be the index such that

$$y_k \in J_h \doteq [p_{h-1}, \, p_h].$$

For each $\nu \geq 1$ consider the function

$$(3.20) \qquad \hat{u}_\nu(x) \doteq \begin{cases} u_\nu(x) & \text{if } \; x \notin \cup_k J_{h(k)}, \\ u_\nu(p_{h(k)-1}) & \text{if } \; x \in [p_{h(k)-1}, \, y_k], \\ u_\nu(p_h) & \text{if } \; x \in]y_k, \, p_{h(k)}]. \end{cases}$$

Observe that each $\hat{u}_\nu$ is continuous at all points $p_0, \ldots, p_r$. Call $\hat{\mu}_{\nu,i}$, $i = 1, \ldots, n$, the corresponding measures, defined as in (2.12)-(2.13) with u replaced by $\hat{u}_\nu$. Clearly $\hat{\mu}_{\nu,i} = \mu_{\nu,i}$ outside the intervals $J_{h(k)}$. By property **(P2)** it follows that

$$Q(\hat{u}_\nu) \leq Q(u_\nu), \qquad V(\hat{u}_\nu)+\kappa_0 Q(\hat{u}_\nu) \leq V(u_\nu)+\kappa_0 Q(u_\nu). \tag{3.21}$$

Since $u_\nu \to u$ pointwise, by (3.13) for each k one has

$$\begin{aligned} &\Big|\mu_i(\{y_k\}) - \hat{\mu}_{\nu,i}(\{y_k\})\Big| \\ &\quad= \Big|E_i\big(u(y_k-), u(y_k+)\big) - E_i\big(u_\nu(p_{h(k)-1}), u_\nu(p_{h(k)})\big)\Big| \\ &\quad\leq C \cdot \Big\{ \big|u(y_k^-) - u(p_{h(k)-1})\big| + \big|u(p_{h(k)-1}) - u_\nu(p_{h(k)-1})\big| \\ &\qquad + \big|u(y_k^+) - u(p_{h(k)})\big| + \big|u(p_{h(k)}) - u_\nu(p_{h(k)})\big| \Big\} \\ &\quad\leq C \cdot \frac{\varepsilon}{N}, \end{aligned} \tag{3.22}$$

for each $k = 1, \ldots, N$ and all ν sufficiently large. By construction we also have

$$|\hat{\mu}_{\nu,i}|\big(J_{h(k)} \setminus \{y_k\}\big) = 0, \qquad |\mu_i|\big(J_{h(k)} \setminus \{y_k\}\big) \leq \frac{\varepsilon}{N}. \tag{3.23}$$

Recalling the definition (2.15) and using (3.17), (3.19), (3.22), (3.23) and (3.9), we obtain an estimate of the form

$$\begin{aligned} Q(u_\nu) &\geq Q(\hat{u}_\nu) \\ &\geq \sum_{i<j} \sum_{h<\ell} \big(|\hat{\mu}_{\nu,j}| \times |\hat{\mu}_{\nu,i}|\big)(J_h \times J_\ell) \\ &\quad + \sum_{i} \sum_{h \neq \ell} \big(\hat{\mu}_{\nu,i}^- \times |\hat{\mu}_{\nu,i}|\big)(J_h \times J_\ell) \\ &\geq \sum_{i<j} \sum_{h<\ell} \big(|\mu_j| \times |\mu_i|\big)(J_h \times J_\ell) \\ &\quad + \sum_{i} \sum_{h \neq \ell} \big(\mu_i^- \times |\mu_i|\big)(J_h \times J_\ell) - C\varepsilon \\ &\geq Q(u) - C'\varepsilon, \end{aligned} \tag{3.24}$$

for suitable constants C, C' and all ν sufficiently large. Since $\varepsilon > 0$ was arbitrary, this establishes (3.1).

By the second inequality in (3.21) together with (3.17), an entirely similar argument yields (3.2). □

By property **(P3)**, the above arguments also imply the lower semicontinuity of the weight functions W_i^u in (2.27).

Lemma 4. *Consider a sequence $u_\nu \in \mathcal{D}_{PL}$ converging to some function $u \in \mathcal{D}_{PL}$ in the $\mathbf{L}^1$ norm. Then for every $x \in \mathbb{R}$ and $i = 1, \ldots, n$ one has*

$$W_i^u(x) \leq \liminf_{\nu \to \infty} W_i^{u_\nu}(x). \tag{3.25}$$

The limit in (3.25) is uniform for x bounded away from the jumps of u. More precisely let J be an open set containing all points $x_1 < \cdots < x_N$ where u has a jump. Then, for each $\varepsilon > 0$, there exists $\rho > 0$ such that every $w \in \mathcal{D}_{PL}$ with $\|w - u\|_{\mathbf{L}^1} < \rho$ satisfies

$$W_i^u(x) \leq W_i^w(x) + \varepsilon \qquad x \notin J, \quad i = 1, \ldots, n. \tag{3.26}$$

Weighted path lengths

In the following we consider the domain $\mathcal{D}$ of functions with small variation, introduced at (2.21), and the weighted norms on generalized tangent vectors, introduced at (2.26). We assume that the constants δ_0, κ_i in (2.21) and (2.27) are chosen so that the properties **(P2)-(P3)** hold.

Theorem 2. *Let $\gamma_\nu : [a, b] \mapsto \mathcal{D}$, $\nu \geq 1$, be a sequence of piecewise regular paths, uniformly converging to a piecewise regular path γ. Then the corresponding weighted lengths satisfy*

$$\|\gamma\|_* \leq \liminf_{\nu \to \infty} \|\gamma_\nu\|_*. \tag{4.1}$$

To prove the theorem, it is not restrictive to assume that γ is a regular path on $]a, b[$. Let the functions $u^\theta \doteq \gamma(\theta)$ have jumps at the points $x_1^\theta < \cdots < x_N^\theta$, and uniform Lipschitz constant L outside the jumps, so that

$$\left|u^\theta(x) - u^\theta(y)\right| \leq L|x - y| \qquad \text{whenever} \quad [x, y] \cap \{x_1^\theta, \ldots, x_N^\theta\} = \emptyset. \tag{4.2}$$

Moreover, by the approximation Lemma 2, we can assume that each γ_ν is a finite concatenation of elementary paths.

Let $\varepsilon > 0$ be given. The proof will be achieved by constructing a family of continuous functionals $\Phi_\theta : \mathcal{D}_{PL} \mapsto \mathbb{R}$, $\theta \in]a,b[$, satisfying the two conditions:

(C1) For some constant C_1, independent of ε, the following holds. For every $\bar\theta$ there exists $\rho_1 = \rho_1(\bar\theta) \in]0,\varepsilon]$ such that

(4.3)
$$\Phi_{\bar\theta}(u^{\bar\theta+\zeta}) - \Phi_{\bar\theta}(u^{\bar\theta}) \geq \int_{\bar\theta}^{\bar\theta+\zeta} \|(v^\theta,\xi^\theta)\|_{u^\theta}\, d\theta - C_1\varepsilon\zeta, \quad \zeta \in [0,\rho_1].$$

(C2) For some constant C_2, independent of ε, the following holds. For every $\bar\theta$, there exist constants $\delta, \rho_2 \in]0,\varepsilon]$ such that

$$\text{(4.4)} \quad \Phi_{\bar\theta}(\tilde u^{\bar\theta+\zeta}) - \Phi_{\bar\theta}(\tilde u^{\bar\theta}) \leq (1 + C_2\varepsilon) \int_{\bar\theta}^{\bar\theta+\zeta} \left\|(\tilde v^\theta, \tilde\xi^\theta)\right\|_{\tilde u^\theta} d\theta,$$

for every $\zeta \in [0,\rho_2]$ and every piecewise elementary path $\tilde\gamma : \theta \mapsto \tilde u^\theta$ with generalized tangent vector $(\tilde v, \tilde\xi)$ satisfying $\left\|u^\theta - \tilde u^\theta\right\|_{\mathbf{L}^1} < \delta$ for all θ.

Let us show that the existence of these functionals implies (4.1). The family of intervals

$$\text{(4.5)} \qquad \left\{[\bar\theta,\ \bar\theta+\zeta];\quad \theta \in]a,b[,\quad 0<\zeta<\min\left\{\rho_1(\bar\theta), \rho_2(\bar\theta)\right\}\right\}$$

is a fine covering of $]a,b[$. Hence by Vitali's theorem [9] there exist finitely many pairwise disjoint intervals $[\theta_j, \theta_j+\zeta_j]$, $j = 1,\ldots,M$, with $\zeta_j \in \left]0, \rho(\theta_j)\right]$, such that

$$\text{(4.6)} \qquad \sum_{j=1}^{M} \int_{\theta_j}^{\theta_j+\zeta_j} \|(v^\theta,\xi^\theta)\|_{u^\theta}\, d\theta > \int_a^b \|(v^\theta,\xi^\theta)\|_{u^\theta}\, d\theta - \varepsilon.$$

Now let $\gamma_\nu : \theta \mapsto u_\nu^\theta$ be a sequence of piecewise elementary paths with tangent vector $(v_\nu^\theta, \xi_\nu^\theta)$, uniformly converging to γ. By the continuity of the functionals Φ_{θ_j} and the uniform convergence $u_\nu^\theta \to u^\theta$, for all ν

sufficiently large we have

$$\left|\Phi_{\theta_j}(u_\nu^{\theta_j}) - \Phi_{\theta_j}(u^{\theta_j})\right| < \frac{\varepsilon}{M}, \qquad \left|\Phi_{\theta_j}(u_\nu^{\theta_j+\zeta_j}) - \Phi_{\theta_j}(u^{\theta_j+\zeta_j})\right| < \frac{\varepsilon}{M}, \tag{4.7}$$

$$\left\|u_\nu^\theta - u^\theta\right\|_{\mathbf{L}^1} < \min\left\{\delta(\theta_1), \ldots, \delta(\theta_M)\right\}, \qquad \theta \in [a,b]. \tag{4.8}$$

Using (4.4), (4.7), (4.3) and (4.6) we now obtain

$$\begin{aligned}
\|\gamma_\nu\|_* &\geq \sum_{j=1}^{M} \int_{\theta_j}^{\theta_j+\zeta_j} \left\|(v_\nu^\theta, \xi_\nu^\theta)\right\|_{u_\nu^\theta} d\theta \\
&\geq \frac{1}{1+C_2\varepsilon} \sum_{j=1}^{M} \left[\Phi_{\theta_j}(u_\nu^{\theta_j+\zeta_j}) - \Phi_{\theta_j}(u_\nu^{\theta_j})\right] \\
&\geq \frac{1}{1+C_2\varepsilon} \sum_{j=1}^{M} \left[\Phi_{\theta_j}(u^{\theta_j+\zeta_j}) - \Phi_{\theta_j}(u^{\theta_j}) - \frac{2\varepsilon}{M}\right] \\
&\geq \frac{1}{1+C_2\varepsilon} \sum_{j=1}^{M} \left[\int_{\theta_j}^{\theta_j+\zeta_j} \|(v^\theta, \xi^\theta)\|_{u^\theta} d\theta - C_1\varepsilon\zeta_j - \frac{2\varepsilon}{M}\right] \\
&\geq \frac{1}{1+C_2\varepsilon} \left[(\|\gamma\|_* - \varepsilon) - C_1\varepsilon(b-a) - 2\varepsilon\right].
\end{aligned} \tag{4.9}$$

Since $\varepsilon > 0$ was arbitrary and the constants C_1, C_2 do not depend on ε, the relation (4.1) is proved.

Completion of the proof

To complete the proof of Theorem 2, we need to construct suitable functionals Φ_θ satisfying the conditions **(C1)**, **(C2)**. Roughly speaking, $\Phi_\theta(w)$ should measure the weighted distance between w and u^θ. Since we do not have an explicit formula for this distance, we resort to a suitable approximation.

Let $\varepsilon > 0$ be given. For any $\bar\theta \in]a,b[$, let $\bar x_1 < \cdots < \bar x_N$ be the jumps in $u^{\bar\theta}$ and define $u_\alpha^\pm \doteq u^{\bar\theta}(x_\alpha\pm)$. Choose

$$\eta = \eta(\bar\theta) \in]0, \varepsilon] \tag{5.1}$$

so that the intervals $I_\alpha \doteq [\bar{x}_\alpha - \eta,\ \bar{x}_\alpha + \eta]$ are mutually disjoint and

$$\sum_{i=1}^{n}\sum_{\alpha=1}^{N}\int_{I_\alpha}\left|v_i^{\bar{\theta}}(x)\right| dx < \varepsilon. \tag{5.2}$$

Define the functional $\Phi_{\bar{\theta}} : \mathcal{D}_{PL} \mapsto \mathbb{R}$ by setting

$$\Phi_{\bar{\theta}}(w) \doteq \int g(x, w(x))\, dx, \tag{5.3}$$

(5.4)

$$g(x,w) \doteq \begin{cases} \sum_{i=1}^{n} J_i^\alpha\big(u^{\bar{\theta}}(x), w\big) W_i^{u^{\bar{\theta}}}(\bar{x}_\alpha) & \text{if } x \in I_\alpha^- \doteq [\bar{x}_\alpha - \eta,\ \bar{x}_\alpha[\,, \\ \sum_{i=1}^{n} J_i^\alpha\big(w, u^{\bar{\theta}}(x)\big) W_i^{u^{\bar{\theta}}}(\bar{x}_\alpha) & \text{if } x \in I_\alpha^+ \doteq]\bar{x}_\alpha,\ \bar{x}_\alpha + \eta]\,, \\ \sum_{i=1}^{n} J_i\big(u^{\bar{\theta}}(x), w\big) W_i^{u^{\bar{\theta}}}(x) & \text{if } x \notin \cup_\alpha I_\alpha, \end{cases}$$

where $J_i(u,v) \doteq \big|E_i(u,v)\big|$ and

- if $\big|E_i(u_\alpha^-, u_\alpha^+)\big| \le \varepsilon$, we let $J_i^\alpha(w_1, w_2) \equiv 0$;
- if $\big|E_i(u_\alpha^-, u_\alpha^+)\big| > \varepsilon$, we define

(5.5)

$$J_i^\alpha(w_1, w_2) \doteq \begin{cases} 0 & \text{if } \operatorname{sgn}\big(E_i(w_1,w_2)\big) \ne \operatorname{sgn}\big(E_i(u_\alpha^-, u_\alpha^+)\big), \\ [\![J_i(w_1,w_2) - \varepsilon]\!]_+ & \text{if } \operatorname{sgn}\big(E_i(w_1,w_2)\big) = \operatorname{sgn}\big(E_i(u_\alpha^-, u_\alpha^+)\big). \end{cases}$$

Here $[\![t]\!]_+$ denotes the positive part of t. Observe that $\Phi_{\bar{\theta}}\big(u^{\bar{\theta}}\big) = 0$.

Proof of property (C1). First of all take $\rho_1(\bar{\theta})$ small so that $x_\alpha(\bar{\theta} + \zeta) \in I_\alpha$ for all $\zeta \in [0, \rho_1]$. Outside the intervals I_α, from part a) of Lemma 1, if ρ_1 is small enough, we have

$$\begin{aligned} &\sum_{i=1}^{n}\int_{\mathbb{R}\setminus\cup_\alpha I_\alpha} J_i\Big(u^{\bar{\theta}}(x), u^{\bar{\theta}+\zeta}(x)\Big) W_i^{u^{\bar{\theta}}}(x)\, dx \\ &\ge \zeta \sum_{i=1}^{n}\int_{\mathbb{R}\setminus\cup_\alpha I_\alpha} \left|v_i^{\bar{\theta}}(x)\right| W_i^{u^{\bar{\theta}}}(x)\, dx - C\int_{\mathbb{R}\setminus\cup_\alpha I_\alpha}\left|u^{\bar{\theta}+\zeta}(x) - u^{\bar{\theta}}(x)\right|^2 dx \\ &\quad - C\int_{\mathbb{R}\setminus\cup_\alpha I_\alpha}\left|u^{\bar{\theta}+\zeta}(x) - u^{\bar{\theta}}(x) - \zeta v^{\bar{\theta}}(x)\right| dx \end{aligned}$$

$$\geq \zeta \sum_{i=1}^{n} \int_{\mathbb{R}\setminus\cup_\alpha I_\alpha} \left|v_i^{\bar\theta}(x)\right| W_i^{u^{\bar\theta}}(x)\,dx - C \sup_{x\in\mathbb{R}\setminus\cup_\alpha I_\alpha} \left|u^{\bar\theta+\zeta}(x) - u^{\bar\theta}(x)\right|$$

$$(5.6)\times \int_{\mathbb{R}\setminus\cup_\alpha I_\alpha} \left[\left|u^{\bar\theta+\zeta}(x) - u^{\bar\theta}(x) - \zeta v^{\bar\theta}(x)\right| + \zeta \sum_{i=1}^{n} \left|v_i^{\bar\theta}(x)\right|\right] dx - C\varepsilon\zeta$$

$$\geq \zeta \sum_{i=1}^{n} \int_{\mathbb{R}\setminus\cup_\alpha I_\alpha} \left|v_i^{\bar\theta}(x)\right| W_i^{u^{\bar\theta}}(x)\,dx - C\varepsilon\zeta.$$

Denote with G_α the set of indices i such that $\left|E_i(u_\alpha^-, u_\alpha^+)\right| > \varepsilon$.

Assume that $\bar\xi_\alpha \doteq \xi_\alpha(\bar\theta) > 0$, the case $\bar\xi_\alpha < 0$ being similar. We can choose $\rho_1(\bar\theta)$ small enough such that $J_i\left(u^{\bar\theta+\zeta}(x), u^{\bar\theta}(x)\right) > \varepsilon$ for all $i \in G_\alpha$, $x \in [\bar x_\alpha, \bar x_\alpha + \bar\xi_\alpha\zeta]$ and $\zeta \leq \rho_1$. By (5.2), and by making ρ_1 smaller (if necessary), it follows that

$$\sum_{i\in G_\alpha} \int_{\bar x_\alpha}^{\bar x_\alpha+\bar\xi_\alpha\zeta} J_i^\alpha\left(u^{\bar\theta+\zeta}(x), u^{\bar\theta}(x)\right) W_i^{u^{\bar\theta}}(\bar x_\alpha)\,dx$$

$$\geq \sum_{i\in G_\alpha} \int_{\bar x_\alpha}^{\bar x_\alpha+\bar\xi_\alpha\zeta} J_i(u_\alpha^-, u_\alpha^+) W_i^{u^{\bar\theta}}(\bar x_\alpha)\,dx$$

$$(5.7)\qquad - C \int_{\bar x_\alpha}^{\bar x_\alpha+\bar\xi_\alpha\zeta} \left(\left|u^{\bar\theta+\zeta}(x) - u_\alpha^-\right| + \left|u^{\bar\theta}(x) - u_\alpha^+\right|\right) dx$$

$$- C\zeta \sum_{i\in G_\alpha} \int_{\bar x_\alpha}^{\bar x_\alpha+\bar\xi_\alpha\zeta} \left|v_i^{\bar\theta}(x)\right| dx - C\varepsilon\zeta$$

$$\geq \zeta\,\bar\xi_\alpha \sum_{i=1}^{n} J_i(u_\alpha^-, u_\alpha^+) W_i^{u^{\bar\theta}}(\bar x_\alpha) - C\zeta \sum_{i=1}^{n} \int_{I_\alpha} \left|v_i^{\bar\theta}(x)\right| dx - C\varepsilon\zeta.$$

By (5.6)-(5.7) it follows that

$$(5.8)\qquad \Phi_{\bar\theta}(u^{\bar\theta+\zeta}) \geq \zeta \left\|(v^{\bar\theta}, \xi^{\bar\theta})\right\|_{u^{\bar\theta}} - C\zeta \sum_{i=1}^{n}\sum_{\alpha=1}^{N} \int_{I_\alpha} \left|v_i^{\bar\theta}(x)\right| dx - C\varepsilon\zeta,$$

hence by (5.2) and part c) of Lemma 1, (4.3) follows. □

Proof of property (C2). It is sufficient to prove that, for some $\rho_2(\bar\theta)$ small enough and all $\theta \in \left[\bar\theta, \bar\theta + \rho_2(\bar\theta)\,\right]$, one has

$$(5.9)\qquad \frac{d\Phi_{\bar\theta}(\tilde u^\theta)}{d\theta} \leq \left\|\left(\tilde v^\theta, \tilde\xi^\theta\right)\right\|_{\tilde u^\theta} (1 + C\varepsilon),$$

provided that the distance $\|\tilde{u}^\theta - u^\theta\|$ remains sufficiently small. For a.e. point θ the map $\theta \mapsto \tilde{u}^\theta$ is an elementary path in a neighborhood of θ. So fix one of these points θ. Assume that $\tilde{\xi}^\theta = (0, \ldots, 0, \tilde{\xi}_\beta, 0, \ldots, 0)$, with $\tilde{\xi}_\beta$ being the shift corresponding to a discontinuity of $\tilde{u}^\theta$ located at point x_β. There are three cases.

Case 1: $x_\beta \in I_\alpha^+$ for some $\alpha = 1, \ldots, N$. Assume $\tilde{\xi}_\beta > 0$. Hence we can see that on a right neighborhood of θ we have

$$\tilde{u}^{\theta+\Delta\theta}(x) = \tilde{u}^\theta(x) + (\tilde{u}^- - \tilde{u}^+)\,\chi_{[x_\beta, x_\beta + \tilde{\xi}_\beta \Delta\theta]}(x), \tag{5.10}$$

for suitable states $\tilde{u}^-, \tilde{u}^+$. Notice that we can also assume that the Riemann problem $(\tilde{u}^-, \tilde{u}^+)$ is solved only by one wave belonging to some characteristic family, say the j-th one. Moreover,

$$\frac{d\Phi_{\bar{\theta}}(\tilde{u}^\theta)}{d\theta} = \tilde{\xi}_\beta \sum_{i=1}^n \Big(J_i^\alpha(\tilde{u}^-, u^{\bar{\theta}}(x_\beta)) - J_i^\alpha(\tilde{u}^+, u^{\bar{\theta}}(x_\beta)) \Big) W_i^{u^{\bar{\theta}}}(\bar{x}_\alpha). \tag{5.11}$$

Observe that $\|(\tilde{v}^\theta, \tilde{\xi}^\theta)\|_{\tilde{u}^\theta} = \tilde{\xi}_\beta\, J_j(\tilde{u}^-, \tilde{u}^+) W_j^{\tilde{u}^\theta}(x_\beta)$, so we will show that

$$\sum_{i=1}^n \Big(J_i^\alpha(\tilde{u}^-, u^{\bar{\theta}}(x_\beta)) - J_i^\alpha(\tilde{u}^+, u^{\bar{\theta}}(x_\beta)) \Big) W_i^{u^{\bar{\theta}}}(\bar{x}_\alpha) \le (1 + C\varepsilon)\, J_j(\tilde{u}^-, \tilde{u}^+) W_j^{\tilde{u}^\theta}(x_\beta). \tag{5.12}$$

Call B_α the set of indices i which belong to G_α and satisfy $J_i^\alpha(\tilde{u}^-, u^{\bar{\theta}}(x_\beta)) - J_i^\alpha(\tilde{u}^+, u^{\bar{\theta}}(x_\beta)) > 0$. Notice that the terms corresponding to indices $i \notin B_\alpha$ do not give any positive contribution to the sum in (5.12). So take $i \in B_\alpha$. Moreover, given $\varepsilon_1 \in]0, \varepsilon]$, it is easy to see that in any fixed small interval there exists a point y^θ such that

$$\Big| \tilde{u}^\theta(y^\theta) - u^{\bar{\theta}}(y^\theta) \Big| \le \varepsilon_1, \qquad \theta \in [\bar{\theta}, \bar{\theta} + \rho_2], \tag{5.13}$$

if $\|\tilde{u}^\theta - u^\theta\|_{\mathbf{L}^1} < \delta$, with ρ_2 and δ sufficiently small. So we can find y_β^+ close to the right of x_β and y_β^- close to the left of $\bar{x}_\alpha$ such that they are continuity points for $\tilde{u}^\theta$, such that $\Big|\tilde{u}^\theta(y_\beta^+) - u^{\bar{\theta}}(x_\beta)\Big| \le \varepsilon_1$,

$\left|\tilde{u}^\theta(y_\beta^-) - u_\alpha^-\right| \le \varepsilon_1$, and

$$\left|E_k\big(\tilde{u}^-, u^{\bar\theta}(x_\beta)\big) - E_k\big(\tilde{u}^-, \tilde{u}^\theta(y_\beta^+)\big)\right| \le \varepsilon/4, \tag{5.14}$$

$$\left|E_k\big(\tilde{u}^+, u^{\bar\theta}(x_\beta)\big) - E_k\big(\tilde{u}^+, \tilde{u}^\theta(y_\beta^+)\big)\right| \le \varepsilon/4, \tag{5.15}$$

for all $k = 1, \ldots, n$. Call $\tilde{u}_\beta^\pm \doteq \tilde{u}^\theta\big(y_\beta^\pm\big)$.

Assume now that $E_i(u_\alpha^-, u_\alpha^+) < 0$. Since i belongs to B_α one has $J_i^\alpha\big(\tilde{u}^-, u^{\bar\theta}(x_\beta)\big) - J_i^\alpha\big(\tilde{u}^+, u^{\bar\theta}(x_\beta)\big) > 0$, hence it follows $E_i\big(\tilde{u}^-, u^{\bar\theta}(x_\beta)\big) < -\varepsilon$. There are two subcases.
Subcase a): $E_i\big(\tilde{u}^+, u^{\bar\theta}(x_\beta)\big) < -\varepsilon/4$. Hence $E_i\big(\tilde{u}^+, \tilde{u}^\theta(y_\beta^+)\big) < 0$ and one can prove that

$$J_i^\alpha\big(\tilde{u}^-, u^{\bar\theta}(x_\beta)\big) - J_i^\alpha\big(\tilde{u}^+, u^{\bar\theta}(x_\beta)\big) \le J_i\big(\tilde{u}^-, u^{\bar\theta}(x_\beta)\big) - J_i\big(\tilde{u}^+, u^{\bar\theta}(x_\beta)\big). \tag{5.16}$$

In connection with $\tilde{u}^\theta$, define the functions

$$u^*(x) \doteq \begin{cases} \tilde{u}^\theta(x) & \text{if } x \le y_\beta^- \text{ or } x \ge y_\beta^+, \\ \tilde{u}^\theta\big(y_\beta^-\big) & \text{if } y_\beta^- \le x < \bar{x}_\alpha, \\ \tilde{u}^\theta\big(y_\beta^+\big) & \text{if } \bar{x}_\alpha < x \le y_\beta^+, \end{cases} \tag{5.17}$$

$$u^*(x) \doteq \begin{cases} \tilde{u}^\theta(x) & \text{if } x \le y_\beta^- \text{ or } x \ge y_\beta^+, \\ \tilde{u}^\theta(x_\beta-) & \text{if } y_\beta^- < x < x_\beta, \\ \tilde{u}^\theta(x_\beta+) & \text{if } x_\beta < x < y_\beta^+. \end{cases} \tag{5.18}$$

By the semicontinuity of the weights and properties **(P1)**-**(P3)**, if we choose δ and ρ_2 sufficiently small, we can see that for every $i = 1, \ldots, n$

$$W_i^{u^{\bar\theta}}(\bar{x}_\alpha) \le W_i^{u^*}(\bar{x}_\alpha) + \varepsilon, \tag{5.19}$$

$$W_i^{u^*}(\bar{x}_\alpha) \le W_i^{u^*}(x_\beta) \le W_i^{\tilde{u}^\theta}(x_\beta). \tag{5.20}$$

Then from (5.11) and (5.19)

(5.21)

$$
\begin{aligned}
&\Big(J_i(\tilde{u}^-, u^{\bar\theta}(x_\beta)) - J_i(\tilde{u}^+, u^{\bar\theta}(x_\beta))\Big) W_i^{u^{\bar\theta}}(\bar{x}_\alpha) \\
&\quad = \tilde{\xi}_\beta \Big(J_i(\tilde{u}^-, u^{\bar\theta}(x_\beta)) - J_i(\tilde{u}^+, u^{\bar\theta}(x_\beta))\Big) W_i^{u^*}(\bar{x}_\alpha) \\
&\qquad + \tilde{\xi}_\beta \Big(J_i(\tilde{u}^-, u^{\bar\theta}(x_\beta)) - J_i(\tilde{u}^+, u^{\bar\theta}(x_\beta))\Big)\Big(W_i^{u^{\bar\theta}}(\bar{x}_\alpha) - W_i^{u^*}(\bar{x}_\alpha)\Big) \\
&\quad \doteq B_1 + B_2,
\end{aligned}
$$

and

$$
B_2 \le C\tilde{\xi}_\beta \big|\tilde{u}^- - \tilde{u}^+\big| \varepsilon \le C \Big\| \Big(\tilde{v}^\theta, \tilde{\xi}^\theta\Big) \Big\|_{\tilde{u}^\theta} \varepsilon, \tag{5.22}
$$

$$
\begin{aligned}
B_1 &= \tilde{\xi}_\beta \Big(J_i\Big(\tilde{u}^-, \tilde{u}^+_\beta\Big) - J_i\Big(\tilde{u}^+, \tilde{u}^+_\beta\Big)\Big) W_i^{u^*}(\bar{x}_\alpha) \\
&\quad + \tilde{\xi}_\beta \Big[\Big(J_i\Big(\tilde{u}^-, u^{\bar\theta}(x_\beta)\Big) - J_i\Big(\tilde{u}^-, \tilde{u}^+_\beta\Big)\Big) \\
&\qquad + \Big(J_i\Big(\tilde{u}^+, \tilde{u}^+_\beta\Big) - J_i\Big(\tilde{u}^+, u^{\bar\theta}(x_\beta)\Big)\Big)\Big] W_i^{u^*}(\bar{x}_\alpha) \\
&\doteq B_1' + B_1''.
\end{aligned} \tag{5.23}
$$

We want to estimate B_1''. First we notice that for $u_1, u_2, u_3 \in \Omega$ and $i = 1, \ldots, n$, we have

(5.24)

$$
\begin{aligned}
J_i(u_1, u_2) - J_i(u_1, u_3) &= \int_0^1 \frac{d}{ds}\Big[J_i\big(u_1, su_2 + (1-s)\,u_3\big)\Big]\,ds \\
&= (u_2 - u_3)\int_0^1 \partial_2 J_i\big(u_1, su_2 + (1-s)\,u_3\big)\,ds.
\end{aligned}
$$

This implies

$$
\begin{aligned}
B_1'' \le{}& \tilde{\xi}_\beta \Big| u^{\bar\theta}(x_\beta) - \tilde{u}^\theta\big(y^+_\beta\big)\Big|\, W_i^{u^*}(\bar{x}_\alpha) \\
&\times \Bigg| \int_0^1 \Big[\partial_2 J_i\Big(\tilde{u}^-, su^{\bar\theta}(x_\beta) + (1-s)\,\tilde{u}^\theta\big(y^+_\beta\big)\Big)
\end{aligned}
$$

$$- \partial_2 J_i \left(\tilde u^+, s u^{\bar\theta}(x_\beta) + (1-s)\,\tilde u^\theta(y_\beta^+)\right)\Big] \, ds \Big|$$

$$= \tilde\xi_\beta \left| u^{\bar\theta}(x_\beta) - \tilde u^\theta(y_\beta^+)\right| \left|\tilde u^- - \tilde u^+\right| W_i^{u^*}(\bar x_\alpha)$$

$$\times \left| \int_0^1 \int_0^1 \partial_1 \partial_2 J_i \left(\sigma \tilde u^- + (1-\sigma)\,\tilde u^+, s u^{\bar\theta}(x_\beta) \right.\right.$$

$$\left.\left. + (1-s)\,\tilde u^\theta(y_\beta^+)\right) ds d\sigma \right|$$

$$\le C\tilde\xi_\beta \left|\tilde u^- - \tilde u^+\right| \left|u^{\bar\theta}(x_\beta) - \tilde u^\theta(y_\beta^+)\right| \le C \left\| \left(\tilde v^\theta, \tilde\xi^\theta\right)\right\|_{\tilde u^\theta} \varepsilon. \tag{5.25}$$

Finally we want to estimate B_1'. Call $E_i \doteq E_i(\tilde u^-, \tilde u_\beta^+)$, $E_i' \doteq E_i(\tilde u^-, \tilde u^+)$, $E_i'' \doteq E_i(\tilde u^+, \tilde u_\beta^+)$, and the same for the J's. Also define $W_i \doteq W_i^{u^*}(\bar x_\alpha)$ and $W_i' \doteq W_i^{u^*}(x_\beta)$. Recall now that $E_i, E_i'' < 0$, $J_i - J_i'' > 0$ and that $E_i' = 0$ for all $i \neq j$. Hence if $i = j$ from Glimm estimates we get $0 < J_j - J_j'' \le -E_j' + C_3\Delta(E', E'')$. If δ_0 is sufficiently small, this implies that $E_j' < 0$ hence $J_j' = -E_j'$. If, instead, $i \neq j$ we get $J_i - J_i'' \le C_3\Delta(E', E'')$.

Subcase b): $E_i(\tilde u^+, u^{\bar\theta}(x_\beta)) > -\varepsilon/4$. Hence $E_i(\tilde u^+, \tilde u^\theta(y_\beta^+)) > -\varepsilon/2$. If it is positive, then

$$\begin{aligned} J_i^\alpha(\tilde u^-, u^{\bar\theta}(x_\beta)) - J_i^\alpha(\tilde u^+, u^{\bar\theta}(x_\beta)) &= J_i(\tilde u^-, u^{\bar\theta}(x_\beta)) - \varepsilon \\ &\le J_i(\tilde u^-, u^{\bar\theta}(x_\beta)). \end{aligned} \tag{5.26}$$

With the notation introduced before, if $i = j$ we have $J_j \le -E_j' - E_j'' + C_3\Delta(E', E'') \le -E_j' + C_3\Delta(E', E'')$ hence $E_j' < 0$, while $J_i \le C_3\Delta(E', E'')$ if $i \neq j$.

If, instead, $0 \ge E_i(\tilde u^+, \tilde u^\theta(y_\beta^+)) > -\varepsilon/2$, we get

$$J_i^\alpha(\tilde u^-, u^{\bar\theta}(x_\beta)) - J_i^\alpha(\tilde u^+, u^{\bar\theta}(x_\beta)) \le J_i(\tilde u^-, u^{\bar\theta}(x_\beta)) - J_i(\tilde u^+, u^{\bar\theta}(x_\beta)), \tag{5.27}$$

so we recover (5.16) and we can conclude as in subcase a).
The case $E_i(u_\alpha^-, u_\alpha^+) > 0$ is treated in a similar way.

In both cases, one can prove that $J_j'(W_j - W_j') \le -\kappa_1\Delta(E', E'')$, if $\kappa_2 > C_3$. By the previous analysis it follows that

$$\sum_{i \in B_\alpha} (J_i - J_i'')W_i - J_j'W_j'$$

$$\text{(5.28)} \qquad \leq \sum_{i\neq j} C_3\Delta(E',E'')W_i + (J_j - J_j'' - J_j')W_j + J_j'(W_j - W_j')$$
$$\leq \Delta(E',E'')\big(C_4 + C_5\kappa_1\delta_0 - \kappa_1\big) < 0,$$

if $\kappa_1 > 2C_4$ and $\delta_0 < (2C_5)^{-1}$. Finally, this together with (5.20) implies (4.4).

Assume now that $\tilde{\xi}_\beta < 0$, hence in a neighborhood of θ

$$\text{(5.29)} \qquad \tilde{u}^{\theta+\Delta\theta}(x) = \tilde{u}^\theta(x) + (\tilde{u}^+ - \tilde{u}^-)\,\chi_{[x_\beta+\tilde{\xi}_\beta\Delta\theta,x_\beta]}(x),$$

and

(5.30)
$$\frac{d\Phi_{\bar\theta}(\tilde{u}^\theta)}{d\theta} = |\tilde{\xi}_\beta|\sum_{i=1}^n \Big(J_i^\alpha(\tilde{u}^+,u^{\bar\theta}(x_\beta)) - J_i^\alpha(\tilde{u}^-,u^{\bar\theta}(x_\beta))\Big)W_i^{u^{\bar\theta}}(\bar{x}_\alpha).$$

Proceeding as before, we will show that

$$\text{(5.31)} \qquad \sum_{i=1}^n \Big(J_i^\alpha(\tilde{u}^+,u^{\bar\theta}(x_\beta)) - J_i^\alpha(\tilde{u}^-,u^{\bar\theta}(x_\beta))\Big)W_i^{u^{\bar\theta}}(\bar{x}_\alpha)$$
$$\leq (1+C\varepsilon)\,J_j(\tilde{u}^-,\tilde{u}^+)W_j^{\tilde{u}^\theta}(x_\beta).$$

We can make an analysis entirely similar to that before and show again that (5.9) holds.

Case 2: $x_\beta \in I_\alpha^-$. Then, if for example $\tilde{\xi}_\beta > 0$, it is easy to show that

(5.32)
$$\frac{d\Phi_{\bar\theta}(\tilde{u}^\theta)}{d\theta} = \tilde{\xi}_\beta\sum_{i=1}^n \Big(J_i^\alpha(u^{\bar\theta}(x_\beta),\tilde{u}^-) - J_i^\alpha(u^{\bar\theta}(x_\beta),\tilde{u}^+)\Big)W_i^{u^{\bar\theta}}(\bar{x}_\alpha).$$

We can treat this case as Case 1.

Case 3: $x_\beta \in \mathcal{F} \doteq \mathbb{R}\setminus\cup_\alpha I_\alpha$. We want to show that also in this case (5.9) holds. Since $u^{\bar\theta}$ and $W_i^{u^{\bar\theta}}(x)$ are continuous on $\mathcal{F}$, by assuming for example that $\tilde{\xi}_\beta > 0$, we get

$$\frac{d\Phi_{\bar\theta}(\tilde{u}^\theta)}{d\theta} \leq \lim_{h\to 0}\frac{1}{h}\int_{x_\beta}^{x_\beta+\tilde{\xi}_\beta h}\sum_{i=1}^n \Big|J_i\Big(u^{\bar\theta}(x),\tilde{u}^-\Big)$$

$$- J_i\Big(u^{\bar\theta}(x), \tilde u^+\Big)\Big| W_i^{u^{\bar\theta}}(x)\,dx \tag{5.33}$$

$$= |\tilde\xi_\beta| \sum_{i=1}^n \Big| J_i\Big(u^{\bar\theta}(x_\beta), \tilde u^-\Big) - J_i\Big(u^{\bar\theta}(x_\beta), \tilde u^+\Big)\Big| W_i^{u^{\bar\theta}}(x_\beta),$$

and the same is true if $\tilde\xi_\beta < 0$ instead.

As in (5.13) we can prove that there exist points $y_\beta^- < x_\beta$ and $y_\beta^+ > x_\beta$ close to x_β such that

$$\Big|\tilde u^\theta(y_\beta^\pm) - u^{\bar\theta}(x_\beta)\Big| \le \varepsilon_1. \tag{5.34}$$

Define u^* and $u^\star$ as in (5.17)-(5.18) with $\bar x_\alpha$ replaced by x_β. By **(P2)-(P3)** and part a) of Lemma 1, if we choose δ and ρ_2 sufficiently small, we can see that

$$W_i^{u^{\bar\theta}}(x_\beta) \le W_i^{u^*}(x_\beta) + \varepsilon, \qquad W_i^{u^*}(x_\beta) \le W_i^{u^\star}(x_\beta) \le W_i^{\tilde u^\theta}(x_\beta). \tag{5.35}$$

We observe that since $u^{\bar\theta}$ is continuous at x_β, if ε_1 is small enough, the Riemann problem $\Big(\tilde u^\theta(y_\beta^-), \tilde u^\theta(y_\beta^+)\Big)$ is solved by waves with small strength, say less than ε. Now call $\tilde u_\beta^\pm \doteq \tilde u^\theta(y_\beta^\pm)$.

By proceeding as in (5.21)-(5.25) one gets

$$\begin{aligned}\frac{d\Phi_{\bar\theta}(\tilde u^\theta)}{d\theta} &\le |\tilde\xi_\beta| \sum_{i=1}^n \Big| J_i\Big(\tilde u_\beta^-, \tilde u^-\Big) - J_i\Big(\tilde u_\beta^-, \tilde u^+\Big)\Big| W_i^{u^*}(x_\beta) \\ &\quad + C\Big\|\Big(\tilde v^\theta, \tilde\xi^\theta\Big)\Big\|_{\tilde u^\theta}\varepsilon.\end{aligned} \tag{5.36}$$

Define $E_i \doteq E^i(\tilde u_\beta^-, \tilde u^+)$, $E_i' \doteq E^i(\tilde u_\beta^-, \tilde u^-)$, $E_i'' \doteq E^i(\tilde u^-, \tilde u^+)$ and the same for the J's. Also call $W_i \doteq W_i^{u^*}(x_\beta)$ and $W_i'' \doteq W_i^{u^\star}(x_\beta)$. Recall that we can assume $J_i'' = 0$ for $i \ne j$. Now we want to show that, for a suitable choice of κ_1 and δ_0, the following holds:

$$\sum_{i=1}^n |J_i' - J_i| W_i \le J_j'' W_j''(1 + C\varepsilon). \tag{5.37}$$

By (2.10) one gets $\big|J_i' - J_i\big| \le J_i'' + C_3\Delta(E', E'')$ for all $i = 1, \ldots, n$, and since the waves in the Riemann problem $\Big(\tilde u^\theta(y_\beta^-), \tilde u^\theta(y_\beta^+)\Big)$ are very small, it follows that

$$J_j''(W_j - W_j'') \leq C\varepsilon J_j'' - \kappa_1\Delta(E', E''). \tag{5.38}$$

By (5.37)-(5.38) we get

(5.39)

$$\begin{aligned}
\sum_{i=1}^{n} &|J_i' - J_i|W_i - J_j''W_j'' \\
&\leq \sum_{i\neq j} |J_i' - J_i|W_i + \Big(|J_j' - J_j| - J_j''\Big)W_j + J_j''(W_j - W_j'') \\
&\leq C\varepsilon J_j''W_j'' + \Delta(E', E'')(C_4 + C_5\kappa_1\delta_0 - \kappa_1) \leq C\varepsilon J_j''W_j'',
\end{aligned}$$

if $\kappa_1 > 2C_4$ and $\delta_0 < (2C_5)^{-1}$. This together with (5.35) implies (5.9). □

References

[1] A. Bressan, A locally contractive metric for systems of conservation laws, *Ann. Scuola Norm. Sup. Pisa* **IV-22** (1995), 109–135.

[2] A. Bressan, The semigroup approach to systems of conservation laws, *Mathematica Contemporanea* **10** (1996), 21–74.

[3] A. Bressan and R. M. Colombo, The semigroup generated by 2×2 conservation laws, *Arch. Rational Mech. Anal.* **133** (1995), 1–75.

[4] A. Bressan and R. M. Colombo, Unique solutions of 2×2 conservation laws with large data, *Indiana Univ. Math. J.* **44** (1995), 677–725.

[5] A. Bressan, G. Crasta and B. Piccoli, Well-posedness of the Cauchy problem for $n \times n$ systems of conservation laws, preprint S.I.S.S.A., Trieste, 1996.

[6] A. Bressan and A. Marson, A variational calculus for discontinuous solutions of systems of conservation laws, *Comm. Part. Diff. Equat.* **20** (1995), 1491–1552.

[7] M. Crandall, The semigroup approach to first-order quasilinear equations in several space variables, *Israel J. Math.* **12** (1972), 108–132.

[8] K. Deimling, "Nonlinear Functional Analysis", Springer-Verlag, Berlin, 1985.

[9] L. C. Evans and R. F. Gariepy, "Measure Theory and Fine Properties of Functions", CRC Press, 1992.

[10] J. Glimm, Solutions in the large for nonlinear hyperbolic systems of equations, *Comm. Pure Appl. Math.* **18** (1965), 697–715.
[11] S. Kruzkov, First order quasilinear equations with several space variables, *Math. USSR Sbornik* **10** (1970), 217–243.
[12] P. Lax, Hyperbolic systems of conservation laws II, *Comm. Pure Appl. Math.* **10** (1957), 537–566.
[13] J. Smoller, "Shock Waves and Reaction-Diffusion equations", Springer-Verlag, New York, 1983.
[14] B. Temple, No $\mathbf{L}^1$-contractive metrics for systems of conservation laws, *Trans. Amer. Math. Soc.* **288** (1985), 471–480.

Paolo Baiti and Alberto Bressan
S.I.S.S.A., Via Beirut 4, Trieste 34014, Italy
E-Mail: baiti@sissa.it
bressan@sissa.it

Time decay of L^p norms for solutions of the wave equation on exterior domains

Michael Beals

1. Introduction and statement of the main results

We consider global estimates for the time dependence of spatial L^p norms of solutions to the wave equation on the exterior of a smooth compact strictly convex obstacle in $\mathbb{R}^n, n \geq 2$, with vanishing Dirichlet data on the boundary:

$$(1.1) \qquad \Box u = 0 \text{ on } \Omega \times \mathbb{R}, \; u(0) = f_0, \; u_t(0) = f_1, \; u\,|_{\partial\Omega\times\mathbb{R}} = 0.$$

The behavior of the norms of solutions to the d'Alembert wave equation in L^p spaces for $p \neq 2$ has been studied for its applications in both linear and nonlinear theory. Estimates have been used to deduce small time existence, long time existence, and scattering properties of solutions to hyperbolic wave equations (see for example Ginibre-Velo [4], Strauss [15], Shatah-Struwe [13], Lindblad-Sogge [8]). Unlike the energy norm, certain other norms of solutions can decay as the time t approaches infinity. If u satisfies the free-space problem

$$(1.2) \qquad \Box u = 0 \text{ on } \mathbb{R}^n \times \mathbb{R}, \; u(0) = f_0, \; u_t(0) = f_1,$$

then as long as the initial data f_0 and f_1 and sufficiently many of their derivatives are in $L^1(\mathbb{R}^n)$, the following time decay holds:

$$(1.3) \qquad \|u(t)\|_\infty \leq Ct^{-(n-1)/2}, \quad \text{as } t \to \infty.$$

(Precise minimal smoothness hypothesis on the data are given in Marshall-Strauss-Wainger [9]). Energy estimates then yield the interpolated results on the norms as maps from $L^p(\mathbb{R}^n)$ to $L^{p'}(\mathbb{R}^n)$, for p' the dual index to p, of the operators $E(t)f$ and $E_0(t)f$ given by

$$(1.4) \qquad E(t)f(x) = \int \frac{\sin t\,|\,\xi\,|}{|\,\xi\,|}\, e^{ix\cdot\xi} \hat{f}(\xi)d\xi,$$

Research supported by NSF grant DMS-9401819

and

$$E_0(t)f(x) = \int \frac{\cos t \mid \xi \mid}{\mid \xi \mid} e^{ix \cdot \xi} \hat{f}(\xi)d\xi. \tag{1.5}$$

An example is the Strichartz estimate [16]: if $n \geq 2$, $\frac{1}{p} = \frac{1}{2} + \frac{1}{n+1}$, $\frac{1}{p'} = \frac{1}{2} - \frac{1}{n+1}$, then

$$\|E(t)f\|_{p'} \leq Ct^{-(n-1)/(n+1)}\|f\|_p. \tag{1.6}$$

The methods in Strauss [14] can then be applied to the long-time existence and scattering for certain nonlinear equations $\Box u = g(u)$ on $\mathbb{R}^n \times \mathbb{R}$, since

$$\begin{aligned} &u(t) = \int E(t-s)g(u(s))ds \text{ satisfies} \\ &\Box u = g(u) \text{ on } \mathbb{R}^n \times \mathbb{R},\ u(0) = 0, u_t(0) = 0. \end{aligned} \tag{1.7}$$

Here we consider the estimates corresponding to (1.3) and (1.6) for the solution of the exterior problem (1.1). In [3] we treated data of support in a fixed compact subset of Ω; here we handle the general case. The compact support hypothesis in [3] meant that norms of the solution near $\partial\Omega$ were known to have sufficient decay, since after finite time the singularities of the solution have passed away from $\partial\Omega$ since Ω is non-trapping. In that case the only large time parameter entered the problem after the interaction of the solution with the boundary, which takes place (essentially) in a bounded time interval. Time estimates involving smoothing operators (such as remainders in the parametrix for the boundary problem and microlocalizations away from the null bicharacteristics passing over the support of the data) were then particularly simple. In the general case considered here, the large time parameter can also be involved before the boundary interaction, and the time behavior of the solution must be analyzed near the boundary as well as far away. The main result is the following estimate, which (from the free space case) is best possible in terms of the time decay and the assumed regularity of the initial data.

Theorem 1.1. *Let $\Omega \subset \mathbb{R}^n$ be the complement of a smooth convex compact obstacle, and assume that the boundary $\partial\Omega$ has nonvanishing* *curvature at each point. Suppose that f_0, f_1 have support in Ω, with $f_0 \in L^{1,\sigma}(\Omega), f_1 \in L^{1,\sigma-1}(\Omega), \sigma > (n+1)/2$. If u is the solution to the initial-boundary problem (1.1), then there is a constant $C = C_\Omega$ such that*

$$\|u(t)\|_\infty \leq C|t|^{-(n-1)/2}(\| \wedge^\sigma f_0\|_1 + \| \wedge^{\sigma-1} f_1\|_1) \quad \text{for} \quad t \geq 1. \tag{1.8}$$

Here $\wedge$ denotes $(1-\Delta)^{1/2}$ for the Laplacian Δ associated with the boundary value problem, and the Sobolev spaces are defined accordingly. If $\sigma = (n+1)/2$ and $f_0 \in h^{1,\sigma}(\Omega), f_1 \in h^{1,\sigma}(\Omega)$, then the estimate (1.8) holds with $\| \ \|_1$ denoting the norm in the local Hardy space $h^1(\Omega)$.

From (1.8), the standard energy estimate, and analytic interpolation of the corresponding family of operators, the optimal decay of the solution to (1.1) for $f_0 = 0$, $f_1 \in L^p(\Omega)$ follows. As in (1.4), let $E_\Omega(t)f$ solve the following Cauchy problem:

$$\begin{gathered} \nu(t) = E_\Omega(t)f \text{ satisfies} \\ \Box\nu = 0 \text{ on } \Omega \times \mathbb{R}, \ \nu(0) = 0, \nu_t(0) = f, \ \nu\,|_{\partial\Omega\times\mathbb{R}} = 0. \end{gathered} \tag{1.9}$$

This operator is of interest for its application in nonlinear problems as in Strauss [14], since the inhomogeneous problem $\Box u = g$ on $\Omega \times \mathbb{R}$, $u(0) = 0$, $u_t(0) = 0$, $u\,|_{\partial\Omega\times\mathbb{R}} = 0$ has solution

$$u(t) = \int_0^t E_\Omega(t-s)g(s)ds. \tag{1.10}$$

Theorem 1.2. *Let Ω be as in Theorem 1.1. Suppose $f \in L^p(\Omega), 1/p = 1/2 + 1/(n+1), 1/p' = 1/2 - /(n+1)$. If $E_\Omega(t)f$ is given by (1.9), then there is a constant $C = C_\Omega$ such that*

$$\|E_\Omega(t)f\|_{p'} \leq C|t|^{-(n-1)/(n+1)}\|f\|_p \text{ for } t \geq 1.$$

These results are an extension to the case of non-compactly supported data of the ones given in [3], where a discussion of earlier and related results is given.

The proof of Theorem 1.1 involves a smooth time-independent decomposition of the solution to (1.1) into pieces αu and βu supported near (respectively away from) the obstacle. The term αu with compact support in space can be analyzed using the Taylor-Melrose parametrix for the correction on the boundary to the solution of the free space problem with the corresponding data. The time-invariance of the equation and the domain means that the time dependence of the parametrix is simple; the spatial localization can be taken near enough to the boundary to yield an explicit representation up to a smoothing operator. Boundedness estimates (somewhat simpler than those given in [3]) coupled with the explicit time dependence handle the term given by the parametrix, and the known local estimates in the compact-data case can be combined with these results to handle the remainder terms. A description of the arguments involved is given subsequently in Section 2.

The remaining term βu satisfies an inhomogeneous equation of the form $\Box\beta u = \alpha_0 Du$ for α_0 a smooth function of compact support near $\partial\Omega$ but identically zero on a smaller neighborhood of $\partial\Omega$. (Here $\alpha_0 Du$ is shorthand for a sum of such terms involving u and its derivatives). Since the solution to the free-space homogeneous problem satisfies the appropriate decay estimates, from (1.4) the primary term remaining to be estimated is

$$\int_0^t E(t-s)\alpha Du(s)ds.$$

Unlike the case of compactly supported data (where the primary contribution to this integral occurs for s in a bounded set), the action of the explicit operator $E(t-s)$ must be analyzed in detail over the entire interval of integration. The term $\alpha_0 Du(s)$ is obtained from the initial data (f_0, f_1) by the action of compositions of Fourier integral operators as in Hörmander [6] and Fourier-Airy operators as in Taylor [17]. Since zeroth order operators of this type are not bounded on $L^p(\mathbb{R}^n)$ for $p \neq 2$ (except when $n = 1$), the estimates on the individual operators in the composition will not alone suffice. As in [3] (and motivated by the considerations in [2]) integral operators of appropriate order again yield an operator with the desired decay estimate. In Section 3 the estimates on the corresponding kernels for the compositions of Fourier and Fourier-Airy operators are derived. Full details will appear elsewhere.

Notation. $\| \|_p$ denotes the norm in the Lebesgue space $L^p(\Omega), 1 \leq p \leq \infty$; $\| \|_1$ will also denote the norm in the local Hardy space $h^1(\Omega)$ when so indicated. This space is defined as follows: fix $\beta \in C^\infty(\mathbb{R}^+)$ with $0 \leq \beta \leq 1, \beta(\rho) \equiv 1$ for $\rho \geq 2, \beta(\rho) \equiv 0$ for $\rho \leq 1$; and let

$\wedge = (1 - \Delta)^{1/2}$ as before. Then $f \in h^1(\Omega)$ is equivalent to requiring $f, \beta(\wedge)f, D_j\beta(\wedge)f \in L^1(\Omega)$. (See Goldberg [5].) $L^{p,\sigma}(\Omega) = \{f : \wedge^\sigma f \in L^p(\Omega)\}$; $h^{1,\sigma}(\Omega) = \{f : \wedge^\sigma f \in h^1(\Omega)\}$.

$\square$ denotes the d'Alembertian $\partial_t^2 - \Delta$. $A \approx B$ means there are constants $C > c > 0$ with $cA \leq B \leq CA$. $\langle a \rangle = \sqrt{1 + a^2}$.

2. Behavior of solutions near $\partial\Omega$

Let $\Omega \subset \mathbb{R}^n$ denote the complement of a smooth compact convex obstacle as in Section 1; we will assume for notational convenience that Ω is contained in the unit ball centered at the origin. Let $\beta \in C^\infty(\mathbb{R}^n)$ be identically one outside a small neighborhood of the complement of Ω, zero on a small neighborhood of $\partial\Omega$, and let $\alpha = 1 - \beta$. If we write

$$u = \alpha u + \beta u, \tag{2.1}$$

then βu is the solution to the following free-space inhomogeneous problem:

(2.2)

$$\square\beta u = -2\nabla\beta \cdot \nabla u - (\Delta\beta)u \text{ on } \mathbb{R}^n \times \mathbb{R},\ \beta u(0) = \beta f_0,\ \beta u_t(0) = \beta f_1.$$

On the other hand, from the expressions for solutions to the free-space problem given in (1.4) and (1.5), we may write αu as

(2.3)

$$\alpha u = \alpha E_0(t)\alpha f_0 + \alpha E(t)\alpha f_1 - \alpha \int_0^1 E_{\partial\Omega}(t-r)[E_0(r)\alpha f_0 + E(r)\alpha f_1]dr.$$

Here $E_{\partial\Omega}$ denotes the operator correcting for the Dirichlet condition on the boundary. Since $\nu = E_0(t)\alpha f_0 + E(t)\alpha f_1$ satisfies the homogeneous free-space problem, the desired decay estimate for holds for ν (Marshall-Strauss-Wainger [9], Beals [2]):

$$\|E_0(t)\alpha f_0 + E(t)\alpha f_1\|_\infty \leq C|t|^{-(n-1)/2}(\|\wedge^\sigma f_0\|_1 + \|\wedge^{\sigma-1} f_1\|_1). \tag{2.4}$$

In order to analyze the remaining term $\alpha \int_0^1 E_{\partial\Omega}(t-r)\nu(r)dr$ in (2.3), we first note that the corresponding kernel can be decomposed as the Taylor-Melrose parametrix $K_{\partial\Omega}$ (Taylor [17]) plus a smoothing remainder term $R_{\partial\Omega}$. (As many terms in the asymptotic expansion for $K_{\partial\Omega}$ can be taken as necessary in order to have $R_{\partial\Omega}$ as smoothing as desired). The remainder term can be written in the form

$$(2.5)\quad \alpha u_1(t,x) = \alpha(x)\iint R_{\partial\Omega,M}(t-r,x,w')\wedge^{-M}\nu(r,w(w'))dw'dr,$$

with $R_{\partial\Omega,M}$ corresponding to having absorbed the factor $\wedge^M$ into the kernel. Here the coordinates on $\partial\Omega$ are taken to be $w = w(w')$ for w' near 0 in $\mathbb{R}^{n-1}$. Since α has compact support, x and w' are in compact sets. The arguments described in [3], using the local time decay for solutions corresponding to compactly supported data (Lax-Phillips [7], Morawetz-Ralston-Strauss [11], Melrose [10], Vainberg [18], Ralston [12]), can be adapted to yield the following estimate:

$$(2.6)\qquad \|\alpha R_{\partial\Omega,M}(t-r,x,w')\|_\infty = C\langle t-r\rangle^{-(n-1)}.$$

Moreover, it follows from (2.6) and the known estimate [2] on the free-space solution ν that, if M is sufficiently large, the remainder term (3.5) then satisfies

$$\begin{aligned}\|\alpha u_1(t)\|_\infty &\leq C\int\langle t-r\rangle^{-(n-1)}\|\wedge^{-M}\nu(r)\|_\infty dr\\ &\leq C(\|\wedge^\sigma f_0\|_1 + \|\wedge^{\sigma-1} f_1\|_1)\int\langle t-r\rangle^{-(n-1)}\langle r\rangle^{-(n-1)/2}dr\\ &\leq C\langle t\rangle^{-(n-1)/2}(\|\wedge^\sigma f_0\|_1 + \|\wedge^{\sigma-1} f_1\|_1.\end{aligned}$$

The last inequality is clear if $n \geq 3$; if $n = 2$ the modification in [3] (using more detailed information about the free-space kernel) can be applied. (See also Section 3, where the details in the $n = 2$ modification are provided for the more complicated estimate involving βu.)

It remains to analyze the term

$$(2.7)\qquad \alpha u_0(t,x) = \alpha(x)\iint K_{\partial\Omega}(t-r,x,w')\nu(r,w(w'))dw'ds$$

corresponding to the Taylor-Melrose parametrix. It is enough to assume that ν is given by

$$\nu_\pm(r,w) = \iint e^{i[(w-y)\cdot\eta\pm r|\eta|]}f(y)d\eta dy,$$

for $f \in L^{1,\sigma}(\Omega)$, $\sigma > (n+1)/2$, or $f \in \mathcal{H}^{1,\sigma}(\Omega)$, $\sigma = (n+1)/2$. If we set

$$g = \Lambda^\sigma f,\ g \in L^1(\Omega) \text{ or } \mathcal{H}^1(\Omega),$$

then this equation can be rewritten as

$$\nu_\pm(r, w) = \iint e^{i[(w-y)\cdot\eta \pm r|\eta|]} \langle\eta\rangle^{-\sigma} g(y) d\eta dy.$$

The argument above allows the analysis on $\partial\Omega$ to be restricted to microlocal neighborhoods of the null bicharacteristics, which fall into two categories — transversal and grazing. The two types of kernels corresponding to the term in (2.7) can be written as follows:

$$\alpha u_0(t,x) = \int K_{tr}(t,x,y) g(y) dy + \int K_{gr}(t,x,y) g(y) dy. \tag{2.8}$$

The following expressions for the kernels are derived in [3], using the results of Taylor [17]. After localization near a fixed point of $\partial\Omega$, say $(w', w_n) = (0,0)$, it can be assumed that coordinates have been chosen so that locally

$$\Omega = \{w_n > \phi(w')\}; \partial\Omega = \{w = w(w') = (w', \phi(w')\},$$

with $\phi(w') = -Q(w') + \mathcal{O}\{| w' |^3\}$ for Q a positive definite quadratic form.

The kernel corresponding to the transversal directions may be written as

(2.9)

$$\begin{aligned} K_{tr}(t,x,y) = & \alpha(x) \int \cdots \int e^{i[\psi_{tr}(x,\zeta') + (t-r)\zeta_n - w'\cdot\zeta' + (w(w')-y)\cdot\eta \pm r|\eta|]} \\ & \times a_{tr}(x,\zeta)\chi(\zeta)\langle\eta\rangle^{-\sigma}\alpha'(w') d\zeta d\eta dr dw'. \end{aligned}$$

The smooth function α' denotes a cutoff near the origin in $\mathbb{R}^{n-1}$, and the zero-order pseudo-differential symbol χ denotes the microlocalizer near the transversal direction, which can be taken to be a conic set where $\zeta_n \neq 0$. The phase function $\psi_{tr}(z, \zeta')$ satisfies the eiconal equation

$$| \nabla_x \psi_{tr} |^2 = \zeta_n^2,\ \psi(x', \phi(x'), \zeta') = z' \cdot \zeta', \tag{2.10}$$

so that on $\partial\Omega \times \mathbb{R}$ it reduces to the usual phase function for the Fourier transform. The amplitude $a_{tr}(x,\zeta) \in S^0$ has principal symbol identically one, and satifies the usual transport equations. By choosing the support of $\alpha(x)$ to be sufficiently close to $\partial\Omega$, we can write (see [3])

$$\psi_{tr}(x,\zeta) = x' \cdot \zeta' - x_n\sqrt{\zeta_n^2 - |\zeta'|^2} + 0(|x|^2|\zeta|). \tag{2.11}$$

The primary term in the kernel corresponding to the grazing directions may be written as

$$\begin{aligned} K_{gr}(t,x,y) = \alpha(x) \int\cdots\int & \\ \times\, e^{i[\psi_{gr}(x,\zeta')+(t-r)|\zeta'|-\psi_0(w',\zeta)+2/3|\zeta'|(\rho_0^{3/2}(x,\zeta)-\theta^{3/2})+(w(w')-y)\cdot\eta\pm r|\eta|]} & \\ \times\, \{\Phi(\rho_0|\zeta'|^{2/3})/\Phi(\theta|\zeta'|^{2/3})\}a_{gr}(x,\zeta)\chi(\zeta)\langle\eta\rangle^{-\sigma}\alpha'(w')d\zeta d\eta dr dw'. & \end{aligned}$$

Here

$$\begin{aligned} \rho_0(x,\zeta) &= \rho(x,\zeta)/|\zeta'|, \\ \theta = -\zeta_n/|\zeta'|, &\text{ and } \Phi(\mu) \sim \mu^{-1/4}(c_0 + c_1\mu^{-3/2} + \cdots). \end{aligned} \tag{2.13}$$

The pair (ψ_{gr},ρ) satisfies the eiconal equation

$$|\nabla_x\psi|^2 + (\rho/|\zeta'|)|\nabla_x\rho|^2 = |\zeta'|^2, \quad \nabla_x\psi\cdot\nabla_x\rho = 0, \tag{2.14}$$

and

$$\psi_0(w',\zeta) = \psi_{gr}(w(w'),\zeta) + \mathcal{O}\{|\zeta'|(\zeta_n/|\zeta'|)^\infty\}. \tag{2.15}$$

The amplitude $a_{gr}(x,\zeta) \in S^0$ satifies the corresponding analogue of the transport equations, and $\chi(\zeta)$ has support on a conic neighborhood of $\zeta = (\xi_1, \xi", \xi_n) = (1,0,0)$, and $\zeta_n < 0$. Again, for $\alpha(x)$ with support near enough to $\partial\Omega$, we have a $\nu' = (\nu_1, \nu") \in \mathbb{R}^{n-1}, \nu_1 > 0$, with (see [3])

$$\begin{aligned} \psi_{gr}(x,\zeta) = x' \cdot [\zeta' + \theta\nu'|\zeta'| + \mathcal{O}\{\theta(\delta+\theta)|\zeta'|\}] \\ + x_n\mathcal{O}\{\theta^2|\zeta'|\} + \mathcal{O}\{|x|^2|\zeta'|\}, \end{aligned} \tag{2.16}$$

and, for $\delta = |\zeta''|/|\zeta'| \ll 1$,

$$\begin{aligned} \rho_0(x,\zeta) = \theta + (x_n - \phi(x'))[\sqrt{2\nu_1} + \mathcal{O}\{(\delta+\theta)\}] \\ + \mathcal{O}\{(x_n - \phi(x'))|x| + \theta^2\}. \end{aligned} \tag{2.17}$$

The following estimates are the key ingredients for handling the term αu supported near $\partial\Omega$ in the proof of Theorem 1.1.

Theorem 2.1. *Let $E_{tr}(t)$ be the operator defined by the kernel $K_{tr}(t,x,y)$ given in* (2.9). *If $\sigma > (n+1)/2$, then*

$$| K_{tr}(t,x,y) | \leq C|t|^{-(n-1)/2} \text{ for } t \geq 1.$$

Thus $E_{tr}(t) : L^1(\Omega) \to L^\infty(\Omega)$ defines a bounded map of norm at most $C|t|^{-(n-1)/2}$. If $\sigma = (n+1)/2$, then $E_{tr}(t) : h^1(\Omega) \to L^\infty(\Omega)$ defines a bounded map with norm at most $C|t|^{-(n-1)/2}$.

Theorem 2.2. *Let $E_{gr}(t)$ be the operator defined by the kernel $K_{gr}(t,x,y)$ given in* (2.12). *If $\sigma > (n+1)/2$, then*

$$| K_{gr}(t,x,y) | \leq C|t|^{-(n-1)/2} \text{ for } t \geq 1.$$

Thus $E_{gr}(t) : L^1(\Omega) \to L^\infty(\Omega)$ defines a bounded map of norm at most $C|t|^{-(n-1)/2}$. If $\sigma = (n+1)/2$, then $E_{gr}(t) : h^1(\Omega) \to L^\infty(\Omega)$ defines a bounded map with norm at most $C|t|^{-(n-1)/2}$.

Proof. From (2.11), the phase function for the oscillating integral defining K_{tr} is given by

(2.18)
$$\psi_{tr}(t-r,x,w',y,\zeta,\eta) = \psi_{tr}(x,\zeta') + (t-r)\zeta_n - w'\cdot\zeta' + (w(w')-y)\cdot\eta \pm r|\eta|.$$

It follows from (2.11) that

$$\nabla_{(w',r)}\psi_{tr} = (-\zeta' + \eta', -\zeta_n + |\eta|) + O(|w'||\eta|). \tag{2.19}$$

Thus repeated integrations by parts in (2.9) yield

$$K_{tr}(t,x,y) = \alpha(x)\int\cdots\int e^{i\psi_{tr}} O(\langle\nabla_{(w',r)}\psi_{tr}\rangle^{-M}) \times a_{tr}(x,\zeta)\chi(\zeta)\langle\eta\rangle^{-\sigma}\alpha'(w')d\zeta d\eta dr dw'. \tag{2.20}$$

Next, (2.11) yields

$$\nabla_\zeta\psi_{tr} = (x'-w', t-r) + O(|x|). \tag{2.21}$$

Since x can be taken to be small by the choice of α, integration by parts gives

$$K_{tr}(t,x,y) = \alpha(x)\int\cdots\int e^{i\psi_{tr}} O(\langle\nabla_{(w',r)}\psi_{tr}\rangle^{-M}) \times O(\langle t-r\rangle^{-(n-1)})\langle\eta\rangle^{-\sigma}\alpha'(w')d\zeta d\eta dr dw'.$$

Moreover,

$$\nabla_\eta\psi_{tr} = (w(w') - y) \pm r\eta/|\eta|. \tag{2.22}$$

As in [3], integration by parts and standard arguments involving the method of stationary phase then imply

$$\begin{aligned} K_{tr}&(t,x,y) \\ &= \alpha(x)\int O(\langle\nabla_{(w',r)}\psi_{tr}\rangle^{-M})O(\langle t-r\rangle^{-(n-1)})O(\langle r\rangle^{-(n-1)/2}) \\ &\quad\times O(\langle r-|y|\rangle^{-(n-1)/2})\langle\eta\rangle^{-\sigma}|\eta|^{-(n-1)/2}\alpha'(w')d\zeta|\eta|^{n-1}d|\eta|dr dw'. \end{aligned}$$

Finally, if we set $\zeta_0 = \nabla_{w',r}\psi_{tr}$, then from (2.11) we have $\det(\partial\zeta_0/d\zeta) = 0(1)$. Consequently, if $\sigma > (n+1)/2$, then

$$\begin{aligned} K_{tr}&(t,x,y) \\ &= O\Big(\int\langle\zeta_0\rangle^{-M}\langle t-r\rangle^{-(n-1)}\langle r\rangle^{-\frac{(n-1)}{2}}\langle r-|y|\rangle^{-\frac{(n-1)}{2}}\langle\eta\rangle^{-\sigma}|\eta|^{-\frac{(n-1)}{2}} \\ &\qquad\times\alpha'(w')d\zeta_0|\eta|^{n-1}d|\eta|dr dw'\Big) \\ &= O\Big(\int\langle t-r\rangle^{-(n-1)}\langle r\rangle^{-\frac{(n-1)}{2}}\langle r-|y|\rangle^{-\frac{(n-1)}{2}}dr\Big) = O(\langle t\rangle^{-\frac{(n-1)}{2}}). \end{aligned} \tag{2.23}$$

This estimate completes the proof in the case of $\sigma > (n+1)/2$; for $\sigma = (n+1)/2$ a modification similar to that in [1] and [3] applies. (After composition with a pseudodifferential operator of order zero (bounded on $h^1(\Omega)$), the resulting composed kernel satisfies the same estimate).

Proof of Theorem 2.2. From (2.16), the phase function for the oscillating integral defining K_{gr} is given by

$$\begin{aligned} \psi_{gr}(t-r,x,w',y,\zeta,\eta) &= \psi_{gr}(x,\zeta') + (t-r)|\zeta| - \psi_0(w',\zeta) \\ &\quad + (2/3)|\zeta'|(\rho_0^{3/2}(x,\zeta) - \theta^{3/2}) \\ &\quad + (w(w')-y)\cdot\eta \pm r|\eta|. \end{aligned} \tag{2.24}$$

From (2.13), (2.16), and (2.17) it follows that

$$\begin{aligned}\nabla_{(w',r)}\psi_{gr} &= (-\zeta' + \eta' + \theta\nu'|\zeta'| + \eta_n|w'|, -|\zeta'| \pm |\eta|) \\ &\quad + O(\{|w'| + \theta(\delta+\theta)\} \mid \zeta' \mid).\end{aligned} \tag{2.25}$$

Integration by parts with respect to (w', r) in (2.12) does not involve the terms ρ_0 or θ, and therefore involves neither the singular part of the phase function nor the delicate part of the amplitude,

$$\Phi(\rho_0 \mid \zeta' \mid^{2/3})/\Phi(\theta \mid \zeta' \mid^{2/3}).$$

Thus we obtain

$$\begin{aligned}K_{gr}(t,x,y) &= \alpha(x)\Big(\int\cdots\int e^{i\psi_{gr}} O(\langle\nabla_{(w',r)}\psi_{gr}\rangle^{-M}) \\ &\quad \times \{\Phi(\rho_0 \mid \zeta' \mid^{2/3})/\Phi(\theta \mid \zeta' \mid^{2/3})\} \\ &\quad \times a_{gr}(x,\zeta)\chi(\zeta)\langle\eta\rangle^{-\sigma}\alpha'(w')d\zeta d\eta dr dw.\end{aligned} \tag{2.26}$$

Next,

$$\partial_{|\zeta'|}\psi_{gr} = (t-r) + O(|x| + |w'|). \tag{2.27}$$

Since x and w' can be taken to be small by the choice of α and of the localizer on the boundary, integration by parts gives

$$\begin{aligned}K_{gr}(t,x,y) &= \alpha(x)\int e^{i\psi_{gr}} O(\langle\nabla_{(w',r)}\psi_{gr}\rangle^{-M}\langle t-r\rangle^{-(n-1)}) \\ &\quad \times \{\Phi(\rho_0 \mid \zeta' \mid^{2/3})/\Phi(\theta \mid \zeta' \mid^{2/3})\} \\ &\quad \times a_{gr}(x,\zeta)\chi(\zeta)\langle\eta\rangle^{-\sigma}\alpha'(w')d\zeta d\eta dr dw.\end{aligned}$$

As in (2.22),

$$\nabla_\eta\psi_{gr} = (w(w') - y) \pm r\eta/|\eta|. \tag{2.28}$$

The argument in [3] involving integration by parts and the method of stationary phase then imply that

$$\begin{aligned}K_{gr}(t,x,y) &= \alpha(x)\int O(\langle\nabla_{(w',r)}\psi_{gr}\rangle^{-M}\langle t-r\rangle^{-(n-1)} \\ &\quad \times \langle r\rangle^{-(n-1)/2}\langle r-|y|\rangle^{-(n-1)/2}) \\ &\quad \times \langle\eta\rangle^{-\sigma}|\eta|^{-(n-1)/2}\alpha'(w')d\zeta|\eta|^{n-1}d|\eta| dr dw'.\end{aligned}$$

Finally, if we set $\zeta_0 = \nabla_{(w',r)}\psi_{gr}$, then from (2.13), (2.16), and (2.17), $\det(\partial\zeta_0/d\zeta) = |\nu_1| + o(1) = O(1)$. Thus the argument is completed exactly as in the case of (2.23).

3. Behavior of solutions away from $\partial\Omega$

It remains to analyze the term "βu" in (2.1), for β smooth and identically zero on a small neighborhood of $\partial\Omega$. Since βu satisfies the inhomogeneous free-space problem, from (2.2), (1.4), and (1.5), it follows that

$$\beta u = E_0(t)\beta f_0 + E(t)\beta f_1 + \int_0^t E(t-s)D\alpha_0 u(s)ds, \tag{3.1}$$

again with α_0 a smooth function of compact support near $\partial\Omega$, identically zero on a smaller neighborhood of $\partial\Omega$, and $\alpha_0 Du$ a shorthand for a sum of such terms involving u and its derivatives. Since $\nu = E_0(t)\beta f_0 + E(t)\beta f_1$ satisfies the homogeneous free-space problem, the desired decay estimate analogous to (2.4) again holds:

$$\|E_0(t)\alpha f_0 + E(t)\alpha f_1\|_\infty \le C|t|^{-(n-1)/2}(\|\wedge^\sigma f_0\|_1 + \|\wedge^{\sigma-1} f_1\|_1). \tag{3.2}$$

The remaining term in (3.1) may be written in terms of kernels involving the Taylor-Melrose parametrix $K_{\partial\Omega}$ and a smoothing remainder $R_{\partial\Omega}$. From (3.1) and (2.5), the remainder term can be written as

$$\begin{aligned}\beta u_1(t,x) = &\int_0^t E(t-s)\wedge^{-M} D\alpha_0 \\ &\times \iint R_{\partial\Omega,2M}(s-r,\cdot,w')\wedge^{-M}\nu(r,w(w'))dw'drds,\end{aligned} \tag{3.3}$$

with $R_{\partial\Omega,2M}$ corresponding to having absorbed two factors of $\wedge^M$ into the kernel. Since α_0 has compact support, z (the variable corresponding to "$\cdot$" in the kernel) and w' are in compact sets. From (2.6) and the known estimates [2] on the free-space solution and the operator

$E(t-s)\wedge^{-M} D$,

$$\begin{aligned}\|\beta u_1(t)\|_\infty &\le C\iint \langle t-s\rangle^{-(n-1)/2}\langle r-s\rangle^{-(n-1)}\|\wedge^{-M}\nu(r)\|_\infty drds\\ &\le C(\|\wedge^\sigma f_0\|_1+\|\wedge^{\sigma-1} f_1\|_1)\\ &\quad\times\iint \langle t-s\rangle^{-(n-1/2)}\langle r-s\rangle^{-(n-1)}\langle r\rangle^{-(n-1/2)}drds\\ &\le C\langle t\rangle^{-(n-1)/2}(\|\wedge^\sigma f_0\|_1+\|\wedge^{\sigma-1} f_1\|_1).\end{aligned}$$

The last inequality is clear if $n\ge 4$; if $n=3$ or 2 more detailed information about the free-space kernel can be used to obtain the same result. The kernel $K(t-s,x-z)$ associated with the operator $E(t-s)\wedge^{-M} D$ satisfies

$$|K(t-s,x-z)|\le C\langle t-s\rangle^{-(n-1)/2}\langle t-s-|x-z|\rangle^{-(n-1)/2}. \tag{3.4}$$

Together with a similar estimate for the kernel $K(r,w(w')-y)$ yielding $\wedge^{-M}\nu(r)$, and the compact integration domain for x and w', the following straightforward estimate suffices:

$$\begin{aligned}\sup_{A,B} C\iint &\langle t-s\rangle^{-(n-1/2)}\langle t-s-A\rangle^{-(n-1/2)}\langle r-s\rangle^{-(n-1)}\langle r\rangle^{-(n-1/2)}\\ &\times\langle r-B\rangle^{-(n-1/2)}drds\le C\langle t\rangle^{-(n-1/2)}.\end{aligned}$$

It remains to analyze the primary term from the Taylor-Melrose parametrix:

$$\begin{aligned}\beta u_0(t,x)=&\int_0^t E(t-s)\wedge^{-M} D\alpha_0\\ &\times\iint K_{\partial\Omega}(s-r,\cdot,w')\wedge^{-M}\nu(r,w(w'))dw'drds.\end{aligned} \tag{3.5}$$

It is again assumed that

$$\nu_\pm(r,w)=\iint e^{i[(w-y)\cdot\eta\pm r|\eta|]}f(y)d\eta dy,$$

for $g=\wedge^\sigma f, g\in L^{1,\sigma}(\Omega)$ or $\mathcal{H}^1(\Omega)$. The kernels involved in (3.5) can then be written as follows:

$$\beta u_0(t,x)=\int K_{tr}(t,x,y)g(y)dy+\int K_{gr}(t,s,r,x,y)g(y)dy. \tag{3.6}$$

From (3.5) and (2.9), the kernel corresponding to the transversal directions may be written as

(3.7)

$$K_{tr}(t,x,y) = \beta(x)$$
$$\times \int_0^t \int \cdots \int e^{i[(x-z)\cdot\xi \pm (t-s)|\xi| + \psi_{tr}(z,\zeta') + (s-r)\zeta_n - w'\cdot\zeta' + (w(w')-y)\cdot\eta \pm r|\eta|]}$$
$$\times \langle\xi\rangle^{-1}\alpha_0(z)\langle\zeta\rangle a_{tr}(z,\zeta)\chi(\zeta)\langle\eta\rangle^{-\sigma}\alpha'(w')dxd\zeta d\eta dzdrdw'ds.$$

And from (3.5) and (2.12), the primary term in the kernel corresponding to the grazing directions may be written as

(3.8)

$$K_{gr}(t,x,y)$$
$$= \beta(x)\int_0^t \int e^{i[(x-z)\cdot\xi \pm (t-s)|\xi|]}$$
$$\times e^{i[\psi_{gr}(z,\zeta') + (s-r)|\zeta'| - \psi_0(w',\zeta) + 2/3|\zeta'|(\rho_0^{3/2}(z,\zeta) - \theta^{3/2}) + (w(w')-y)\cdot\eta \pm r|\eta|]}$$
$$\times \langle\xi\rangle^{-1}\alpha_0(x)\langle\zeta\rangle\{\Phi(\rho_0|\zeta'|^{2/3})/\Phi(\theta|\zeta'|^{2/3})\}a_{gr}(z,\zeta)\chi(\zeta)\langle\eta\rangle^{-\sigma}$$
$$\times \alpha'(w')d\xi d\zeta d\eta drdw'dzds.$$

The argument above coupled with the estimates below handle the term βu supported away from $\partial\Omega$ and (2.1), and together with the results of Section 2 yield Theorem 1.1.

Theorem 3.1. *Let $E_{tr}(t)$ be the operator defined by the kernel $K_{tr}(t,x,y)$ given in* (3.7). *If $\sigma > (n+1)/2$, then*

$$|K_{tr}(t,x,y)| \leq C|t|^{-(n-1)/2} \text{ for } t \geq 1.$$

Thus $E_{tr}(t) : L^1(\Omega) \to L^\infty(\Omega)$ defines a bounded map of norm at most $C|t|^{-(n-1)/2}$. If $\sigma = (n+1)/2$, then $E_{tr}(t) : h^1(\Omega) \to L^\infty(\Omega)$ defines a bounded map with norm at most $C|t|^{-(n-1)/2}$.

Theorem 3.2. *Let $E_{gr}(t)$ be the operator defined by the kernel $K_{gr}(t,x,y)$ given in* (3.8). *If $\sigma > (n+1)/2$, then*

$$|K_{gr}(t,x,y)| \leq C|t|^{-(n-1)/2} \text{ for } t \geq 1.$$

Thus $E_{gr}(t) : L^1(\Omega) \to L^\infty(\Omega)$ defines a bounded map of norm at most $C|t|^{-(n-1)/2}$. If $\sigma = (n+1)/2$, then $E_{gr}(t) : h^1(\Omega) \to L^\infty(\Omega)$ defines a bounded map with norm at most $C|t|^{-(n-1)/2}$.

Proof of Theorem 3.1. From (3.7), the phase function for the oscillating integral defining K_{tr} is given by

$$\begin{aligned}\psi_{tr}&(t-s,x,z,w',y,\zeta,\eta)\\ &= (x-z)\cdot\xi \pm (t-s)|\xi| + \psi_{tr}(z,\zeta') + (s-r)\zeta_n\\ &\quad - w'\cdot\zeta' + (w(w')-y)\cdot\eta \pm r|\eta|.\end{aligned} \tag{3.9}$$

As in [3], it follows from (2.11) that

$$\begin{aligned}\nabla_{(z,w',r)}\psi_{tr} &= (-\xi'+\zeta', -\xi_n + \sqrt{\zeta_n^2 - |\zeta'|^2}, -\zeta'+\eta', -\zeta_n + |\eta|)\\ &\quad + O(|z||\zeta|) + O(|w'||\eta|).\end{aligned} \tag{3.10}$$

Repeated integrations by parts in (3.7) then yield

$$\begin{aligned}K_{tr}(t,x,y) &= \beta(x)\int_0^t\int\cdots\int e^{i\psi_{tr}}O(\langle\nabla_z\psi_{tr}\rangle^{-M})O(\langle\nabla_{(w',r)}\psi_{tr}\rangle^{-M})\\ &\quad\times\langle\xi\rangle^{-1}\alpha_0(z)\langle\zeta\rangle a_{tr}(z,\zeta)\chi(\zeta)\langle\eta\rangle^{-\sigma}\alpha'(w')d\xi d\zeta d\eta dz dr dw' ds.\end{aligned} \tag{3.11}$$

As in (2.21), integration by parts with respect to ζ gives

$$\begin{aligned}K_{tr}(t,x,y) = \beta(x)&\int_0^t\int\cdots\int e^{i\psi_{tr}}O(\langle\nabla_z\psi_{tr}\rangle^{-M})\\ &\times O(\langle\nabla_{w',r}\psi_{tr}\rangle^{-M})O(\langle s-r\rangle^{-(n-1)})\\ &\times\langle\xi\rangle^{-1}\alpha_0(z)\langle\zeta\rangle a_{tr}(z,\zeta)\chi(\zeta)\langle\eta\rangle^{-\sigma}\alpha'(w')d\xi d\zeta d\eta dz dr dw' ds.\end{aligned}$$

Now,

$$\nabla_\xi\psi_{tr} = (x-z)\pm(t-s)\xi/|\xi|,\ \nabla_\eta\psi_{tr} = (w(w')-y)\pm r\eta/|\eta|. \tag{3.12}$$

Since z and $w(w')$ are small, it follows that if $(t-s)\geq t/2$ we can use integration by parts and stationary phase in ξ as in [3] and integration by parts in η, and if $s\geq t/2$ we can use integration by parts and stationary phase in η as in (2.22) and integration by parts in ξ, to obtain

$$
\begin{aligned}
K_{tr}(t,x,y) = \beta(x) & \int_0^{t/2} \int \cdots \int O(\langle \nabla_z \psi_{tr}\rangle^{-M}) O(\langle \nabla_{w',r}\psi_{tr}\rangle^{-M}) \\
& \times O(\langle s-r\rangle^{-(n-1)}) O(\langle t-s\rangle^{-(n-1)/2}) \\
& \times O(\langle t-s-|x|\rangle^{-(n-1)/2}) O(\langle \xi\rangle^{-\sigma}) |\xi|^{-(n-1)/2} \\
& \times O(\langle r-|y|\rangle^{-(n-1)}) \alpha_0(z)\alpha'(w') |\xi|^{n-1} d|\xi| d\zeta d\eta dz dr dw' ds \\
& + \beta(x) \int_{t/2}^{t} \int \cdots \int O(\langle \nabla_z \psi_{tr}\rangle^{-M}) O(\langle \nabla_{w',r}\psi_{tr}\rangle^{-M}) \\
& \times O(\langle s-r\rangle^{-(n-1)}) O(\langle t-s-|x|\rangle^{-(n-1)/2}) \\
& \times O(\langle \eta\rangle^{-\sigma}) |\eta|^{-(n-1)/2} O(\langle r\rangle^{-(n-1)/2}) \\
& \times O(\langle r-|y|\rangle^{-(n-1)/2}) \alpha_0(z)\alpha'(w') d\xi d\zeta |\eta|^{n-1} d|\eta| dz dr dw' ds
\end{aligned}
$$

Finally, if we set $\zeta_0 = \nabla_z \psi_{tr}$ and $\eta_0 = \nabla_{(w',r)}\psi_{tr}$ in the first integral, and $\xi_0 = \nabla_z \psi_{tr}$ and $\zeta_0 = \nabla_{(w',r)}\psi_{tr}$ in the second integral, then, from (3.10) we have $\det(\partial\zeta_0/d\zeta) = O(1), \det(\partial\eta_0/d\eta) = O(1)$ and $\det(\partial\xi_0/d\xi) = O(1)$. Consequently, if $\sigma > (n+1)/2$, then

(3.13)

$$
\begin{aligned}
K_{tr}(t,x,y) &= O\Big(\int_0^{t/2}\int \langle\zeta_0\rangle^{-M}\langle\eta_0\rangle^{-M}\langle s-r\rangle^{-(n-1)}\langle t-s\rangle^{(n-1)/2} \\
&\quad \times \langle t-s-|x|\rangle^{(n-1)/2}\langle r-|y|\rangle^{-(n-1)}\alpha_0(z)\alpha'(w') \\
&\quad \times \langle\xi\rangle^{-\sigma}|\xi|^{(n-1)/2} d|\xi| d\zeta d\eta dz dr dw' ds \\
&\quad + \int_{t/2}^{t}\int \langle\xi_0\rangle^{-M}\langle\zeta_0\rangle^{-M}\langle s-r\rangle^{-(n-1)} \\
&\quad \times \langle t-s-|x|\rangle^{-(n-1)}\langle r\rangle^{-(n-1)/2}\langle r-|y|\rangle^{-(n-1)/2} \\
&\quad \times \alpha_0(z)\alpha'(w') d\xi d\zeta \langle\eta\rangle^{-\sigma}|\eta|^{(n-1)/2} d|\eta| dz dr dw' ds\Big) \\
&= O\Big(\int_0^{t/2}\int \langle s-r\rangle^{-(n-1)}\langle t-s\rangle^{(n-1)/2} \\
&\quad \times \langle t-s-|x|\rangle^{(n-1)/2}\langle r-|y|\rangle^{-(n-1)} dr ds \\
&\quad + \int_{t/2}^{t}\int \langle s-r\rangle^{-(n-1)}\langle t-s-|x|\rangle^{-(n-1)} \\
&\quad \times \langle r\rangle^{-(n-1)/2}\langle r-|y|\rangle^{-(n-1)/2} dr ds\Big) \\
&= O(\langle t\rangle^{-(n-1)/2}).
\end{aligned}
$$

This estimate completes the proof, with the usual modification in the case of $\sigma = (n+1)/2$.

Proof of Theorem 3.2. From (2.16), the phase function for the oscillating integral defining K_{gr} is given by

(3.14)
$$\begin{aligned}\psi_{gr}&(t-r,x,w',y,\zeta,\eta)\\ &= (x-z)\cdot\xi \pm (t-s)|\xi|\psi_{gr}(z,\zeta') + (s-r)|\zeta| - \psi_0(w',\zeta)\\ &\quad + (2/3)|\zeta'|(\rho_0^{3/2}(z,\zeta) - \theta^{3/2}) + (w(w')-y)\cdot\eta \pm r|\eta|.\end{aligned}$$

As in [3], it follows from (2.13), (2.16), and (2.17) that

(3.15)
$$\begin{aligned}\nabla_{(z,w',r)}&\psi_{gr}\\ &= (-\xi' + \zeta' + \theta\nu'|\zeta'|, -\xi_n + \rho_0^{1/2}\sqrt{2\nu_1}, -\zeta'\\ &\quad + \eta' - \theta\nu'|\zeta'| + \eta_n|w'|, -|\zeta'| \pm |\eta|)\\ &\quad + O(\{|z| + \theta(\delta+\theta)\} \mid \zeta' \mid) + O(\{|w'| + \theta(\delta+\theta)\} \mid \zeta' \mid).\end{aligned}$$

As in Section 2, we then obtain

(3.16)
$$\begin{aligned}K_{gr}(t,x,y) = \beta(x)\int\cdots\int & e^{i\psi_{gr}}O(\langle\nabla_z\psi_{gr}\rangle^{-M})O(\langle\nabla_{w',r}\psi_{gr}\rangle^{-M})\langle\xi\rangle^{-1}\\ &\times \alpha_0(x)\langle\zeta\rangle\{\Phi(\rho_0 \mid \zeta' \mid^{2/3})/\Phi(\theta \mid \zeta' \mid^{2/3})\}\\ &\times a_{gr}(z,\zeta)\chi(\zeta)\langle\eta\rangle^{-\sigma}\alpha'(w')d\zeta d\eta dr dw'.\end{aligned}$$

As in (2.27), integration by parts with respect to $|\zeta'|$ gives

$$\begin{aligned}K_{gr}(t,x,y) = \beta(x)\int\cdots\int & e^{i\psi_{gr}}O(\langle\nabla_z\psi_{gr}\rangle^{-M})\\ &\times O(\langle\nabla_{(w',r)}\psi_{gr}\rangle^{-M})O(\langle t-r\rangle^{-(n-1)})\\ &\times \{\Phi(\rho_0 \mid \zeta' \mid^{2/3})/\Phi(\theta \mid \zeta' \mid^{2/3})\}a_{gr}(x,\zeta)\chi(\zeta)\langle\eta\rangle^{-\sigma}\\ &\times \alpha'(w')d\zeta d\eta dr dw.\end{aligned}$$

The analysis after (3.12), depending on whether $s \le t/2$ or $s \ge t/2$, again applies. The estimates in the case $s \le t/2$ are derived as at the end of the proof of Theorem 2.2, and in the case $s \ge t/2$ are derived in [3]. The argument is then completed exactly as above for (3.13).

References

[1] M. Beals, L^p boundedness of Fourier integral operators, *Memoirs AMS* **38** (264), (1982).

[2] ———, Optimal L^∞ decay for solutions to the wave equation with a potential, *Comm. in PDE* **19** (1994), 1319–1369.

[3] ———, Global time decay of the amplitude of a reflected wave, Partial Differential Equations and Mathematical Physics, The Danish-Swedish Analysis Seminar 1995, *Progress in Nonlinear Diff. Eqs. and Their Appl.* **21** (1996), 25–44.

[4] J. Ginibre and G. Velo, The global Cauchy problem for the non-linear Klein-Gordon equation, *Math. Zeit.* **189** (1985), 487–505.

[5] D. Goldberg, A local version of real Hardy spaces, *Duke Math. J.* **46** (1979), 27–42.

[6] L. Hörmander, Fourier integral operators I, *Acta Math.* **127** (1971), 79–183.

[7] P.D. Lax and R.S. Phillips, *Scattering Theory*, Academic Press, New York and London, (1967).

[8] H. Lindblad and C.D. Sogge, On existence and scattering with minimal regularity for semilinear wave equations, *J. Funct. Analysis* **130** (1995), 357–426.

[9] B. Marshall, W. Strauss and S. Wainger, $L^p - L^q$ estimates for the Klein-Gordon equations, *J. Math. Pures Appl.* (A) **59** (1980), 417–440.

[10] R.B. Melrose, Singularities and energy decay in acoustical scattering, *Duke Math. J.* **46** (1979), 43–59.

[11] C.S. Morawetz, J.V. Ralston and W.A. Strauss, Decay of solutions of the wave equation outside nontrapping obstacles, *Comm. Pure Appl. Math.* **30** (1977), 447–508.

[12] J. Ralston, Note on the decay of acoustic waves, *Duke Math. J.* **46** (1979), 799–804.

[13] J. Shatah and M. Struwe, Regularity results for nonlinear wave equations, *Annals of Math.* **138** (1993), 503–518.

[14] W.A. Strauss, Nonlinear scattering at low energy, *J. Funct. Analysis* **41** (1981), 110–133.

[15] W.A. Strauss, Nonlinear Wave Equations, *CBMS Reg. Conf. Series in Math.* **73**, Amer. Math. Soc., Providence, (1989).

[16] R.S. Strichartz, A priori estimates for the wave equation and some applications, *J. Funct. Anal.* **5** (1970), 218–235.

[17] M. Taylor, *Pseudodifferential Operators*, Princeton University Press, Princeton, (1981).

[18] B.R. Vainberg, On the short wave asymptotic behavior of solutions of stationary problems, *Russ. Math. Surveys* **30:2** (1975), 1–58.

Department of Mathematics
Rutgers University
New Brunswick, New Jersey

Sobolev embeddings in Weyl-Hörmander calculus

Jean-Yves Chemin and Chao-Jiang Xu[1]

Introduction

The motivation for studying problems of Sobolev embeddings in Weyl-Hörmander calculus was the proof of Sobolev embeddings for Sobolev spaces associated to subelliptic systems of order two, see [4]. In this text, we focus on L^p embeddings for abstract Sobolev spaces in the context of the Weyl-Hörmander calculus. The structure of this text will be the following: in the first section, we briefly present Weyl-Hörmander calculus, then we state the abstract Sobolev embedding; in the second section, we introduce smoothing operators with respect to an Hörmander's metric g and a g-weight m; in the third and last section, we prove the abstract embedding theorem.

1. Weyl-Hörmander calculus

In this section, we follow [5] and [3]. Weyl's quantization associates to $a \in \mathcal{S}(\mathbf{R}^n)$ the operator a^w defined by

$$a^w u(x) \stackrel{\text{def}}{=} (2\pi)^{-n} \int_{\mathbf{R}^{2n}} e^{i\langle x-z,\zeta\rangle} a\left(\frac{x+z}{2},\zeta\right) u(z)dzd\zeta.$$

Let us denote by $[X, Y]$ the standard symplectic form

$$[X,Y] = [(x,\xi),(y,\eta)] = \langle y,\xi\rangle - \langle x,\eta\rangle.$$

We have the following composition formula $a^w \circ b^w = (a\#b)^w$ with

$$(a\#b)(X) = \pi^{-2n} \int_{\mathbf{R}^{2n}\times\mathbf{R}^{2n}} e^{-2i[X-Y_1,X-Y_2]} a(Y_1)b(Y_2)dY_1dY_2.$$

Let us define the concept of Hörmander's metric.

Definition 1.1 *Let g be a measurable map from $\mathbf{R}^{2n}$ to the set of positive definite quadratic forms on $\mathbf{R}^{2n}$. The metric g is an Hörmander's metric if and only if the following three conditions are satisfied:*

$$g_X(X-Y) \le \frac{1}{C_0} \Rightarrow C_0^{-1} g_X \le g_Y \le C_0 g_X \tag{1}$$

[1]The second author is partly suported by China's NNSF.

$$g_X \leq g_X^\sigma \quad with \quad g_X^\sigma(T) \stackrel{\text{def}}{=} \sup_{W\neq 0} \frac{[T,W]^2}{g_X(W)}\ ; \tag{2}$$

There exists an integer N_0 so that for any (X,Y) of $\mathbf{R}^{2n}\times\mathbf{R}^{2n}$,

$$C_0^{-1}(1+g_Y^\sigma(X-Y))^{-N_0}g_X \leq g_Y \leq C_0(1+g_Y^\sigma(X-Y))^{N_0}g_X. \tag{3}$$

Considering an Hörmander's metric g, let us fix a strictly positive real number r strictly smaller than C_0^{-1}, where the constant C_0 is that of condition (1). In all that follows, let us denote by U_X the g_X-ball of center X and of radius $r^{1/2}$, i.e.

$$U_X \stackrel{\text{def}}{=} \{Y\in\mathbf{R}^{2n}\ /\ g_X(Y-X)<r\}.$$

Let us define the following function Δ:

$$\begin{aligned}\Delta(X,Y) &\stackrel{\text{def}}{=} 1+\max\{g_X^\sigma(U_X-U_Y), g_Y^\sigma(U_X-U_Y)\} \text{ with } (4)\\ g_X^\sigma(U_X-U_Y) &= \inf_{(X',Y')\in U_X\times U_Y} g_X^\sigma(X'-Y').\end{aligned}$$

In all that follows, we drop the fact that this function depends on r. As proved in [3], we may substitute conditions (1) and (3) by

$$\frac{1}{C_0}\Delta(X,Y)^{-N_0}g_X \leq g_Y \leq C_0\Delta(X,Y)^{N_0}g_X. \tag{5}$$

One of the key properties of function Δ, obviously symmetric, is the following lemma proved in [3].

Lemma 1.2 *There exists an integer N_1 so that*

$$\sup_{X\in\mathbf{R}^{2n}}\int_{Y\in\mathbf{R}^{2n}}\Delta(X,Y)^{-N_1}|g_Y|^{1/2}dY<\infty, \tag{6}$$

where $|g_Y|$ denotes the determinant of the quadratic form g_Y in any symplectic basis of $\mathbf{R}^{2n}$.

An Hörmander metric describes a localization procedure in the phase space. Let us notice that if a and b are smooth and compactly supported functions on $\mathbf{R}^{2n}$, there is no reason why $a\#b$ should be so. The correct notion is the following.

Definition 1.3 *Let γ be a strictly positive definite quadratic form on $\mathbf{R}^{2n}$ such that $\gamma^\sigma\geq\gamma$ and Y a point of $\mathbf{R}^{2n}$. Let us define on $\mathcal{S}(\mathbf{R}^{2n})$ the following semi-norms*

$$\|a\|_{k,Conf(\gamma,Y)} \stackrel{\text{def}}{=} \sup_{\substack{X\in\mathbf{R}^{2n}\\ j\leq k,\gamma(T_j)\leq 1}} (1+\gamma^\sigma(X-B_\gamma(Y,r)))^{k/2}|\partial_{T_1}\cdots\partial_{T_j}a(X)|.$$

Let g be an Hörmander's metric and $(a_Y)_{Y\in\mathbf{R}^{2n}}$ a family of functions of $\mathcal{S}(\mathbf{R}^{2n})$. This family is uniformly confined if and only if, for any integer k,

$$\|(a_Y)\|_{k,Conf(g)} \stackrel{\text{def}}{=} \sup_{Y\in\mathbf{R}^{2n}} \|a_Y\|_{k,Conf(g_Y,Y)} < \infty.$$

The key estimate, proved in [3], is the following.

Theorem 1.4 *Let g be an Hörmander's metric and a and b two functions of $\mathcal{S}(\mathbf{R}^{2n})$. For any pair of integers (k, N), an integer ℓ and a constant C exist such that, for any pair (Y, Z) of $\mathbf{R}^{2n} \times \mathbf{R}^{2n}$, we have*

$$\|a\#b\|_{k,Conf(g_Y,Y)} + \|a\#b\|_{k,Conf(g_Z,Z)} \\ \leq C\Delta(Y,Z)^{-N}\|a\|_{\ell,Conf(g_Y,Y)}\|b\|_{\ell,Conf(g_Z,Z)}.$$

In all that follows, we shall assume, for the sake of simplicity, that the metric g is strongly temperate (see [2] for a precise definition). The hypothesis implies in particular the following theorem.

Theorem 1.5 *There exist two uniformly confined families (φ_Y) and (ψ_Y) so that, for any $X \in \mathbf{R}^{2n}$,*

$$\int_{Y\in\mathbf{R}^{2n}} \varphi_Y(X)|g_Y|^{1/2}dY = \int_{Y\in\mathbf{R}^{2n}} (\psi_Y\#\varphi_Y)(X)|g_Y|^{1/2}dY = 1. \tag{7}$$

As proved in [2], the above Theorem 1.5 is always true with series of integrals instead of a single integral. All the results that follow are true in this case. In the proofs, just substitute a single integral on $\mathbf{R}^{2n}$ by a series of integrals on $\mathbf{R}^{2n}$.

Convention In all that follows, we denote by (φ_Y) and (ψ_Y) any two uniformly confined families satisfying (7).

Let us define the concepts of g-weight and of symbols associated to some g-weight.

Definition 1.6 *Let g be an Hörmander's metric. Then a measurable function m defined on $\mathbf{R}^{2n}$ with value in $\mathbf{R}_+^\star$ is a g-weight if and only if*

$$\exists\tilde{C} \,/\, \left(\frac{m(X)}{m(Y)}\right)^{\pm 1} \leq \tilde{C}\Delta(X,Y)^{\tilde{N}}. \tag{8}$$

Definition 1.7 *Let m be a g-weight. Let us denote by $S(m, g)$ the set of all smooth functions a so that, for any integer k,*

$$\|a\|_{k;S(m,g)} \stackrel{\text{def}}{=} \sup_{\substack{j\leq k, X\in\mathbf{R}^{2n} \\ g_X(T_j)\leq 1}} \frac{|\partial_{T_1}\cdots\partial_{T_j}a(X)|}{m(X)} < \infty$$

where $\partial_T a$ denotes the map $\langle da, T\rangle$.

As we have good localization procedure in the phase space, we introduce, following R. Beals's paper [1] (see also [6] and [2]), a "Littlewood-Paley" definition of Sobolev spaces.

Definition 1.8 *Let g be an Hörmander's metric and m be a g-weight. The space $H(m,g)$ is the set of tempered distributions u so that*

$$\|u\|_{H(m,g)} \stackrel{\text{def}}{=} \left(\int m(Y)^2 \|\varphi_Y^w u\|_{L^2}^2 |g_Y|^{1/2} dY \right)^{1/2} < \infty.$$

As it is proved for instance in [2], the space $H(m,g)$ is the set of tempered distributions u on $\mathbf{R}^n$ such that for any $a \in S(m,g)$, $a^w u \in L^2$. In [2], it is also proved that the space $H(1,g)$ is L^2. Moreover the space $H(m,g)$ is "almost independent" of the metric g. And it is also proved that, for any g-weight m, there exist a symbol $\mathcal{M}$ belonging to $S(m,g)$ and a constant C such that $u \in H(m,g) \Leftrightarrow \mathcal{M}^w u \in L^2$ and

$$C^{-1}\|u\|_{H(m,g)} \leq \|\mathcal{M}^w u\|_{L^2} \leq C\|u\|_{H(m,g)}.$$

From Theorem 18.6.6 of [5], we deduce immediately the following theorem.

Theorem 1.9 *Let m and m_2 be two g-weights so that*

$$\lim_{X \to \infty} \frac{m_1(X)}{m_2(X)} = +\infty.$$

Then the space $H(m_1,g)$ is compactly included in $H(m_2,g)$.

We are interested in Sobolev embeddings. So it is natural to do the following hypothesis on the metric g.

Definition 1.10 *Let g be an Hörmander's metric. It is a split one if and only if we have, for any X of $\mathbf{R}^{2n}$,*

$$g_X(dx, d\xi)^2 = g_{1,X}(dx^2) + g_{2,X}(d\xi^2).$$

Let us notice that if g is split, then the metric g^σ defined by (2) satisfies

$$g_X^\sigma(dx, d\xi)^2 = g_{2,X}^{-1}(dx^2) + g_{1,X}^{-1}(d\xi^2).$$

The uncertainty principle $g \leq g^\sigma$ implies that

$$g_{1,X} \leq g_{2,X}^{-1} \quad \text{and obviously} \quad g_{2,X} \leq g_{1,X}^{-1}. \tag{9}$$

Convention In all that follows, we denote by g a strongly temperate split Hörmander's metric and denote $H(m,g)$ by $H(m)$.

In Weyl-Hörmander calculus, lemmas of the Cotlar type are very important, see for instance [5] and [3]. We prove here a Cotlar-type lemma, which takes into account the localization in x-space.

Lemma 1.11 (localized Cotlar) *Let $(\theta_Y)_{Y\in\mathbf{R}^{2n}}$ be a uniformly confined family and $(u_Y)_{Y\in\mathbf{R}^{2n}}$ a family of functions of $L^2(\mathbf{R}^n)$ such that*

$$\int \|u_Y\|_{L^2}^2 |g_Y|^{1/2} dY < +\infty. \tag{10}$$

Then, for any N, a constant C and an integer k exist such that

$$\begin{aligned}\left\|\int_{\mathbf{R}^{2n}} \theta_Y^w u_Y |g_Y|^{1/2} dY\right\|_{L^2}^2 &\leq C\|(\theta_Y)\|_{k,Conf(g)}^2 \\ &\quad\times \int_{Y\in\mathbf{R}^{2n}} \left\|\left(1+g_{2,Y}^{-1}(\cdot - U_Y)\right)^{-N} u_Y\right\|_{L^2}^2 |g_Y|^{1/2} dY.\end{aligned} \tag{11}$$

This lemma is proved in [4]. We give here only a sketch of the proof.

$$I = \int_{\mathbf{R}^{2n}\times\mathbf{R}^{2n}} (\theta_Y^w u_Y | \theta_Z^w u_Z)_{L^2} |g_Y|^{1/2} |g_Z|^{1/2} dY dZ.$$

Defining $\Theta_{Y,Z} \stackrel{\text{def}}{=} \overline{\theta}_Z \# \theta_Y$, we have

$$I = \int_{\mathbf{R}^{2n}\times\mathbf{R}^{2n}} I_{Y,Z} |g_Y|^{1/2} |g_Z|^{1/2} dY dZ \quad \text{with} \quad I_{Y,Z} = \left(\Theta_{Y,Z}^w u_Y | u_Z\right)_{L^2}.$$

Let us define the constant coefficients differential operator $L_{Y,Z}$ by

$$L_{Y,Z} f(\xi) = f(\xi) - \sum_{1\leq i,j\leq n} (g_{2,Y} + g_{2,Z})_{i,j}^{-1} \frac{\partial^2}{\partial\xi_i \partial\xi_j} f(\xi).$$

This operator $L_{Y,Z}$ is a finite sum of derivations of $g_{2,Y} + g_{2,Z}$-length smaller than 1. Using integration by parts with respect to $L_{Y,Z}$ and Theorem 1.4, we get

$$\begin{aligned}|I_{Y,Z}| &\leq C\Delta(Y,Z)^{-N_1} \\ &\quad\times \int \left(1 + g_Y^\sigma\left(\left(\frac{x+t}{2},\tau\right) - U_Y\right)\right)^{-N+n/2} \\ &\quad\quad \times (1 + (g_{2,Y}^{-1}(x-t))^{-N+n/2} |u_Y(t)| \\ &\quad\quad \times \left(1 + g_Z^\sigma\left(\left(\frac{x+t}{2},\tau\right) - U_Z\right)\right)^{-N+n/2} \\ &\quad\quad \times (1 + (g_{2,Z}^{-1}(x-t))^{-N+n/2} |u_Z(x)| dx dt d\tau.\end{aligned} \tag{12}$$

The metric g is split and satisfies (2), so standard computations on quadratic forms imply that we have, for any $Y = (y,\eta)$ of $\mathbf{R}^{2n}$,

$$(1+g_{2,Y}^{-1}(x-t))^{-N+n/2}\left(1+g_Y^\sigma\left(\left(\frac{x+t}{2},\tau\right)-U_Y\right)\right)^{-N+n/2}$$
$$\leq C(1+g_{2,Y}^{-1}(t-U_Y))^{-N/2}(1+g_{2,Y}^{-1}(x-t))^{-n/2}(1+g_{2,Y}(\tau-\eta))^{-n/2}.$$

Using inequality (12) and the Schwarz inequality, we get

$$\begin{aligned}|I_{Y,Z}| \;\leq\; & C\Delta(Y,Z)^{-N_1}\int(1+g_{2,Y}^{-1}(x-t))^{-n/2}(1+g_{2,Y}(\tau-\eta))^{-n/2}\mathcal{U}_Y(t)\\ &\times(1+g_{2,Z}^{-1}(x-t))^{-n/2}(1+g_{2,Z}(\tau-\zeta))^{-n/2}\mathcal{U}_Z(x)dxdtd\tau\end{aligned}$$

where $\mathcal{U}_Y$ is defined by $\mathcal{U}_Y(t)\stackrel{\text{def}}{=}(1+g_{2,Y}^{-1}(t-U_Y))^{-N/2}|u_Y(t)|$. The Schwarz inequality implies that

$$\begin{aligned}|I_{Y,Z}| \;&\leq\; C\Delta(Y,Z)^{-N_1}J_Y^{1/2}J_Z^{1/2}\quad\text{with}\\ J_Y \;&\stackrel{\text{def}}{=}\; \int(1+g_{2,Y}^{-1}(x-t))^{-n}(1+g_{2,Y}(\tau-\eta))^{-n}\\ &\qquad\times(1+g_{2,Y}^{-1}(t-U_Y))^{-N}|u_Y(t)|^2dxdtd\tau.\end{aligned}$$

With the change of variables $x'=g_{2,Y}^{-1/2}(x-t)$, $\tau'=g_{2,Y}^{1/2}(\tau-\eta)$ and $t'=t$, whose Jacobian is 1, we find that

$$J_Y\leq C\|(1+g_{2,Y}^{-1}(\cdot-U_Y))^{-N}u_Y\|_{L^2}^2.$$

So, we have proved that $|I_{Y,Z}|$ is smaller than

$$C\Delta(Y,Z)^{-N_1}\|(1+g_{2,Y}^{-1}(\cdot-U_Y))^{-N}u_Y\|_{L^2}\|(1+g_{2,Z}^{-1}(\cdot-U_Z))^{-N}u_Z\|_{L^2}.$$

By the Schwarz inequality with measure $\Delta(Y,Z)^{-N_1}|g_Y|^{1/2}|g_Z|^{1/2}dYdZ$, we get

$$I^2\leq\int\|(1+g_{2,Y}^{-1}(\cdot-U_Y))^{-N}u_Y\|_{L^2}^2\Delta(Y,Z)^{-N_1}|g_Y|^{1/2}|g_Z|^{1/2}dYdZ.$$

Condition (6) on Δ implies the lemma.

Now, let us state the main results of the paper. First of all, let us recall Theorem 4.7 of [2].

Theorem 1.12 *Let m be a g-weight, and let us denote by Ω_∞ the set of all x of $\mathbf{R}^n$ so that*

$$\Pi_\infty(x)\stackrel{\text{def}}{=}\int_{\mathbf{R}^n}m^{-2}(x,\xi)d\xi<\infty. \tag{13}$$

Then, for any x of Ω_∞, the linear form on $\mathcal{S}(\mathbf{R}^n)$ defined by

$$\left\{\begin{array}{ccc}\mathcal{S}(\mathbf{R}^n) & \rightarrow & \mathbf{C}\\ u & \mapsto & u(x)\end{array}\right.$$

can be extended in a continuous linear form on $H(m)$ and we have

$$\forall u\in H(m,g),\ \forall x\in\Omega_\infty,\ |u(x)|\leq C\,\Pi_\infty(x)\|u\|_{H(m)}. \tag{14}$$

Sobolev embeddings in L^p were then obtained by interpolation, so it was impossible to catch critical cases. The aim of this paper is to prove the two following theorems. The first one is already proved in [4].

Theorem 1.13 *Let m be a g-weight greater than 1 and A_0 a real number strictly greater than 1. Let us denote by Ω_p the set of all x so that*

$$\Pi_p^2(x) \stackrel{\text{def}}{=} \sup_{A\geq A_0} \Pi_p^2(A,x) \quad \textit{with}$$
$$\Pi_p^2(A,x) \stackrel{\text{def}}{=} A^{-\frac{4}{p-2}} \int_{\{\xi\ /\ m(x,\xi)\leq A\}} m(x,\xi)^{-2} d\xi < \infty.$$

If the Lebesgue measure of Ω_p is strictly positive, the map from $\mathcal{S}(\mathbf{R}^n)$ to L^∞_{loc} defined by $u \mapsto \dfrac{u}{\Pi_p}$ can be extended in a linear continuous map from $H(m)$ to $L^p(\Omega_p, d\mu_p)$, where $d\mu_p$ denotes the measure $\Pi_p^2(x)dx$.

Theorem 1.14 *Consider m a g-weight as in Theorem 1.13 above. Assume moreover that*

$$\lim_{A\to\infty} \frac{\Pi_p(A,x)}{\Pi_p(x)} = 0\,, \textit{uniformly in } x \textit{ and that} \quad \lim_{(x,\xi)\to\infty} m(x,\xi) = +\infty.$$

Then $H(m)$ is compactly embedded in $L^p(\Omega_p, d\mu_p)$.

Let us notice that when $m(x,\xi) = (1+|\xi|^2)^s$, we recover the usual Sobolev embeddings. For corollaries involving subelliptic systems, see [4].

2. Smoothing operators in Weyl-Hörmander calculus

Before introducing the concept of smoothing operators, let us present a proof of usual Sobolev embeddings which is

$$\|f\|_{L^p} \leq C\left(\int_{\mathbf{R}^n} |\xi|^{2s} |\widehat{f}(\xi)|^2 d\xi\right)^{1/2} \quad \text{with} \quad p = \frac{2d}{d-2s}.$$

This will justify the definition of smoothing operators given in a while. This proof is based on classical real interpolation ideas. Let us write

$$\|f\|_{L^p}^p = p\int_0^\infty \lambda^{p-1} \mu(|f| > \lambda) d\lambda.$$

We use the following decomposition in a low and high frequency part. We write $f = f_{1,A} + f_{2,A}$ with $f_{1,A} = \mathcal{F}^{-1}(\mathbf{1}_{B(0,A)}\widehat{f})$ and $f_{2,A} = \mathcal{F}^{-1}(\mathbf{1}_{B^c(0,A)}\widehat{f})$. It is obvious that

$$\|f_{1,A}\|_{L^\infty} \leq \|\widehat{f_{1,A}}\|_{L^1} \leq \frac{C}{d-2s} A^{d/2-s} |f|_{\dot{H}^s}. \tag{15}$$

From the above inequality (15), we have

$$A = A_\lambda \stackrel{\text{def}}{=} \left(\frac{\lambda(d-2s)}{4C|f|_{\dot{H}^s}}\right)^{p/d} \Rightarrow \mu\left(|f_{1,A}| > \frac{\lambda}{2}\right) = 0.$$

As $(|f| > \lambda) \subset (2|f_{1,A}| > \lambda) \cup (2|f_{2,A}| > \lambda)$, we infer that

$$\|f\|_{L^p}^p \leq p \int_0^\infty \lambda^{p-1} \mu(2|f_{2,A_\lambda}| > \lambda) d\lambda.$$

Using the Bienaymé-Tchebychev inequality, we get

$$\|f\|_{L^p}^p \leq 4p \int_0^\infty \lambda^{p-3} \|f_{2,A_\lambda}\|_{L^2}^2 d\lambda. \tag{16}$$

Using the Fourier-Plancherel theorem, we have from inequality (16) that

$$\|f\|_{L^p}^p \leq 4p(2\pi)^d \int_{\mathbf{R}_+ \times \mathbf{R}^d} \lambda^{p-3} \mathbf{1}_{\{(\lambda,\xi)\,/\,|\xi| \geq A_\lambda\}}(\lambda, \xi) |\widehat{f}(\xi)|^2 d\xi d\lambda.$$

But by the definition of A_λ, we have

$$|\xi| \geq A_\lambda \Leftrightarrow \lambda \leq C_\xi \stackrel{\text{def}}{=} \frac{4C|f|_{\dot{H}^s}}{d-2s} |\xi|^{d/p}.$$

Using Fubini's theorem, we have

$$\|f\|_{L^p}^p \leq 4\frac{p(2\pi)^d}{p-2} \left(\frac{4C}{d-2s}\right)^{p-2} |f|_{\dot{H}^s}^{p-2} \int_{\mathbf{R}^d} |\xi|^{d(p-2)/p} |\widehat{f}(\xi)|^2 d\xi.$$

As $2s = \dfrac{d(p-2)}{p}$, the usual Sobolev embedding is proved.

The problem here is to substitute $m(x,\xi)$ to $|\xi|$ in the proof written above. We are going to define a smoothing operator. Let us choose a cut-off function $\chi \stackrel{\text{def}}{=} \mathbf{1}_{[0,1]}$. For any g-weight m, any strictly positive real number A and any $Y = (y, \eta)$ in $\mathbf{R}^{2n}$, we define

$$\chi_{Y,m,A}(x) \stackrel{\text{def}}{=} \chi\left(\frac{m(x,\eta)}{A}\right).$$

Let us define the operator $S_{m,A}$ by

$$S_{m,A} u \stackrel{\text{def}}{=} \int_{\mathbf{R}^{2n}} \psi_Y^w (\chi_{Y,m,A} \varphi_Y^w u) |g_Y|^{1/2} dY. \tag{17}$$

This operator is analogous to the frequency cut-off used in the preceding proof. It is a smoothing operator in the spaces $H(m^\sigma m_1)$ in the following sense.

Theorem 2.1 *Let m and m_1 be two g-weights and $\sigma > 0$. There exists a constant C such that, for any real number $A > 1$, we have*

$$\|S_{m,A}\|_{\mathcal{L}(H(m_1),H(m^\sigma m_1))} \le CA^\sigma.$$

This theorem is proved in [4].

The following L^∞-estimate, which is the equivalent of inequality (15), will be very important for us. An analogous statement is proved in [4].

Theorem 2.2 *Let m be a g-weight and (δ, σ) a pair of real numbers so that δ is positive. Let us denote by $\Omega_{\delta,\sigma}$ the set of all x of $\mathbf{R}^n$ such that, for some strictly positive real number A_0, we have*

$$\begin{aligned} \Pi^2(x) &\stackrel{\text{def}}{=} \sup_{A>A_0} \Pi^2(A,x) < +\infty \quad \textit{with} \\ \Pi^2(A,x) &\stackrel{\text{def}}{=} A^{-2\delta} \int_{m(x,\eta)\le A} m(x,\eta)^{-2} d\eta. \end{aligned} \tag{18}$$

Then there exists a constant C so that, for any $A \ge A_0$ and for any x of $\Omega_{\delta,\sigma}$, the linear form defined on $\mathcal{S}(\mathbf{R}^n)$ by

$$\begin{cases} \mathcal{S}(\mathbf{R}^n) & \to \mathbf{C} \\ u & \mapsto S_{m,A}u(x) \end{cases}$$

can be extended to a continuous linear form on $H(m)$ so that

$$\forall u \in H(m), \forall x \in \Omega_{\delta,\sigma}, \ |S_{m,A}(u)(x)| \le CA^\delta \|u\|_{H(m)} \sup_{B\ge A} \Pi(B,x). \tag{19}$$

The proof of this theorem relies on the following pointwise estimate, proved in [4] and implicitly contained in the proof of Lemma 4.8 of [2].

Lemma 2.3 *For any pair of integers (N, N'), a constant C and an integer k exist so that for any function $\theta \in \mathcal{S}(\mathbf{R}^{2n})$ and for any split quadratic form*

$$\gamma(dx^2, d\xi^2) = \gamma_1(dx^2) + \gamma_2(d\xi^2) \quad \textit{with} \quad \gamma \le \gamma^\sigma,$$

for any $Y \in \mathbf{R}^{2n}$ and for any $x \in \mathbf{R}^n$, we have,

$$\begin{aligned} &|(1 + \gamma_2^{-1}(x - U_{1,Y}))^N \theta^w u(x)| \\ &\qquad \le C\|\theta\|_{k,Conf(\gamma,Y)} |\gamma_2|^{-1/2} ((1 + \gamma_2^{-1}(\cdot))^{-N'} * |u|)(x). \end{aligned}$$

Let us go back to the proof of Theorem 2.2. From Lemma 2.3, for any pair of integers (N, N'), an integer k and a constant C exist such that

$$
\begin{aligned}
|S_{m,A}u(x)| \;\le\; & C\int_{(Y,z)\in\mathbf{R}^{2n}\times\mathbf{R}^n}(1+g_{2,Y}^{-1}(x-U_Y))^{-N}|g_{2,Y}|^{-1/2}\\
& \times(1+g_{2,Y}^{-1}(x-z))^{-N'}m(Y)\chi_{Y,m,A}(z)m\\
& \times(Y)^{-1}|(\varphi_Y^w u)(z)||g_Y|^{1/2}dY\,dz.
\end{aligned}
$$

As the weight m is temperate, applying (8), we infer the existence of an integer N_2 so that

$$m(Y)^{-1}\le Cm(z,\eta)^{-1}(1+g_{2,Y}^{-1}(z-U_{1,Y}))^{N_2}.$$

As we know that

$$g_{2,Y}^{-1}(z-U_{1,Y})\le 2g_{2,Y}^{-1}(x-z)+2g_{2,Y}^{-1}(x-U_{1,Y}),$$

we deduce from this that

$$
\begin{aligned}
S_{m,A}u(x)^2 \;\le\; & C\int_{\mathcal{I}_m(A)}(1+g_{2,Y}^{-1}(x-U_Y))^{N_2-N}(1+g_{2,Y}^{-1}(x-z))^{N_2-N'}\\
& \times m(z,\eta)^{-1}|g_{2,Y}|^{-\frac{1}{2}}m(Y)|(\varphi_Y^w u)(z)||g_Y|^{\frac{1}{2}}dY\,dz
\end{aligned}
$$

with

$$\mathcal{I}_m(A)\stackrel{\text{def}}{=}\{(Y,z)\in\mathbf{R}^{2n}\times\mathbf{R}^n\;/\;m(z,\eta)\le A\}.$$

The Schwarz inequality for the measure $|g_Y|^{1/2}dY\,dz$ implies that

$$
\begin{aligned}
S_{m,A}u(x)^2 \;&\le\; C\|u\|_{H(m)}^2 I(x)\quad\text{with}\\
I(x) \;&\stackrel{\text{def}}{=}\; \int_{\mathcal{I}_m(A)} m(z,\eta)^{-2}(1+g_{2,Y}^{-1}(x-U_Y))^{2N_2-2N}\\
&\qquad\times(1+g_{2,Y}^{-1}(x-z))^{2N_2-2N'}|g_{1,Y}|^{1/2}|g_{2,Y}|^{-1/2}dY\,dz.
\end{aligned}
$$

The metric g is temperate, so, stating $X=(x,\eta)$, we have

$$|g_{1,Y}|^{1/2}|g_{2,Y}|^{-1/2}\le|g_{1,X}|^{1/2}|g_{2,X}|^{-1/2}(1+g_{2,Y}^{-1}(x-U_Y))^{N_0 n}\quad\text{and}$$

$$
\begin{aligned}
(1+g_{2,Y}^{-1}(x-z))^{2N_2-2N'} \le\; & (1+g_{2,X}^{-1}(x-z))^{2N_2-2N'}\\
& \times(1+g_{2,Y}^{-1}(x-U_Y))^{2N_0(N'-N_2)}.
\end{aligned}
$$

Using the uncertainty principle and the fact that g is temperate, we have

$$(1+g_{2,Y}^{-1}(x-U_Y))^{-n}\le C(1+g_{1,X}(x-y))^{-n}(1+g_{2,Y}^{-1}(x-U_Y))^{N_0 n}.$$

Then we claim that the quantity

$$(1+g_{2,Y}^{-1}(x-U_Y))^{2N_2-2N}(1+g_{2,Y}^{-1}(x-z))^{2N_2-2N'}|g_{1,Y}|^{1/2}|g_{2,Y}|^{-1/2}$$

is smaller than

$$C(1+g_{1,X}(x-y))^{-n}(1+g_{2,X}^{-1}(x-z))^{2N_2-2N'}|g_{1,X}|^{1/2}|g_{2,X}|^{-1/2}$$
$$\times(1+g_{2,Y}^{-1}(x-U_Y))^{2N_0(N'-N_2)+n(N_0+1)+2(N_2-N)}.$$

So choosing $N=(1-N_0)N_2+N'+n\left[\frac{N_0+1}{2}\right]$, we find that

$$I(x)\leq C\int_{\mathcal{I}_m(A)} m(z,\eta)^{-2}(1+g_{2,X}^{-1}(x-z))^{2N_2-2N'}$$
$$\times(1+g_{1,X}(x-y))^{-n}|g_{1,X}|^{1/2}|g_{2,X}|^{-1/2}dYdz.$$

As m is a g-weight, we have, by the definition of $\mathcal{I}_m(A)$, that

$$I(x)\leq \int_{\mathcal{J}_m(A)} m(x,\eta)^{-2}(1+g_{2,X}^{-1}(x-z))^{4N_2-2N'}$$
$$\times(1+g_{1,X}(x-y))^{-n}|g_{1,X}|^{1/2}|g_{2,X}|^{-1/2}dYdz \quad \text{with}$$

$$\mathcal{J}_m(A)\stackrel{\text{def}}{=}\{(y,\eta,z)\in\mathbf{R}^{2n}\times\mathbf{R}^n\ /\ m(x,\eta)\leq CA(1+g_{2,X}^{-1}(x-z))^{N_0}\}.$$

With the change of variables $\eta'=\eta$, $y'=g_{1,X}^{1/2}(x-y)$ and $z'=g_{2,X}^{-1/2}(x-z)$, whose Jacobian is $|g_{1,X}|^{\frac{1}{2}}|g_{2,X}|^{-\frac{1}{2}}$, we get

$$I(x)\leq C\int_{\mathcal{K}_m(A)} m(x,\eta')^{-2}(1+|z'|^2))^{2N_2-2N'}(1+|y'|^2)^{-n}dy'd\eta'dz'$$

with $\mathcal{K}_m(A)=\{(y',\eta',z')\ /\ m(x,\eta')\leq CA(1+|z'|^2)^{N_0}\}$. As it is obvious that $A\leq CA(1+|z'|^2)^{N_0}$, we have, by the definition of functions $\Pi(A,\cdot)$,

$$\int_{\{\eta/\ m(x,\eta)\leq A(1+|z|^2)^{N_0}\}} m(x,\eta)^{-2}d\eta\leq C\sup_{B\geq A}\Pi^2(B,x)A^{2\delta}(1+|z|^2)^{2N_0\delta}.$$

So we get

$$I(x)\leq CA^{2\delta}\sup_{B\geq A}\Pi^2(B,x)\int(1+|z|^2))^{4N_2+2N_0\delta-2N'}(1+|y|^2)^{-n}dydz.$$

Choosing for instance $N'=[N_2\delta]+2N_2+n+1$, we conclude the proof of the theorem.

3. Proof of the embedding theorems in L^p

We begin the proof exactly as in the classical case by writing

$$\begin{aligned}\left\|\frac{u}{\Pi_p}\right\|^p_{L^p(\Omega_p,d\mu_p)} \leq\ & p\int_{\lambda\leq\lambda_0}\lambda^{p-1}\mu_p\left(\frac{|u|}{\Pi_p}>\lambda\right)d\lambda \\ & +p\int_{\lambda\geq\lambda_0}\lambda^{p-1}\mu_p\left(\frac{|S_{m,A}u|}{\Pi_p}>\frac{\lambda}{2}\right)d\lambda \\ & +p\int_{\lambda\geq\lambda_0}\lambda^{p-1}\mu_p\left(\frac{|u-S_{m,A}u|}{\Pi_p}>\frac{\lambda}{2}\right)d\lambda.\end{aligned}$$

If C is the constant given by inequality (19) of Theorem 2.2, let us define

$$A_\lambda=\left(\frac{\lambda}{4C\|u\|_{H(m)}}\right)^{(p-2)/2} \quad \text{and} \quad \lambda_0=4C\|u\|_{H(m)}A_0^{2/(p-2)}.$$

Theorem 2.2 applied with $\delta=\dfrac{2}{p-2}$ and $A=A_\lambda$ gives

$$\left(\frac{|S_{m,A}u|}{\Pi_p}>\frac{\lambda}{2}\right)=\emptyset.$$

So we get

$$\begin{aligned}\left\|\frac{u}{\Pi_p}\right\|^p_{L^p(\Omega_p,d\mu_p)} \leq\ & p\int_{\lambda\leq\lambda_0}\lambda^{p-1}\mu_p\left(\frac{|u|}{\Pi_p}>\lambda\right)d\lambda \\ & +p\int_{\lambda\geq\lambda_0}\lambda^{p-1}\mu_p\left(\frac{|u-S_{m,A_\lambda}u|}{\Pi_p}>\frac{\lambda}{2}\right)d\lambda.\end{aligned}$$

We infer that

$$\begin{aligned}\left\|\frac{u}{\Pi_p}\right\|^p_{L^p(\Omega_p,d\mu_p)} \leq\ & \lambda_0^{p-2}p\int_{\lambda\leq\lambda_0}\lambda\mu_p\left(\frac{|u|}{\Pi_p}>\lambda\right)d\lambda \\ & +p\int_{\lambda\geq\lambda_0}4\lambda^{p-3}\left\|\frac{u-S_{m,A_\lambda}u}{\Pi_p}\right\|^2_{L^2(\Omega_p,d\mu_p)}d\lambda.\end{aligned}$$

Let us notice that, for any function v of $L^2(\mathbf{R}^n,dx)$, we have

$$\left\|\frac{v}{\Pi_p}\right\|^2_{L^2(\Omega_p,d\mu_p)}=\|v\|^2_{L^2(\Omega_p,dx)};$$

so we have

$$\lambda_0^{p-2} p \int_{\lambda \le \lambda_0} \lambda \mu_p \left(\frac{|u|}{\Pi_p} > \lambda \right) d\lambda \le \lambda_0^{p-2} \|u\|^2_{L^2(\Omega_p, dx)} \le (4C)^{p-2} A_0^2 \|u\|^p_{H(m)}$$

and then

$$\left\| \frac{u}{\Pi_p} \right\|^p_{L^p(\Omega_p, d\mu_p)} \le (4C)^{p-2} A_0^2 \|u\|^p_{H(m)} + p \int_{\lambda \ge \lambda_0} 4\lambda^{p-3} \|u - S_{m,A_\lambda} u\|^2_{L^2(\mathbf{R}^n, dx)} d\lambda.$$

Now let us estimate $\|u - S_{m,A_\lambda} u\|^2_{L^2}$. We apply the localized Cotlar Lemma 1.11 with $\theta_Y = \psi_Y$ and $u_Y(x) = (1 - \chi_{Y,m,A_\lambda})(x) \varphi_Y^w u(x)$; so we get

$$\|u - S_{m,A_\lambda} u\|^2_{L^2} \le C \int (1 + g_{2,Y}^{-1}(x - U_{1,Y}))^{-N} (1 - \chi_{Y,m,A_\lambda}(x))^2 \times |\varphi_Y^w u(x)|^2 |g_Y|^{1/2} dY dx.$$

So we infer that

$$p \int_{\lambda \ge \lambda_0} \lambda^{p-3} \|u - S_{m,A_\lambda} u\|^2_{L^2} d\lambda \le C \int (1 + g_{2,Y}^{-1}(x - U_{1,Y}))^{-N} \times (1 - \chi_{Y,m,A_\lambda})(x))^2 |\varphi_Y^w u(x)|^2 \lambda^{p-3} |g_Y|^{1/2} dY dx d\lambda.$$

By the definition of χ_{Y,m,A_λ} and A_λ, we have

$$\lambda \le 4C \|u\|_{H(m)} m(x,\eta)^{2/(p-2)} = \lambda(x,\eta).$$

So we get

$$\begin{aligned} I_p &\stackrel{\text{def}}{=} p \int_{\lambda \ge \lambda_0} \lambda^{p-3} \|u - S_{m,A_\lambda} u\|^2_{L^2} d\lambda \\ &\le \frac{p}{p-2} \int (1 + g_{2,Y}^{-1}(x - U_{1,Y}))^{-N_0} \lambda(x,\eta)^{p-2} dY dx \\ &\le C_N^p \|u\|^{p-2}_{H(m)} \int (1 + g_{2,Y}^{-1}(x - U_{1,Y}))^{-N} m^2(x,\eta) |\varphi_Y^w u(x)|^2 dY dx. \end{aligned}$$

As m is a g-weight, we have $m^2(x,\eta) \le C m^2(y,\eta)(1 + g_{2,Y}^{-1}(x - U_{1,Y}))^{N_0}$; so we get

$$p \int_{\lambda \ge \lambda_0} \lambda^{p-3} \|u - S_{m,A_\lambda} u\|^2_{L^2} d\lambda \le C \|u\|^p_{H(m)}.$$

This proves Theorem 1.13.

Now let us prove Theorem 1.14. The main step consists in the proof of the following interpolation lemma.

Lemma 3.1 *Let us assume the hypotheses of Theorem 1.14. For any strictly positive real number ε, there exists a constant C_ε such that, for any u in $H(m)$, we have*

$$\left\| \frac{u}{\Pi_p} \right\|_{L^p(\Omega_p;d\mu_p)} \leq C_\varepsilon \|u\|_{L^2(\Omega_p;dx)} + \varepsilon \|u\|_{H(m)}.$$

Let us start from inequality (20) and assume that $\varepsilon < 1$. There exists a strictly positive real number A_ε so that

$$\forall x \in \Omega_p, \sup_{B \geq A_\varepsilon} \Pi_p(B,x) \leq \varepsilon \Pi_p(x).$$

Consider the constant C of inequality (19) of Theorem 2.2 and let us define

$$A_{\lambda,\varepsilon} = \left(\frac{\lambda}{4C\|u\|_{H(m)}\varepsilon} \right)^{(p-2)/2} \quad \text{and} \quad \lambda_{0,\varepsilon} = 4C\varepsilon \|u\|_{H(m)} A_\varepsilon^{2/p-2}.$$

Theorem 2.2 implies that $\|S_{m,A_{\lambda,\varepsilon}}\|_{L^\infty} \leq \lambda/2$, so, we have

$$\left\| \frac{u}{\Pi_p} \right\|^p_{L^p(\Omega_p;d\mu_p)} \leq \lambda_{0,\varepsilon}^{p-2} p \|u\|^2_{L^2(\Omega_p,d\mu_p)} + p \int_{\lambda \geq \lambda_0} \lambda^{p-3} \|u - S_{m,A_{\lambda,\varepsilon}} u\|^2_{L^2} d\lambda.$$

Applying the localized Cotlar Lemma 1.11, and using the fact that, by definition of $\chi_{Y,m,A}$, we have

$$\begin{aligned} \left\| \frac{u}{\Pi_p} \right\|^p_{L^p(\Omega_p;d\mu_p)} \leq\; & \lambda_{0,\varepsilon}^{p-2} p \|u\|^2_{L^2(K,d\mu_p)} \\ & + C_{p,N} \int_{\lambda \leq \lambda_\varepsilon(x,\eta)} (1 + g_{2,Y}^{-1}(x - U_{1,Y}))^{-N} \\ & \times |\varphi_Y^w u(x)|^2 \lambda^{p-3} |g_Y|^{1/2} dY\, dx\, d\lambda \end{aligned}$$

with $\lambda_\varepsilon(x,\xi) \stackrel{\text{def}}{=} 4C\varepsilon\|u\|_{H(m)} m(x,\eta)^{2/p-2}$. Along the same lines as the proof of Theorem 1.13 and using the definition of $\lambda_{0,\varepsilon}$ we prove that

$$\left\| \frac{u}{\Pi_p} \right\|^p_{L^p(\Omega_p;d\mu_p)} \leq C_p A_\varepsilon^2 \|u\|^{p-2}_{H(m)} \|u\|^2_{L^2(K,d\mu_p)} + C_p \varepsilon^{p-2} \|u\|^p_{H(m)}.$$

This implies the lemma. Using Theorem 1.9, the theorem comes from the above Lemma 3.1 with standard functional analysis arguments.

References

[1] R. Beals, Weighted distribution spaces and pseudodifferential operators, *Journal d'Analyse Mathématique*, **39**, 1981, pages 130–187.

[2] J.-M. Bony and J.-Y. Chemin, Espaces fonctionnels associés au calcul de Weyl-Hörmander, *Bulletin de la Société Mathématique de France* **122** (1994), 77–118.

[3] J.-M. Bony and N. Lerner, Quantification asymptotique et microlocalisation d'ordre supérieur, *Annales de l'École Normale Supérieure* **22** (1989), 377–433.

[4] J.-Y. Chemin and C.-J. Xu, Inclusions de Sobolev en calcul de Weyl-Hörmander et systèmes sous-elliptiques, *prépublication du Laboratoire d'Analyse Numérique de l'Université Paris 6, à paraître aux Annales de l'École Normale Supérieure.*

[5] L. Hörmander, *The analysis of linear partial differential equations*, tome 3, Springer Verlag, 1985.

[6] N. Lerner, Sur les espaces de Sobolev généraux associés aux classes récentes d'opérateurs pseudo-différentiels, *Notes aux Comptes-Rendus de l'Académie des Sciences de Paris* **289** (1979), 663–666.

Jean-Yves Chemin
Analyse Numérique, Tour 55-65, 5ème étage
Université Pierre and Marie CURIE, 4 Place Jussieu
75230 Paris Cedex 05, France

Chao-Jiang Xu
Institut de Mathématiques, Université de Wuhan
Wuhan 430072, R. P. China

About the Cauchy problem for a system of conservation laws

C. Cheverry

I. Introduction

This note summarizes some recent developments in the field of hyperbolic systems of conservation laws (in one-dimensional space). More precisely, we consider the inital value problem for data of small amplitude. The question is to find a relevant criterion leading to global existence whatever the variation of the initial condition.

Progress in this direction was first achieved by way of geometric optics. This asymptotic analysis has allowed us to isolate the key concepts. New tools were then introduced to tackle the Cauchy problem with large BV-initial data. This article should be considered as a complement to [2], [3] and [4]. The main ideas present in these works are developed and illustrated by some new examples. For details of the proofs, the reader should refer to the cited literature.

Consider a $n \times n$ system of nonlinear conservation laws:

$$u_t + [f(u)]_x = 0 \quad ; \quad -\infty < x < \infty, \quad u \in \mathbb{R}^n, \tag{1.1}$$

such that, in a neighborhood of some basic state $\bar{u}$, the function f is sufficiently smooth, the system is strictly hyperbolic, and each field is genuinely nonlinear in the sense of Lax [11].

Let $\lambda_j(u)$ be the jth eigenvalue of the matrix $Df(u)$. Let $r^j(u)$ and $l_j(u)$ denote the corresponding right and left eigenvectors:

$$Df(u).r^j(u) = \lambda_j(u)\, r^j(u) \quad ; \quad l_j(u).Df(u) = \lambda_j(u)\, l_j(u). \tag{1.2}$$

By changing the state variable u, we may always arrange

$$\Gamma^i_{ii} := \nabla\lambda_i(\bar{u}).r^i(\bar{u}) > 0 \quad ; \quad 1 \le i \le n. \tag{1.3}$$

The global existence of weak solutions was proved in 1965, in the fundamental paper of Glimm [6]. Glimm's existence theorem says that the system (1.1) with initial data

$$u(0,x) = u_0(x) \tag{1.4}$$

Typeset by $\mathcal{A}\mathcal{M}\mathcal{S}$-TEX

has a weak solution for all time provided that

$$\| u_0(x) - \bar{u} \|_{L^\infty(\mathbb{R})} \le \varepsilon_1 \ll 1, \tag{1.5}$$

$$TV_{\mathbb{R}}\big(u_0(.)\big) := \sum_{x_1 < \ldots < x_N} | u_0(x_{i+1}) - u_0(x_i) | \le \varepsilon_2 \ll 1 \tag{1.6}$$

where the ε_i are small positive constants which depend only on f.

The solution $u(t,x)$ of the Cauchy problem (1.1)–(1.4) is obtained as the limit in L^1_{loc} of a sequence $\{ u_\nu(t,x) \}_\nu$ of piecewise constant approximate solutions. For each ν, the constructive procedure for $u_\nu(t,x)$ starts at $t = 0$, choosing a piecewise constant function $u_\nu(0,x)$ such that

$$\| u_\nu(0,x) - \bar{u} \|_{L^\infty(\mathbb{R})} \le \varepsilon_1 \ll 1 . \tag{1.7}$$

Unless we add supplementary hypotheses on the flux function $f(u)$, the request (1.5) (which in turn implies (1.7)) is necessary. Indeed, it allows us to solve the Riemann problems at every point where the function $u_\nu(0,x)$ has a jump. A local approximate solution of (1.1) – (1.4) can thus be obtained by piecing together the solutions of the different Riemann problems.

In order to extend this local solution, it is important to handle the successive wave interactions. It is on this level that condition (1.6) arises. Several authors have tried to relax constraint (1.6). This program has been undertaken under very restrictive assumptions (see [16] and more recently [20] for the Euler equations; see [18] for explanations concerning the rich systems). Nevertheless, no systematic study had ever been made. The question approached here is this: Given a system of conservation laws, do we necessarily have recourse to (1.6)?

Let us recall how the information (1.6) usually occurs. It is useful for bounding the total variation of the approximation $u_\nu(t,x)$ at any time $t > 0$. Then, by Helly's theorem, a subsequence which converges to a solution of (1.1), can be selected.

The classical method is the following: a nonlinear functional F is introduced. It measures the total variation plus a quadratic potential Q for wave interaction:

$$F\big(u_\nu(t,.)\big) := TV_{\mathbb{R}}\big(u_\nu(t,.)\big) + C_1 \; Q\big(u_\nu(t,.)\big) . \tag{1.8}$$

Two waves α and β are approaching if they will interact at some later time. After their interaction, their strengths may be modified and new waves may be created. It induces at worst an increase in Q by

$$C_2 \; |\alpha| \; |\beta| \; TV_{\mathbb{R}}\big(u_\nu(t,.)\big) . \tag{1.9}$$

On the other hand, waves α and β are no longer approaching after the interaction. Thus, we have to cut the product $|\alpha|\,|\beta|$ from Q. Briefly, the increase (1.5) is offset by the corresponding decrease if

$$C_2 \; TV_{\mathbb{R}}\big(u_\nu(t,.)\big) \; < \; 1\,, \tag{1.10}$$

which is precisely the requirement of a small total variation.

When condition (1.10) is fulfilled, it's over. Indeed, by choosing the weight C_1 appropriately, the functional F is easily shown to decrease. It guarantees that the variation of $u_\nu(t,x)$ is still bounded at any time $t > 0$.

The preceding argument is clearly no longer valid when the total variation grows (when (1.10) is no longer verified). This remark does not mean that a weak solution does not exist for all time. Indeed, as long as the nearby states in $u_\nu(t,x)$ are close enough, the approximation may be defined further, without regard to the size of the total variation. This only indicates that the preceding global analysis, which estimates the contribution given by two interacting waves at time t by the product of the strengths of the corresponding waves at the initial time, generally lacks precision in considering solutions with a large BV–norm.

In fact, the problem that matters is to understand the phenomena that locally induce the increasing or decreasing of the variation. Intuitively, the genuine nonlinearity gives rise to decay, whereas the presence of nonzero interaction coefficients implies the increasing. It is important to measure and compare these different effects. Indeed, two main problems lie in hiding beyond these preoccupations:

(i) The well-posedness of the Cauchy problem (1.1)–(1.4) with large initial data (when only the L^∞-bound (1.5) is required).

(ii) The compactness properties from L^∞ to L^1_{loc} of the solution operator associated to the hyperbolic system (1.1).

The plan of this lecture follows the historical development of the subject. Section 2 states the point of view of weakly nonlinear geometric optics. The complex problem (1.1) is reduced to the simpler one (but non-trivial) which consists in the modulation equations. In Section 3, we recall how to investigate the BV-stability properties of the modulation equations. The method is then applied to the important special case of nonisentropic gas dynamics. It allows us to improve the estimate (2.16) given on page 212 in the work of Majda-Rosales-Shonbek [13]. In Section 4, we first recall one of the most significant results exposed in [4]. We then carefully explain the main ideas underlying the proof.

II. Weakly nonlinear geometric optics

It is a long-standing problem to establish the global existence of solutions to the Cauchy problem (1.1)–(1.4) with large BV-data. We remark that a natural way to prove existence in general for a nonlinear evolution equation

$$\dot{u}(t) = F\big(u(t)\big) \tag{2.1}$$

relies on finding uniform bounds for the natural metrics associated with the differential identity (2.1).

The discontinuous solutions of system (1.1) are controlled by the amplitude as measured by the L^∞-norm and the total amount of wave magnitude as measured by the total variation norm. Thus, the stability of (1.1) may be expressed by the two following inequalities:

$$\| u(t,.) - \bar{u} \|_{L^\infty(\mathbb{R})} \leq C_3 \; \| u_0(.) - \bar{u} \|_{L^\infty(\mathbb{R})} \; ; \quad t \in \mathbb{R}_*^+ , \tag{2.2}$$

$$TV_{[a\,;\,b]}\big(u(t,.)\big) \leq C_3 \; TV_{[a-\lambda_t t\,;\,b+\lambda_t t]}\big(u_0(.)\big) \; ; \tag{2.3}$$

$; -\infty < a < b < \infty \quad ; \quad t \in \mathbb{R}_*^+$, where we have introduced the constant λ_t, which bounds the speed of propagation before time t :

$$\lambda_t := \max_{1 \leq i \leq n} \; \sup_{(s,y) \in [0,t] \times \mathbb{R}} \; |\, \lambda_i\big(u(s,y)\big) \,| \, . \tag{2.4}$$

The two estimates (2.2) and (2.3) must be uniform.

In particular, they must be achieved by the solutions $u^\varepsilon(t,x)$ of the hyperbolic system (1.1) that correspond to small amplitude rapidly oscillating initial data:

$$u^\varepsilon(0,x) = u_0^\varepsilon(x) = \bar{u} + \varepsilon \sum_{j=1}^{n} U_0^j(x/\varepsilon) \; r^j(\bar{u}) \, . \tag{2.5}$$

The perturbation in (2.5) of the constant state $\bar{u}$ is built with the help of n profiles:

$$U_0^j(x+1) = U_0^j(x) \quad ; \quad U_0^j(.) \in BV(\mathbb{T}) \quad ; \quad 1 \leq j \leq n \, . \tag{2.6}$$

The function $u_0^\varepsilon(x)$ thus constructed is periodic with period ε. It is nearly constant. For ε sufficiently small, the condition (1.5) is automatically satisfied. On the other hand, the total variation becomes infinite as soon as one of the profiles $U_0^j(x)$ is non-trivial. Therefore, the constraint (1.6) is not verified by the expression $u^\varepsilon(0,x)$. This particularity grounds the interest of our study. Indeed, when condition (1.6) is fulfilled, the validity of the two estimates (2.2) and (2.3) goes without saying [6].

Nevertheless, our framework here concerns functions with large total variation. In that case, the estimate

$$(2.7)\qquad \begin{aligned} TV_{[a\,;b]}\big(u^\varepsilon(t,.)\big) \leq C\ TV_{[a-\lambda_t t\,;b+\lambda_t t]}\big(u_0^\varepsilon(.)\big)\,; \\ -\infty < a < b < \infty \quad ; \quad t \in \mathbb{R}_*^+, \end{aligned}$$

is not at all guaranteed.

The purpose of weakly nonlinear geometric optics is to describe the leading order behaviour of the sequence $\{u^\varepsilon(t,x)\}_\varepsilon$ when the parameter ε tends to zero. The theory is accomplished in the general setting of weak solutions, even after shock waves have formed (see [2] and [16]). The exact solution $u^\varepsilon(t,x)$ is known to be suitably described by an asymptotic expansion of the form

(2.8)

$$u^\varepsilon(t,x) \sim \tilde u^\varepsilon(t,x) := \bar u + \varepsilon \sum_{j=1}^{n} U^j\Big(t\,,\big(x-\lambda_j(\bar u)\,t\big)/\varepsilon\Big)\, r^j(\bar u)\,.$$

A formal analysis [12] reveals the constraints to impose on the different components $U^j(t,y)$ of the profile $U(t,y)$. These are Burgers equations coupled through terms incorporating nonlocal resonant wave interactions.

The system of integro-partial-differential equations in question assumes the canonical form

$$(2.9)\qquad \begin{aligned} &(\partial_t U^i)(t,y) + (1/2)\,\Gamma^i_{ii}\,\partial_y\,\big[U^i(t,y)^2\big] + \\ &+ \sum_{i\neq p<q\neq i} \Lambda^i_{pq}\,\partial_y\Big\{\int_0^1 U^p(t,y-s)\,U^q(t,s)\,ds\Big\} = 0\,, \\ &\qquad\qquad ;\text{in } \mathbb{R}^+\times\mathbb{T} \text{ for } 1\leqslant i\leqslant n\,. \end{aligned}$$

It is completed by prescribing the initial conditions

$$(2.10)\qquad U^i(0,y) = U_0^i(y) = U_0^i(y+1) \ \text{ in } \mathbb{T} \text{ for } 1\leqslant i\leqslant n\,.$$

The interaction coefficients Λ^i_{pq} introduced in (2.9) are defined as

$$(2.11)\qquad \Lambda^i_{pq} := l_i(\bar u)\,.\,\big[D\,r^q(\bar u)\,.\,r^p(\bar u) - D\,r^p(\bar u)\,.\,r^q(\bar u)\big]\,.$$

According to the preceding definition, these coefficients describe the Lie-algebra structure of the eigenvector fields $\{r^j(u)\}_{1\leq j\leq n}$. Concerning the solutions of (1.1), they take into account the effects of interactions between the pth and q^{th} families, as seen in the ith family. They will play a very important role in what follows.

We have said that the solutions $u^\varepsilon(t,x)$ would be approached with precision by the expansions $\tilde{u}^\varepsilon(t,x)$. By this assertion, we mean that we have the following local L^1-convergence result [2]:

$$(2.12) \qquad \| u^\varepsilon(t,.) - \tilde{u}^\varepsilon(t,.) \|_{L^1([a,b])} = o(\varepsilon) \quad ; \quad ; -\infty < a < b < \infty \quad , \quad t \in \mathbb{R}_*^+ .$$

With regard to (2.12), it seems natural to replace the expression $u^\varepsilon(t,x)$ in (2.7) by its corresponding model $\tilde{u}^\varepsilon(t,x)$. Passing to the limit when the parameter ε tends to zero, we see, after some elementary manipulations, that an inequality similar to

$$(2.13) \qquad TV_{\mathbb{T}}\big(U(t,.)\big) \leq C_4 \; TV_{\mathbb{T}}\big(U_0(.)\big) \quad ; \quad t \in \mathbb{R}_*^+ ,$$

must be achieved for some suitable constant C_4. Thus, we may assert that a prerequisite for the well-posedness of the hyperbolic system (1.1) is the BV-stability expressed on the level of (2.13).

Following this idea, Joly, Métivier and Rauch [10] constructed a smooth solution to (2.9), for a particular 3×3 system of conservation laws which blows up in finite time. They have concluded that the corresponding solution to (1.1) must blow up in L^1-norm (and also in BV-norm) in finite time.

We adopt a different point of view here. We look for a relevant criterion which guarantees global existence for (2.9). Then we ask about the stability of the initial system (1.1).

The contents of this second section can be summarized by the following figure:

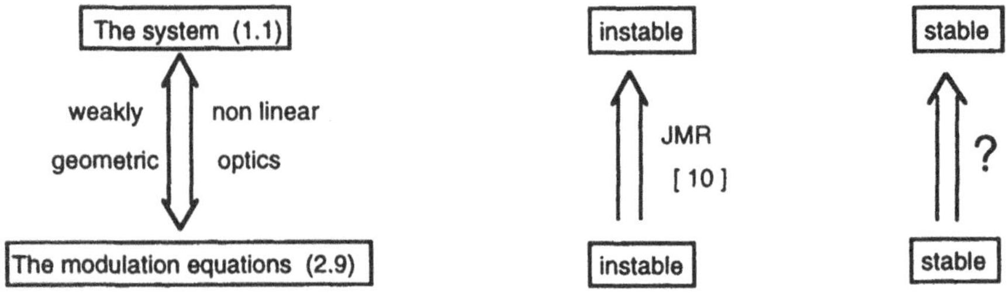

III. Stability of the modulation equations

In this section, we review the technicalities developed in [3]. We take care of explaining the notions underlying the proofs given in [3]. This step is important. Indeed, when discussing with the Cauchy problem (1.1)–(1.4), we shall take inspiration from the ideas expressed here.

The strategy for obtaining information about the large time behavior of a given solution $U(t,y)$ to system (2.9) follows standard arguments. We first introduce an expression $G_U(t)$, which is equivalent to the local variation of the profile $U(t,y)$:

$$(3.1) \qquad G_U(t) \sim TV_{\mathbb{T}}\big(U(t,.)\big) \qquad ; \qquad t \in \mathbb{R}^+_* .$$

Then we seek a bound on $G_U(t)$, uniformly valid for all times $t \geq 0$:

$$(3.2) \qquad \sup_{t \in \mathbb{R}^+} G_U(t) \leq M < \infty .$$

The quantity $G_U(t)$ is obtained by a rather general procedure. We build an operator Op which acts on BV-periodic functions:

$$(3.3) \qquad Op : BV(\mathbb{T}) \longrightarrow \mathbb{R}^+$$

Next we define

$$(3.4) \qquad G_U(t) := Op\big(U(t,.)\big) .$$

The operator Op presents some very nice properties in connection with the periodic solutions of a Burgers law:

$$(3.5) \qquad V_t + (\Gamma\ V^2/2)_y = 0 \qquad ; \qquad \Gamma > 0 ,$$

with

$$(3.6) \qquad V(t,y+1) = V(t,y) .$$

Indeed, it takes into account the manner by which the genuine non-linearity in (3.5) acts on the function $V(t,y)$. This requirement finds expression in the differential inequality

$$(3.7) \qquad dG_V(t)/dt \leqslant -\Gamma\ G_V(t)^2 .$$

Let us now describe the ideas involved in the construction of $G_V(t)$.

A solution to (3.5) inherits naturally the BV-regularity. Therefore, the partial derivative $\partial_y V(t,y)$ is a Radon measure which may be decomposed into its positive part plus its negative part

$$\mu(t,.) := \partial_y V(t,.) = \mu_+(t,.) + \mu_-(t,.) \in M_b(\mathbb{T}), \tag{3.8}$$

with

$$< \mu_+(t,.) \, ; \, \varphi(.) > \, \geq 0, \tag{3.9}$$

for all positive continuous functions $\varphi(y)$.

As the profile $V(t,y)$ is periodic in y, we have the identity

$$TV_{\mathbb{T}}\big(V(t,.)\big) = 2 \, \| \mu_+(t,.) \|_{M_b(\mathbb{T})} \, . \tag{3.10}$$

The equality (3.10) shows that we may concentrate our analysis on the expressions $\mu_+(t,y)$ without losing control of the local variation. Now, it is more convenient to handle $\mu_+(t,y)$ than $\mu(t,y)$. Let us explain why this is so.

It is a well known fact that the difficulties when dealing with (3.5) come from the presence of shocks. The solution $V(t,y)$ is in general discontinuous. At any point z of discontinuity, the entropy condition imposes the relation

$$V(t,z+) < V(t,z-) \, . \tag{3.11}$$

The preceding inequality implies that all the singularities of order zero are contained in the negative part $\mu_-(t,y)$. They are not seen on the level of the positive part $\mu_+(t,y)$. It follows that the quantity $\mu_+(t,y)$ recovers better properties of regularity. This fact has already been observed in the work of Dafermos [5], who has proved the following result:

$$\mu^+(t,.) \in L^\infty(\mathbb{T}) \quad ; \quad t \in \mathbb{R}_*^+ \, . \tag{3.12}$$

Working with solutions of (3.5), people used to consider the local variation. Unfortunately, the expression $TV_{\mathbb{T}}\big(V(t,.)\big)$ fails to satisfy (3.7). Indeed, the variation of a solution to Burgers law may very well remain the same during some nontrivial time interval.

A different approach is proposed in [3]. Because of (3.12), it makes sense to talk about the value taken by the function $\mu^+(t,.)$ at any point y. We may study how these values are distributed along the space line. As the expression $\mu^+(t,y)$ is the solution to

$$\mu_t^+ + \Gamma \; V \; \mu_y^+ = - \; \Gamma \; (\mu^+)^2 \; < 0, \tag{3.13}$$

it is decreasing along each characteristic curve. Intuitively, the values of $\mu^+(t,y)$ become less and less concentrated under the effects of genuine nonlinearity. The construction of $G_V(t)$ draws inspiration from this principle. Roughly speaking, it is defined by assigning a convenient weight (bigger as the support of the function $\mu^+(t,y)$ is getting smaller) as a factor of the expression $TV_{\mathbb{T}}\left(V(t,.)\right)$. The technical details can be found in [3].

We stress the fact that identity (3.7) is remarkable. Our natural expectation should have been to find an inhomogeneous differential inequality whose second member should have been a locally finite Borel Measure. What makes (3.7) possible here is the regularity property (3.12) (or the fact that our analysis gets around the difficulty induced by the presence of shocks).

Now we shall implement the two relations (3.1) and (3.7) in order to study the stability of (2.9). At this stage, it only remains to understand the contribution brought by the integral expressions placed as a factor of the interaction coefficients.

For this purpose, we perform a rough analysis. By differentiating (2.9) with regard to y and then integrating in y the expression thus obtained, we ascertain that the convolution terms give rise to an increase in Q. More precisely, the BV-norm can grow at a rate at most proportional to

$$\sum_{i\neq p<q\neq i} |\Lambda^i_{pq}| \; \left\| \int_0^1 \partial_y U^p(t,.-s)\; \partial_y U^q(t,s)\; ds \right\|_{L^1(\mathbb{T})} \tag{3.14}$$

which, because of (3.1), is equivalent to

$$\sum_{i\neq p<q\neq i} |\Lambda^i_{pq}| \; G_{U^p}(t)\; G_{U^q}(t)\,. \tag{3.15}$$

In turn, the method leads to the following system of n differential inequalities:

$$\begin{aligned} dG_{U^i}(t)\,/\,dt \;\leq\; &-\;\Gamma^i_{ii}\;G_{U^i}(t)^2 + \\ &+\;16\;e\sum_{i\neq p<q\neq i} |\Lambda^i_{pq}|\;G_{U^p}(t)\;G_{U^q}(t)\,;\quad 1\leq i\leq n\,. \end{aligned} \tag{3.16}$$

These identities do not take into account all the effects. When considering realistic systems, physical assumptions lead to extra conditions on the interaction coefficients. For example, the existence of a convex entropy function for which an additional conservation law can be derived imposes constraints on the signs of the Λ^i_{pq}'s. Now, the presence of Λ^i_{pq}'s with well-adjusted signs may induce cancellations. Nevertheless, the exploitation of these sign properties (even at the level of (2.9)) is still an open problem.

For the present, we may regard (3.16) as a system describing rather well the phenomena that govern the solutions of (2.9). Indeed, in despite of the preceding objection, the analysis is sharp: it expresses very well the competition that occurs between the influence of genuine nonlinearity and the de-stabilizing amplification effects due to resonance. Keeping this in mind, we investigate the large time stability properties of system (3.16).

Obviously, a uniform bound is guaranteed for the functionals $G_{U^i}(t)$ as soon as we have

$$\left(\inf_{1 \leq i \leq n} \Gamma^i_{ii} \right) \geq \kappa \left(\sup_{i \neq p < q \neq i} |\Lambda^i_{pq}| \right), \tag{3.17}$$

for a sufficiently large constant κ.

The preceding assumption is sufficient but not necessary. It is introduced for the sake of simplicity. A more precise and intrinsic criterion of stability may be formulated (see [3]).

In practice, each type of equation requires a specific treatment. In particular, it is sometimes also possible to consider systems that admit a linearly degenerate field (which is excluded by the condition (3.17)). As an illustration of this assertion, the rest of this first chapter is devoted to the important special case of nonisentropic gas dynamics. The reader is invited to consult the references [8], [13] and [14] for an introduction to the subject.

The simplified asymptotic equations associated with the system of compressible fluid flow are composed by the identity

$$\begin{aligned} (\partial_t U^1)(t,y) + (1/2)\ \partial_y \left[U^1(t,y)^2 \right] + \\ + \int_0^1 (\partial_y K)(y-z)\ U^2(t,z)\ dz = 0, \end{aligned} \tag{3.18}$$

coupled with

$$\begin{aligned} (\partial_t U^2)(t,y) - (1/2)\ \partial_y \left[U^2(t,y)^2 \right] + \\ - \int_0^1 (\partial_y K)(-y+z)\ U^1(t,z)\ dz = 0. \end{aligned} \tag{3.19}$$

It is completed by prescribing the initial conditions

$$U^i(0,y) = U^i_0(y) = U^i_0(y+1) \quad \text{in } \mathbb{T} \quad \text{for } i = 1, 2. \tag{3.20}$$

In (3.18)–(3.19), the kernel $K(y)$ is a specified BV-function (independent of time t) with period one. By applying our general procedure to system (3.18)–(3.19), we obtain

$$dG_{U^1}(t)\,/\,dt \;\leq\; -\,G_{U^1}(t)^2 \;+\; 16\;e\;G_K(0)\;G_{U^2}(t) \tag{3.20}$$

and

$$dG_{U^2}(t)\,/\,dt \;\leq\; -\,G_{U^2}(t)^2 \;+\; 16\;e\;G_K(0)\;G_{U^1}(t)\,. \tag{3.21}$$

The phase portrait for the autonomous system corresponding to the inequalities (3.20)–(3.21) is depicted below.

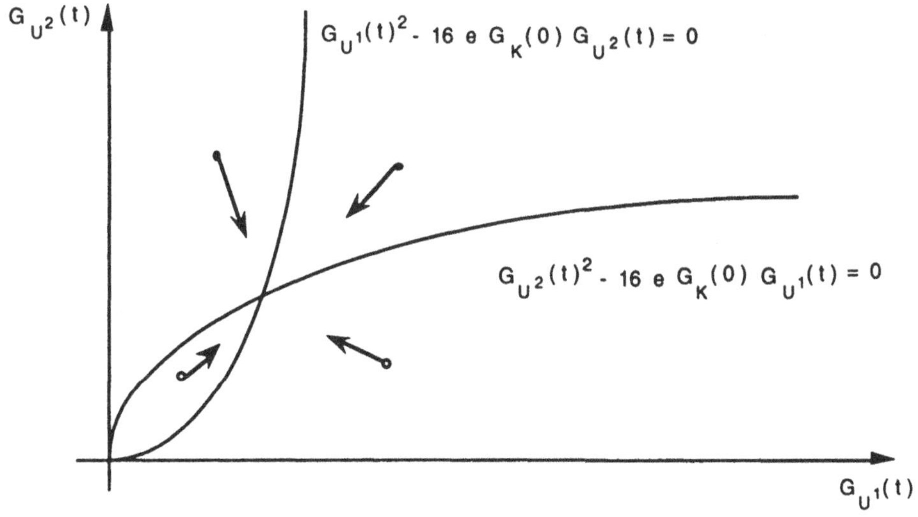

We see in this picture that the flow associated with (3.20)–(3.21) remains bounded for all times $t \geq 0$. It follows:

Corollary. *Global existence for the system* (3.18)–(3.19). *Moreover, the solution* $U(t,y) = \big(U^1(t,y)\,,\,U^2(t,y)\big)$ *is subjected to the uniform following control*

$$\sup_{t\in\mathbb{R}^+} \;\; TV_{\mathbb{T}}\big(U(t,.)\big) \;\leq\; C_5 \;\; \|\,K(.)\,\|_{M_b(\mathbb{T})} \;<\; \infty \tag{3.22}$$

for some constant C_5 *independant of the initial data.*

IV. The Cauchy problem with small amplitude initial data

The purpose of this section is to present an introduction to the central aspects of the Cauchy problem (1.1)−(1.4) with small amplitude initial data. We have attempted to state the most significant results to date. Surely, many improvements can still be found.

The strategy consists of bringing out the study from the modulation equations. We saw in the previous section that system (2.9) is well understood when condition (3.17) is fulfilled. The principle by which the effects of the genuine nonlinearity and of the interaction coefficients can be measured and are comparable should be recovered at the level of system (1.1). In any case, it presents a very good starting point in view of a general theory.

The reader should have already guessed that we will adopt the reverse approach of this followed by weakly nonlinear geometric optics. We mean gradually to modify the asymptotic equations (2.9) in order to obtain the complete system (1.1).

To this end, we proceed by stages in accordance with the following picture.

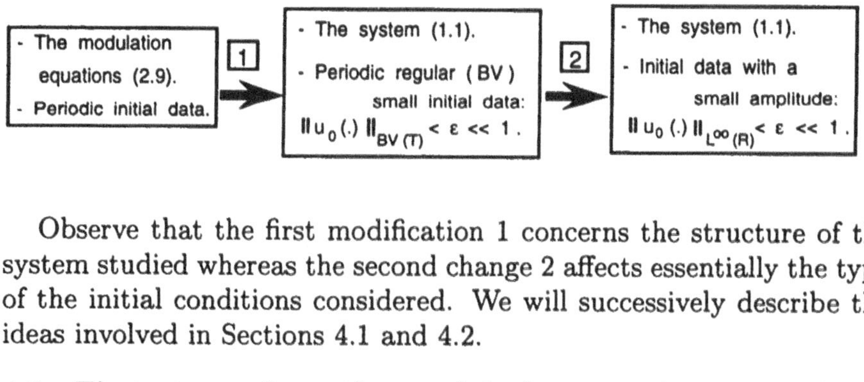

Observe that the first modification 1 concerns the structure of the system studied whereas the second change 2 affects essentially the type of the initial conditions considered. We will successively describe the ideas involved in Sections 4.1 and 4.2.

4.1. First stage: from the modulation equations to periodic global solutions of system (1.1).

Let us first recall the original purpose: the study of the BV-stability properties of system (1.1). Keeping this goal in mind, we draw a parallel between (2.9) and (1.1). Of course, such a comparison would be irrelevant when considering other problems (like the regularity of the solutions or their uniqueness). On the other hand, it is indeed relevant when discussing the mechanisms of exchange that govern the BV-norm.

In order to put (2.9) and (1.1) together, we have to specify clearly what is possible to compare. This point runs into a lot of difficulties. We intend to make a formal analysis, which allows us to understand the broad outline of the reasoning. A rigorous self-contained approach is given in [4].

The analysis of (2.9) is based on the study of the expression $G_{U^i}(t)$. We wish to transfer the information contained about the quantity $G_{U^i}(t)$ on to the system (1.1). One possible approach is prompted by the formula (2.8). It consists in making the association

$$(4.1.1) \qquad U^i(t,y) \longleftrightarrow l_i(\bar{u}) \,.\, u(t,x) \,.$$

Such a comparison is useful in order to justify weakly nonlinear geometric optics. On the other hand, it is not convenient for our purpose. Indeed, we cannot carry out the study of exact solutions to (1.1) with precision by freezing the state variable in the vector $l_i(u)$.

In order to have a more precise idea of the subject, we have to remember that the construction of the quantity $G_{U^i}(t)$ makes principal use of the derivative $(\partial_y U^i)(t,y)$ (rather than $U^i(t,y)$). Therefore, the question is: To what does each component $(\partial_y U^i)(t,y)$ correspond to in (1.1)?

Several replies have already been developed in the literature. They mainly depend on the regularity of the solutions involved. Consult Fritz John [9] in the case of regular solutions and Tai-Ping-Liu [19] in the case of piecewise constant functions. In fact, these approaches are very close to each other. It is always profitable to consider them simultaneously. In particular, each of them brings part of the information concerning the BV-stability of (1.1).

We begin our discussion by assuming the point of view of regular solutions. To this end, we have to introduce some preliminary notation.

Before everything else, observe that, by performing an affine change on the variables (t,x), one may fix the speed of propagation at the ground state. We impose here

$$(4.1.2) \qquad -\lambda_1(\bar{u}) = \lambda_n(\bar{u}) = 1/2 \,.$$

Now, let $u(t,x)$ be a piecewise Lipschitz periodic (with period 1) solution of (1.1). Then the derivative $u_x(t,x)$ exists almost everywhere and one can define its ith component to be

$$(4.1.3) \qquad u_x^i(t,x) = l_i\big(u(t,x)\big) \,.\, u_x(t,x) \qquad ; \quad 1 \le i \le n \,.$$

With this notation, for almost every $(t\,,\,x\,)$, one has

$$(4.1.4) \qquad u_x(t,x) = \sum_{i=1}^{n} u_x^i(t,x)\; r^i\big(u(t,x)\big) \,.$$

Moreover, the equation (1.1) may be rewritten as

$$(4.1.5) \qquad u_t(t,x) + \sum_{i=1}^{n} \lambda_i\big(u(t,x)\big)\ u_x^i(t,x)\ r^i\big(u(t,x)\big) = 0\,.$$

Differentiating the identity (4.1.4) with respect to t and the expression (4.1.5) with respect to x (see [9] for the details), one obtains the following semilinear system of n scalar equations:

(4.1.6)

$$(u_x^i)_t + \big(\lambda_i(u)\ u_x^i\big)_x + \sum_{p<q} G_{pq}^i(u)\ u_x^p\ u_x^q = 0\,;$$

$$;\ (t,x) \in \mathbb{R}^+ \times \mathbb{T}\quad,\quad 1 \le i \le n\,,$$

where

$$(4.1.7) \qquad G_{pq}^i(u) := \big(\lambda_q(u) - \lambda_p(u)\big)\ \Lambda_{pq}^i(u)\,.$$

Taking the spatial derivative of (2.9) we get the identity

$$(4.1.8) \qquad \begin{aligned} &(U_y^i)_t + (\Gamma_{ii}^i\ U^i\ U_y^i)_y + \sum_{p<q} \Lambda_{pq}^i\ U_y^p * U_y^q = 0 \\ &\qquad (t,x) \in \mathbb{R}^+ \times \mathbb{T}\quad,\quad 1 \le i \le n\,. \end{aligned}$$

The similarity between the two equations (4.1.6) and (4.1.8) is obvious. In particular, the expression

$$(4.1.9) \qquad \lambda_i(u)\ (u_x^i)_x + \nabla\lambda_i(u)\,.\,r^i(u)\ (u_x^i)^2$$

plays a part completely similar to the quantity

$$(4.1.10) \qquad (\Gamma_{ii}^i\ U^i\ U_y^i)_y = \Gamma_{ii}^i\ U^i\ (U_y^i)_y + \Gamma_{ii}^i\ (U_y^i)^2\,.$$

This comparison is useful in order to understand the manner by which genuine nonlinearity acts in (4.1.6). It means that the instantaneous rate of decreasing induced by the presence of the coefficients Γ_{ii}^i on periodic solutions of (1.1) is again expressed by the decay law (3.7). Of course some supplementary technicalities must be introduced in order to deal rigorously with exact solutions of the system (1.1), but in fact the ideas are the same. The result (expressed by a rule analogous to (3.7)) is also the same.

The analogy between the expressions placed used as factors of the interaction coefficients in (4.1.6) and in (4.1.8) is more difficult to achieve. Indeed, there is a gap of regularity between the convolution terms

$$(U_y^p * U_y^q)(t,y) \in M_b(\mathbb{T}) \tag{4.1.11}$$

and the products

$$u_x^p(t,x)\ u_x^q(t,x) \in ? \tag{4.1.12}$$

The first quantity (4.1.11) is always well defined. It inherits the regularity of a Radon measure. This fact facilitates the study of (4.1.8) in the sense that the amount indicated by formula (3.15) appears directly.

On the contrary, the product in (4.1.12) has no sense for the usual solutions of bounded variation of system (1.1). Now, piecewise constant functions naturally arise in the study of (1.1), and it is essential to understand the contribution given by (4.1.12) in this framework.

To this end, we look at the integral version of the identity (4.1.6). Let $y^i(t,z)$ be

$$\dot{y}_i(t,z) = \lambda_i\big(u(t,y_i(t,z))\big)\,,\quad y_i(0,z) = z\ ;\ 1 \le i \le n\,, \tag{4.1.13}$$

the characteristic line of the ith family starting at the point z.

The component $u_x^i(t,x)$ is absolutely continuous and satisfies

$$\begin{aligned} u_x^i\big(t,y_i(t,z)\big) &= u_x^i(0,z) \\ &\quad - \int_0^t \big(\nabla\lambda_i(u)\,.\,r^i(u)\ (u_x^i)^2\big)(\tau,y_i(\tau,z))\ d\tau \\ &\quad - \sum_{p<q} \int_0^t \big(G_{pq}^i(u)\ u_x^p\ u_x^q\big)(\tau,y_i(\tau,z))\ d\tau \\ &\quad - \sum_{j\neq i} \int_0^t \big(\nabla\lambda_i(u)\,.\,r^j(u)\ u_x^j\ u_x^i\big)(\tau,y_i(\tau,z))\ d\tau\,. \end{aligned} \tag{4.1.14}$$

The flow

$$T_i(t)\ :\ z \longmapsto T_i(t)(z) = y_i(t,z) \tag{4.1.15}$$

is a Lipschitz-continuous transformation whose Jacobian is bounded away from zero. Hence, it admits an inverse denoted by $T_i^{-1}(t)$. Our purpose is to evaluate

$$\int_{\mathbb{T}} |u_x^i(t,x)|\ dx\,. \tag{4.1.16}$$

The peculiarity of the sums in the identity (4.1.14) comes from the fact that they only involve products $u_x^p\ u_x^q$ with different indices. We thus have to control for p not equal to q an expression similar to

$$(4.1.17)\qquad \int_{\mathbb{T}}\int_0^t \big(G_{pq}^i(u)\ u_x^p\ u_x^q\big)\big(\tau\,,\,y_i(\tau,T_i^{-1}(\tau)(x))\big)\ d\tau\ dx\,.$$

In (4.1.17), we cut the time axis into intervals of size one. After an appropriate change of variables, we see that it remains to bound the quantities

$$(4.1.18)\qquad \int_n^{n+1}\int_{\mathbb{T}} |(u_x^p\ u_x^q)(\tau,x)|\ d\tau\, dx\ \ ;\ \ p\neq q\ \ ,\ \ 0\leqslant n\leqslant t\,.$$

Observe that the partition of $\mathbb{R}^+\times\mathbb{R}$ performed here (into subdomains whose diameters are equivalent to one) is not arbitrary. This choice is linked to condition (4.1.2). The existence of a particular spatial length induces the appearance of a characteristic time interval (whose length is equivalent to a period). It implies the presence of a local structure at the level of which it is convenient to measure the exchanges of variation. It motivates the setting of an analysis which achieves a compromise between the global computations of Glimm [6] (which raises the objections recalled in the introduction) and the differential calculus (which loses some information because it isolates the solution from its general context). At this stage lies the key observation and the reason why periodic solutions have called our attention.

The two expressions $u_x^p(t,x)$ and $u_x^q(t,x)$ have transversal wave fronts in the sense that we formally have

$$(4.1.19)\qquad \big(\partial_t + \lambda_p(u)\ \partial_x\big)\, u_x^p(t,x)\ \sim\ 0\,.$$

This property indicates that (4.1.18) should again be controlled by the product of the local variation of the components $u_x^p(t,x)$ and $u_x^q(t,x)$. In order to make this idea more clear, let us suppose that the different jumps in u are located at the points $x_1(t)<\ldots<x_N(t)<x_1(t)+1$. Assume that the jump at $x_j(t)$ occurs in the k_jth characteristic family and has strength measured by the scalar $\alpha_j(t)$.

Following [1] and [15], we regard $|u_x^i(x)|\ dx$ as the strength of an infinitesimal $\mathrm{i}-$wave in u located at x. Thus, the local variation of the k_j^{th} characteristic and ith component of $u(t,x)$ is evaluated as

$$(4.1.20)\qquad G_{u^i}(t) := \int_{\mathbb{T}} |u_x^i(t,x)|\ dx\ +\ \sum_{\{j\ ;\ k_j=i\}} |\alpha_j(t)|\,.$$

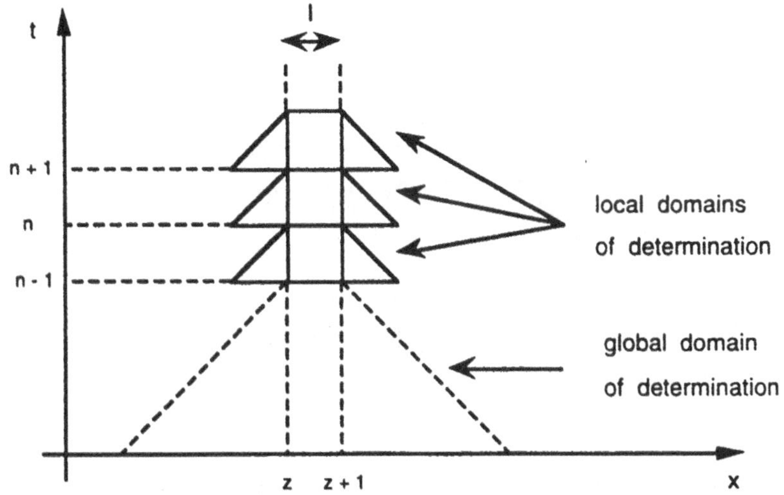

The waves that interact between the times n and $n+1$ inside a specified interval I of size one must be present inside the domain of determination associated with I (which is determined by the finite speed of propagation (4.1.2)). The new feature here is that the interaction terms are given by wave strengths just before the interaction (at the time $t = n$, in agreement with the preceding partition) rather than the initial wave strengths.

This principle leads to the introduction of local domains of determination (see the preceding figure). To these domains correspond local interaction potentials $Q_n(\mathbb{T})$, which depends on the selected instant n:

$$
\begin{aligned}
Q_n(\mathbb{T}) := {} & \sum_{p \le q} \iint_{0<x<y<3} |u_x^p(n,x)|\,|u_x^q(n,y)|\,dx\,dy + \\
& + \sum_{\{k_i \le k_j;\, x_j < x_i\}} |\,\alpha_i(n)\,|\,|\alpha_j(n)| + \\
& + \sum_i \Bigg[\sum_{j \le k_i} |\alpha_i(n)| \int_{x_\alpha}^{\infty} |u_x^j(n,x)|\,dx + \\
& \qquad + \sum_{j \ge k_i} |\alpha_i(n)| \int_{-\infty}^{x_\alpha} |u_x^j(n,x)|\,dx \Bigg].
\end{aligned}
\tag{4.1.21}
$$

Next, consider the interaction of two waves at the point $x_l(s) =$

$x_{l+1}(s)$ with $n \leq s \leq n+1$. The leading effect of such an interaction has been carefully studied in a paper by R. Young [21].

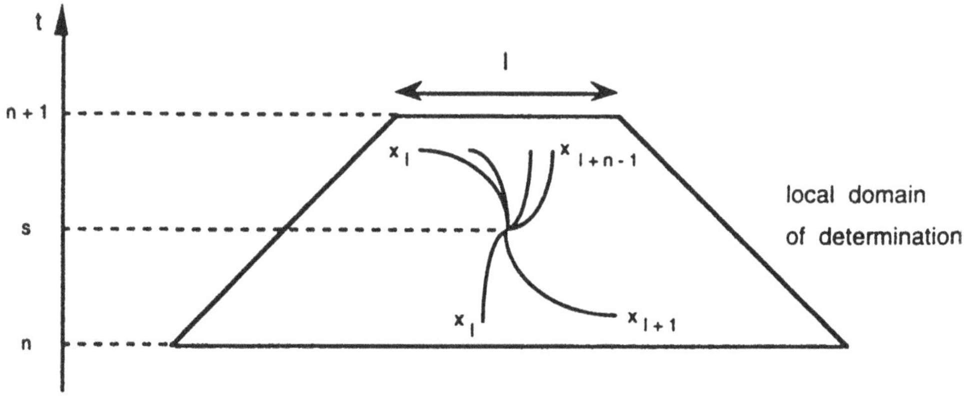

There are n waves that are produced. These waves are located at the different points

$$x_l(r) < x_{l+1}(r) < \dots < x_{l+n-1}(r) \quad ; \quad r > s. \tag{4.1.22}$$

They have, respectively, the strengths $\alpha_l(r), \dots, \alpha_{l+n-1}(r)$, with

$$\begin{aligned} \alpha_i(s+) = \; & \delta_{i\,k_l}\ \alpha_l(s-) + \delta_{i\,k_{l+1}}\ \alpha_{l+1}(s-) + \\ & + \Lambda^i_{k_l\,k_{l+1}}\ \alpha_l(s-)\ \alpha_{l+1}(s-) + \\ & + O\big(|\alpha_l(s-)|\ |\alpha_{l+1}(s-)|\ (|\alpha_l(s-)| + |\alpha_{l+1}(s-)|)\big). \end{aligned} \tag{4.1.23}$$

This formula suggests that the increase of the local variation during the time interval $[n\,;\,n+1]$ is approximately controlled by

$$\big(\sup_{i\neq p<q\neq i} |\Lambda^i_{pq}|\big) \times Q_n(\mathbb{T}). \tag{4.1.24}$$

In (4.1.21), we can separate the variables x and y in the integrand. By applying Fubini's theorem, we find a contribution at most equal to

$$\big(\sup_{i\neq p<q\neq i} |\Lambda^i_{pq}|\big) \times \big(\sum_{p\leq q} G_{u^p}(n)\ G_{u^q}(n)\big). \tag{4.1.25}$$

In conclusion, the hypothesis of periodicity allows us to reduce the study of the solution $u(t,x)$ to its local behavior on a period. Its main features are contained in the expressions $\{G_{u^i}(t)\}_{1\leq i\leq n}$, which binds both the local variation and the amplitude of the function $u(t,x)$. Indeed, we have

$$\max_{1 \le i \le n} G_{u^i}(t) \sim TV_{\mathbb{T}}\big(u(t,.)\big), \tag{4.1.26}$$

and because the mean of a periodic solution to (1.1) is a conserved quantity, we also have

$$\| u(t,.) - \bar{u} \|_{L^\infty(\mathbb{T})} \le C_6 \,\big(\max_{1 \le i \le n} G_{u_i}(t) \big). \tag{4.1.27}$$

Moreover, the preceding formal analysis shows that the time evolution of the $G_{u^i}(t)$'s is controlled by an approximate version of the differential inequalities (3.16):

$$\begin{aligned} G_{u^i}(n+1) \le & - \Gamma^i_{ii} \; G_{u^i}(n)^2 + \\ & + C_7 \sum_{i \ne p < q \ne i} |\Lambda^i_{pq}| \, G_{u^p}(n) G_{u^q}(n) \,; \quad 1 \le i \le n. \end{aligned} \tag{4.1.28}$$

Again, it follows from the criterion (3.17) and the inequality (4.1.28) that the n sequences $\{ G_{u^i}(n) \}_{n \in \mathbb{N}}$ are clearly bounded under the condition (3.17).

These explanations indicate how the proof of the statement given below works:

Theorem: (see [4] for the rigorous details). *Let the equation* (1.1) *be strictly hyperbolic, genuinely nonlinear and smooth in a neighborhood of* $\bar{u}$. *Then, there is a* $\kappa_0 > 0$ *and a* $\delta_0 > 0$ *with the following properties.*

If $\kappa > \kappa_0$ *and if the initial data* $u_0(x)$ *are periodic of period* 1 *and are given so that*

$$TV_{\mathbb{T}}\big(u_0(.)\big) \le \delta_0, \tag{4.1.29}$$

then there is a weak solution $u(t,x)$ *of* (1.1) *defined for all* x *and all* $t \ge 0$ *with initial data* $u_0(x)$ *and such that*

$$TV_{\mathbb{T}}\big(u(t,.)\big) \le (C_8/t) \quad ; \quad t \in \mathbb{R}^+_*. \tag{4.1.30}$$

Remark 4.1. This result contains and deeply generalizes the preceding work of Glimm and Lax [7] concerning the decay of periodic solutions to 2×2 systems.

4.2. Second stage: from global periodic solutions of system (1.1) to global solutions with small amplitude.

Our interest now turns toward initial data with small amplitude:

$$\| u_0(.) \|_{L^\infty(\mathbb{R})} \leq \varepsilon \ll 1 . \tag{4.2.1}$$

It is not clear whether the preceding approach can successfully be implemented in this new framework. The difficulty consists in recovering at the level of any function $g(x)$ some of the characteristics that were useful in the study of periodic solutions to (1.1). In the preceding approach, we have made use of two properties linked with the periodicity:

(i) The repetition of a given structure.

(ii) The existence of a particular spatial and time length (equivalent to a period), which allows us to observe the exchanges of local variation with precision.

Working with any solution of (1.1), it does not seem possible to use the first point (i), which is too specific. In particular, an L^∞-estimate cannot be deduced from a local BV-control (as in (4.1.27)). That is the reason for which we now introduce a supplementary (very restrictive) hypothesis.

We suppose the existence of an *a priori* estimate for the amplitude of the approximate solutions associated with (1.1). For example, such a property is guaranteed when the flux function f admits an invariant domain for the Riemann problem.

Even under this favourable context, a major difficulty (that built the compensated compactness) subsists. In order to construct global approximate solutions and to obtain their convergence, one has to control the evolution of their (local) variation. Precisely, the analysis that we have presented here gives a complete reply to this question.

From that time, it remains only to study the evolution of the variation. This may be done by applying principle (ii). The special length L is determined by imposing the relation

$$\sup_{\{I\,;\,|I|=L\}} TV_I\big(u_0(.)\big) \sim \| u_0(.) \|_{L^\infty(\mathbb{R})} . \tag{4.2.2}$$

It indicates the size of the domains on which it is advantageous to measure the exchanges of variation. Under condition (3.17), the local variation is again proving to be decreasing (after some relaxation time, which, of course, depends on L). A weak solution may be constructed for all times (see [4]). Moreover, the solution operator

$$(4.2.3) \qquad S(t) \,:\, \{\, u_0(x) \,;\, \| \, u_0(.) \, \|_{L^\infty(\mathbb{R})} \leq \varepsilon \,\} \longrightarrow L^1_{loc}(\mathbb{R})$$

$$u_0(x) \longmapsto S(t)\big(u_0(.)\big) = u(t,x)$$

satisfies the following compactness property:

$$(4.2.4) \qquad TV_I\big(u(t,.)\big) \leq C_9 \, |I| \,/\, t \,,$$

for every spatial interval I whose length $|I|$ is bigger than t.

Remark 4.2. In the case where no *a priori* L^∞ – estimate is available, the method proposed here allows us to improve the life span of solutions associated with large BV-initial conditions (see [4]).

References

[1] A. Bressan, The Semigroup Approach to Systems of Conservation laws, S.I.S.S.A., Via Beirut 4, Trieste 34014, Italy. Ref. 135/95/M.

[2] C. Cheverry, Justification de l'optique géométrique nonlinéaire pour un système de lois de conservation. Séminaire EDP de l'école Polytechnique. Exposé n° XV. 21 Mars 1995, *Duke Math. Jour.* **87** (13) (1997), 1–51.

[3] C. Cheverry, The modulation equations of non linear geometric optics, *Comm. in Part. Diff. Eq.* **21** (1996) (7-8), 1119–1140.

[4] C. Cheverry, Systèmes de lois de conservation et stabilité BV, Prépublication 96-08. Institut de Recherche Mathématique de Rennes (1996). Submitted in *Bulletin et Memoires de la Societé Mathematique de France.*

[5] C.M. Dafermos, Generalized characteristics and the structure of solutions of hyperbolic conservation laws, *Indiana U. Math. Jour.* **26** (1977) No 6., 1097–1119.

[6] J. Glimm, Solutions in the large for nonlinear hyperbolic systems of equations, *Comm. Pure. Appl. Math.* **18** (1965), 697–715.

[7] J. Glimm and P. Lax, Decay of solutions of systems of hyperbolic conservation laws, *Amer. Math. Soc. Memoir* **101**, Providence, 1970.

[8] J. Hunter, Hyperbolic waves and nonlinear geometric acoustics, *Transactions of the Sixth Army Conference on Applied Mathematics and Computing* (1989), No 2, 527–569.

[9] F. John, Formation of singularities in One Dimensional Nonlinear wave propagation, *J. Math. Anal. and Appl.* **45** (1974), 375–381.

[10] Joly-Métivier-Rauch, A Nonlinear Instability for 3×3 systems of Conservation Laws, *Commun. Math. Phys.* **162** (1994), 47–59.

[11] P. Lax, Hyperbolic systems of conservation laws II, *Comm. Pure. Appl. Math.* **10** (1957), 537–566.

[12] A. Majda, Non linear geometric optics for hyperbolic systems of Conservation Laws, *Oscillation Theory, Computation, and Methods of compensated compactness* (C. Dafermos, J. Erikson, D. Kinderlehrer, and M. Slemrod, eds.), IMA Volumes in Mathematics and Its Applications 2, Springer Verlag, New-York, 1986, pp 115–165.

[13] Majda-Rosales-Schonbek, A canonical system of integrodifferential equations arising in resonant nonlinear acoustics, *Studies in Applied Mathematics* **79** (1988), 205–262.

[14] L. Pego, Some explicit resonanting Waves in weakly nonlinear gas dynamics, *Studies in Applied Mathematics* **79** (1988), 263–270.

[15] M. Schatzman, Continuous Glimm functionals and uniqueness of solutions of the Riemann problem, *Indiana Univ. Math. J.* **34** (1985), 533–589.

[16] S. Schochet, Glimm's scheme for systems with almost-planar interactions, *Comm. in Part. Diff. Eq.* **16** (1991), 423–1440.

[17] S. Schochet, Resonant nonlinear geometrical optics for weak solutions of conservation laws *J. Diff. Eq.* **113** (1994), 473–504.

[18] D. Serre, Richness and the classification of quasilinear hyperbolic systems, *IMA volumes* **29**, Springer Verlag, 1991, 315–333.

[19] Tai-Ping-Liu, Decay to N-waves of solutions of general systems of nonlinear hyperbolic conservation laws, *Comm. in Part. Diff. Eq.* **30** (1977), 585–610.

[20] R. Young, The large time stability of sound waves, To appear in *Comm. Math. Physics* (1996).

[21] R. Young, Sup-norm stability for Glimm's scheme, *Comm. Pure. Appl. Math.* **46** (1993), 903–948.

C.N.R.S URA 305
Institut de Recherche Mathématique de Rennes
Campus de Beaulieu. Av du Gl Leclerc
35 042 RENNES cedex

Global existence of the solutions and formation of singularities for a class of hyperbolic systems

Daniele Del Santo, Vladimir Georgiev**, Enzo Mitidieri**

0. Introduction

In this paper we prove some results concerning existence and nonexistence of global solutions of the Cauchy problem for a class of semilinear hyperbolic systems of the form

$$\begin{aligned} \partial_t^2 u - \Delta u &= H_v(u,v), \\ \partial_t^2 v - \Delta v &= H_u(u,v), \end{aligned} \qquad \text{in } \mathbf{R}^n \times [0,+\infty[, \tag{0.1}$$

with smooth compactly supported initial data in $\mathbf{R}^n$. Here $n \geq 1$ and $H : \mathbf{R}^2 \to \mathbf{R}$ is a given C^2 function. We shall call (0.1) a hyperbolic system of Hamiltonian type (see [5]). For the sake of simplicity, we shall concentrate our attention to the special case

$$H(u,v) = \frac{1}{q+1} u|u|^q + \frac{1}{p+1} v|v|^p, \tag{0.2}$$

with p, $q > 1$.

A big effort has been devoted in the last few years to the study of optimal conditions for the solvability of problems of type

$$\begin{aligned} Tu &= H_v(u,v), \\ Tv &= H_u(u,v), \end{aligned}$$

for different kinds of operators T and functions H. Let us mention two canonical examples. Consider the Lane–Emden system

$$\begin{aligned} -\Delta u &= |v|^{p-1}v, \\ -\Delta v &= |u|^{q-1}u, \end{aligned} \qquad \text{in } \mathbf{R}^n. \tag{0.3}$$

It is known that if p, $q > 0$, $pq > 1$ and $n \geq 3$ then (0.3) has no global positive solutions (see [17] and [19]) if

$$\max\{\frac{p+1}{pq-1}, \frac{q+1}{pq-1}\} \geq \frac{n-2}{2},$$

while (0.3) possesses infinitely many global positive solutions in $\mathbf{R}^n$ (see [16] and [18]) if

$$\frac{1}{p+1}+\frac{1}{q+1}\leq\frac{n-2}{n}.$$

We note that if

$$\max\{\frac{p+1}{pq-1},\frac{q+1}{pq-1}\}<\frac{n-2}{2} \qquad \text{and} \qquad \frac{1}{p+1}+\frac{1}{q+1}>\frac{n-2}{n},$$

the problem of nonexistence of positive solutions for (0.3) has been solved only for radially symmetric solutions and it is still open in the general case (see [17]). For $n=1,\ 2$, the problem (0.3) does not have any positive solutions for all $p,\ q>0$.

The parabolic system associated to (0.3), i.e.

$$\begin{aligned}\partial_t u-\Delta u&=|v|^{p-1}v,\\ \partial_t v-\Delta v&=|u|^{q-1}u,\end{aligned}\qquad \text{in } \mathbf{R}^n\times[0,+\infty[, \tag{0.4}$$

has been considered, among others, in [4]. For this problem, nonexistence of global positive solutions has been proved, if $p,\ q>0$, $pq>1$ and $n\geq 1$, under the assumption

$$\max\{\frac{p+1}{pq-1},\frac{q+1}{pq-1}\}\geq\frac{n}{2}.$$

This theorem is sharp (see [4] for the details).

In this paper we show that a similar qualitative result holds for (0.1) when H is the function defined in (0.2). More precisely we prove that if

$$p,\ q>1 \qquad \text{and} \qquad \max\{\frac{p+2+q^{-1}}{pq-1},\frac{q+2+p^{-1}}{pq-1}\}>\frac{n-1}{2}, \tag{0.5}$$

then the Cauchy problem for (0.1) has no global solutions. This result admits a partial converse: in fact, at least for (p,q) belonging to a subset of the pq plane, if

$$p,\ q>1, \quad \text{and} \quad \max\{\frac{p+2+q^{-1}}{pq-1},\frac{q+2+p^{-1}}{pq-1}\}<\frac{n-1}{2},$$

then the Cauchy problem for (0.1) has a global solution, provided the initial data are sufficiently small.

It is interesting to compare these results with the corresponding ones for the hyperbolic equation

$$\partial_t^2 u-\Delta u=|u|^p \qquad \text{in } \mathbf{R}^n\times[0,+\infty[. \tag{0.6}$$

It is well known (see [12], [10], [20]) that the Cauchy problem associated to (0.6) has no global solutions if

$$1 < p < p_c, \tag{0.7}$$

where p_c is the positive root of the quadratic polynomial

$$(n-1)x^2 - (n+1)x - 2.$$

We note that (0.7) can be rewritten equivalently as

$$p > 1 \qquad \text{and} \qquad \frac{1+p^{-1}}{p-1} > \frac{n-1}{2};$$

in this way it is easily seen that we obtain exactly (0.7) taking $p = q$ in (0.5). If $p > p_c$ then the Cauchy problem for (0.6) with small initial data has a global solution (see [12], [11], [15], [9]). Considering the system (0.1), we can say that, at least in a neighborhood of (p_c, p_c), the curve

$$\max\left\{\frac{p+2+q^{-1}}{pq-1}, \frac{q+2+p^{-1}}{pq-1}\right\} = \frac{n-1}{2}$$

plays the same role as the value p_c for the equation (0.6).

Our results reveal a peculiar phenomenon of systems containing Hamiltonian nonlinearities. More precisely, in dealing with questions related to existence, nonexistence and regularity of the solutions, we are faced with new aspects which are not present in the scalar case: concerning the blow–up condition on the parameter involved (in our case p, q and n) we see that the region C defined by

$$C = \{(p,q) \ : \ (0.1) \text{ has no global solutions } \}$$

is unbounded (at least for $n = 2,\ 3$). Roughly speaking, we may say that for systems with Hamiltonian nonlinearities the unboundedness of C is due to the fact that there exists a sort of balance of growth between the components of the solution.

1. Existence results

Theorem 1 *Let $2 \le n \le 3$ and $p,\ q > 1$. Consider the following Cauchy problem :*

$$\begin{aligned} \partial_t^2 u - \Delta u &= |v|^p, \\ \partial_t^2 v - \Delta v &= |u|^q, \end{aligned} \qquad \textit{in } \mathbf{R}^n \times [0, +\infty[, \tag{1.1}$$

$$\begin{aligned} u(x,0) = \varepsilon f_1(x), \ \partial_t u(x,0) = \varepsilon g_1(x), \\ v(x,0) = \varepsilon f_2(x), \ \partial_t v(x,0) = \varepsilon g_2(x), \end{aligned} \quad \text{in } \mathbf{R}^n, \tag{1.2}$$

where f_i, $g_i \in C_0^\infty(\mathbf{R}^n)$ *and* supp f_i, supp $g_i \subset \{|x| \leq R\}$, *for* $i = 1,\ 2$ *(*$R > 0$*);*

If $p,\ q \in]5-n, 6-n]$,

$$\max\{\frac{p+2+q^{-1}}{pq-1}, \frac{q+2+p^{-1}}{pq-1}\} < \frac{n-1}{2}, \tag{1.3}$$

and $\varepsilon > 0$ *is sufficiently small, then the problem* (1.1), (1.2) *has a unique global classical solution.*

Theorem 2 *Let* $n \geq 4$ *and* $p,\ q > 1$. *If*

$$\min\{\frac{p-1}{pq-1}, \frac{q-1}{pq-1}\} > \frac{n-1}{2(n+1)}, \tag{1.4}$$

$$\max\{\frac{p+2+q^{-1}}{pq-1}, \frac{q+2+p^{-1}}{pq-1}\} < \frac{n-1}{2},$$

and $\varepsilon > 0$ *is sufficiently small, then the problem* (1.1), (1.2) *has a global (weak) solution.*

Remark 1.1 The restriction $p,\ q \leq 6-n$ for the case of space dimensions $n = 2,\ 3$ can also be used to show that the solutions of (1.1) have the same rate of decay of the solutions of the free wave equation. One can relax this condition by using the estimates in [1] and [21].

Remark 1.2 The assumption (1.4) seems to be optimal for the existence of a local (in time) solution in L^p- spaces. This assumption corresponds to requiring in the scalar case that the exponent of the nonlinearity be bounded by $(n+3)/(n-1)$. The bound $(n+3)/(n-1)$, called the conformal exponent, appears in the local existence theorem for the scalar case discussed in [20]. Even in the scalar case it is an open problem to establish the existence of a local solution above the conformal exponent for arbitrarily high space dimension, without making any assumptions about the size of the initial data.

1.1. Proof of Theorem 1

In this section we shall prove the existence of global solutions to (1.1), (1.2) for $\varepsilon > 0$ small enough. Our main tool in doing so will be some integral estimates on the solution of a nonhomogeneous linear wave equation. These results, which we collect in the lemma below, have been proved by Glassey [11] if $n = 2$ and by John [12] for $n = 3$.

Lemma 1.1 *Let $n = 2,\ 3$ and let u be a smooth solution of*

$$\partial_t^2 u - \Delta u = h \qquad \text{in } \mathbf{R}^n \times [0, +\infty[, \tag{1.5}$$
$$u(x,0) = 0,\ \partial_t u(x,0) = 0 \qquad \text{in } \mathbf{R}^n, \tag{1.6}$$

where $h \in C^\infty(\mathbf{R}^n \times [0, +\infty[)$ and supp $h \subset \{|x| \leq t + R\}$. *Let $\tau_\pm = 1 + |t \pm |x||$.*

Then

$$\|\tau_+^\alpha \tau_-^\beta u\|_{L^\infty(\mathbf{R}^n \times [0,+\infty[)} \leq C \|\tau_+^\gamma \tau_-^\delta h\|_{L^\infty(\mathbf{R}^n \times [0,+\infty[)},$$

where α, β, γ, δ are nonnegative parameters such that $\alpha < (n-1)/2$ and

$$\beta = \gamma - \frac{n+3}{2} + \min\{1, \delta\} < \frac{n-1}{2}.$$

Let us now briefly describe the method of proof of Theorem 1.

Let (u_0, v_0) be the solution of

$$\begin{aligned} \partial_t^2 u_0 - \Delta u_0 &= 0, \\ \partial_t^2 v_0 - \Delta v_0 &= 0, \end{aligned} \qquad \text{in } \mathbf{R}^n \times [0, +\infty[,$$

and, for $j \geq 1$, let (u_j, v_j) be the solution of

$$\begin{aligned} \partial_t^2 u_j - \Delta u_j &= |v_{j-1}|^p, \\ \partial_t^2 v_j - \Delta v_j &= |u_{j-1}|^q, \end{aligned} \qquad \text{in } \mathbf{R}^n \times [0, +\infty[,$$

where in both cases initial conditions are given by (1.2). We will show that if ε is sufficiently small, the sequence (u_j, v_j) converges to a global classical solution of (1.1), (1.2). Since (u_0, v_0) is a solution of the free wave equation with compactly supported inital data, it follows that for all integers $N \geq 0$ there exists $C_N > 0$ such that

$$\|\tau_+^{(n-1)/2} \tau_-^N u_0(t,x)\|_{L^\infty} + \|\tau_+^{(n-1)/2} \tau_-^N v_0(t,x)\|_{L^\infty} \leq C_N \varepsilon. \tag{1.7}$$

Our method consists in the construction of two sequences of real numbers, $(\beta_j^{(1)})$, $(\beta_j^{(2)})$, with the property that there exists $j_0 \geq 0$ such that for all $j \geq j_0$, $\beta_j^{(i)} = \beta_{j_0}^{(i)}$ for $i = 1,\ 2$, and moreover the following inequalities holds

$$\|\tau_+^\alpha \tau_-^{\beta_j^{(1)}} u_j\|_{L^\infty} + \|\tau_+^\alpha \tau_-^{\beta_j^{(2)}} v_j\|_{L^\infty} \leq C\varepsilon, \tag{1.8}$$

and

$$\|\tau_+^\alpha \tau_-^{\beta_j^{(1)}}(u_j - u_{j-1})\|_{L^\infty} + \|\tau_+^\alpha \tau_-^{\beta_j^{(2)}}(v_j - v_{j-1})\|_{L^\infty} \leq \frac{C\varepsilon}{2^j}, \tag{1.9}$$

for some $0 \leq \alpha < (n-1)/2$ and for all $j \geq 1$.

For the sake of brevity we shall present only the proof of (1.8), leaving to the reader the verification of (1.9).

Our reasoning will be by induction. Let (1.8) be fulfilled for some j. By (1.7), the estimate from Lemma 1.1 and the inductive assumption (1.8), we get

$$\|\tau_+^\alpha \tau_-^{\beta_{j+1}^{(1)}} u_{j+1}\|_{L^\infty} \leq C\varepsilon + C\varepsilon^p \|\tau_+^{\gamma^{(1)} - p\alpha} \tau_-^{\delta^{(1)} - p\beta_j^{(2)}}\|_{L^\infty}$$

and

$$\|\tau_+^\alpha \tau_-^{\beta_{j+1}^{(2)}} v_{j+1}\|_{L^\infty} \leq C\varepsilon + C\varepsilon^q \|\tau_+^{\gamma^{(2)} - q\alpha} \tau_-^{\delta^{(2)} - q\beta_j^{(1)}}\|_{L^\infty}.$$

We choose

$$\gamma^{(1)} = p\alpha, \quad \delta^{(1)} = p\beta_j^{(2)},$$
$$\gamma^{(2)} = q\alpha, \quad \delta^{(2)} = q\beta_j^{(1)}.$$

Having in mind the relation between β, γ and δ in Lemma 1.1, we obtain

$$\begin{aligned} \beta_{j+1}^{(1)} &= A + \min\{1, p\beta_j^{(2)}\} \\ \beta_{j+1}^{(2)} &= B + \min\{1, q\beta_j^{(1)}\} \end{aligned} \tag{1.10}$$

where

$$A = p\alpha - \frac{n+3}{2}, \quad B = q\alpha - \frac{n+3}{2}. \tag{1.11}$$

We now need the following lemma.

Lemma 1.2 *Let $A, B > -1$. Let (x_j), (y_j) be two sequences of real numbers defined by*

$$x_0 = A + 1, \qquad x_{j+1} = A + \min\{1, py_j\},$$
$$y_0 = B + 1, \qquad y_{j+1} = B + \min\{1, qx_j\}.$$

Suppose that one of the following conditions holds:

a) $p(B+1) > 1$ *and* $q(A+1) > 1$;

b) $p(B+1) > 1$, $q(A+1) \leq 1$ *and* $p(B + q(A+1)) > 1$;

c) $p(B+1) \leq 1$, $q(A+1) > 1$ *and* $q(A+p(B+1)) > 1$.

Then the sequences (x_j), (y_j) *are constant for* $j \geq 1$.

Proof. If *a)* holds, then it is evident that

$$x_j = A+1, \quad y_j = B+1,$$

for all $j \geq 0$. If *b)* is fulfilled, then $x_1 = A+1$, $y_1 = B+q(A+1)$. But then, as $py_1 = p(B+q(A+1)) > 1$, it follows that $x_2 = x_1$, $y_2 = y_1$ and consequently

$$x_j = A+1, \quad y_j = B+q(A+1),$$

for all $j \geq 1$. The case *c)* is similar. This completes the proof of the lemma.

We apply Lemma 1.2 to the sequences $(\beta_j^{(1)})$, $(\beta_j^{(2)})$ since (1.10) holds and A, $B > -1$ if α is taken very close to $(n-1)/2$. The conditions *a)*, *b)*, *c)* of Lemma 1.2 take the form

$$pq\frac{n-1}{2} - p\frac{n+1}{2} - 1 > 0 \text{ and } pq\frac{n-1}{2} - q\frac{n+1}{2} - 1 > 0, \tag{1.12}$$

$$\begin{aligned} &pq\frac{n-1}{2} - p\frac{n+1}{2} - 1 > 0, \quad pq\frac{n-1}{2} - q\frac{n+1}{2} - 1 \leq 0, \\ &\text{and } p^2q\frac{n-1}{2} - pq - p\frac{n+3}{2} - 1 > 0, \end{aligned} \tag{1.13}$$

and

$$\begin{aligned} &pq\frac{n-1}{2} - p\frac{n+1}{2} - 1 \leq 0, \quad pq\frac{n-1}{2} - q\frac{n+1}{2} - 1 > 0, \\ &\text{and } p^2q\frac{n-1}{2} - pq - q\frac{n+3}{2} - 1 > 0, \end{aligned} \tag{1.14}$$

It is easy to see that the set obtained as the union of the sets where (1.12), (1.13) and (1.14) are satisfied is given by (1.3). It remains to check the requirement $0 \leq \beta_j^{(1)}$, $\beta_j^{(2)} < (n-1)/2$ in Lemma 1.1. To assure the nonnegativeness of $\beta_j^{(1)}$, $\beta_j^{(2)}$ it is sufficient to show that $B+q(A+1) \geq 0$ and $A+p(B+1) \geq 0$, i.e.

$$pq\frac{n-1}{2} - q > \frac{n+3}{2}, \quad \text{and} \quad pq\frac{n-1}{2} - p > \frac{n+3}{2}.$$

Both inequalities follow from (1.3). To verify $\beta_j^{(1)}$, $\beta_j^{(2)} < (n-1)/2$ we remark that by the proof of Lemma 1.2 we have $\beta_j^{(1)} \leq A+1$ and $\beta_j^{(2)} \leq$

$B+1$ for all j, consequently it will be sufficient that $A,\ B < (n-3)/2$; these last inequalities are implied by

$$p, q \leq \frac{2n}{n-1}.$$

Finally we notice that $2n/(n-1) = 6 - n$ for $n = 2,\ 3$.

The regularity of the solution is proved by adapting to our situation the arguments introduced in [12] and [11]. The details are left to the reader.

1.2. Proof of Theorem 2

In this section we shall sketch the proof of the Theorem 2. The key point is the following weighted Strichartz estimate established in [9] (see also [6], [7], [8]).

Lemma 1.3 *Let $n \geq 3$ and let u be a smooth solution of* (1.5), (1.6), *where $h \in C^\infty(\mathbf{R}^n \times [0, +\infty[)$ and* $\operatorname{supp} h \subset \{|x| \leq t + R\}$.

Suppose that $1 < \lambda < \mu$ satisfy

$$\frac{1}{\lambda} + \frac{1}{\mu} = 1, \quad \frac{n-3}{2} < \frac{n}{\mu} - \frac{1}{\lambda}.$$

Then

$$\|\tau_+^\alpha \tau_-^\beta u\|_{L^\mu(\mathbf{R}^n \times [0,+\infty[)} \leq C \|\tau_+^\gamma \tau_-^\delta h\|_{L^\lambda(\mathbf{R}^n \times [0,+\infty[)},$$

where α, β, γ, δ are nonnegative parameters such that $\alpha \in [0, (n-1)/2 - n/\mu[$ and

$$\beta = \gamma - \alpha - 2 + \frac{n}{\lambda} - \frac{n+1}{\mu} + \min\{1, \delta + \frac{1}{\lambda}\}.$$

We will prove that, taking (u_j, v_j) constructed as in the proof of Theorem 1, the sequence $((\tau_+^{\alpha^{(1)}} \tau_-^{\beta_j^{(1)}} u_j, \tau_+^{\alpha^{(2)}} \tau_-^{\beta_j^{(2)}} v_j))$ converges in $L^{\mu^{(1)}} \times L^{\mu^{(2)}}$ for suitable values of the parameters. We start with the following version of (1.7):

$$\|\tau_+^{\alpha^{(1)}} \tau_-^N u_0\|_{L^{\mu^{(1)}}} + \|\tau_+^{\alpha^{(2)}} \tau_-^N v_0\|_{L^{\mu^{(2)}}} \leq C\varepsilon, \tag{1.15}$$

where

$$\alpha^{(1)} = \frac{n-1}{2} - \frac{n}{\mu^{(1)}} - \sigma,$$
$$\alpha^{(2)} = \frac{n-1}{2} - \frac{n}{\mu^{(2)}} - \sigma.$$

Here N is an arbitrary large integer and $\sigma > 0$ is sufficiently small. We shall find $\mu^{(1)}, \mu^{(2)}, \lambda^{(1)}, \lambda^{(2)}$ and we shall construct the sequences $(\beta_j^{(1)})$, $(\beta_j^{(2)})$ such that there exists $j_0 \geq 0$ such that for all $j \geq j_0$, $\beta_j^{(i)} = \beta_{j_0}^{(i)}$ for $i = 1,\ 2$, and

$$\|\tau_+^{\alpha^{(1)}} \tau_-^{\beta_j^{(1)}} u_j\|_{L^{\mu^{(1)}}} + \|\tau_+^{\alpha^{(2)}} \tau_-^{\beta_j^{(2)}} v_j\|_{L^{\mu^{(2)}}} \leq C\varepsilon, \tag{1.16}$$

$$\|\tau_+^{\alpha^{(1)}} \tau_-^{\beta_j^{(1)}} (u_j - u_{j-1})\|_{L^{\mu^{(1)}}} + \|\tau_+^{\alpha^{(2)}} \tau_-^{\beta_j^{(2)}} (v_j - v_{j-1})\|_{L^{\mu^{(2)}}} \leq \frac{C\varepsilon}{2^j}. \tag{1.17}$$

We verify (1.16) by an iterative argument and in the meantime we will determine all the parameters. Applying the estimate from Lemma 1.3, we get

$$\|\tau_+^{\alpha^{(1)}} \tau_-^{\beta_{j+1}^{(1)}} u_{j+1}\|_{L^{\mu^{(1)}}} \leq C\varepsilon + C\varepsilon^p \|\tau_+^{\gamma^{(1)} - p\alpha^{(2)}} \tau_-^{\delta^{(1)} - p\beta_j^{(2)}}\|_{L^\infty},$$
$$\|\tau_+^{\alpha^{(2)}} \tau_-^{\beta_{j+1}^{(2)}} v_{j+1}\|_{L^{\mu^{(2)}}} \leq C\varepsilon + C\varepsilon^q \|\tau_+^{\gamma^{(2)} - q\alpha^{(1)}} \tau_-^{\delta^{(2)} - q\beta_j^{(1)}}\|_{L^\infty}.$$

We choose

$$\gamma^{(1)} = p\alpha^{(2)}, \quad \delta^{(1)} = p\beta_j^{(2)}, \quad \lambda^{(1)} p = \mu^{(2)},$$
$$\gamma^{(2)} = q\alpha^{(1)}, \quad \delta^{(2)} = q\beta_j^{(1)}, \quad \lambda^{(2)} q = \mu^{(1)}.$$

Recalling the relation between β, γ and δ from Lemma 1.3, we get (modulo O(σ)–terms)

$$\beta_{j+1}^{(1)} = p\frac{n-1}{2} - \frac{n+3}{2} - \frac{1}{\mu^{(1)}} + \min\{1, p\beta_j^{(2)} + \frac{1}{\lambda^{(1)}}\},$$
$$\beta_{j+1}^{(2)} = q\frac{n-1}{2} - \frac{n+3}{2} - \frac{1}{\mu^{(2)}} + \min\{1, q\beta_j^{(1)} + \frac{1}{\lambda^{(2)}}\},$$

and

$$\frac{1}{\lambda^{(1)}} = \frac{p(q-1)}{pq-1}, \qquad \frac{1}{\lambda^{(2)}} = \frac{q(p-1)}{pq-1},$$
$$\frac{1}{\mu^{(1)}} = \frac{p-1}{pq-1}, \qquad \frac{1}{\mu^{(2)}} = \frac{q-1}{pq-1}.$$

We define

$$A = p\frac{n-1}{2} - \frac{n+3}{2} - \frac{1}{\mu^{(1)}} + \frac{1}{\lambda^{(1)}},$$
$$B = q\frac{n-1}{2} - \frac{n+3}{2} - \frac{1}{\mu^{(2)}} + \frac{1}{\lambda^{(2)}},$$

and we set $a = 1/\mu^{(1)}$, $b = 1/\mu^{(2)}$. To verify that $(\beta_j^{(1)})$, $(\beta_j^{(2)})$ are constant after a certain j_0, we shall make use of the following version of Lemma 1.2.

Lemma 1.4 *Let $A > -a$ and $B > -b$. Let (x_j), (y_j) be two sequences of real numbers defined by*

$$x_0 = A + a, \qquad x_{j+1} = A + \min\{a, py_j\},$$
$$y_0 = B + b, \qquad y_{j+1} = B + \min\{b, qx_j\}.$$

Suppose that one of the following conditions holds:

a) $p(B+b) > a$ *and* $q(A+a) > b$;

b) $p(B+b) > 1$, $q(A+a) \leq 1$ *and* $p(B + q(A+a)) > a$;

c) $p(B+b) \leq 1$, $q(A+a) > b$ *and* $q(A + p(B+b)) > b$.

Then the sequences (x_j), (y_j) are constant for $j \geq 1$.

Also in this case the conditions *a), b), c)* of Lemma 1.4 take the form of (1.12), (1.13), (1.14). The fact that $\beta_j^{(1)}, \beta_j^{(2)} \geq 0$ follows from $B + q(A+1) \geq 0$ and $A + p(B+1) \geq 0$. A simple computation shows that the condition $B + q(A+1) \geq 0$, for example, is equivalent to

$$pq\frac{n-1}{2} - q \geq \frac{n+1}{2} + \frac{q-1}{pq-1}.$$

Having in mind that $(q-1)/(pq-1) < 1$, we see that the above inequality is a consequence of

$$pq\frac{n-1}{2} - q \geq \frac{n+3}{2},$$

and this inequality is implied by (1.3). Finally it is immediate to check that (1.4) is equivalent to

$$\frac{n-3}{2} < \frac{n}{\mu^{(1)}} - \frac{1}{\lambda^{(1)}} \quad \text{and} \quad \frac{n-3}{2} < \frac{n}{\mu^{(2)}} - \frac{1}{\lambda^{(2)}}.$$

The proof of Theorem 2 is complete.

2. Nonexistence results

In this section we present the main nonexistence results of this paper; we shall show that the functional approach of Glassey and Sideris may also be used in our situation.

Theorem 3 *Let $1 \leq n \leq 3$, $T \in]0,+\infty]$ and let $(u,v) \in C^2(\mathbf{R}^n \times [0,T[)^2$ be a solution of the problem*

$$\begin{aligned} \partial_t^2 u - \Delta u &= |v|^p, \\ \partial_t^2 v - \Delta v &= |u|^q, \end{aligned} \qquad \text{in } \mathbf{R}^n \times [0,T[, \tag{2.1}$$

$$\begin{aligned} u(x,0) = f_1(x),\ \partial_t u(x,0) = g_1(x), \\ v(x,0) = f_2(x),\ \partial_t v(x,0) = g_2(x), \end{aligned} \qquad \text{in } \mathbf{R}^n, \tag{2.2}$$

where f_i, $g_i \in C_0^\infty(\mathbf{R}^n)$ and $\operatorname{supp} f_i$, $\operatorname{supp} g_i \subset \{|x| \leq R\}$, for $i = 1,\ 2$ $(R > 0)$. Suppose that

i) $\int_{\mathbf{R}^n} g_i(x)\,dx > 0$, *for* $i = 1,\ 2$;

ii) $p,\ q > 1$ *and*

$$\max\{\frac{p+2+q^{-1}}{pq-1}, \frac{q+2+p^{-1}}{pq-1}\} > \frac{n-1}{2}. \tag{2.3}$$

Then $T < +\infty$.

For all integers n we define $\eta = \eta(n)$ to be 0 if n is odd and 1/2 if n is even.

Theorem 4 *Let $n \geq 4$, $T \in]0,+\infty]$ and let $(u,v) \in C^2(\mathbf{R}^n \times [0,T[)^2$ be a solution of the problem* (2.1), (2.2).

Suppose that

i') $\int_{\mathbf{R}^n} |x|^{\eta-1} f_i(x)\,dx > 0$, $\int_{\mathbf{R}^n} g_i(x)\,dx > 0$, $\int_{\mathbf{R}^n} |x|^{\eta} g_i(x)\,dx > 0$, *for* $i = 1,\ 2$;

ii) $p,\ q > 1$ and the condition (2.3) *holds.*

Then $T < +\infty$.

Remark 2.1 If (u, v) is a C^2 solution of the problem (2.1), (2.2), the assumption on the support of the initial data implies that $\operatorname{supp} u$, $\operatorname{supp} v \subset \{|x| \le R + t\}$ (see e. g. [13, Theorem 4a]).

Remark 2.2 We know that if $n = 1$ then the problem (2.1), (2.2) has a unique classical solution defined locally in time, for every value of p and q. If $n \ge 2$, at least in the case when $1 \le p,\ q \le (n+3)/(n-1)$ it is possible to show that (2.1), (2.2) admit a unique local (in time) generalized solution (see [20, Par.1]; related local existence results can be found in [2]). Using the technique of Sideris [20] it is possible to adapt our proof of nonexistence of global classical solutions also to generalized solutions.

Remark 2.3 If $n = 3$ the result of Theorem 3 holds without requiring *i)*. To see this it is sufficient to modify in a suitable way the proof of John's result [12, Theorem 2] (see [3]).

Remark 2.4 Using a functional technique similar to that of Theorems 3 and 4 we can prove nonexistence of global solutions for hyperbolic systems which are more general than (2.1). In particular, rephrasing Kato's result [14], the Lemma 2.2 below permits us to show the blow–up of the generalized solutions of the following nonlinear problem.

Let us consider

$$\begin{aligned} &\partial_t^2 u + A_1 u = H_1(x,t,u,v), \\ &\partial_t^2 v + A_2 v = H_2(x,t,u,v), \end{aligned} \qquad \text{in } \mathbf{R}^n \times [0,T[\tag{2.4}$$

$$\begin{aligned} &u(x,0) = f_1(x),\ \partial_t u(x,0) = g_1(x), \\ &v(x,0) = f_2(x),\ \partial_t v(x,0) = g_2(x), \end{aligned} \qquad \text{in } \mathbf{R}^n, \tag{2.5}$$

where A_1, A_2 are elliptic operators of type

$$A_i(\cdot) = -\sum_{j,k=1}^{n} \partial_{x_j}(a^i_{j,k}(x,t)\partial_{x_k}(\cdot)) - \sum_{j=1}^{n} \partial_{x_j}(a^i_j(x,t)(\cdot)), \qquad i = 1,\ 2.$$

The functions $a^i_{j,k}$, a^i_j are supposed to be smooth, and moreover the operator A_i must satisfy the property that $A_i^*1 = 0$, where A_i^* is the formal adjoint of A_i, for $i = 1, 2$. The functions H_1, H_2 satisfy the following conditions:

$$H_1(x,t,u,v) \geq c|v|^p,$$

$$H_2(x,t,u,v) \geq c|u|^q,$$

for some $c > 0$ and for all $(x,t,u,v) \in \mathbf{R}^n \times [0,T[\times \mathbf{R} \times \mathbf{R}$. Let (u,v) be a generalized solution of (2.4), (2.5) defined in $\mathbf{R}^n \times [0,T[$, i. e.

$$u,\ v,\ H_1(x,t,u,v),\ H_2(x,t,u,v) \in C^0([0,T[, \mathrm{L}^1_{\mathrm{loc}}(\mathbf{R}^n)),$$

and

$$\begin{aligned} &\tfrac{d^2}{dt^2}\langle u,\varphi\rangle + \langle u, A_1^*\varphi\rangle = \langle H_1(x,t,u,v),\varphi\rangle, \\ &\tfrac{d^2}{dt^2}\langle v,\varphi\rangle + \langle v, A_2^*\varphi\rangle = \langle H_2(x,t,u,v),\varphi\rangle, \end{aligned} \qquad \text{for all } \varphi \in C_0^\infty(\mathbf{R}^n).$$

The result is the following.

Suppose that the hypotheses i) and ii) of Theorem 4 hold, $\operatorname{supp} \mathrm{u},\ \operatorname{supp} \mathrm{v} \subset \{|x| \leq R+t\}$ *and*

$$\max\left\{\frac{p+1}{pq-1},\ \frac{q+1}{pq-1}\right\} \geq \frac{n-1}{2}.$$

Then $T < +\infty$.

2.1. Some systems of differential inequalities

In the proofs of Theorems 3 and 4 we shall use the following version of [20, Lemma 4].

Lemma 2.1 *Let* $a,\ b \in [0,+\infty]$, *with* $a < b$ *and let* $F,\ G \in C^2([a,b[,\mathbf{R})$. *Suppose that, for all* $t \in [a,b[$, *we have*

$$F(t) \geq C(R+t), \tag{2.6}$$

$$G(t) \geq C(R+t)^s, \tag{2.7}$$

$$F''(t) \geq C(R+t)^{-\alpha}(G(t))^p, \tag{2.8}$$

$$G''(t) \geq C(R+t)^{-\beta}(F(t))^q, \tag{2.9}$$

where $C,\ R > 0$, $s \geq 1$, $\alpha,\ \beta \geq 0$, *and* $p,\ q > 0$. *Suppose moreover that* $pq > 1$ *and*

$$q(\alpha-2)+\beta-2 < s(pq-1). \tag{2.10}$$

Then $b < +\infty$.

Proof. We shall argue by contradiction. Suppose that F and G are defined on $[a, +\infty[$. By (2.6), (2.7) it follows that F, $G > 0$ on $[a, +\infty[$ and

$$\lim_{t\to+\infty} F(t) = \lim_{t\to+\infty} G(t) = +\infty. \tag{2.11}$$

Inequalities (2.8), (2.9) imply that F'', $G'' > 0$, hence F and G are convex functions on $[a, +\infty[$. This fact together with (2.11) implies that there exists $T_0 \geq a$ such that $F'(t)$, $G'(t) > 0$, for $t \geq T_0$. Multiplying (2.8) by $G'(t)$ and integrating the resulting inequality by parts, we obtain

$$\begin{aligned} F'(t)G'(t) - F'(T_0)G'(T_0) \quad &- \int_{T_0}^{t} F'(s)G''(s)\, ds \\ &\geq \frac{C(R+t)^{-\alpha}}{p+1}[(G(t))^{p+1} - (G(T_0))^{p+1}]. \end{aligned}$$

Consequently there exist $C > 0$ and $T_0 \geq a$ such that, for $t \geq T_0$,we have

$$F'(t)G'(t) \geq C(R+t)^{-\alpha}(G(t))^{p+1}. \tag{2.12}$$

Similarly, multiplying (2.12) by $G'(t)$ and integrating by parts, it follows that

$$F(t) \geq \frac{C(R+t)^{-\alpha}(G(t))^{p+2}}{(G'(t))^2}, \tag{2.13}$$

where $C > 0$ and t is taken to be large. From (2.9) and (2.13) we get

$$(G'(t))^{2q}G''(t) \geq C(R+t)^{-\beta-q\alpha}(G(t))^{q(p+2)}, \tag{2.14}$$

hence, as before there exist $C > 0$ and $T_0 \geq a$ such that, for $t \geq T_0$,

$$G'(t) \geq C(R+t)^{-\frac{\beta+q\alpha}{2(q+1)}}(G(t))^{\frac{q(p+2)+1}{2(q+1)}}. \tag{2.15}$$

Now, by (2.10) we can choose $\varepsilon > 0$ such that

$$\frac{pq-1}{2(q+1)} > \varepsilon > \frac{1}{s} \cdot \frac{q(\alpha-2)+\beta-2}{2(q+1)},$$

that is,

$$\frac{q(p+2)+1}{2(q+1)} - \varepsilon > 1 \quad \text{and} \quad -\frac{\beta+q\alpha}{2(q+1)} + \varepsilon s > -1. \tag{2.16}$$

Finally from (2.7) and (2.15) we obtain

$$G'(t) \geq C(R+t)^{-\frac{\beta+q\alpha}{2(q+1)}+\varepsilon s}(G(t))^{\frac{q(p+2)+1}{2(q+1)}-\varepsilon}.$$

Integrating this last inequality and using (2.16) we obtain a contradiction. This completes the proof of the lemma.

For the sake of completeness we state a result similar to the previous one which can be considered to be a vector version of Kato's Lemma [14, Lemma 2].

Lemma 2.2 *Let $a, b \in [0, +\infty]$, with $a < b$ and let F, $G \in C^2([a, b[, \mathbf{R})$. Suppose that, for all $t \in [a, b[$, we have*

$$F(t) \geq C(R+t), \tag{2.17}$$
$$G(t) \geq C(R+t), \tag{2.18}$$
$$F''(t) \geq C(R+t)^{-\alpha}(G(t))^p, \tag{2.19}$$
$$G''(t) \geq C(R+t)^{-\beta}(F(t))^q, \tag{2.20}$$

where C, $R > 0$, α, $\beta \geq 0$, and p, $q > 0$. Suppose moreover that $pq > 1$, $\alpha \leq p + 1$ and

$$q(\alpha - 2) + \beta - 2 \leq pq - 1.$$

Then $b < +\infty$.

Proof. In view of the result of Lemma 2.1 it is sufficient to consider only the case

$$q(\alpha - 2) + \beta - 2 = pq - 1. \tag{2.21}$$

We shall obtain a contradiction by assuming that F and G are defined in $[a, +\infty[$. Arguing as in the first part of the proof of Lemma 2.1, we obtain that there exist $C > 0$ and $T_0 \geq a$ such that, for all $t \geq T_0$,

$$G'(t) \geq C(R+t)^{-\frac{\beta+q\alpha}{2(q+1)}}(G(t))^{\frac{q(p+2)+1}{2(q+1)}}. \tag{2.22}$$

By (2.21) we have that

$$\frac{\beta + q\alpha}{2(q+1)} = \frac{q(p+2)+1}{2(q+1)},$$

and consequently (2.22) reads as

$$G'(t) \geq C(R+t)^{-\gamma}(G(t))^{\gamma}, \tag{2.23}$$

where

$$\gamma = 1 + \frac{pq-1}{2(q+1)} > 1.$$

Since by hypothesis $-\alpha + p \leq -1$, we deduce that

$$F(t) \geq C(R+t)^{-\alpha+p+2}.$$

Consequently, it follows from (2.20) that

$$G''(t) \geq C(R+t)^{-\beta-q\alpha+pq+2q} = C(R+t)^{-1}.$$

Integrating this last inequality twice it follows that there exist $C > 0$ and $T_0 \geq a$ such that

$$G(t) \geq C(R+t)\log(1+R+t), \tag{2.24}$$

for all $t \geq T_0$. From (2.24) and (2.23) we deduce that

$$G'(t) \geq C(R+t)^{-1}(\log(1+R+t))^{\gamma-1}G(t).$$

Hence it follows that for all $\delta > 0$ there exist $C > 0$ and $T_0 \geq a$ such that

$$G(t) \geq C(R+t)^{\delta},$$

for all $t \geq T_0$. Now, by choosing $\delta = 2\gamma/(\gamma-1)$, from (2.23) we get that

$$G'(t) \geq C(R+t)^{-\gamma}(G(t))^{\frac{\gamma-1}{2}}(G(t))^{\frac{\gamma+1}{2}} \geq C(G(t))^{\frac{\gamma+1}{2}}. \tag{2.25}$$

Since $(\gamma-1)/2 > 1$, from (2.25) the claim easily follows. This concludes the proof of the lemma.

2.2. Proof of the Theorem 3

First we observe that by the symmetry of our problem, in what follows we may suppose that $1 < q \leq p$. Letting (u, v) be a solution to (2.1), (2.2), we define

$$F(t) := \int_{\mathbf{R}^n} u(x,t)\,dx,$$
$$G(t) := \int_{\mathbf{R}^n} v(x,t)\,dx.$$

Since (u, v) are C^2 functions with support in $\{|x| \leq R+t\}$ it follows that F and G are C^2 functions defined in $[0, T[$. By Hölder's inequality we have that

$$\begin{aligned} |\int_{\mathbf{R}^n} v(x,t)\,dx|^p &\leq (\int_{|x|\leq R+t} dx)^{p-1}\int_{|x|\leq R+t}|v(x,t)|^p\,dx \\ &\leq C(R+t)^{n(p-1)}\int_{\mathbf{R}^n}|v(x,t)|^p\,dx. \end{aligned}$$

On the other hand, from the divergence theorem it follows that

$$F''(t) = \int_{\mathbf{R}^n} \partial_t^2 u(x,t)\,dx = \int_{\mathbf{R}^n} (\partial_t^2 - \Delta) u(x,t)\,dx = \int_{\mathbf{R}^n} |v(x,t)|^p\,dx. \tag{2.26}$$

Hence

$$F''(t) \geq C(R+t)^{-n(p-1)}|G(t)|^p, \tag{2.27}$$

where $C > 0$. Similarly,

$$G''(t) \geq C(R+t)^{-n(q-1)}|F(t)|^q. \tag{2.28}$$

From (2.26) we deduce that F is a convex function and consequently $F(t) \geq F'(0)t + F(0)$, where $F'(0) = \int_{\mathbf{R}^n} g_1(x)dx$. Hence by hypothesis *i)* there exists $T_0 > 0$ such that, for all $t \geq T_0$,

$$F(t) \geq C(R+t), \tag{2.29}$$

for some $C > 0$. Arguing in the same way,

$$G(t) \geq C(R+t). \tag{2.30}$$

If $n = 1$ the nonexistence result follows via Lemma 2.1 by choosing $s = 1$, $\alpha = p - 1$ and $\beta = q - 1$.

Let $n = 2$. In this case (2.3) is equivalent to

$$(pq-1)(3 - q/2) > q(2p-4) + 2q - 4,$$

and consequently there exists $q^* \in]q, 4[$ such that

$$(pq-1)(3 - \frac{q^*}{2}) > q(2p-4) + 2q - 4. \tag{2.31}$$

Suppose that u_0 is the solution of

$$\begin{aligned} \partial_t^2 u_0 - \Delta u_0 &= 0 && \text{in } \mathbf{R}^2 \times [0,+\infty[, \\ u_0(x,0) = f_1(x),\ \partial_t u_0(x,0) &= g_1(x) && \text{in } \mathbf{R}^2. \end{aligned}$$

We know by [10, Lemma 2] that

$$\lim_{t\to+\infty} \frac{1}{\sqrt{t\log t}} \int_{t-\log t<|x|<R+t} u_0(x,t)\,dx = \sqrt{2}\int_{\mathbf{R}^2} g_1(x)\,dx.$$

Since $\int g_1(x)\,dx > 0$, we deduce that there exist C, $T_0 > 0$ such that, for all $t \geq T_0$,

$$\int_{t-\log t<|x|<R+t} u_0(x,t) \geq C((R+t)\log(1+R+t))^{1/2}.$$

By the positivity of the fundamental solution of the wave equation for space dimension $n = 2,\ 3$, we have that $u(x,t) \geq u_0(x,t)$ for all $(x,t) \in \mathbf{R}^2 \times [0,+\infty[$. It follows that

$$\begin{aligned} C((R+t)\log(1+R+t))^{1/2} &\leq \int_{t-\log t<|x|<R+t} u_0(x,t)\,dx \\ &\leq \int_{t-\log t<|x|<R+t} u(x,t)\,dx \\ &\leq C'\Big(\int_{\mathbf{R}^2} |u(x,t)|^q\Big)^{1/q} dx((R+t) \\ &\qquad \times \log(1+R+t))^{(q-1)/q}. \end{aligned}$$

Hence there exist C, $T_0 \geq 0$ such that, for all $t > T_0$,

$$\int_{\mathbf{R}^2} |u(x,t)|^q\,dx \geq C((R+t)\log(1+R+t))^{1-q/2}, \tag{2.32}$$

and consequently

$$G''(t) = \int_{\mathbf{R}^2} |u(x,t)|^q\,dx \geq C(R+t)^{1-q^*/2},$$

for all $t \geq T_0$. Integrating this last inequality twice (notice that $q^* < 4$), we get

$$G(t) \geq C(R+t)^{3-q^*/2}. \tag{2.33}$$

In order to conclude the proof we apply Lemma 2.1 with the choice $s = 3 - q^*/2$, $\alpha = 2(p-1)$ and $\beta = 2(q-1)$.

Suppose finally that $n = 3$. In this case (2.3) reads as

$$(pq-1)(4-q) > q(3p-5) + 3q - 5. \tag{2.34}$$

As before, let u_0 be the solution of

$$\begin{aligned} \partial_t^2 u_0 - \Delta u_0 &= 0 && \text{in } \mathbf{R}^3 \times [0,+\infty[, \\ u_0(x,0) = f_1(x),\ \partial_t u_0(x,0) &= g_1(x) && \text{in } \mathbf{R}^3. \end{aligned}$$

Since $\frac{d^2}{dt^2}\int u_0(x,t)dx = 0$, it follows that $\int u_0(x,t)dx = t\int g_1(x)dx + \int f_1(x)dx$. By the strong Huygens' principle we have that supp $u_0 \subset$

$\{t - R \le |x| \le R + t\}$. Hence using the fact that $u(x,t) \ge u_0(x,t)$, we get

$$C(R+t) \le \int_{t-R<|x|<R+t} u_0(x,t)\,dx \le \int_{t-R<|x|<R+t} u(x,t)\,dx.$$

Then, by Hölder's inequality,

$$C(R+t) \le \Big(\int_{t-R<|x|<R+t} |u(x,t)|^q\,dx\Big)^{1/q}(R+t)^{2(q-1)/q},$$

which implies

$$G''(t) = \int_{\mathbf{R}^3} |u(x,t)|^q\,dx \ge C(R+t)^{2-q}.$$

Integrating the above inequality twice (remark that $q < 3$) we deduce that there exist C, $T_0 > 0$ such that, for $t \ge T_0$, we have

$$G(t) \ge C(R+t)^{4-q}. \tag{2.35}$$

An application of Lemma 2.1 completes the proof of the theorem.

2.3. Proof of Theorem 4

We start by recalling two results due to Sideris [20, Lemma 5, Lemma 6] that we will use in what follows.

Lemma 2.3 *Let $n \ge 4$. Let $h \in C^1(\mathbf{R}^n \times [0,T[)$, with $h(x,t) \ge 0$ for all $(x,t) \in \mathbf{R}^n \times [0,T[$. Let $w_1 \in C^2(\mathbf{R}^n \times [0,T[)$ be the solution of*

$$\begin{aligned} &\partial_t^2 w_1 - \Delta w_1 = h && \text{in } \mathbf{R}^n \times [0,T[,\\ &w_1(x,0) = 0,\ \partial_t w_1(x,0) = 0 && \text{in } \mathbf{R}^n. \end{aligned}$$

Consider

$$W(x,t) = \int_0^t (t-s)^m w_1(x,s)\,ds,$$

where $m = (n-5)/2$ if n is even, or $m = (n-4)/2$ if n is odd.

Then $W(x,t) \ge 0$ for all $(x,t) \in \mathbf{R}^n \times [0,T[$.

Lemma 2.4 *Let $n \ge 4$. Let w_0 be the solution of*

$$\begin{aligned} &\partial_t^2 w_0 - \Delta w_0 = 0 && \text{in } \mathbf{R}^n \times [0,+\infty[,\\ &w_0(x,0) = f(x),\ \partial_t w_0(x,0) = g(x) && \text{in } \mathbf{R}^n, \end{aligned}$$

where $f,\ g \in C_0^\infty(\mathbf{R}^n)$. *Suppose that* $\operatorname{supp} f$, $\operatorname{supp} g \subset \{|x| \leq R\}$ *and* $\int_{\mathbf{R}^n} |x|^{\eta-1} f(x)\,dx > 0$, $\int_{\mathbf{R}^n} |x|^{\eta} g(x)\,dx > 0$, *where* $\eta = 0$ *if* n *is odd or* $\eta = 1/2$ *if* n *is even. Then there exist* C, $T_0 > 0$ *such that, for* $t > T_0$,

$$\int_{t-R}^{t} (t-s)^m \int_{|x|>t} w_0(x,s)\,dx\,ds \geq C(R+t)^{(n-1)/2},$$

where m *is defined in the Lemma* 2.3.

Coming to the proof of Theorem 4 we can suppose that $1 < q \leq p$ without loss of generality. As $\operatorname{supp} u$, $\operatorname{supp} v \in \{|x| \leq R+t\}$, we can define

$$a_1(t) := \int_{\mathbf{R}^n} u(x,t)\,dx,$$
$$a_2(t) := \int_{\mathbf{R}^n} v(x,t)\,dx.$$

It follows that $a_1,\ a_2 \in C^2([0,+\infty[)$, and since

$$a_1''(t) = \int_{\mathbf{R}^n} |v(x,t)|^p\,dx \geq 0,$$
$$a_2''(t) = \int_{\mathbf{R}^n} |u(x,t)|^q\,dx \geq 0,$$

$a_1,\ a_2$ are convex functions. Moreover $a_i(0) = \int_{\mathbf{R}^n} f_i(x)dx$ and $a_i'(0) = \int_{\mathbf{R}^n} g_i(x)dx$, for $i = 1,\ 2$. Then we deduce from *i')* that there exist C, $T_0 > 0$ such that, for $t \geq T_0$,

$$a_i(t) \geq C(R+t), \tag{2.36}$$

for $i = 1,\ 2$. Let us now set, for $t \geq R$,

$$F(t) := \int_{t-R}^{t} (t-s)^m a_1(s)\,ds,$$
$$G(t) := \int_{t-R}^{t} (t-s)^m a_2(s)\,ds,$$

where $m = (n-5)/2$ if n is even, or $m = (n-4)/2$ if n is odd. Integrating by parts we obtain that

$$\begin{aligned} F(t) &= \frac{R^{m+1} a_1(t-R)}{m+1} + \frac{R^{m+2} a_1'(t-R)}{(m+1)(m+2)} \\ &\quad + \frac{1}{(m+1)(m+2)} \int_{t-R}^{t} (t-s)^{m+2} a_1''(s)\,ds, \\ G(t) &= \frac{R^{m+1} a_2(t-R)}{m+1} + \frac{R^{m+2} a_2'(t-R)}{(m+1)(m+2)} \\ &\quad + \frac{1}{(m+1)(m+2)} \int_{t-R}^{t} (t-s)^{m+2} a_2''(s)\,ds. \end{aligned}$$

Differentiating these identities twice with respect to t, we get

$$F''(t) = \int_{t-R}^{t} (t-s)^m a_1''(s)\, ds = \int_{t-R}^{t} (t-s)^m \int_{\mathbf{R}^n} |v(x,s)|^p \, dx\, ds,$$
$$G''(t) = \int_{t-R}^{t} (t-s)^m a_2''(s)\, ds = \int_{t-R}^{t} (t-s)^m \int_{\mathbf{R}^n} |u(x,s)|^q \, dx\, ds.$$

By Hölder's inequality, using the fact that $\operatorname{supp} u$, $\operatorname{supp} v \subset \{|x| \leq R+t\}$ we have

$$F''(t) \geq C(R+t)^{-n(p-1)} |G(t)|^p, \tag{2.37}$$
$$G''(t) \geq C(R+t)^{-n(q-1)} |F(t)|^q, \tag{2.38}$$

for some $C > 0$. On the other hand, from (2.36) it is easy to prove that there exist C, $T_0 > 0$ such that, for $t \geq T_0$,

$$F(t) \geq C(R+t), \tag{2.39}$$
$$G(t) \geq C(R+t). \tag{2.40}$$

Consider now the first component of the solution, i. e. the function u; we know that $u(x,t) = u_0(x,t) + u_1(x,t)$ where

$$\begin{aligned} &\partial_t^2 u_0 - \Delta u_0 = 0 && \text{in } \mathbf{R}^n \times [0,+\infty[, \\ &u_0(x,0) = f_1(x),\ \partial_t u_0(x,0) = g_1(x) && \text{in } \mathbf{R}^n, \end{aligned}$$

and

$$\begin{aligned} &\partial_t^2 u_1 - \Delta u_1 = |v|^p && \text{in } \mathbf{R}^n \times [0,T[, \\ &u_1(x,0) = 0,\ \partial_t u_1(x,0) = 0 && \text{in } \mathbf{R}^n. \end{aligned}$$

Then, by Lemma 2.3,

$$\int_0^t (t-s)^m u(x,s)\, ds \geq \int_0^t (t-s)^m u_0(x,s)\, ds, \tag{2.41}$$

for all $(x,t) \in \mathbf{R}^n \times [0,T[$. We integrate both side of (2.41) on $\{|x| > t\}$ and we reverse the order of integration. We get

$$\int_0^t (t-s)^m \int_{|x|>t} u(x,s)\, dx\, ds \geq \int_0^t (t-s)^m \int_{|x|>t} u_0(x,s)\, dx\, ds,$$

and since $\operatorname{supp} u$ and $\operatorname{supp} u_0$ are contained in $\{|x| \leq R+s\}$, this last inequality is equivalent to

$$\int_{t-R}^t (t-s)^m \int_{|x|>t} u(x,s)\, dx\, ds \geq \int_{t-R}^t (t-s)^m \int_{|x|>t} u_0(x,s)\, dx\, ds.$$

By Hölder's inequality and Lemma 2.4 we deduce, for $t \geq T_0$,

$$(R+t)^{(n-1)(q-1)/q}\left(\int_{t-R}^{t}(t-s)^m \int_{\mathbf{R}^n} |u(x,s)|^q dx ds\right)^{1/q} \geq C(R+t)^{(n-1)/2}.$$

Consequently there exist C, $T_0 > 0$ such that, for $t \geq T_0$,

$$G''(t) \geq C(R+t)^{n-1-q(n-1)/2}.$$

We remark that $n-1-q(n-1)/2 > -1$, and, integrating this last inequality twice, we have

$$G(t) \geq C(R+t)^{n+1-q(n-1)/2}. \tag{2.42}$$

Recalling (2.39), (2.42), (2.37), (2.38) and observing that (2.3) implies that

$$(pq-1)(n+1-q\frac{(n-1)}{2}) > q(n(p-1)-2)+n(q-1)-2,$$

we can use Lemma 2.1 with $s = n+1-q(n-1)/2$, $\alpha = n(p-1)$ and $\beta = n(q-1)$. The proof is complete.

References

[1] F.Asakura, Existence of a global solution to a semilinear wave equation with slowly decreasing initial data, *Comm. Partial Differential Equations* **11** (1986), 1459–1484.

[2] Y. Choquet–Bruhat, Global existence for solutions of $\Box u = A|u|^p$, *J. Differential Eq.* **92** (1989), 98–108.

[3] D. Del Santo, Global existence and blow-up for a hyperbolic system in three space dimension, preprint 1997.

[4] M. Escobedo, M. A. Herrero, Boundedness and blow up for a semilinear reaction–diffusion system, *J. Differential Eq.* **89** (1991), 176–202.

[5] P. Felmer, D. G. de Figueiredo, Superquadratic elliptic systems, *Trans. Amer. Math. Soc.* **343** (1994), 99–116.

[6] V. Georgiev, Weighted estimate for the wave equation, preprint University of Tsukuba, n. 95-010, 1995.

[7] V. Georgiev, Une estimation avec poids pour l'équation des ondes, *C. R. Acad. Sci. Paris , Série I* **322** (1996), 829–834.

[8] V. Georgiev, Existence of global solution to supercritical semilinear wave equation, to appear in *Serdica.*

[9] V. Georgiev, H. Lindblad, C. Sogge, Weighted Strichartz estimate and global existence for semilinear wave equation, to appear in *Amer. J. Math.*

[10] R. Glassey, Finite–time blow–up for solutions of nonlinear wave equations, *Math. Z.* **177** (1981), 323–340.

[11] R. Glassey, Existence in the large for $\Box u = F(u)$ in two space dimensions, *Math. Z.* **178** (1981), 233–261.

[12] F. John, Blow–up of solutions of nonlinear wave equations in three space dimension, *Manuscripta Math.* **28** (1979), 235–268.

[13] F. John, Blow–up for quasi–linear wave equations in three space dimension, *Comm. Pure Appl. Math.* **34** (1981), 29–51.

[14] T. Kato, Blow–up of solutions of some nonlinear hyperbolic equations, *Comm. Pure Appl. Math.* **33** (1980), 501–505.

[15] H. Lindblad, C. Sogge, On existence and scattering with minimal regularity for semilinear wave equations, *J. Funct. Anal.* **130** (1995), 357–426.

[16] E. Mitidieri, A Rellich type identity and applications, *Comm. Partial Differential Equations* **18** (1993), 125–151.

[17] E. Mitidieri, Nonexistence of positive solutions of semilinear elliptic systems in $\mathbf{R}^n$, *Diff. Integral Eqs.* **9** (1996), 465–469.

[18] J. Serrin, H. Zou, Existence of positive solutions of the Lane–Emden system, preprint 1996.

[19] J. Serrin, H. Zou, Nonxistence of positive solutions of Lane–Emden systems, preprint 1996.

[20] T. Sideris, Nonexistence of global solutions to semilinear wave equations in high dimension, *J. Differential Eq.* **52** (1984), 378–406.

[21] K. Tsutaya, Global existence and the life span of solutions of semilinear wave equations with data of noncompact support in three space dimensions, to appear in *Funkcialaj Ekvacioj*.

Daniele Del Santo and Enzo Mitidieri
Dipartimento di Scienze Matematiche
Università di Trieste
Piazzale Europa 1, 34127
Trieste, Italy

Vladimir Georgiev
Institute of Mathematics
Bulgarian Academy of Science
Sofia 1113,
Acad. G. Bonchev bl. 8
Bulgaria

A class of solvable operators

Nils Dencker

I . Introduction

In this paper we shall study the operator

$$P = D_t + iF(t, x, D_x) \tag{1.1}$$

where $F \in C^2(\mathbf{R}, S^1_{1,0}(\mathbf{R}^n))$ has real principal symbol $\sigma(F) = f(t, x, \xi)$. We shall assume condition $(\overline{\Psi})$:

(1.2)

$t \mapsto f(t, x, \xi)$ does not change sign from $+$ to $-$ with increasing t.

Then the adjoint operator P^* satisfies condition (Ψ), that is, $t \mapsto f(t, w)$ does not change sign from $-$ to $+$ with increasing t.

It was conjectured by Nirenberg and Treves [9] that condition (Ψ) was equivalent to local solvability for operators of principal type, and they proved this in several cases. Local solvability for P^* means that the equation

$$P^* u = v \tag{1.3}$$

has a local solution $u \in \mathcal{D}'$ for any $v \in C^\infty$ satisfying a finite number of compatibility conditions. Local solvability in L^2 means that (1.3) has a local solution $u \in L^2$ for any $v \in L^2$ satisfying a finite number of compatibility conditions.

The necessity of (Ψ) for local solvability in the C^∞ category was proved by Moyers in two dimensions and by Hörmander in general, see Corollary 26.4.8 in [3]. In the analytic category, the sufficiency of condition (Ψ) for solvability of microdifferential operators acting on microfunctions was proved by Trépreau [10]. The sufficiency of (Ψ) for local L^2 solvability for first order pseudodifferential operators in two dimensions was proved by Lerner [5].

For differential operators, condition (Ψ) is equivalent to condition (P), which rules out any sign changes of $t \mapsto f(t, x, \xi)$. The sufficiency of (P) for local L^2 solvability for first order pseudodifferential operators was proved by Nirenberg and Treves [9] when the principal symbol is real analytic, and by Beals and Fefferman [1] in the general case.

Typeset by $\mathcal{A}\mathcal{M}\mathcal{S}$-TEX

Lerner [6] constructed a counterexample to the sufficiency of (Ψ) for local L^2 solvability. It was proved in [2] that Lerner's counterexamples were locally solvable with a loss of at most one derivative. Now the questions are which extra condition is needed to prove local L^2 solvability, and is condition (Ψ) sufficient for local solvability in general?

In this paper we shall prove, assuming condition $(\overline{\Psi})$ and (1.6), an estimate for P which implies local solvability for P^* with a loss of at most one derivative. We shall use the Weyl calculus notation:

$$|f|_1^g = (|\partial_x f|^2 + |\partial_\xi f|^2 \langle \xi \rangle^2)^{1/2} \tag{1.4}$$

where $g_{x,\xi} = |dx|^2 + |d\xi|^2/\langle\xi\rangle^2$ is the homogeneous metric. This satisfies $g/g^\sigma = h^2$ where $h^{-1} = \langle\xi\rangle = (|\xi|^2+1)^{1/2}$. We make the assumption

$$\begin{aligned}(\langle\xi\rangle^{-1/2}|\partial_x f| + \langle\xi\rangle^{1/2}|\partial_\xi f| + 1)(\langle\xi\rangle^{-1/2}|\partial_x f'| + \langle\xi\rangle^{1/2}|\partial_\xi f'|)\\ \le C(f' + \langle\xi\rangle^{-1}|\partial_x f|^2 + \langle\xi\rangle|\partial_\xi f|^2 + 1) \qquad \text{when } f = 0\end{aligned} \tag{1.5}$$

where $f' = \partial_t f$. Observe that, because of $(\overline{\Psi})$, we find that $f' \ge 0$ when $f = 0$. Condition (1.5) can be written as

$$h|f'|_1^g|f|_1^g + h^{1/2}|f'|_1^g \le C(f' + h(|f|_1^g)^2 + 1) \qquad \text{when } f = 0. \tag{1.6}$$

This condition is not satisfied for Lerner's counterexamples in [6].

Observe that Lerner [7] proved that P^* is locally L^2 solvable under condition $(\overline{\Psi})$ and the condition

$$\langle\xi\rangle^{-1}|\partial_x f|^2 + \langle\xi\rangle|\partial_\xi f|^2 \le Cf' \qquad \text{when } f = 0, \tag{1.7}$$

that is, $h(|f|_1^g)^2 \le Cf'$ when $f = 0$. Menikoff [8] proved local L^2 solvability for P^* when $(\overline{\Psi})$ is satisfied, and

$$\langle\xi\rangle^{-1}|\partial_x f|^2 + \langle\xi\rangle|\partial_\xi f|^2 \le C|f|, \tag{1.8}$$

that is, $h(|f|_1^g)^2 \le C|f|$. Hörmander [4] proved local L^2 solvability for P^* satisfying $(\overline{\Psi})$ in the case the sign changes of f are independent of ξ, i.e., $f(t,x,\xi)f(t,x,\eta) \ge 0$, $\forall t, x, \xi, \eta$.

Condition (1.6) is not homogeneous, and by Proposition 2.2 it is stable, i.e., it is satisfied in a G neighborhood of $f^{-1}(0)$. Here $G = Hg/h$, with $H^{-1} = |f| + h(|f|_1^g)^2 + 1$, is the Beals–Fefferman metric. We shall prove in Proposition 2.3, that (1.6) implies that the metric G only varies with a constant factor in t near the sign changes of f. It also implies that $f \neq 0$ in G neighborhoods where $f' \not\gtrsim -C(|f|+1)$–see Lemma 3.2.

Observe that condition (1.6) is not invariant under changes of scale in the t variable. In fact, a scale change replaces $f(t,w)$ by $\varrho f(\varrho t, w)$ for $\varrho > 0$, which replaces (1.6) by

(1.9)
$$(\varrho h^{1/2}|f|_1^g + 1)\varrho^2 h^{1/2}|f'|_1^g \leq C(\varrho^2 f' + \varrho^2 h(|f|_1^g)^2 + 1) \qquad \text{when } f = 0.$$

By dividing by ϱ^3 and taking $\varepsilon = \varrho^{-1} \leq 1$, we obtain that

$$(1.10)\quad h|f'|_1^g|f|_1^g + \varepsilon h^{1/2}|f'|_1^g \leq C\varepsilon(f' + h(|f|_1^g)^2 + 1) \qquad \text{when } f = 0$$

for any chosen $0 < \varepsilon \leq 1$. Observe that f still satisfies (1.6), but the constant in the estimate below and the size of the time interval are changed.

Before stating the result, we shall introduce more general operators. Let

$$(1.11)\qquad P = D_t + if^w(t,x,D_x) + r^w(t,x,D_x)$$

where $f(t,w) \in C^2(\mathbf{R}, S(h^{-1},g))$ and $r(t,w) \in C(\mathbf{R}, S(1,g))$, $w = (x,\xi)$. Here we use the Weyl notation

(1.12)
$$a^w u(x) = (2\pi)^{-n} \iint \exp(i\langle x-y,\xi\rangle) a\left(\tfrac{x+y}{2},\xi\right) u(y)\, dy d\xi \quad u \in \mathcal{S}.$$

We assume that g is slowly varying and σ temperate, $\sup g/g^\sigma = h^2 \leq 1$ and g is independent of t. As usual, we denote the seminorms of f in $S(h^{-1},g)$ by

$$(1.13)\qquad |f|_k^g = \sup_{T_j} \frac{|f^{(k)}(T_1,\ldots,T_k)|}{\prod_1^k g(T_j)^{1/2}} \qquad \text{for } k \geq 0.$$

Let $\|u\|(t) = \langle u,u\rangle^{1/2}(t)$ be the L^2 norm of u in the x variables for fixed t.

Theorem 1.1. *Assume that P is of the form* (1.11), *where f satisfies condition* $(\overline{\Psi})$ *and* (1.6). *There exist positive constants T_0 and C_0, a metric $G_{0,t}$ on $T(T^*\mathbf{R}^n)$, which is uniformly σ temperate $\forall\, t$ such that $\sup G_0/G_0^\sigma = H_0^2 \geq ch^2$, and there exists $b(t,w) \in C(\mathbf{R}, S(H_0^{-1}, G_0))$ such that*

$$(1.14)\qquad \int \|u\|^2(t)\, dt \leq C_0 T^2 \int \operatorname{Im}\langle b^w Pu, u\rangle(t) + \|Pu\|^2(t)\, dt$$

for $u(t,x) \in \mathcal{S}(\mathbf{R}\times\mathbf{R}^n)$ having support where $|t| \leq T \leq T_0$.

Remark 1.2. By the Cauchy–Schwarz inequality, we obtain from (1.14) that

$$\int \|u\|^2(t)\,dt \le C\int T^4\|b^w Pu\|^2(t) + T^2\|Pu\|^2(t)\,dt. \tag{1.15}$$

Since g is σ temperate, we find $h^{-1}(x,\xi) \le C_N\langle\xi\rangle^N$, which gives $|b(t,x,\xi)| \le C_N\langle\xi\rangle^N$ and thus

$$\int \|u\|^2\,dt \le C'_N T^2 \int \|Pu\|^2_{(N+n+1)}\,dt \tag{1.16}$$

for $u \in C_0^\infty$ having support where $|t| \le T$. Here

$$\|u\|^2_{(s)}(t) = (2\pi)^{-n}\int |\hat{u}(t,\xi)|^2\langle\xi\rangle^{2s}\,d\xi \qquad s \in \mathbf{R} \tag{1.17}$$

is the square of the Sobolev norm in the x variables, for fixed t. This gives local solvability of P^* near $t = 0$. Observe that (1.14) can be microlocalized with respect to the metric g, for small enough T_0.

In the case $g_{x,\xi} = |dx|^2 + |d\xi|^2/\langle\xi\rangle^2$ we immediately obtain the following result by conjugation with $\langle D_x\rangle^s$.

Corollary 1.3. *Assume that P is of the form* (1.1), *where f satisfies condition* ($\overline{\Psi}$) *and* (1.5). *For $T_s > 0$ small enough, we find*

$$\begin{aligned}\int \|u\|^2_{(s)}(t)\,dt &\le C\int T^4\|Pu\|^2_{(s+1)}(t) + T^2\|Pu\|^2_{(s)}(t)\,dt \\ &\le C_1 T^2\int \|Pu\|^2_{(s+1)}(t)\,dt\end{aligned} \tag{1.18}$$

for $u(t,x) \in \mathcal{S}(\mathbf{R}\times\mathbf{R}^n)$ having support where $|t| \le T \le T_s$.

This gives local solvability for P^* near $t = 0$, with a loss of at most one derivative.

The idea of the proof of Theorem 1.1 is to make an approximate factorization $f = ab + r$, where a is a positive and bounded symbol, b is "almost" increasing in the sense that $\partial_t b \ge -cb - C$ and r is bounded. If $\partial_t f \ge cf - C$ we may take $a \equiv 1$ and $b \equiv f$. Condition (1.6) implies that $f \ne 0$ when $\partial_t f \not\ge -C(|f| + 1)$. In this case, we are going to take $|b| \ge |f|$ so that b is non-decreasing. Essentially, we are going to use $b = \pm\sup_t H^{-1}$ over a suitable t interval. This means that we replace the Beals–Fefferman metric $G = Hg/h$ with a monotone metric in these neighborhoods. This metric must be regularized and special care must be taken when constructing the cut-off functions.

The plan of the paper is as follows. In Section 2 we introduce the Beals–Fefferman metric and prove that condition (1.6) is locally stable with respect to this metric. In Section 3 we localize in neighborhoods where either $f' \geq cf - C$ or not. By condition (1.6) we find that $f \neq 0$ in the latter case, and then we modify the Beals–Fefferman metric to get a monotone metric in t, which is done in Section 4. In Section 5 we construct suitable monotone cut-off functions. These are used in Section 6 to factor $f = ab + r$, where $b' \geq cb - C$, a is positive and bounded symbol and r is bounded. Finally, we prove Theorem 1.1 in Section 7 by using this factorization and Theorem A.2 in Appendix A. We shall use the Wick quantization, which we define in Appendix B.

The author wishes to thank Nicolas Lerner and Anders Melin for some valuable comments.

2. The metrics

We shall use the Beals–Fefferman metric $G = Hg/h$, for fixed t, where

$$H^{-1} = 1 + |f| + h(|f|_1^g)^2. \tag{2.1}$$

We shall also use the metric $\widetilde{G} = \widetilde{H}g/h$ where

$$\widetilde{H}^{-1} = 1 + |f'| + h(|f'|_1^g)^2. \tag{2.2}$$

These metrics are easily seen to be uniformly slowly varying and σ temperate, $\sup G/G^\sigma = H^2 \leq 1$ and $\sup \widetilde{G}/\widetilde{G}^\sigma = \widetilde{H}^2 \leq 1$. We also have $f \in S(H^{-1}, G)$ and $f' \in S(\widetilde{H}^{-1}, \widetilde{G})$ uniformly in t — see for example Lemma 26.10.2 in [3]. By replacing g with an equivalent metric, we may assume that $g(w) \in S(g(w), g)$ for any $w \in T^*\mathbf{R}^n$, and that $h \in S(h, g)$. Observe that H^{-1} and $\widetilde{H}^{-1}$ are continuous functions of (t, w). Since $f'' \in S(h^{-1}, g)$ we also find that H^{-1} and $\widetilde{H}^{-1}$ only changes with a term $O(\varepsilon)$ when $\Delta t = O(\varepsilon h)$, thus H^{-1} and $\widetilde{H}^{-1}$ only vary with a fixed factor in t for small ε. This implies that $dt^2/h^2 + G$ is slowly varying.

Lemma 2.1. *If $h(|f|_1^g)^2 \geq \varrho(|f| + 1)$ at (t, w_0), we find $h(|f|_1^g)^2 \geq \frac{\varrho}{1+\varrho} H^{-1}$ and $|f| + 1 \leq \frac{1}{1+\varrho} H^{-1}$ at (t, w_0). If ϱ is large enough, we find that $f\big|_t$ changes sign in the neighborhood $\left\{ w : G_{t,w_0}(w - w_0) \leq \frac{4}{\varrho^2} \right\}$.*

Proof. Since $H^{-1} = 1 + |f| + h(|f|_1^g)^2 \leq (1 + \frac{1}{\varrho}) h(|f|_1^g)^2$ at (t, w_0), we easily get the first statement. By considering $-f$ we may assume

$f(t,w_0) \geq 0$. If $\Delta f = f(t,w_0) - f(t,w)$ and $H^{-1} = H^{-1}(t,w_0)$ we find for some w satisfying $G_{t,w_0}(w-w_0) = \varepsilon^2$ that

$$\Delta f \geq \varepsilon |f|_1^G(t,w_0) - C_0\varepsilon^2 H^{-1} \geq \varepsilon\left(\sqrt{\frac{\varrho}{1+\varrho}} - C_0\varepsilon\right) H^{-1}. \tag{2.3}$$

Thus, when $\varepsilon \leq \sqrt{\frac{\varrho}{(1+\varrho)}}/(2C_0)$ we find $\Delta f \geq \frac{\varepsilon}{2}\sqrt{\frac{\varrho}{(1+\varrho)}}H^{-1}$. This gives

$$f(t,w_0) < \frac{H^{-1}}{1+\varrho} \leq \frac{2}{\varepsilon\sqrt{\varrho+\varrho^2}}\Delta f < \frac{2}{\varepsilon\varrho}\Delta f \leq \Delta f \tag{2.4}$$

if $\frac{2}{\varrho} \leq \varepsilon \leq \sqrt{\frac{\varrho}{(1+\varrho)}}/(2C_0)$. For large ϱ we obtain $f(t,w) < 0$, for some w satisfying $G_{t,w_0}(w-w_0) = 4/\varrho^2$, which proves the lemma. □

Next, we shall prove that condition (1.10) is essentially stable, that is, holds in $dt^2/h^2 + G$ neighborhoods of $f^{-1}(0)$.

Proposition 2.2. *Assume that f satisfies*

$$h|f'|_1^g|f|_1^g + c_0\varrho h^{1/2}|f'|_1^g \leq \varrho(f' + h(|f|_1^g)^2 + 1) \qquad \textit{when } f = 0 \tag{2.5}$$

for some positive constants c_0 and ϱ. If $f(s,w) = 0$ for some (s,w) such that $w \in \omega_0 = \left\{w : G_{t,w_0}(w-w_0) \leq \varepsilon_0^2\right\}$ and $|s-t| \leq \varepsilon_0 h(w_0)$ where $0 < \varepsilon_0 \ll \varrho$ is small enough, then we find

$$h|f'|_1^g|f|_1^g + c_0\varrho h^{1/2}|f'|_1^g \leq 2\varrho f' + C_0\varrho(h(|f|_1^g)^2 + 1) \qquad \textit{at } (t,w_0) \tag{2.6}$$

for some positive constant C_0. Here the sizes of ε_0 and C_0 only depend on c_0 and ϱ in (2.5), and on the seminorms of f, f' and f''.

Proof. Let $G_s = G\big|_{t=s}$, then $G_s \cong G_t$ when $|s-t| \leq \varepsilon h$ for small ε. Now the terms in (2.6) only change with $O(\varepsilon)$ when $|s-t| \leq \varepsilon h$. Thus, it suffices to consider $s = t$ for ε_0 small.

We may assume, by the slow variation, that ε_0 is chosen so that

$$1/C \leq H(t,w)/H(t,w_0) \leq C \tag{2.7}$$

when $w \in \omega_0$. Since $h \in S(h,g)$ we find that $|h(w) - h(w_0)| \leq C\varepsilon_0 H^{-1/2}(t,w_0)h^{3/2}(w_0)$ when $w \in \omega_0$ for small ε_0. By Taylor's formula we have

$$|f'(t,w) - f'(t,w_0)| \leq \varepsilon_0|f'|_1^G(t,w_0) + C\varepsilon_0^2 H^{-1}(t,w_0) \tag{2.8}$$

when $w \in \omega_0$.

Since $g = hG/H$, we have that $|f|_1^g$ and $|f'|_1^g$ only vary with terms that are of size $O(\varepsilon_0 h^{-1/2}(w_0)H^{-1/2}(t,w_0))$ in ω_0, and $h(|f|_1^g)^2 = H(|f|_1^G)^2$ only varies with terms that are $O(\varepsilon_0 H^{-1}(t,w_0))$ in ω_0. Since $H^{-1} = h(|f|_1^g)^2 + 1$ when $f = 0$ we obtain that $H^{-1} \le C_1(1 + h(|f|_1^g)^2)$ in $\{t\} \times \omega_0$ for small ε_0. We find that

$$
\begin{aligned}
&\big|h|f'|_1^g|f|_1^g(t,w) - h|f'|_1^g|f|_1^g(t,w_0)\big| \\
&\qquad \le |h(w) - h(w_0)||f'|_1^g|f|_1^g(t,w) \\
&\qquad\quad + h(w_0)|f|_1^g(t,w)\big||f'|_1^g(t,w) - |f'|_1^g(t,w_0)\big| \\
&\qquad\quad + h|f'|_1^g(t,w_0)\big||f|_1^g(t,w) - |f|_1^g(t,w_0)\big| \\
&\qquad \le C\varepsilon_0(H^{-1/2}h^{1/2}|f'|_1^g(t,w_0) + H^{-1}(t,w_0)).
\end{aligned} \tag{2.9}
$$

We find from (2.5) and (2.9) that

$$
\begin{aligned}
(1 - C_2\varepsilon_0)h|f'|_1^g|f|_1^g + (c_0\varrho - C_2\varepsilon_0)h^{1/2}|f'|_1^g& \\
\le \varrho f' + C_3(\varrho + \varepsilon_0)(1 + h(|f|_1^g)^2) \qquad &\text{at } (t,w_0).
\end{aligned} \tag{2.10}
$$

When $\varepsilon_0 \le \min(1, c_0\varrho)/2C_2$ we obtain (2.6). □

The next proposition shows that the metric G only varies with a fixed factor near the sign changes of f, by condition (1.6).

Proposition 2.3. *Assume that*

$$2h|f'|_1^g|f|_1^g \le f' + CH^{-1} \tag{2.11}$$

on $I = [t_1, t_2] \times \{w\}$. *If* $f \ge 0$ *on* I *we find that*

$$G_{t_2,w} \le C_0 G_{t_1,w}. \tag{2.12}$$

If $f \le 0$ *on* I *we find*

$$G_{t_1,w} \le C_0 G_{t_2,w}. \tag{2.13}$$

Observe that (2.11) follows from (1.10) when $C\varepsilon \le \frac{1}{2}$. If $f\big|_t$ changes sign in a small G_t neighborhood of w, for $t \in [t_1, t_2]$, we find from Propositions 2.2 and 2.3 that $1/C \le G_{t_1,w}/G_{t_2,w} \le C$.

Proof. It suffices to consider the case $f \ge 0$ on $[t_1,t_2] \times \{w\}$. In fact, by choosing $-t$ as variable and considering $-f$ instead of f, we get the corresponding result for the case $f \le 0$. When $f \ge 0$ on I, we find $H^{-1} = 1 + h(|f|_1^g)^2 + f$, which implies

$$\partial_t H^{-1} \ge f' - 2h|f'|_1^g|f|_1^g \ge -CH^{-1} \tag{2.14}$$

on $]t_1, t_2[\times \{w\}$. This gives by continuity that

$$H^{-1}(t_2,w) \ge \exp(-C(t_2 - t_1))H^{-1}(t_1,w) \tag{2.15}$$

and proves the result. □

3. The localizations

We shall now make some localizations, and we shall only consider $t \in [-1,1]$. We define $H_t = H\big|_t$ and similarly $\widetilde{G}_t = \widetilde{G}\big|_t$ and $\widetilde{H}_t = \widetilde{H}\big|_t$. For each t we shall localize in small G_t neighborhoods. Let

$$\omega^{\varepsilon}_{t,w_0} = \left\{ w : G_{t,w_0}(w - w_0) < \varepsilon^2 \right\} \tag{3.1}$$

for $\varepsilon > 0$. In what follows, we shall assume that we have chosen the neighborhoods $\omega^{\varepsilon}_{t,w}$ so that G and g only vary with a fixed factor in $\omega^{\varepsilon}_{t,w}$. We shall compare H_t^{-1} with $\widetilde{H}_t^{-1}$ in these neighborhoods. We say that the neighborhoods $\omega^{\varepsilon}_{t,w_0}$ are of

$$\begin{aligned} &\text{type A:} \qquad \text{if } H_t/\widetilde{H}_t \le \mu \ \text{ or } f' \ge 0 \text{ in } \omega^{\varepsilon}_{t,w_0}, \\ &\text{type B:} \qquad \text{if } \sup H_t/\widetilde{H}_t > \mu \ \text{ and } \inf f' < 0 \text{ in } \omega^{\varepsilon}_{t,w_0}, \end{aligned} \tag{3.2}$$

for $1 \le \mu$. The parameter μ is going to be fixed later. Observe that $\widetilde{H}_t^{-1} \le \mu H_t^{-1}$ if and only if $\mu \widetilde{G}_t \ge G_t$ since $\widetilde{G} = \widetilde{H} G/H$. Since we are going to regularize in t intervals of length $O(\varepsilon h)$, we shall look at the behavior in such intervals. Then, we shall use the metric $dt^2/h + G$.

Now $\widetilde{H}_t$ may vary with more than a fixed factor in the neighborhoods $\omega^{\varepsilon}_{t,w_0}$ for arbitrarily small ε. But in the type B cases we have $\widetilde{G} < G/\mu$ which implies that the condition is stable in uniform $dt^2/h^2 + G$ neighborhoods.

Lemma 3.1. *We find that there exist $c_0 > 0$ and $C_0 > 0$, with the property that if $\widetilde{H}_t^{-1}(w') > \mu H_t^{-1}(w')$ for some $w' \in \omega^{\varepsilon}_{t,w_0}$ with $\varepsilon^2/\mu \le c_0$, then $\widetilde{H}_s^{-1}(w) > \mu H_s^{-1}(w)/C_0$ when $|s - t| \le \varepsilon h(w_0)$ and $w \in \omega^{\varepsilon}_{t,w_0}$.*

Proof. If $\widetilde{H}_t^{-1}(w') > \mu H_t^{-1}(w')$ at $w' \in \omega^{\varepsilon}_{t,w_0}$ we find that

$$H_t^{-1}(w_0) \le C H_t^{-1}(w') < C\mu^{-1}\widetilde{H}_t^{-1}(w'). \tag{3.3}$$

Thus, $\widetilde{G}_{t,w'}(w' - w_0) \le C_1\varepsilon^2/\mu$, which gives $\widetilde{H}_t^{-1}(w') \le C_2\widetilde{H}_t^{-1}(w_0)$ when $\varepsilon^2/\mu \ll 1$, and implies that $\widetilde{H}_t^{-1}(w_0) > \mu H_t^{-1}(w_0)/C'$. By the slow variation, we find that $\widetilde{H}_t^{-1}(w)$ only varies with a fixed factor in $\omega^{\varepsilon}_{t,w_0}$ when $\varepsilon^2/\mu \ll 1$. Also H^{-1} and $\widetilde{H}^{-1}$ only varies with a term which is $O(\varepsilon) = o(H^{-1})$ when $|\Delta t| \le \varepsilon h$ and ε is small. This proves the lemma. □

The next proposition shows that condition (1.6) implies that f has constant sign in uniform dt^2/h^2+G neighborhoods of the type B neighborhoods.

Proposition 3.2. *Assume that f satisfies condition* (1.6). *Then, there exist $\mu_1 \geq 1$ and $\varepsilon_1 > 0$, such that if $\omega^\varepsilon_{t,w_0}$ is of type B with $\varepsilon \leq \varepsilon_1$ and $\mu \geq \mu_1$, then $f(s,w) \neq 0$ when $|s-t| \leq \varepsilon h(w_0)$ and $G_{t,w'}(w-w') < \varepsilon^2$ for any $w' \in \omega^\varepsilon_{t,w_0}$. Here μ_1 and ε_1 only depend on the seminorms of f'.*

Proof. Actually, it suffices to prove that $f(s,w) \neq 0$ when $|s-t| \leq \varepsilon h(w_0)$ and $w \in \omega^\varepsilon_{t,w_0}$ when $\omega^\varepsilon_{t,w_0}$ is of type B, for $\varepsilon \ll 1$ and $\mu \gg 1$. In fact, if $G_{t,w'}(w-w') < \varepsilon^2$ for some $w' \in \omega^\varepsilon_{t,w_0}$, we find $G_{t,w_0}(w-w_0) < 2(1+C)\varepsilon^2$ since $G_{t,w_0} \leq CG_{t,w'}$. Thus $w \in \omega^{C_1\varepsilon}_{t,w_0}$ for some $C_1 \geq 1$, and this neighborhood is also of type B. We then get the result by shrinking ε. Observe that by Lemma 3.1 we may assume that $\widetilde{H}_t^{-1} > \mu H_t^{-1}/C_0$ in $\omega^\varepsilon_{t,w_0}$, which implies that $\widetilde{G}_t < C_0 G_t/\mu$.

It is clear that if $|f'| \geq \widetilde{H}_t^{-1}/3$ at (t,w_0), we obtain that $f'(t,w)$ has constant sign for w in a (uniform) $\widetilde{G}_t$ neighborhood of w_0, since $|f'(t,w)-f'(t,w_0)| \leq (\delta+C\delta^2)\widetilde{H}_t^{-1}(w_0)$ when $\widetilde{G}_{t,w_0}(w-w_0) \leq \delta^2$ and $\delta > 0$ is small enough. This contains the neighborhood $G_{t,w_0}(w-w_0) \leq \varepsilon^2$ if $C_0\varepsilon^2/\mu \leq \delta^2$. Also, we find $|f'(s,w)| \geq c\widetilde{H}^{-1}(s,w)$ for some positive constant c when $|s-t| \leq \varepsilon h(w_0)$ and $w \in \omega^\varepsilon_{t,w_0}$. Since $f' \geq 0$ when $f = 0$ by condition $(\overline{\Psi})$, we get a contradiction with type B, if $f(s,w) = 0$ for $|s-t| \leq \varepsilon h(w_0)$ and $w \in \omega^\varepsilon_{t,w_0}$ when $\varepsilon^2/\mu \ll \delta^2 \ll 1$. Thus, we get the result in this case.

If $h(|f'|^g_1)^2 \geq \widetilde{H}_t^{-1}/3$ at (t,w_0) we obtain that $h(|f'|^g_1)^2 \geq c\widetilde{H}_t^{-1}$ in a uniform $\widetilde{G}_t$ neighborhood, since $h(|f'|^g_1)^2 = \widetilde{H}(|f'|^{\widetilde{G}}_1)^2$ only varies with $O(\delta\widetilde{H}_t^{-1}(w_0))$ when $\widetilde{G}_{t,w_0}(w-w_0) \leq \delta^2$ is small enough. If $f(s,w) = 0$ for some (s,w) satisfying $G_{t,w_0}(w-w_0) < C\varepsilon^2$ and $|s-t| \leq \varepsilon h(w_0)$ and ε is small enough, we find that $H^{-1} \leq C(1+h(|f|^g_1)^2)$ at (t,w_0). Also Proposition 2.2 and condition (1.6) give for small ε that

$$\begin{aligned} c\widetilde{H}_t^{-1/2}H_t^{-1/2} &\leq h|f'|^g_1|f|^g_1 + h^{1/2}|f'|^g_1 \\ &\leq C_1 f' + C_2 H_t^{-1} \leq C_1 f' + C_3\widetilde{H}_t^{-1/2}H_t^{-1/2}/\sqrt{\mu} \end{aligned} \tag{3.4}$$

at (t,w_0), which gives $(c-C_3/\sqrt{\mu})\widetilde{H}_t^{-1/2}(w_0)H_t^{-1/2}(w_0) \leq C_1 f'(t,w_0)$. Thus, for large μ we obtain $c_1\widetilde{H}_t^{-1/2}(w_0)H_t^{-1/2}(w_0) \leq f'(t,w_0)$ for some $c_1 > 0$. Since

$$\begin{aligned} |f'(t,w)-f'(t,w_0)| &\leq C\widetilde{G}_{t,w_0}(w-w_0)^{1/2}\widetilde{H}_t^{-1}(w_0) \\ &\leq C'\varepsilon\widetilde{H}_t^{-1/2}(w_0)H_t^{-1/2}(w_0) \end{aligned} \tag{3.5}$$

when $G_{t,w_0}(w-w_0) \le \varepsilon^2 \ll \mu$, this implies $f' > 0$ when $G_{t,w'}(w-w') < \varepsilon^2 \ll \mu$, and again gives a contradiction with type B.

Finally, if $1 \ge \widetilde{H}_t^{-1}/3$ at w_0, we also get a contradiction for large enough μ, since $\mu/C_0 \le \mu H_t^{-1}/C_0 \le \widetilde{H}_t^{-1}$. Since $1 + h(|f'|_1^g)^2 + |f'| = \widetilde{H}^{-1}$, one of these cases must hold. This gives the proposition. □

Proposition 3.3. *There exists $\varepsilon_2 > 0$ such that if $\omega_{t,w_0}^\varepsilon$ is of type A with $\varepsilon \le \varepsilon_2$, we find that*

$$f'(s,w) \ge c(s,w)f(s,w) - C, \tag{3.6}$$

when $w \in \omega_{t,w_0}^\varepsilon$ and $|s-t| < \varepsilon h(w)$. Here C is constant, and $c \in C(\mathbf{R}, S(1,G))$ uniformly.

Proof. When $\omega_{t,w_0}^\varepsilon$ is of type A, we find that either $\widetilde{H}_t^{-1}(w) \le \mu H_t^{-1}(w)$ or $f'(t,w) \ge 0$ for any $w \in \omega_{t,w_0}^\varepsilon$. In the latter case we obtain (3.9) with $c = 0$, after perturbing in t. In the former, we consider the cases $h(|f|_1^g)^2 \gtrless \varrho(1+|f|)$ at (t,w_0), for large fixed $\varrho \ge 1$, which we shall determine later.

When $h(|f|_1^g)^2 \le \varrho(1+|f|)$ at (t,w_0), we find $H_t^{-1} \le (1+\varrho)(1+|f|)$ there. Since H_t^{-1} only varies with a fixed factor in the neighborhood, we find $H_t^{-1} \le C(1+\varrho)$ if $f = 0$ in the neighborhood $\omega_{t,w_0}^\varepsilon$. Then $\partial_t f \ge -\widetilde{H}_t^{-1} \ge -\mu H_t^{-1} \ge -\mu C(1+\varrho)$ in $\omega_{t,w_0}^\varepsilon$. We find that $\partial_t f(s,w) \ge -C_{\mu\varrho}$ when $w \in \omega_{t,w_0}^\varepsilon$ and $|s-t| < \varepsilon h(w)$ for ε small, thus we obtain the result in this case with $c \equiv 0$. If $f \not\equiv 0$ in the neighborhood we find $1 + |f| = 1 \pm f$. Now, by assumption we have

$$\begin{aligned}|f(t,w) - f(t,w_0)| &\le \varepsilon|f|_1^G(t,w_0) + C\varepsilon^2 H_t^{-1}(w_0)\\ &\le \varepsilon H_t^{-1/2}(w_0)\varrho^{1/2}(1+|f(t,w_0)|)^{1/2} + C_0\varepsilon^2 H_t^{-1}(w_0)\\ &\le C_1\varepsilon\varrho(1+|f(t,w_0)|) + C_1\varepsilon H_t^{-1}(w_0)\end{aligned} \tag{3.7}$$

when $w \in \omega_{t,w_0}^\varepsilon$, since $H^{1/2}|f|_1^G = h^{1/2}|f|_1^g$. If $\varepsilon\varrho$ is small enough, we find that $1 + |f(t,w_0)| \le C_2(1 + |f(t,w)| + \varepsilon H_t^{-1}(w_0))$ and

$$\begin{aligned}H_t^{-1}(w) &\le CH_t^{-1}(w_0) \le C(1+\varrho)(1+|f(t,w_0)|)\\ &\le C_3(1+\varrho)(1+|f(t,w)| + \varepsilon H_t^{-1}(w))\end{aligned} \tag{3.8}$$

when $w \in \omega_{t,w_0}^\varepsilon$. Thus, if we put $\varepsilon\varrho = c_0$, we find for small enough c_0 that $H_t^{-1} \le C_4(1+\varrho)(1 \pm f)$ in $\omega_{t,w_0}^\varepsilon$. We obtain $\partial_t f \ge -\widetilde{H}_t^{-1} \ge -\mu H_t^{-1} \ge -\mu C(1+\varrho)(1 \pm f)$ in $\omega_{t,w_0}^\varepsilon$, and since $\Delta(\partial_t f) = O(\varepsilon)$ and

$\Delta(\varrho f) = O(c_0)$ when $\Delta t = O(\varepsilon h)$, we obtain the result in this case too, with c constant.

In the case $h(|f|_1^g)^2 > \varrho(1+|f|)$ at (t, w_0), we find that $1 \le H^{-1} \le \frac{1+\varrho}{\varrho} h(|f|_1^g)^2$ at (t, w_0), which implies that $h(|f|_1^g)^2 > \varrho(1+|f|)/2$ at (s, w_0) when $|s-t| < \varepsilon h(w_0)$ for $\varepsilon = c_0/\varrho$ with small enough c_0. In fact, we find that $\Delta(\varrho f) = O(c_0)$ when $\Delta t = O(\varepsilon h)$. Since $\varrho = c_0/\varepsilon$ we find by Lemma 2.1 that, for large enough ϱ, $f\big|_s$ changes sign in the neighborhood $\omega_{t,w_0}^{C\varepsilon}$, for some constant C, when $|s-t| < \varepsilon h(w_0)$. By condition $(\overline{\Psi})$ we find $\partial_t f \ge 0$ when $f = 0$. By shrinking ε, we may introduce G_{t,w_0} orthonormal coordinates and then choose $Hf\big|_s$ as a local coordinate uniformly in $\omega_{t,w_0}^{C\varepsilon}$ when $|s-t| < \varepsilon h(w)$. By Taylor's formula and the fact that $|f'|_1^G = \sqrt{\widetilde{H}/H}|f'|_1^{\widetilde{G}} \le CH^{-1/2}\widetilde{H}^{-1/2} \le C\sqrt{\mu}H^{-1}$ when $w \in \omega_{t,w_0}^{\varepsilon}$ and $|s-t| < \varepsilon h(w_0)$, we find $\partial_t f \ge c_\mu f$ in that domain, for some $c_\mu \in S(1, G_t)$ which is continuous in t. □

Proposition 3.4. *There exists ε_3 such that if $f'(t,w) \ge 0$ in $\omega_{t,w_0}^{\varepsilon}$ for $\varepsilon \le \varepsilon_3$, and $\widetilde{H}^{-1}(t,w_0) \ge 3H^{-1}(t,w_0)$, then*

$$1/4 \le f'(t,w)/f'(t,w_0) \le 2 \tag{3.9}$$

when $w \in \omega_{t,w_0}^{\varepsilon/2}$.

Proof. We have by Taylor's formula

$$0 \le f'(t,w) \le f'(t,w_0) + \langle w-w_0, df'(t,w_0)\rangle + c_0\widetilde{H}_t^{-1}(w_0)\widetilde{G}_{t,w_0}(w-w_0) \tag{3.10}$$

when $G_{t,w_0}(w-w_0) \le \varepsilon^2$ is small enough, since $\widetilde{G}_{t,w_0} \le \frac{1}{3}G_{t,w_0}$. By choosing suitable w satisfying $G_{t,w_0}(w-w_0) = \varepsilon^2$, we find

$$\varepsilon h^{1/2}H^{-1/2}|f'|_1^g \le f'(t,w_0) + c_0\varepsilon^2 H^{-1} \qquad \text{at } (t,w_0) \tag{3.11}$$

since $\widetilde{H}^{-1}\widetilde{G} = H^{-1}G$ and $g_{w_0}(w-w_0) = \varepsilon^2 h(w_0)H^{-1}(t,w_0)$. Thus $f'(t,w_0) \le \varepsilon H^{-1}(t,w_0)$ implies

$$h^{1/2}|f'|_1^g \le (1+c_0\varepsilon)H^{-1/2} \qquad \text{at } (t,w_0), \tag{3.12}$$

which gives $\widetilde{H}^{-1} \le ((1+c_0\varepsilon)^2 + \varepsilon + 1)H^{-1} < 3H^{-1}$ at (t,w_0) when ε is small enough, giving a contradiction in this case. Thus, we may assume that $f'(t,w_0) > \varepsilon H^{-1}(t,w_0)$, and then (3.10) gives

$$|\langle w-w_0, df'(t,w_0)\rangle| < (1+c_0\varepsilon)f'(t,w_0) \tag{3.13}$$

when $G_{t,w_0}(w-w_0) \le \varepsilon^2$. If $G_{t,w_0}(w-w_0) \le \varepsilon^2/4$, we find that

$$|\langle w-w_0, df'(t,w_0)\rangle| < (1+c_0\varepsilon)f'(t,w_0)/2, \tag{3.14}$$

and

$$\begin{aligned} c_0\widetilde{H}_t^{-1}(w_0)\widetilde{G}_{t,w_0}(w-w_0) &= c_0H_t^{-1}(w_0)G_{t,w_0}(w-w_0) \\ &\le c_0\varepsilon^2 H_t^{-1}(w_0)/4 < c_0\varepsilon f'(t,w_0)/4. \end{aligned} \tag{3.15}$$

By (3.10) this implies

$$\begin{aligned} f'(t,w_0)(1-(1+c_0\varepsilon)/2 - c_0\varepsilon/4) &< f'(t,w) \\ &< f'(t,w_0)(1+(1+c_0\varepsilon)/2 + c_0\varepsilon/4). \end{aligned} \tag{3.16}$$

We find that $f'(t,w_0)/4 \le f'(t,w) \le 2f'(t,w_0)$ when $G_{t,w_0}(w-w_0) < \varepsilon^2/4$ and $3c_0\varepsilon/4 \le 1/4$. □

Corollary 3.5. *If $\omega^\varepsilon_{t,w_0}$ is of type A for $\varepsilon \le \varepsilon_3$ and $\mu \ge \mu_2$, then there exists a positive constant C such that*

$$|f'(s,w)| \le 2f'(t,w_0) + C\mu H_t^{-1}(w_0) \tag{3.17}$$

when $w \in \omega^{\varepsilon/2}_{t,w_0}$ and $|s-t| \le \varepsilon h(w_0)$.

Proof. Since the terms only change with a bounded term when $\Delta t = O(h(w_0))$, we only have to consider $s = t$. Since $\omega^\varepsilon_{t,w_0}$ is of type A, we have either $\widetilde{H}^{-1} \le \mu H^{-1}$ or $\widetilde{H}^{-1}(t,w) > \mu H^{-1}(t,w)$ for some $w \in \omega^\varepsilon_{t,w_0}$ and then $f' \ge 0$ in $\omega^\varepsilon_{t,w_0}$.

In the latter case, we find $\widetilde{H}_t^{-1} > \mu H_t^{-1}/C_0$ in $\omega^\varepsilon_{t,w_0}$ by Lemma 3.1 if $\varepsilon^2/\mu \le c_0$. If we choose $\mu_2 = 3C_0$ we then find $\widetilde{H}_t^{-1} > 3H_t^{-1}$ in $\omega^\varepsilon_{t,w_0}$ when $\mu \ge \mu_2$. Since $f' \ge 0$ in $\omega^\varepsilon_{t,w_0}$, we find by Proposition 3.4 that

$$0 \le f'(t,w) \le 2f'(t,w_0) \qquad \text{when } w \in \omega^{\varepsilon/2}_{t,w_0}. \tag{3.18}$$

In the case $\widetilde{H}^{-1} \le \mu H^{-1}$ we find

$$|f'(t,w)| \le \widetilde{H}^{-1}(t,w) \le \mu H^{-1}(t,w) \le C_1\mu H^{-1}(t,w_0) \tag{3.19}$$

when $w \in \omega^\varepsilon_{t,w_0}$. By combining these estimates, we obtain the corollary with $C = 3C_1$. □

In what follows we assume that we have made a scale change in t to make (1.10) hold with $C\varepsilon \le \frac{1}{4}$. Then, we fix ε and μ in (3.2) so that the conclusions of Lemma 3.1, Propositions 2.2 and 3.2–3.4 and Corollary 3.5 hold. Thus, by Proposition 2.2 we have

$$2h|f'|_1^g|f|_1^g + \varrho h^{1/2}|f'|_1^g \le f' + C(h(|f|_1^g)^2 + 1) \qquad \text{in } \omega_{t,w_0}^\varepsilon \tag{3.20}$$

in the case $f(t,w) = 0$ for some $w \in \omega_{t,w_0}^\varepsilon$.

Now we shall make the preparation. Since we have problems estimating f' in the type B cases, we shall redefine the metric and the symbol in these cases, to make them monotone. Now $f \ne 0$ in the type B cases by Proposition 3.2, which is essential for the construction. We shall consider the neighborhoods of type B in which $\pm f > 0$, and we call these neighborhoods type $B^\pm$.

Let $\Xi_t^\pm = \bigcup_{B^\pm} \omega_{t,w_0}^\varepsilon$ be the union of neighborhoods of type $B^\pm$ for fixed $t \in [-1,1]$, and let

$$\Theta_t^+ = \bigcup_{-1\le s\le t} \Xi_s^+, \quad \Theta_t^- = \bigcup_{t\le s\le 1} \Xi_s^-. \tag{3.21}$$

Then $\Theta_s^+ \subseteq \Theta_t^+$ and $\Theta_s^- \supseteq \Theta_t^-$ when $-1 \le s \le t \le 1$, and by $(\overline{\Psi})$ we find $\pm f\big|_t \ge 0$ on $\Theta_t^\pm$. We also let

$$\Theta^\pm = \bigcup_{t\in[-1,1]} \{t\} \times \Theta_t^\pm. \tag{3.22}$$

In order to make the cut-offs, we need to take $dt^2/h^2 + G$ neighborhoods of the sets $\Theta^\pm$. First, we define

$$\Gamma_t^\pm(\delta) = \left\{ w : \exists w' \in \Theta_t^\pm \text{ such that } G_{t,w'}(w - w') < \delta^2 \right\}, \tag{3.23}$$

for some $\delta > 0$ to be chosen later. Then, we define

$$\Omega^\pm(\delta) = \left\{ (t,w) : w \in \Gamma_s^\pm(\delta) \quad \text{for some } \pm(s-t) < \delta h(w) \right\} \tag{3.24}$$

which is a $dt^2/h^2 + G$ neighborhood of $\Theta^\pm$. We let $\Omega_t^\pm(\delta) = \Omega^\pm(\delta)\big|_t$, then since h is constant in t we find $\Omega_s^\pm(\delta) \subseteq \Omega_t^\pm(\delta)$ when $\pm(t-s) \ge 0$. Since the condition of being of type B involves strict inequalities, and the metric is continuous, we find that $\Theta^\pm$ and $\Omega^\pm$ are open sets. Observe that

(3.25)

$$\bigcup_{s\le t} \Gamma_s^+(\delta) = \left\{ w : \exists w' \in \Theta_t^+ \text{such that} \inf_{t_1(w')<s\le t} G_{s,w'}(w-w') < \delta^2 \right\}$$

where $t_1(w') = \inf\{r : w' \in \Theta_r^+\} < t$ since $w' \in \Theta_t^+$, and correspondingly for $\bigcup_{t \le s} \Gamma_s^-(\delta)$. This implies that

$$(3.26)\quad \Omega_t^+(\delta) = \Big\{w : \quad \exists s < t + \delta h(w) \quad \exists w' \in \Theta_s^+, \\ \text{such that} \inf_{t_1(w') < s < t + \varepsilon h(w)} G_{s,w'}(w - w') < \delta^2\Big\}$$

We shall later prove that $\inf_{t_1(w') < s \le t} G_s$ gives a slowly varying metric on Θ^+, and correspondingly on Θ^- (see Proposition 4.2). The next proposition shows that the positivity of f on Θ^+ is extended to $\Omega^+(\delta)$ for small δ.

Proposition 3.6. *There exists* $0 < \delta_0$ *such that* $\pm f \ge 0$ *on* $\Omega^\pm(\delta_0)$, *and*

$$(3.27)\qquad H^{-1} \le c_0(1 \pm f)$$

on $\Omega^\pm(\delta_0)$.

Proof. We shall prove that $\pm f(t,w) \ge 0$ when $w \in \omega_{t,w_0}^\delta$ and $w_0 \in \bigcup_{\pm(s-t)\le 0} \Gamma_s^\pm(\delta)$ for small enough $\delta > 0$. Then we find $H^{-1} < (1 + \frac{2}{\delta})(1 + |f|) = (1 + \frac{2}{\delta})(1 \pm f)$ at (t, w_0) for small δ. In fact, by Lemma 2.1 we find that $f\big|_t$ changes sign in ω_{t,w_0}^δ, if $h(|f|_1^g)^2 \ge \frac{2}{\delta}(1 + |f|)$ at (t, w_0) for δ small enough. Since f and H^{-1} only vary with a term that is $O(\delta)$ when $\Delta t = O(\delta h)$ we immediately obtain (3.27) and the proposition for small enough δ. As before, we only have to consider the positive case.

Thus assume that $f(t,w) < 0$ for some $w \in \omega_{t,w_0}^\delta$ when $w_0 \in \bigcup_{s \le t} \Gamma_s^+(\delta)$. Since $w_0 \in \bigcup_{s\le t} \Gamma_s^+(\delta)$, we find that there exists $w' \in \Theta_t^+$ for which $\inf_{t_1(w') < r \le t} G_{r,w'}(w_0 - w') \le \delta^2$, where as before $t_1(w') = \inf\{r : w' \in \Theta_r^+\} < t$. Clearly, for any $c > 1$ we find that $w' \in \Theta_\sigma^+$ for some $\sigma \le t$ for which $G_{\sigma,w'}(w_0 - w') < c\delta^2$. By definition, $w' \in \Xi_s^+$ for some $s \le \sigma$. Then $f(s, w') > 0$, so we find from $(\overline{\Psi})$ that $f(\tau, w') \ge 0$ when $\tau \ge s$, and $f(\tau, w) \le 0$ when $\tau \le t$. We have $G_{t,w_0}(w - w_0) < \delta^2$ and $G_{\sigma,w'}(w_0 - w') < c\delta^2$. By the slow variation we have $G_{\sigma,w_0}(w_0 - w') < C\delta^2$ when δ is small enough. Now by considering the cases where $H(\sigma, w_0) \gtrless H(t, w_0)$ (i.e., $G_{\sigma,w_0} \gtrless G_{t,w_0}$), we obtain that w and $w' \in \omega_{r,w_0}^{c\delta}$ for some $r \in [s,t]$ and $c \ge 1$. By shrinking δ we obtain from Lemma 3.7 below that

$$(3.28)\qquad 1/C_0 \le G_{\tau,w_0}/G_{r,w_0} \le C_0 \qquad \text{for } s \le \tau \le t.$$

We find $G_{s,w_0} \le CG_{\sigma,w_0}$, which gives for small δ that $G_{s,w'} \le CG_{s,w_0}$, and thus

$$\begin{aligned}(3.29)\quad G_{s,w'}(w-w') &\le CG_{s,w_0}(w-w') \\ &\le 2C'(G_{r,w_0}(w_0-w')+G_{r,w_0}(w-w_0)) \le C_2\delta^2.\end{aligned}$$

When $\delta \ll \varepsilon_1$ we obtain that $w \in \omega^{\varepsilon_1}_{s,w'}$, which by Proposition 3.2 gives $f(s,w) > 0$. By condition $(\overline{\Psi})$ this contradicts the assumption that $f(t,w) < 0$, and proves the proposition. □

Lemma 3.7. *Assume that f satisfies condition $(\overline{\Psi})$ and condition (3.20). There exist $\delta_0 > 0$ and $C_0 > 0$ such that if $f(r,w_1) > 0$ and $f(t,w_2) < 0$, where $r < t$, w_1, $w_2 \in \omega^{\delta}_{s,w_0}$ for some $s \in [r,t]$ and $\delta \le \delta_0$, then*

$$(3.30)\qquad 1/C_0 \le G_{\sigma,w}/G_{s,w} \le C_0$$

for $w \in \omega^{\delta}_{s,w_0}$ and $\sigma \in [r,t]$.

Proof. First we note that by condition $(\overline{\Psi})$ we find $f(\sigma,w_1) \ge 0$ and $f(\sigma,w_2) \le 0$ for $\sigma \in [r,t]$. We shall prove (3.30) for $r \le \sigma \le s$, by choosing $-t$ as new t variable and $-f$ as new symbol, we obtain (3.30) for $s \le \sigma \le t$.

For $0 < \delta \le \varepsilon$ we put

$$(3.31)\qquad \sigma_\delta = \inf\left\{\sigma : G_{\tau,w_0} \le \frac{\varepsilon^2}{\delta^2}G_{s,w_0} \quad \text{for } \sigma \le \tau \le s\right\} \le s.$$

Observe that σ_δ decreases as δ decreases, and that by continuity we have $G_{\sigma,w_0}(w_j - w_0) < \varepsilon^2$ when $\sigma_\delta \le \sigma \le s$, thus $w_j \in \omega^{\varepsilon}_{\sigma,w_0}$ then. This implies that $f(\sigma,w) = 0$ for some $w \in \omega^{\varepsilon}_{\sigma,w_0}$ when $\max(r,\sigma_\delta) \le \sigma \le s$. By condition (3.20) we find that

$$(3.32)\qquad 2h|f'|^g_1|f|^g_1 \le f' + CH^{-1}$$

at (σ, w_j) for $j = 1$, 2, and $\max(r,\sigma_\delta) \le \sigma \le s$, since $\varepsilon \le \varepsilon_0$. Then Proposition 2.3 implies that

$$(3.33)\qquad \begin{aligned} G_{s,w_1} &\le CG_{\sigma,w_1} \\ G_{\sigma,w_2} &\le CG_{s,w_2}\end{aligned} \qquad \text{for } \max(r,\sigma_\delta) \le \sigma \le s.$$

By the slow variation, we find $1/C_1 \le G_{\sigma,w}/G_{s,w} \le C_1$ for $w \in \omega^{\delta}_{s,w_0}$ when $\max(r,\sigma_\delta) \le \sigma \le s$, since then $G_{\sigma,w_0}(w-w_0) < \varepsilon^2$. In the case $\sigma_\delta \le r$, we obtain (3.30).

If we choose $\delta \le \varepsilon/\sqrt{2C_1}$ we obtain a contradiction when $\sigma_\delta > r$, since we find that

$$G_{\sigma_\delta,w_0} \le C_1 G_{s,w_0} \le \frac{\varepsilon^2}{2\delta^2} G_{s,w_0}. \tag{3.34}$$

Thus $G_{\tau,w_0} \le \frac{\varepsilon^2}{\delta^2} G_{s,w_0}$ for $\sigma \le \tau \le \sigma_\delta$ for some $\sigma < \sigma_\delta$. This contradicts the definition of σ_δ, proves that $\sigma_\delta \le r$ and gives the lemma. □

4. The modified metric

Next, we choose $\delta > 0$ small enough in the definitions of $\Gamma_t^\pm(\delta)$ and $\Omega_t^\pm(\delta)$, so that the conclusions in Proposition 3.6 hold and that G_t and g only vary with a fixed factor in the neighborhoods $\omega_{t,w}^\delta$ defined in (3.1), and when $\Delta t = O(\delta h(w))$. In the following, except when explicitly needed, we shall suppress the dependency on δ and write $\Omega^\pm = \Omega^\pm(\delta)$.

Now, $\pm H^{-1}(t,w)$ is neither C^∞ nor monotone on $\Omega^\pm$. We shall modify $\pm H^{-1}$ so that it is non-decreasing and C^∞ on $\Omega^\pm$. Let $t_0(w) = \inf\{\, s : w \in \Omega_s^+ \,\} < t$ for $w \in \Omega_t^+$ and $t_0(w) = \sup\{\, s : w \in \Omega_s^- \,\} > t$ for $w \in \Omega_t^-$. One way to get a decreasing metric on Ω^+ is by using the metric $\inf_{t_0(w)<s\le t} G_s$, but we shall instead use an equivalent metric. Recall that $H^{-1} \le c_0(1 \pm f)$ on $\Omega^\pm$ by Proposition 3.6. Therefore, we put

$$H_0^{-1}(t,w) = \begin{cases} \sup_{t_0(w)<s\le t}(1+f(s,w)) & \text{on } \Omega_t^+ \\ \sup_{t\le s<t_0(w)}(1-f(s,w)) & \text{on } \Omega_t^- \\ H^{-1}(t,w) & \text{elsewhere,} \end{cases} \tag{4.1}$$

put $G_0 = H_0 g/h = H_0 G/H$ and $G_{0,t} = G_0\big|_t$. Then $\pm H_0^{-1}(t,w)$ is increasing on $\Omega^\pm$, but is not C^∞. Since $H^{-1} \le c_0(1 \pm f)$ on $\Omega^\pm$ by Proposition 3.6, we find that $H^{-1} \le C_0 H_0^{-1}$. We shall later regularize H_0^{-1}, but first we need the following proposition.

Proposition 4.1. *We have that $dt^2/h^2 + G_{0,t}$ is slowly varying.*

Proof. When $|\Delta t| \le \varepsilon h$ we find that f and H^{-1} only vary with a term which is $O(\varepsilon)$, for small ε. Since $1 \pm f \cong H^{-1}$ on $\Omega^\pm$, we only have to consider t constant.

Assuming that $G_{0,t,w}(w'-w) < \varrho^2$, we shall prove that $G_{0,t,w'} \le C G_{0,t,w}$ or equivalently $H_0^{-1}(t,w) \le C H_0^{-1}(t,w')$. We shall assume ϱ chosen small enough so that g only varies with a fixed factor when

$G_{0,t,w}(w'-w) < \varrho^2$. We shall consider the cases where $(t,w) \in \Omega^\pm$ or not.

First we consider the case where $(t,w) \notin \Omega^\pm$. Then $G_{0,t,w} = G_{t,w}$, and we find $G_{t,w'} \le CG_{t,w}$ for small enough ϱ, by the uniform slow variation of G. Since $H^{-1} \le C_0 H_0^{-1}$ we find $G_0 \le C_0 G$, which proves the estimate in this case.

It remains to consider the case where $(t,w) \in \Omega^\pm$, and as before we shall only consider $(t,w) \in \Omega^+$. We may assume that $H_0^{-1}(t,w) \ge 2$, otherwise $G_{0,t,w'}/G_{0,t,w} < 2C$. Thus we may assume $2 \le H_0^{-1}(t,w) = 1 + f(s,w)$ for some $t_0(w) < s \le t$ and $w \in \Omega_s^+$. This gives $1 \le f(s,w)$ and implies that $H^{-1}(t,w) \le C_0 H_0^{-1}(t,w) \le 2C_0 f(s,w)$. We obtain

$$f(s,w') \ge f(s,w) - C\varrho H^{-1}(s,w) \ge f(s,w)/2 \tag{4.2}$$

when $G_{s,w}(w'-w) < \varrho^2$ is small enough. In fact, $H^{-1}(s,w) \le c_0(1 + f(s,w))$ when $(s,w) \in \Omega^+$ by Proposition 3.6. Since $f(s,w') > 0$, we find $w' \notin \Omega_s^-$. If $w' \in \Omega_s^+$ we find $H_0^{-1}(t,w') \ge 1 + f(s,w')$ which proves the estimate in this case. If not, we consider the cases $w' \in \Omega_t^+$ or not. If $w' \notin \Omega_t^+$ we find that $I = [s,t] \times \{w'\}$ is disjoint from Ω^+. By Proposition 3.3 we find that $\partial_t f \ge -c(f+1)$ on I, since $\omega^\varepsilon_{\sigma,w'}$ cannot be of type B when $s \le \sigma \le t$. This implies that $H_0^{-1}(t,w') \ge f(t,w') + 1 \ge c_1(f(s,w')+1)$ for some positive constant c_1, which proves the estimate in this case. If $w' \in \Omega_t^+$ we find that $I_0 = [s, t_0(w')] \times \{w'\}$ is disjoint from Ω^+. As before, we find that $H_0^{-1}(t,w') \ge f(t_0(w'),w') + 1 \ge c_1(f(s,w')+1)$ for some positive constant c_1, which proves the estimate in the final case. □

The following proposition shows that the metric G_0 is equivalent to G on $\Omega^\pm \setminus \Theta^\pm$.

Proposition 4.2. *We find that*

$$cH_0^{-1} \le 1 \pm f \le H_0^{-1} \qquad \text{on } \Omega^\pm \setminus \Theta^\pm \tag{4.3}$$

for some positive constants c and C. We also find that $\Omega_t^\pm$ is a union of $G_{0,t}$ balls with radius $\ge c\delta$, for some $c > 0$, in which $G_{0,t}$ only varies with a fixed constant. There exists $c_3 > 0$ with the property that if $0 < c_1 < c_2 < c_3$, then there exists $\lambda > 0$ such that $\{\,(t,w) : \delta_0(t,w) < \lambda\delta\,\} \subseteq \Omega^\pm(c_2\delta)$, where

$$\delta_0(t,w) = \inf\left\{\, \sqrt{G_{0,t,w}(w_0 - w)} : w_0 \in \Omega_t^\pm(c_1\delta) \,\right\} \tag{4.4}$$

is the G_0 distance to $\Omega^\pm(c_1\delta)$.

Proof. The upper estimate in (4.3) is trivial. As before, we shall only consider $(t,w) \in \Omega^+ \setminus \Theta^+$. We have by definition $H_0^{-1}(t,w) = (1 + f(s,w))$ for some $s \le t$ such that $w \in \Omega_s^+ \setminus \Theta_s^+$, which implies $I = [s,t] \times w \subset \Omega^+ \setminus \Theta^+$. Thus, since $f \ge 0$ on I, Proposition 3.3 gives $\partial_t f \ge -c(f+1)$ on I, where c is constant. This implies that

$$f(t,w) + 1 \ge c(f(s,w)+1) = cH_0^{-1}(t,w), \tag{4.5}$$

and proves (4.3). In fact, $\omega_{\sigma,w}^{\varepsilon}$ is not of type B when $\sigma \in [s,t]$. Now assume $(t,w) \in \Omega^+(\delta)$, we shall prove that there exists $w_0 \in \Theta_t^+$ such that

$$w \in \{\, z : G_{0,t,w_0}(z - w_0) < c^2\delta^2 \,\} \subseteq \Omega_t^+(\delta) \tag{4.6}$$

for some positive constant c. Since $w \in \Gamma_s^+$ for some $s < t + \delta h(w)$, we find by (3.25) that $\exists w_0 \in \Theta^+_{t+\delta h(w)}$ such that

$$w \in B_{t,w_0}^{\delta} = \left\{ z : \inf_{t_1(w_0) < r < t+\delta h(w)} G_{r,w_0}(z - w_0) < \delta^2 \right\} \subseteq \Omega_t^+(\delta) \tag{4.7}$$

where $t_1(w_0) = \inf\{\, r : w_0 \in \Theta_r^+ \,\}$ as before. Since $H(t + \delta h(w), w) \cong H(t,w)$ and $G_s = H_s G_t / H_t$, we find that

$$B_{t,w_0}^{\delta} = \left\{ z : \inf_{t_1(w_0) < r \le t} G_{r,w_0}(z - w_0) < c^2\delta^2 \right\} \subseteq \Omega_t^+(\delta) \tag{4.8}$$

for some constant c. Now (4.3) implies that

$$\sup_{t_0(w_0) < \sigma \le r} (1 + f(\sigma, w_0)) = H_0^{-1}(r, w_0) \le C(1 + f(r, w_0)) \le CH^{-1}(r, w_0) \tag{4.9}$$

when $t_0(w_0) < r < t_1(w_0)$. When $r = t_1(w_0)$ we obtain by continuity that

$$\sup_{t_0(w_0) < s \le t} (1 + f(s, w_0)) \le C \sup_{t_1(w_0) < s \le t} (1 + f(s, w_0)) \tag{4.10}$$

which implies that

$$H_0^{-1}(t, w_0) \le C' \sup_{t_1(w_0) < s \le t} (1 + f(s, w_0)) \le C' \sup_{t_1(w_0) < s \le t} H^{-1}(s, w_0) \tag{4.11}$$

for $w_0 \in \Theta_t^+$. On the other hand we have

$$\sup_{t_1(w_0) < s \le t} H^{-1}(s, w_0) \le C \sup_{t_1(w_0) < s \le t} (1 + f(s, w_0)) \le C' H_0^{-1}(t, w_0). \tag{4.12}$$

Thus, we find for some constant C'' that

$$G_{0,t,w_0}/C'' \le \inf_{t_1(w_0)<s\le t} G_{s,w_0} \le C'' G_{0,t,w_0} \tag{4.13}$$

when $w_0 \in \Theta_t^+$. This gives

$$w \in B_{t,w_0}^{\delta} = \{\, z : G_{0,t,w_0}(z - w_0) < c_0^2\delta^2 \,\} \subseteq \Omega_t^+(\delta) \tag{4.14}$$

for some $c_0 > 0$. By the choice of δ we obtain that $G_{0,t}$ only varies with a fixed factor in the balls.

Assume that $\delta_0(t,w) < \lambda\delta$. Then there exists $(t, w_0) \in \Omega^+(c_1\delta)$ such that $G_{0,t,w}(w_0 - w) < \lambda^2\delta^2$. By definition we have $\overline{G}_{t_0,w'}(w_0 - w') \le c_1^2\delta^2$ where $w' \in \Theta_{t_0}^+$, $t_0 = t + c_1\delta h(w')$ and $\overline{G}_{t_0,w'} = \inf_{t_1(w')<s\le t_0} G_{s,w'}$. By (4.13) we find that $G_{0,t_0,w'}(w_0 - w') \le Cc_1^2\delta^2$, thus when $Cc_1^2\delta^2 < Cc_3^2\delta^2$ and λ are small enough, we find that $G_{0,t_0,w'} \le C' G_{0,t_0,w_0} \le C'' G_{0,t_0,w} \le C''' G_{0,t,w}$. By (4.13) we obtain $\overline{G}_{t_0,w'}(w - w')^{1/2} \le \overline{G}_{t_0,w'}(w - w_0)^{1/2} + \overline{G}_{t_0,w'}(w_0 - w')^{1/2} \le C_0\lambda\delta + c_1\delta \le c_2\delta$, if $\lambda \ll 1$. This implies that $(t,w) \in \Omega^+(c_2\delta)$ and completes the proof of the proposition. □

We find from (4.13) that $G_{0,t,w}$ is equivalent to $\inf_{t_1(w)<s\le t} G_{s,w}$ on Θ^+, and it is equivalent to G on $\Omega^+\setminus\Theta^+$ by (3.27) and (4.3). Thus, $\inf_{t_1(w)<s\le t} G_{s,w}$ is slowly varying on Θ^+.

Now, $\pm H_0^{-1}$ is increasing on Ω^+ but not C^∞. We shall next regularize $\pm H_0^{-1}$ in the t variables, so that it is still increasing. Let $0 \le \phi(s) \in C_0^\infty$ such that $\int \phi(s)\,ds = 1$ and $\phi(s) = 0$ when $s \ge 1$. For $\lambda > 0$ we put

$$\mathcal{H}_0^{-1}(t,w) = h^{-1}(w)\lambda^{-1} \int \phi((t-s)/\lambda h(w)) H_0^{-1}(s,w)\,ds, \tag{4.15}$$

and $\mathcal{G}_0 = \mathcal{H}_0 g/h = \mathcal{H}_0 G_0/H_0$. We are going to use the neighborhoods $\Omega^\pm(\delta/k\varrho)$ for $k \in \mathbf{Z}_+$ and $\varrho \ge 1$.

Lemma 4.3. *We find that $t \mapsto \mathcal{H}_0(t,w)$ is C^∞ and that $\partial_t \mathcal{H}_0^{-1} \le Ch^{-1}\mathcal{H}_0^{-1}$. We have $1/C \le H_0/\mathcal{H}_0 \le C$ and*

$$\pm\partial_t \mathcal{H}_0^{-1} \ge 0 \tag{4.16}$$

on $\Omega^\pm(\delta/2)$, when $\lambda = c\delta$ and $c > 0$ is small enough. For any $\varrho \ge 1$ there exists c_ϱ such that

$$|\partial_t \mathcal{H}_0^{-1}| \le 2f' + C\mu\mathcal{H}_0^{-1} \tag{4.17}$$

on $\complement(\Omega^+(\delta/9\varrho) \cup \Omega^-(\delta/9\varrho))$ when $\lambda = c_\varrho\delta$.

Proof. We have

$$(4.18)\quad \partial_t \mathcal{H}_0^{-1}(t,w) = \lambda^{-2}h^{-2}(w)\int \phi'((t-s)/\lambda h(w))H_0^{-1}(s,w)\,ds$$

which immediately gives

$$(4.19)\qquad |\partial_t \mathcal{H}_0^{-1}| \le Ch^{-1}\mathcal{H}_0^{-1}.$$

Since H_0^{-1} only varies with a term that is $O(\lambda)$ when $|s-t| \le \lambda h(w)$ and λ is small, we easily get the first statements. Recall that

$$(4.20)\quad \Omega^{\pm}(\delta) = \left\{\, (t,w) : w \in \Gamma_s^{\pm}(\delta) \quad \text{for some } \pm(s-t) < \delta h(w) \,\right\}.$$

Thus for every $0 < c_0 < c_1 \le 1$ there exists $\varrho > 0$ such that

$$(4.21)\quad \left\{\, (t,w) : (s,w) \in \Omega^{\pm}(c_0\delta) \text{ for some } \pm(s-t) \le \varrho h(w) \,\right\} \subseteq \Omega^{\pm}(c_1\delta).$$

We find that $|\partial_t H_0^{-1}| \le |f'|$ almost everywhere in t, for fixed w. In fact, $t \mapsto f(t,w)$ may only change sign once. Thus in $\complement(\Omega^+(\delta/9\varrho) \cup \Omega^-(\delta/9\varrho))$ we find by Corollary 3.5 that

$$(4.22)\quad \begin{aligned} |\partial_t \mathcal{H}_0^{-1}(t,w)| &\le h^{-1}(w)\lambda^{-1}\int \phi((t-s)/\lambda h(w))|\partial_t H_0^{-1}(s,w)|\,ds \\ &\le h^{-1}(w)\lambda^{-1}\int \phi((t-s)/\lambda h(w)) \\ &\qquad \times (2f'(t,w) + C\mu H^{-1}(t,w))\,ds \\ &= 2f'(t,w) + C\mu H^{-1}(t,w) \end{aligned}$$

when $\lambda = c\delta$ and $c > 0$ is small enough. In fact, when c is sufficiently small, we find by (4.21) that the integrand in (4.15) is supported outside $\Theta^{\pm}$, which implies that $\omega_{t,w}^{\varepsilon}$ is of type A.

Now $t \mapsto \pm H_0^{-1}(t,w)$ is non-decreasing on $\Omega^{\pm}(\delta)$. Thus, when $(t,w) \in \Omega^{\pm}(\delta/2)$ we find for small $\lambda = c\delta$ that $t \mapsto \pm H_0^{-1}$ is non-decreasing in the support of the integrand by (4.21). This gives $t \mapsto \pm\mathcal{H}_0^{-1}$ is non-decreasing on $\Omega^{\pm}(\delta/2)$, which implies that $\pm\partial_t\mathcal{H}_0^{-1} \ge 0$ on $\Omega^{\pm}(\delta/2)$. □

We shall next regularize $\pm\mathcal{H}_0^{-1}$ in w. Let $0 \le \phi(s) \in C^\infty$ such that $\phi(s) = 1$ when $s \le \frac{1}{2}$, $\phi(s) = 0$ when $s \ge 1$ and $\phi' \le 0$. Put

$$(4.23)\qquad E_\lambda(t,w) = h^{-n}(w)|g_w|^{1/2}\int_{T^*\mathbf{R}^n} \phi(\mathcal{G}_{0,t,z}(w-z)/\lambda^2)\,dz.$$

Observe that the integrand is supported where $\mathcal{G}_{0,t,z}(w-z) \leq \lambda^2$, which implies that $\mathcal{G}_{0,t,w}(w-z) \leq C\lambda^2$ for small λ, by the slow variation.

Proposition 4.4. *For $\lambda = c\delta$ where $0 < c$ is small enough, we find that $E_\lambda \in S(\mathcal{H}_0^{-n}, \mathcal{G}_0)$ with seminorms depending on λ, and*

$$1/C_\lambda \leq E_\lambda(t,w)\mathcal{H}_0^n(t,w) \leq C_\lambda \qquad \forall t, w. \tag{4.24}$$

We find $|\partial_t E_\lambda| \leq Ch^{-1}E_\lambda$, $\forall$ t, w. We also find that for small λ

$$\pm\partial_t E_\lambda \geq c_\lambda C_0 E_\lambda \qquad \text{at } (t,w) \tag{4.25}$$

if $\pm\partial_t\mathcal{H}_0^{-1}(t,z) \geq -C_0\mathcal{H}_0^{-1}(t,z)$ when $\mathcal{G}_{0,t,w}(z-w) \leq c\lambda^2$ for sufficiently large $c \geq 1$. For sufficiently large ϱ, there exists $c_\varrho > 0$ such that $\pm\partial_t E_\lambda \geq 0$ in $\Omega^\pm(\delta/3\varrho)$ and $|\partial_t E_\lambda| \leq C'(f'\mathcal{H}_0 + C\mu)E_\lambda$ in $\complement\,(\Omega^+(\delta/8\varrho) \cup \Omega^-(\delta/8\varrho))$, when $\lambda = c_\varrho\delta$.

Proof. Since $\mathcal{G}_0 = \mathcal{H}_0 g/h$, we find

(4.26)
$$E_\lambda(t,w) \leq h^{-n}(w)|g_w|^{1/2}\int_{\mathcal{G}_{0,t,w}(w-z)\leq C\lambda^2} dz \leq C^n\lambda^{2n}\mathcal{H}_0^{-n}(t,w)$$

for small λ. Then we also have $\mathcal{G}_{0,t,z} \leq C\mathcal{G}_{0,t,w}$ in the support of the integrand, which gives

$$\begin{aligned}E_\lambda(t,w) &\geq h^{-n}(w)|g_w|^{1/2}\int_{\mathcal{G}_{0,t,w}(w-z)\leq\lambda^2/2C} dz \\ &\geq C^{-n}2^{-n}\lambda^{2n}\mathcal{H}_0^{-n}(t,w)\end{aligned} \tag{4.27}$$

and proves that $1/C_\lambda \leq E_\lambda(t,w)\mathcal{H}_0^n(t,w) \leq C_\lambda$ for small λ.

Next we prove that $E_\lambda \in S(\mathcal{H}_0^{-n}, \mathcal{G}_0)$. By the assumptions on the metric g, we have $h^{-n}|g|^{1/2} \in S(h^{-n}|g|^{1/2}, g)$. Thus, we only have to differentiate under the integral sign. Take $w_0 \in T(T^*\mathbf{R}^n)$ such that $\mathcal{G}_{0,t,w}(w_0) \leq 1$, which implies $\mathcal{G}_{0,t,z}(w_0) \leq C$ for small λ. Then, differentiating with respect to w_0 for small λ give terms

(4.28)
$$\begin{aligned}h^{-n}(w)|g_w|^{1/2}\int_{T^*\mathbf{R}^n} \phi'(\mathcal{G}_{0,t,z}(w-z)/\lambda^2)2\mathcal{G}_{0,t,z}(w-z,w_0)/\lambda^2\,dz \\ = O(H_0^{-n}(t,w))\end{aligned}$$

where $\mathcal{G}_{0,t,z}(w-z,w_0) = O(\mathcal{G}_{0,t,z}(w-z)^{1/2})$ is the polarized form of $\mathcal{G}_{0,t,z}$. Since repeated differentiations produce similar terms, we obtain $E_\lambda \in S(\mathcal{H}_0^{-n}, \mathcal{G}_0)$ with seminorms depending on λ.

It remains to estimate $\partial_t E_\lambda$. Since $\phi' \le 0$ we find for small enough λ that

(4.29)

$$
\begin{aligned}
\pm\, \partial_t E_\lambda(t,w) &= \pm h^{-n}(w)|g_w|^{1/2} \int_{T^*R^n} \phi'(\mathcal{G}_{0,t,z}(w-z)/\lambda^2)\partial_t \mathcal{G}_{0,t,z}(w-z)/\lambda^2\, dz \\
&\ge h^{-n}(w)|g_w|^{1/2} \int_{T^*R^n} \phi'(\mathcal{G}_{0,t,z}(w-z)/\lambda^2) C_0 \mathcal{G}_{0,t,z}(w-z)/\lambda^2\, dz \\
&\ge -C_0 C E_\lambda(t,w)
\end{aligned}
$$

if $\mp\partial_t \mathcal{G}_0 = \mp\partial_t \mathcal{H}_0 g/h = \pm \partial_t \mathcal{H}_0^{-1}\mathcal{H}_0^2 g/h \ge -C_0\mathcal{G}_0$ in the support of the integrand. This proves (4.25). We also have $|\partial_t \mathcal{G}_0| = |\partial_t \mathcal{H}_0 g/h| \le Ch^{-1}\mathcal{G}_0$ by Lemma 4.3, since $\partial_t\mathcal{H}_0 = -\mathcal{H}_0^2\partial_t\mathcal{H}_0^{-1}$. In view of (4.29), we obtain that $|\partial_t E_\lambda| \le Ch^{-1}E_\lambda$.

We shall prove that $\pm\partial_t E_\lambda \ge 0$ in $\Omega^\pm(\delta/3\varrho)$ for large ϱ when $\lambda = c_\varrho\delta$ for c_ϱ small enough. As before, we shall only consider $\Omega^+(\delta/3\varrho)$. Now, we find that the integrand in (4.23) is supported where $G_{0,t,w}(z-w) \le C\lambda^2$ where $(t,w) \in \Omega^+(\delta/3\varrho)$. By Proposition 4.2 we find that $(t,z) \in \Omega^+(\delta/2\varrho)$, for ϱ large and $\lambda = c_\varrho\delta$ for $c_\varrho \ll 1$. There $\pm\partial_t\mathcal{H}_0^{-1} \ge 0$ by Lemma 4.3, which implies $\pm\partial_t E_\varrho \ge 0$.

In a similar way, one proves that integrand is supported outside $\Omega^\pm(\delta/9\varrho)$ when (t,w) is in the complement of $\Omega^+(\delta/8\varrho) \cup \Omega^-(\delta/8\varrho)$ for ϱ large and λ small. In view of Lemma 4.3 and Corollary 3.5, this gives $|\partial_t\mathcal{H}_0(t,z)| \le C'(f'(t,w)\mathcal{H}_0^2(t,w) + C\mu\mathcal{H}_0(t,w))$ in the support of the integrand, for small enough c_ϱ. By (4.29) we find $|\partial_t E_\lambda| \le C'(f'\mathcal{H}_0 + C\mu)E_\lambda$ in the complement of $\Omega^+(\delta/8\varrho)\cup\Omega^-(\delta/8\varrho)$, for small enough c_ϱ. □

Now we fix the values of λ and ϱ so that $\frac{1}{\varrho} \le c_3$ in Proposition 4.2 and that the results of Proposition 4.4 hold, and we put $E = E_\lambda$ for this choice of λ.

Remark 4.5. Let $\mathcal{H} = E^{-1/n} \in S(\mathcal{H}_0, \mathcal{G}_0)$, where $E = E_\lambda$ is defined by (4.23) for the above choice of λ. Then $\mathcal{H} \cong \mathcal{H}_0$ and $\partial\mathcal{H} = -\frac{1}{n}\mathcal{H}^{1+n}\partial E_\lambda$, which implies that $\pm\partial_t\mathcal{H}^{-1} \ge 0$ in $\Omega^\pm(\delta/3\varrho)$, $|\partial_t\mathcal{H}^{-1}| \le C'(f' + C\mu\mathcal{H}^{-1})$ and $|\partial_t\mathcal{H}| \le -C'(f'\mathcal{H}^2 + C\mu\mathcal{H})$ in the complement of $\Omega^+(\delta/8\varrho) \cup \Omega^-(\delta/8\varrho)$. We have $|f| \le c_1\mathcal{H}^{-1}$ for some constant c_1. Finally, we have $|\partial_t\mathcal{H}| \le Ch^{-1}\mathcal{H}$ which gives $|\partial_t\mathcal{H}^{-1}| \le Ch^{-1}\mathcal{H}^{-1}$.

5. The cut-off functions

In this section we shall construct suitable monotone cut-off functions, which we shall use to localize in $\Omega^\pm$.

Assume that we have measurable sets $\Omega_t \subseteq T^*\mathbf{R}^n$, for $t \in \mathbf{R}$. We assume that $G_{t,w} = H(t,w)g_w/h(w) \geq cg_w$, $t \in \mathbf{R}$, where g is a uniformly slowly varying metric on $T^*\mathbf{R}^n$, which is constant in t and satisfies $\sup g/g^\sigma = h^2$. We assume that g and G are C^∞ in (t,w), and in fact that $g(w) \in S(g(w),g)$ and $G(w) \in S(G(w),G)$ for any $w \in T^*\mathbf{R}^n$, and also that $H \in S(H,G)$ and $h \in S(h,g)$. We shall use

$$\delta_1(t,w) = \inf\left\{ \sqrt{G_{t,w}(w-w')} : w' \in \Omega_t \right\} \tag{5.1}$$

which is the $G_{t,w}$ distance to Ω_t. We shall first construct a smooth approximation to this distance function. Let

$$\begin{aligned} F(t,w) &= H(t,w)^n h(w)^{-n} |g_w|^{1/2} \int_{\Omega_t} \frac{dw'}{G_{t,w}(w'-w)^{n+1}} \\ &= H(t,w)^{-1} h(w) |g_w|^{1/2} \int_{\Omega_t} \frac{dw'}{g_w(w'-w)^{n+1}}. \end{aligned} \tag{5.2}$$

For sets which are not too irregular we obtain nice properties for F.

Proposition 5.1. *Assume that Ω_t is a union of $G_{t,w}$ balls $\omega^\varrho_{t,w_0} = \{ w : G_{t,w_0}(w-w_0) < \varrho^2 \}$ of radius $\varrho \geq c > 0$, such that G_t only varies with a fixed factor in ω^ϱ_{t,w_0}. Then there exist positive constants c_j, $0 \leq j \leq 2$, such that*

$$0 < c_1 \leq \delta_1^2(t,w)F(t,w) \leq c_2 \qquad \text{when } \delta_1 \leq c_0\varrho. \tag{5.3}$$

When $\delta_1(t,w) \geq \lambda > 0$ we find $F(t,w) \in S(1,G_t)$ uniformly in t with semi-norms depending on λ. If $\pm\partial_t H^{-1}(t,w) \geq -C_0 H^{-1}(t,w)$ and $\Omega_t \subseteq \Omega_{t+h}$ for $\pm h > 0$ small, we find

$$\pm\partial_t F(t,w) \geq -C_0 F(t,w) \tag{5.4}$$

in the distribution sense.

Proof. We choose $g_w/h(w)$ orthogonal coordinates, so that $G_{t,w}(z) = H(t,w)|z|^2$. By definition we have that $|w-w'| \geq \delta_1(t,w)H^{-1/2}$ when $w' \in \Omega_t$, where $H^{-1/2} = H^{-1/2}(t,w)$, and

$$F(t,w) = H(t,w)^{-1} \int_{\Omega_t} \frac{dw'}{|w'-w|^{2n+2}}. \tag{5.5}$$

Since the integrand is positive we find by using polar coordinates

(5.6)

$$F(t,w) \le H^{-1}\int_{|w-w'|\ge\delta_1 H^{-1/2}} \frac{dw'}{|w-w'|^{2n+2}}$$
$$= c_n H^{-1}\int_{\delta_1 H^{-1/2}}^{\infty} \frac{r^{2n-1}dr}{r^{2n+2}} = c_n H^{-1}\int_{\delta_1 H^{-1/2}}^{\infty} \frac{dr}{r^3} = \frac{c_n}{2{\delta_1}^2}.$$

This proves the upper bound of the estimate.

Next, we prove the lower bound. From the definition of $\delta_1(t,w)$, we find that there exists $w_1 \in \overline{\Omega}_t$ such that $\delta_1(t,w)H^{-1/2} < |w-w_1| \le 2\delta_1(t,w)H^{-1/2}$, where $H^{-1} = H^{-1}(t,w)$. By the condition that Ω_t is a union of G_t balls of radius $\varrho \ge c > 0$ there exists $w_2 \in \Omega_t$ such that $G_{t,w_2}(w_2-w_1) < \varrho^2$, and

$$\omega = \left\{ w' : G_{t,w_2}(w'-w_2) < \varrho^2 \right\} \subseteq \Omega_t. \tag{5.7}$$

If δ_1 is small enough, we obtain by the assumptions and the slow variation that $1/C \le G_{t,w}/G_{t,w_2} \le C$, which gives $c\varrho H^{-1/2} \le |w_1 - w_2| \le C\varrho H^{-1/2}$ and $(c\varrho - 2\delta_1(t,w))H^{-1/2} \le |w-w_2| \le (C\varrho + 2\delta_1(t,w))H^{-1/2}$. In $g_w/h(w)$ orthogonal coordinates, we find that ω is an ellipsoid with half-axes of length $\cong \varrho H^{-1/2}$. If we take $\delta_1(t,w) \le c_0\varrho$ with $c_0 \ll 1$, we find that there exist $1 < c_3 < c_4 < 3$ such that when $c_3\delta_1 H^{-1/2} \le \varepsilon \le c_4\delta_1 H^{-1/2}$, the intersection of ω with the sphere of radius ε centered at w contains the intersection with a cone with angle $\theta > 0$ and vertex at w. By using polar coordinates with center at w, we find that

(5.8)

$$F(t,w) \ge H^{-1}\int_{\omega} \frac{dw'}{|w-z|^{2n+2}} \ge c_n H^{-1} c \int_{c_3\delta_1 H^{-1/2}}^{c_4\delta_1 H^{-1/2}} \frac{dr}{r^3} \ge c_5/\delta_1^2.$$

This gives the lower bound.

Next, we shall prove that $F(t,w) \in S(1, G_t)$. It is clear that

$$H(t,w)^n h(w)^{-n}|g_w|^{1/2} \in S(H^n h^{-n}|g|^{1/2}, G),$$

thus we only have to differentiate under the integral sign. This gives terms

$$-(n+1)H(t,w)^n h(w)^{-n}|g_w|^{1/2}\int_{\Omega_t} \frac{\langle w_0, \partial_w\rangle G_{t,w}(w'-w)dw'}{G_{t,w}(w'-w)^{n+2}}. \tag{5.9}$$

If $G_{t,w}(w_0) \leq 1$ we find $|\langle w_0, \partial_w\rangle G_{t,w}(w'-w)| \leq C(G_{t,w}(w'-w) + G_{t,w}(w'-w)^{1/2})$, which in the chosen coordinates gives terms bounded by

$$CH^{-1}\int_{\Omega_t}\left(\frac{1}{|w'-w|^{2n+2}} + \frac{1}{|w'-w|^{2n+3}}\right)dw'. \tag{5.10}$$

This gives

$$|F(t,w)|_1^G \leq (C + C_n\lambda^{-1}H^{1/2}(t,w))|F(t,w)| \cdot \tag{5.11}$$

when $\delta_1 \geq \lambda > 0$, since then $|w - w'| \geq \lambda H^{-1/2}(t,w)$ in the integration domain. Continued differentiation produces the same type of terms, which proves that $F(t,w) \in S(1, G_t)$ uniformly when $\delta_1 \geq \lambda > 0$.

Finally, we shall prove (5.4). Since, in the chosen $g_w/h(w)$ orthogonal coordinates,

$$\begin{aligned} F(t+h,w) - F(t,w) &= H^{-1}(t+h,w)\int_{\Omega_{t+h}\setminus\Omega_t}\frac{dw'}{|w'-w|^{2n+2}} \\ &+ (H^{-1}(t+h,w) - H^{-1}(t,w))\int_{\Omega_t}\frac{dw'}{|w'-w|^{2n+2}} \end{aligned} \tag{5.12}$$

we find that $\partial_t F(t,w) \geq \partial_t H^{-1}(t,w)\int_{\Omega_t}\frac{dw'}{|w'-w|^{2n+2}} \geq -C_0F(t,w)$ in the distribution sense, if $\partial_t H^{-1}(t,w) \geq -C_0H^{-1}(t,w)$ and $\Omega_t \subseteq \Omega_{t+h}$ for small $h > 0$. By choosing $-t$ as time variable, we obtain the corresponding result when $-\partial_t H^{-1}(t,w) \geq -C_0H^{-1}(t,w)$ and $\Omega_t \subseteq \Omega_{t+h}$ for $h < 0$. □

Next, we construct the cut-off functions. Let $\chi(s) \in C^\infty(\mathbf{R})$ such that $0 \leq \chi \leq 1$, $\chi(s) = 0$ when $s \leq 1$, $\chi(s) = 1$ when $s \geq 2$, such that $0 \leq \chi' \leq 2$. Let

$$\Psi_\lambda(t,w) = \chi(F(t,w)\lambda^2) \qquad \lambda > 0 \tag{5.13}$$

and $\Psi_\lambda = 1$ on Ω_t.

Lemma 5.2. *We find that $\Psi_\lambda \in S(1, G_t)$ with seminorms depending on λ. Also Ψ_λ is supported where $\delta_1(t,w) \leq \sqrt{c_2}\lambda$ and $\Psi_\lambda = 1$ when $\delta_1(t,w) \leq \sqrt{\frac{c_1}{2}}\lambda$. We find*

$$\pm\partial_t\Psi_\lambda \geq -4C_0 \tag{5.14}$$

in the distribution sense, independently of λ, if $\pm\partial_t H^{-1} \geq -C_0H^{-1}$ when $\sqrt{\frac{c_1}{2}}\lambda \leq \delta_1(t,w) \leq \sqrt{c_2}\lambda$ and $\Omega_t \subseteq \Omega_{t+h}$ for $\pm h > 0$ small. Here $0 < c_1 < c_2$ are given as in Proposition 5.1.

Proof. When $\delta_1 \geq \sqrt{c_2}\lambda$, we find $F\lambda^2 \leq F{\delta_1}^2/c_2 \leq 1$, thus $\Psi_\lambda = 0$. When $\delta_1 \leq \sqrt{c_1/2}\lambda$, we find $F\lambda^2 \geq 2F{\delta_1}^2/c_1 \geq 2$, which implies $\Psi_\lambda = 1$.

If $G_{t,w}(w_0) \leq 1$, we have $\langle w_0, \partial_w \rangle \Psi_\lambda = \chi'(F\lambda^2)\langle w_0, \partial_w \rangle F\lambda^2 = O(1)$, since $\chi'(F\lambda^2)$ is supported where $\delta_1 \geq \lambda\sqrt{c_1/2}$ and $\langle w_0, \partial_w \rangle F = O(1)$ there. Since repeated differentiation gives the same type of terms we find that $\Psi_\lambda \in S(1, G_t)$ with seminorms depending on λ.

It remains to prove (5.14). It follows from Proposition 5.1 that

$$\pm\partial_t\Psi_\lambda = \pm\chi'(F\lambda^2)\partial_t F\lambda^2 \geq -C_0\chi'(F\lambda^2)F\lambda^2 \geq -4C_0 \tag{5.15}$$

independently of λ, if $\pm\partial_t H^{-1} \geq -C_0 H^{-1}$ in $\operatorname{supp}\chi'(F\lambda^2)$, i.e., when $\sqrt{\frac{c_1}{2}}\lambda \leq \delta_1 \leq \sqrt{c_2}\lambda$. In fact, we then have $\pm\partial_t F \geq -C_0 F$ and $1 \leq F\lambda^2 \leq 2$ in $\operatorname{supp}\partial_t\Psi_\lambda$. □

Recall that $\mathcal{H} = E^{-1/n} \in S(H_0, G_0)$, where $E = E_\lambda$ is defined by (4.23) for the choice of $\lambda > 0$ above. By Remark 4.5 we find that $\pm\partial_t\mathcal{H}^{-1} \geq 0$ in $\Omega^\pm(\delta/3\varrho)$, $|\partial_t\mathcal{H}| \leq -C'(f'\mathcal{H}^2 + C\mu\mathcal{H})$ in the complement of $\Omega^+(\delta/8\varrho) \cup \Omega^-(\delta/8\varrho)$ for the above choice of ϱ. Let

$$\psi_\pm(t, w) = \chi(F_\pm(t, w)\lambda^2) \tag{5.16}$$

and $\psi_\pm \equiv 1$ on $\Omega_t^\pm(\delta/7\varrho)$. Here $\lambda = c\delta$, $\chi(s)$ is as in Lemma 5.2, $F_\pm$ is defined in (5.2) with $\mathcal{G} = \mathcal{H}G/H = \mathcal{H}g/h$ instead of G and $\Omega_t^\pm(\delta/7\varrho)$ instead of Ω_t, i.e.,

$$F_\pm(t, w) = \mathcal{H}^{-n}(t, w)h^{-n}(w)|g_w|^{1/2} \int_{\Omega_t^\pm(\delta/7\varrho)} \frac{dw'}{\mathcal{G}_{t,w}(w'-w)^{n+1}}. \tag{5.17}$$

We also define

$$\widetilde{\psi}_\pm(t, w) = \psi(\widetilde{F}_\pm(t, w)\lambda^2) \tag{5.18}$$

and $\widetilde{\psi}_\pm \equiv 1$ on $\Omega_t^\pm(\delta/5\varrho)$. Here we have replaced Ω^+ with $\Omega^\pm(\delta/5\varrho)$ in the definition of $\widetilde{F}_\pm$.

Lemma 5.3. *We have* $0 \leq \psi_\pm$, $0 \leq \widetilde{\psi}_\pm$, $\psi_\pm$ *and* $\widetilde{\psi}_\pm \in S(1, G_0)$. *For* $\lambda = c\delta$ *with* $c \ll 1$ *in (5.16) and (5.18) we find that* $\pm\partial_t\psi_\pm \geq 0$ *and* $\pm\partial_t\widetilde{\psi}_\pm \geq 0$ *in the distribution sense,* $\psi_\pm \equiv 1$ *on* $\Omega^\pm(\delta/7\varrho)$, $\operatorname{supp}\psi_\pm \subseteq \Omega^\pm(\delta/6\varrho)$, $\widetilde{\psi}_\pm \equiv 1$ *on* $\Omega^\pm(\delta/5\varrho)$ *and* $\operatorname{supp}\widetilde{\psi}_\pm \subseteq \Omega^\pm(\delta/4\varrho)$.

Proof. The first statements are obvious in view of Lemma 5.2. By Lemma 5.2, $\psi_\pm$ is supported where $\delta_2^\pm \le \sqrt{c_2}\lambda = \sqrt{c_2}c\delta$. Here

$$\delta_2^\pm(t,w) = \inf\left\{ \sqrt{G_{0,t,w}(w_0 - w)} : w_0 \in \Omega^\pm(\delta/7\varrho) \right\} \tag{5.19}$$

is the G_0 distance to $\Omega^\pm(\delta/7\varrho)$. Since $\frac{1}{\varrho} \le c_3$, we find from Proposition 4.2 that $\operatorname{supp}\psi_\pm \subseteq \Omega^\pm(\delta/6\varrho)$ for small enough c.

We find by the definition that $\widetilde{\psi}_\pm \equiv 1$ on $\Omega^\pm(\delta/5\varrho)$, and as before we find $\operatorname{supp}\widetilde{\psi}_\pm \subseteq \Omega^\pm(\delta/4\varrho)$ for small c. Since $\pm\partial_t\mathcal{H}^{-1} \ge 0$ in $\Omega^\pm(\delta/3\varrho)$ we obtain as before that $\pm\partial_t\psi_\pm \ge 0$ and $\pm\partial_t\widetilde{\psi}_\pm \ge 0$ for small c. □

Next, we regularize in t and put

$$\chi_\pm(t,w) = h^{-1}(w)\lambda^{-1}\int \phi((s-t)\lambda^{-1}h^{-1}(w))\psi_\pm(s,w)\,ds \tag{5.20}$$

and

$$\widetilde{\chi}_\pm(t,w) = h^{-1}(w)\lambda^{-1}\int \phi((s-t)\lambda^{-1}h^{-1}(w))\widetilde{\psi}_\pm(s,w)\,ds \tag{5.21}$$

where $0 \le \phi(s) \in C^\infty$ such that $\int \phi(s)\,ds = 1$ and $\phi(s) = 0$ when $s \ge 1$.

Proposition 5.4. *We have $0 \le \chi_\pm$, $0 \le \widetilde{\chi}_\pm$, and that $\chi_\pm, \widetilde{\chi}_\pm \in C^\infty(\mathbf{R}, S(1,G_0))$. When $\lambda = c\delta$ and $0 < c$ is small enough, we find $\pm\partial_t\widetilde{\chi}_\pm \ge 0$, $\pm\partial_t\widetilde{\chi}_\pm \ge 0$, $\chi_\pm \equiv 1$ on $\Omega^\pm(\delta/8\varrho)$, $\widetilde{\chi}_\pm \equiv 1$ on $\operatorname{supp}\chi_\pm$ and $\operatorname{supp}\widetilde{\chi}_\pm \subseteq \Omega^\pm(\delta/3\varrho)$. We also find $|\partial_t\chi_\pm| + |\partial_t\widetilde{\chi}_\pm| \le Ch^{-1}$.*

Proof. Since $\psi_\pm$ and $\widetilde{\psi}_\pm \in S(1,G_0)$ and G_0 only vary with a fixed factor when $\Delta t = O(\lambda h)$ for small λ, we find that $\chi_\pm$ and $\widetilde{\chi}_\pm \in S(1,G_0)$. We find

$$\begin{aligned}\pm\partial_t\chi_\pm(t,w) &= \mp\lambda^{-2}h^{-2}(w)\int \phi'((s-t)\lambda^{-1}h^{-1}(w))\psi_\pm(s,w)\,ds\\ &= h^{-1}(w)\lambda^{-1}\int \phi((s-t)\lambda^{-1}h^{-1}(w))(\pm\partial_t\psi_\pm(s,w))\,ds\end{aligned} \tag{5.22}$$

which gives $|\partial_t\chi_\pm| + |\partial_t\widetilde{\chi}_\pm| \le C_\lambda h^{-1}$. By (4.21), we obtain the other statements by choosing $\lambda = c\delta$ with $0 < c$ small enough. □

6. The preparation

Now we shall make the preparation, and for that purpose we need that $\pm f \ge cH_0^{-1} > 0$ on $\operatorname{supp} d\chi_\pm$. Therefore, we put

$$f_\kappa = f + \kappa\mathcal{H}(\widetilde{\chi}_+ - \widetilde{\chi}_-) \qquad \kappa > 0 \tag{6.1}$$

which is continuous in t. Observe that this construction only changes P with a term in $\operatorname{Op} S(H_0, G_0) \subset \operatorname{Op} S(H, G)$, and we find $\pm f_\kappa \geq \kappa\mathcal{H}$ on $\operatorname{supp}\chi_\pm$.

Lemma 6.1. *We find that $f_\kappa \in S(H_0^{-1}, G_0)$ and $\pm f_\kappa \geq cH_0^{-1} > 0$ on $\operatorname{supp} d_{\chi_\pm}$. For small enough κ we find*

$$c_\kappa f_\kappa - C_\kappa \leq \partial_t f_\kappa \leq C_\kappa h^{-1} \qquad \text{outside } \Omega^\pm(\delta/8\varrho) \tag{6.2}$$

for some constant C_κ, and $c_\kappa \in S(1, G_0)$ is continuous in t.

Proof. First we prove $f \in S(H_0^{-1}, G_0)$, which is obvious in $\complement\Theta^\pm$. We have $|f| \leq cH_0^{-1}$, and $H_0^{1/2}|f|_1^{G_0} = H^{1/2}|f|_1^G \leq CH^{-1/2} \leq C'H_0^{-1/2}$ by Proposition 3.6 on $\Omega^\pm$. When $k \geq 2$, we find $|f|_k^{G_0} = (h/H_0)^{k/2}|f|_k^g \leq C_k H_0^{-1}(h/H_0)^{k/2-1} \leq C'_k H_0^{-1}$ since $H_0^{-1} \leq ch^{-1}$. This proves that f and $f_\kappa \in S(H_0^{-1}, G_0)$.

Clearly $\pm f \geq c_0\mathcal{H}^{-1} - 1$ on $\Omega^\pm(\delta) \setminus \Theta^\pm$ for some constant $c_0 > 0$, by Proposition 4.2. Thus, when $\mathcal{H}^{-1} \gg 1$ we find $\pm f \geq c_1\mathcal{H}^{-1}$. Else, we find that $\pm f_\kappa \geq \kappa\widetilde{\chi}_\pm\mathcal{H} \geq c_2\mathcal{H}^{-1} > 0$ on $\operatorname{supp} d_{\chi_\pm}$.

By Proposition 3.3, (6.2) holds for $f_\kappa = f$ outside $\Omega^\pm(\delta/8\varrho)$ since $G_0 \cong G$ there. In fact, if $\{\phi_j(t,w)\} \in S(1, dt^2/h^2 + G)$ is a partition of unity such that $\phi_j \geq 0$ and

$$\operatorname{supp}\phi_j \subseteq \left\{ (t,w) : w \in \omega_{w_j}^\lambda \text{ and } |t - t_j| \leq \lambda h(w_j) \right\}, \tag{6.3}$$

then for small λ we find that $\omega_{w_j}^\lambda$ is of type A, if $\operatorname{supp}\phi_j$ intersects $\complement(\Omega^+(\delta/8\varrho) \cup \Omega^-(\delta/8\varrho))$. Thus, we may apply Proposition 3.3 to get $\partial_t f \geq c_j f - C$ in $\operatorname{supp}\phi_j$. By taking $c = \sum_j \phi_j c_j$, we obtain (6.2) for f. Outside $\Omega^\pm(\delta/8\varrho)$ we find $|\partial_t\mathcal{H}| \leq C'(f'\mathcal{H}^2 + C\mu\mathcal{H})$ by Remark 4.5, which gives for small κ

$$\begin{aligned} \partial_t f_\kappa &= \partial_t f + \kappa\mathcal{H}\partial_t(\widetilde{\chi}_+ - \widetilde{\chi}_-) + \kappa(\widetilde{\chi}_+ - \widetilde{\chi}_-)\partial_t\mathcal{H} \\ &\geq (1 - \kappa(\widetilde{\chi}_+ + \widetilde{\chi}_-)C\mathcal{H}^2)f' - C \geq -c_\kappa f_\kappa - C' \end{aligned} \tag{6.4}$$

at (t,w). Here $c_\kappa \in S(1, G_0)$ is continuous in t.

Finally, we find $|\partial_t f| + |\partial_t\mathcal{H}| + |\partial_t\widetilde{\chi}_\pm| \leq ch^{-1}$ by Remark 4.5 and Proposition 5.4, which proves that $|\partial_t f_\kappa| \leq ch^{-1}$, and completes the proof of the lemma. □

Let f_κ be defined by (6.1). Since $\mathcal{H}^{-1} \geq cH_0^{-1} \geq CH^{-1}$, we find $|f_\kappa| \leq c_1\mathcal{H}^{-1}$ for some constant c_1. Now we define

$$b_\kappa(t,w) = c_1\mathcal{H}^{-1}(\chi_+ - \chi_-) + f_\kappa(1 - \chi_+ - \chi_-) \in S(H_0^{-1}, G_0), \tag{6.5}$$

which is a real symbol, and is continuous in t. Here f_κ is defined by (6.1) and $\chi_\pm$ by (5.20).

Proposition 6.2. *For κ small enough, we obtain*

$$0 \le f_\kappa/b_\kappa \le 1 \tag{6.6}$$

$$a_\kappa = f_\kappa/b_\kappa \in S(1, G_0) \tag{6.7}$$

$$c_\kappa b_\kappa - C_\kappa \le \partial_t b_\kappa \le C_\kappa h^{-2}, \tag{6.8}$$

for some constant C_κ, and $c_\kappa \in S(1, G_0)$ uniformly in t, having support where $a_\kappa \ge c > 0$. We find that a_κ and c_κ are continuous in t.

Proof. Since $\pm f_\kappa \ge 0$ in $\Omega^\pm$ we find $f_\kappa/b_\kappa \ge 0$. Since $|f_\kappa| \le c_1\mathcal{H}^{-1}$, we obtain $f_\kappa/b_\kappa \le 1$. Since $\pm f_\kappa \ge cH_0^{-1}$ on $\operatorname{supp}\chi_\pm$ by Lemma 6.1, we find $|b_\kappa| \ge c'H_0^{-1}$ on $\operatorname{supp}\chi_\pm$. Thus $b_\kappa^{-1} \in S(H_0, G_0)$ on $\operatorname{supp}\chi_\pm$, which gives $a_\kappa \in S(1, G_0)$ there. Since $a_\kappa \equiv 1$ outside $\operatorname{supp}\chi_\pm$, we obtain (6.7). We also see that $a_\kappa \ge c_1 > 0$ on $\operatorname{supp} d\chi_\pm$.

We have

$$\begin{aligned}\partial_t b_\kappa = c_1(\chi_+ - \chi_-)\partial_t\mathcal{H}^{-1} + \partial_t f_\kappa(1-\chi)\\ + (c_1\mathcal{H}^{-1} - f_\kappa)\partial_t\chi_+ - (c_1\mathcal{H}^{-1} + f_\kappa)\partial_t\chi_-\end{aligned} \tag{6.9}$$

where $\chi = \chi_+ + \chi_-$. Now $c_1\mathcal{H}^{-1} \pm f_\kappa \ge 0$ and $\pm\partial_t\chi_\pm \ge 0$. Also, $\pm\partial_t\mathcal{H}^{-1} \ge 0$ on $\operatorname{supp}\chi_\pm$ by Remark 4.5, which gives

$$\partial_t b_\kappa \ge \partial_t f_\kappa(1-\chi) \ge (-c_\kappa f_\kappa - C)(1-\chi) \tag{6.10}$$

for some constant C and $c_\kappa \in S(1, G_0)$ which is continuous in t by Lemma 6.1. Since $\pm f_\kappa \ge 0$ on $\operatorname{supp}\chi_\pm$, we may take $\pm c_\kappa > 0$ on $\operatorname{supp}\chi_\pm$. In fact, we may replace c_κ by $\widetilde{c}_\kappa = c_\kappa + \varrho(\widetilde{\chi}_+ - \widetilde{\chi}_-)$ in (6.10) with $\varrho \gg 1$. Since $\pm(b_\kappa - f_\kappa) \ge 0$ on $\Omega^\pm$, $\pm\widetilde{c}_\kappa > 0$ on $\operatorname{supp}\chi_\pm$, and $b_\kappa = f_\kappa$ on $\complement\operatorname{supp}\chi$, we obtain $\partial_t b_\kappa \ge -\overline{c}_\kappa b_\kappa - C$, where $\overline{c}_\kappa = \widetilde{c}_\kappa(1-\chi) \in S(1, G_0)$ is supported in $\operatorname{supp}(1-\chi)$ and is continuous in t. Thus, we find $a_\kappa \ge c_1 > 0$ on $\operatorname{supp}\overline{c}_\kappa$, since $a_\kappa \ge c_1$ on $\operatorname{supp} d\chi_\pm$ and $a_\kappa = 1$ outside $\operatorname{supp}\chi_\pm$.

Since $|\partial_t f_\kappa| + |\partial_t\chi_\pm| \le Ch^{-1}$ and $|\partial_t\mathcal{H}^{-1}| \le Ch^{-1}\mathcal{H}^{-1} \le C'h^{-2}$, we obtain $|\partial_t b_\kappa| \le C''h^{-2}$. □

7. Proof of Theorem 1.1

First, we note that we may assume that $\operatorname{Im} r \in S(h, g) \subset S(H_0, G_0)$ in (1.11). In fact, by conjugating with the operator E^w, where $E =$

$\exp(\int^t \operatorname{Im} r(s)\,ds)$, r is replaced by $\operatorname{Re} r + \int^t \{ f(t), \operatorname{Im} r(s) \}\, ds$, modulo $S(h,g)$.

In the following we shall use Wick operators as well as Weyl operators, see Section B in the Appendix. We shall write $A \geq B$ when we have $\langle Au, u\rangle \geq \langle Bu, u\rangle \Leftrightarrow \langle (A-B)u, u\rangle \geq 0$ for $u \in \mathcal{S}$. We shall use the notation $\operatorname{Re} B = \frac{1}{2}(B + B^*)$ and $\operatorname{Im} B = \frac{1}{2i}(B - B^*)$ on $\mathcal{S}$. As before, we shall suppress the dependence on t.

By choosing κ small enough, we find from Proposition 6.2 that

$$f = ab + r_0 \tag{7.1}$$

where $b \in S(H_0^{-1}, G_0)$, $0 \leq a \in S(1, G_0)$, and $r_0 \in S(H_0, G_0)$. Also

$$-c_0 b - C \leq \partial_t b \leq Ch^{-2} \tag{7.2}$$

for some constant C, and $c_0 \in S(1, G_0)$ continuous is t, with support where $a \geq c > 0$.

Now let

$$B = b^{Wick}(t, x, D_x) \tag{7.3}$$

as a Wick operator (see the definition in Appendix B). Then $B^* = B$ on $\mathcal{S}$, and by Proposition B.2 and Remark B.3 we find that $\partial_t B = (\partial_t b)^{Wick}$ is well-defined and bounded $\mathcal{S} \mapsto L^2$, since $|\partial_t b| \leq h^{-2}$, and we find $(\partial_t b)^{Wick} \geq -(c_0 b + C)^{Wick}$. We also have $q^{Wick} \cong q^w$ modulo an L^2 bounded operator, when $q \in S(H_0^{-1}, G_0)$ by Proposition B.2. This implies

$$\partial_t B \geq -\operatorname{Re}(FB) - C_0, \tag{7.4}$$

where $F = F^* = c_0^w \in \operatorname{Op} S(1, G_0)$. In fact, we have $\operatorname{Re}(FB) = \frac{1}{2}(FB + BF) = FB - \frac{1}{2}[F, B] \cong FB$, where $[F, B] \cong [c_0^w, b^w] \cong 0$ and $FB \cong c_0^w b^w \cong (c_0 b)^w \cong (c_0 b)^{Wick}$, modulo L^2 bounded operators, for fixed t.

Since $a \geq 0$, we find by the Fefferman-Phong inequality that $a^w \geq c_1^w$ for some $c_1 \in S(H_0^2, G_0)$. Put $A = a^w - c_1^w \geq 0$, then we obtain from (7.1) that

$$f^w(t, x, D_x) = A(B + R_0) + R_1 \tag{7.5}$$

where R_0 is an L^2 bounded operator, $R_1 \in \operatorname{Op} S(1, G_0)$ and $\operatorname{Re} R_1 \in \operatorname{Op} S(H_0, G_0)$, $\forall t$. In fact, $R_0 = b^w - b^{Wick}$ is bounded in L^2, and $a^w b^w = f^w + r_1^w$, where $r_1 \cong \frac{i}{2}\{a, b\}$ modulo $S(H_0, G_0)$, and $c_1^w b^w \in S(H_0, G_0)$, which proves $\operatorname{Re} R_1 \in \operatorname{Op} S(H_0, G_0)$.

Lemma 7.1. *We have*

$$F = AR_3 + R_4 \tag{7.6}$$

for some $R_3 \in \operatorname{Op} S(1, G_0)$ *and* $R_4 \in \operatorname{Op} S(H_0^2, G_0)$, $\forall t$.

Proof. Since $F = c_0^w$ and $a \geq c > 0$ on $\operatorname{supp} c_0$ we may for any t take $0 \leq a_0 \in S(1, G_0)$ such that $a_0 = a^{-1}$ on $\operatorname{supp} c_0$. Then we have $a^w a_0^w = 1 + r_0^w$ by the calculus, where $r_0 \in S(H_0^2, G_0)$ on $\operatorname{supp} c_0$. This gives $F = AR_3 + R_4$, with $R_3 = a_0^w c_0^w$ and $R_4 = -r_0^w c_0^w + c_1^w a_0^w c_0^w$, for any t. □

We have by (1.11) and (7.5) that

$$P = D_t + iAB + R \tag{7.7}$$

where $R = AR_0 + R_1$, R_0 is bounded in L^2, $R_1 \in \operatorname{Op} S(1, G_0)$ and $\operatorname{Im} R_1 \in \operatorname{Op} S(H_0, G_0)$.

Lemma 7.2. *We find that*

$$\partial_t B + 2\operatorname{Im}(BR) \geq -BAB - C \qquad \forall\, t, \tag{7.8}$$

for some constant C.

Proof. We have $R = AR_0 + R_1$ and $\partial_t B \geq -\operatorname{Re}(FB) - C_0 = -\operatorname{Re}(BF) - C_0$ for any t, where $F = AR_3 + R_4$. Here R_3 and $R_1 \in \operatorname{Op} S(1, G_0)$, R_0 is L^2 bounded and R_4 and $\operatorname{Im} R_1 \in \operatorname{Op} S(H_0, G_0)$, $\forall t$. Clearly, we find $\|B\overline{R}\| \leq C$, when $\overline{R} \in \operatorname{Op} S(H_0, G_0)$ since $B \cong b^w$ modulo $\operatorname{Op} S(1, G_0)$. We also find that $2\operatorname{Im}(B \operatorname{Re} R_1) = \frac{1}{i}[B, \operatorname{Re} R_1] \geq -C$. Finally, for $u \in \mathcal{S}$, we have

$$|\langle BA\widetilde{R}u, u\rangle| = |\langle ABu, \widetilde{R}u\rangle| \leq \frac{1}{4}\langle ABu, Bu\rangle + C'\|u\|^2 \tag{7.9}$$

when $\widetilde{R}$ is bounded in L^2, since $0 \leq A \leq C$. This gives by (7.4)

$$\partial_t B + 2\operatorname{Im}(BR) \geq -\operatorname{Re}(BF) - C_0 + 2\operatorname{Im}(BR) \geq -BAB - C'', \tag{7.10}$$

which proves the lemma. □

Now Theorem A.2 in Appendix A gives for small T that

$$\int \|u\|^2(t)\, dt \leq C_0 T^2 \int \operatorname{Im}\langle BPu, u\rangle(t) + \|Pu\|^2(t)\, dt \tag{7.11}$$

for some positive constant C_0, if $u \in \mathcal{S}(\mathbf{R} \times \mathbf{R}^n)$ has support where $|t| \leq T$. Since $B = b^{Wick} \cong b^w$ modulo $\operatorname{Op} S(1, G_0)$, we find

$$\operatorname{Im}\langle BPu, u\rangle \leq \operatorname{Im}\langle b^w Pu, u\rangle + C\|Pu\|^2 \tag{7.12}$$

which gives (1.14), and proves Theorem 1.1.

Appendix A. The estimate

In this appendix we shall prove an *a priori* estimate, which is needed to prove Theorem 1.1. We assume that

$$P = D_t + iA(t)B(t) + R(t) \qquad t \in \mathbf{R} \tag{A.1}$$

where $\{ A(t) \}$ and $\{ R(t) \}$ are weakly continuous families of bounded operators on $L^2(\mathbf{R}^n)$, such that $A(t) = A^*(t) \geq 0$, $\forall t$. We assume that $B(t)$ is a strongly continuous family of operators: $\mathcal{S}(\mathbf{R}^n) \mapsto L^2(\mathbf{R}^n)$ such that $\langle B(t)u, v\rangle = \langle u, B(t)v\rangle$ when u, $v \in \mathcal{S}(\mathbf{R}^n)$. Also, we assume that $t \mapsto \langle B(t)u, v\rangle$ is differentiable for any u, $v \in \mathcal{S}(\mathbf{R}^n)$, with derivative

$$\partial_t\langle B(t)u, v\rangle = \langle B'(t)u, v, \rangle \tag{A.2}$$

where $\{ B'(t) \}$ is a weakly continuous family of bounded operators: $\mathcal{S}(\mathbf{R}^n) \mapsto L^2(\mathbf{R}^n)$. We also assume that there exists $\gamma < 2$ such that

$$\langle B'(t)u, u\rangle + 2\,\mathrm{Im}\langle R(t)u, B(t)u\rangle + \gamma\langle A(t)B(t)u, B(t)u\rangle \geq -C\|u\|^2 \tag{A.3}$$

for all $|t| < c$, when $u \in \mathcal{S}(\mathbf{R}^n)$. Here $\langle u, v\rangle$ is the scalar product in $L^2(\mathbf{R}^n)$, and we let $\|u\| = \langle u, u\rangle^{1/2}$ be the L^2 norm.

Remark A.1. It follows from (A.2) that if u, $v \in \mathcal{S}(\mathbf{R} \times \mathbf{R}^n)$ then $t \mapsto \langle B(t)u, v\rangle$ is differentiable and

$$\partial_t\langle B(t)u, v\rangle = \langle B'(t)u, v\rangle + \langle B(t)u', v\rangle + \langle B(t)u, v'\rangle. \tag{A.4}$$

In particular we obtain $\partial_t\langle B(t)u, u\rangle = \langle B'(t)u, u\rangle + 2\,\mathrm{Re}\langle B(t)u, u'\rangle$. If $u \in \mathcal{S}(\mathbf{R} \times \mathbf{R}^n)$ we find that $t \mapsto Ru(t)$ is weakly and $t \mapsto Bu(t)$ is strongly continuous. Observe that by Banach-Steinhaus' theorem, we find that $\|A(t)\|$ and $\|R(t)\|$ are locally bounded.

Theorem A.2. *If P in* (A.1) *satisfies conditions* (A.2) *and* (A.3), *then there exists C_0 and $T_0 > 0$ such that*

$$\int \|u\|^2\, dt \leq C_0T^2 \int \mathrm{Im}\langle Pu, Bu\rangle + \|Pu\|^2\, dt \tag{A.5}$$

for $u \in \mathcal{S}(\mathbf{R}\times\mathbf{R}^n)$ having support where $|t| \leq T \leq T_0$. The constants T_0 and C_0 only depend on $\|A(t)\|$, $\|R(t)\|$, γ and the constant C in (A.3).

Proof. We shall prove (A.5) by first using that

$$\|u\|^2(t) = \int_{-T}^{t} 2\operatorname{Re}\langle \partial_t u, u\rangle\, dt \tag{A.6}$$

if $u \in \mathcal{S}(\mathbf{R}\times\mathbf{R}^n)$ and $|t| \le T$ in $\operatorname{supp} u$, which gives after integation in t

$$\int \|u\|^2(t)\, dt = \int 2(T-t)\operatorname{Re}\langle \partial_t u, u\rangle\, dt \tag{A.7}$$

if $u \in \mathcal{S}(\mathbf{R}\times\mathbf{R}^n)$ and $|t| \le T$ in $\operatorname{supp} u$. Now

$$\begin{aligned} \operatorname{Re}\langle \partial_t u, u\rangle &= \operatorname{Re}\langle iPu, u\rangle + \operatorname{Re}\langle ABu, u\rangle + \operatorname{Re}\langle -iRu, u\rangle \\ &= -\operatorname{Im}\langle Pu, u\rangle + \operatorname{Re}\langle ABu, u\rangle + \operatorname{Im}\langle Ru, u\rangle, \end{aligned} \tag{A.8}$$

which gives

$$\int \|u\|^2\, dt = \int 2(T-t)\big(\operatorname{Im}\langle u, Pu\rangle + \operatorname{Re}\langle ABu, u\rangle + \operatorname{Im}\langle Ru, u\rangle\big)\, dt \tag{A.9}$$

if $u \in \mathcal{S}(\mathbf{R}\times\mathbf{R}^n)$ and $|t| \le T$ in $\operatorname{supp} u$. Since $R(t)$ is locally bounded, we obtain that

$$\int \|u\|^2\, dt \le 4\int (T-t)\big(\operatorname{Im}\langle u, Pu\rangle + \operatorname{Re}\langle ABu, u\rangle\big)\, dt \tag{A.10}$$

if $u \in \mathcal{S}$ and $8T\|R(t)\| \le 1$ in $\operatorname{supp} u$.

Next, we use the Cauchy–Schwarz inequality

$$\operatorname{Re}\langle ABu, u\rangle \le (4\varrho)^{-1}\|ABu\|^2 + \varrho\|u\|^2 \qquad \forall\, \varrho > 0 \tag{A.11}$$

if $u \in \mathcal{S}(\mathbf{R}^n)$. Since $A(t) \ge 0\ \forall\, t$, we can use spectral theory to construct $A^{1/2}(t) \ge 0$ such that $A^{1/2}(t)A^{1/2}(t) = A(t)$ and $\|A^{1/2}(t)\| = \|A(t)\|^{1/2}$ for any t. Then we obtain

$$\|ABu\|^2 \le \|A\|\|A^{1/2}Bu\|^2 = \|A\|\langle ABu, Bu\rangle \qquad u \in \mathcal{S}(\mathbf{R}^n). \tag{A.12}$$

This gives

$$\begin{aligned} \operatorname{Re}\langle ABu, u\rangle &\le \tfrac{1}{4\varrho}\|ABu\|^2 + \varrho\|u\|^2 \\ &\le \tfrac{1}{4\varrho}|\!|\!| A |\!|\!| \langle ABu, Bu\rangle + \varrho\|u\|^2, \quad |t| < c \end{aligned} \tag{A.13}$$

$u \in \mathcal{S}(\mathbf{R}^n)$, where $|||A||| = \sup_{|t|<c} \|A\|$. By taking $\varrho = \frac{1}{16T}$ we obtain from (A.10)

$$\int \|u\|^2\,dt \le \int 8(T-t)\operatorname{Im}\langle u, Pu\rangle + 64T^2|||A|||\langle ABu, Bu\rangle\,dt \tag{A.14}$$

for $u \in \mathcal{S}(\mathbf{R}\times\mathbf{R}^n)$ having support where $|t| \le T \le c$. In order to estimate the $\langle ABu, Bu\rangle$ term we use the following

Lemma A.3. *If P satisfies conditions* (A.1)–(A.3) *we find*

$$(2-\gamma)\int \langle ABu, Bu\rangle\,dt \le 2\int \operatorname{Im}\langle Pu, Bu\rangle + C\|u\|^2\,dt \tag{A.15}$$

for $u \in \mathcal{S}(\mathbf{R}\times\mathbf{R}^n)$ having support where $|t| < c$.

By using Lemma A.3 we find

$$\int \|u\|^2\,dt \le \int 8(T-t)\operatorname{Im}\langle u, Pu\rangle + \tfrac{128}{2-\gamma}T^2|||A|||\operatorname{Im}\langle Pu, Bu\rangle + \tfrac{64}{2-\gamma}CT^2\|u\|^2\,dt \tag{A.16}$$

if $|t| \le T \ll 1$ in $\operatorname{supp} u$. Now by the Cauchy–Schwarz inequality we have

$$8(T-t)\operatorname{Im}\langle u, Pu\rangle \le 16T|\langle u, Pu\rangle| \le 256T^2\|Pu\|^2 + \frac{1}{4}\|u\|^2. \tag{A.17 }$$

When $\frac{64}{2-\gamma}CT^2 \le \frac{1}{4}$ this gives (A.5). □

Proof of Lemma A.3. Define

$$M_t(u) = \langle Bu, u\rangle \tag{A.18}$$

for $u \in \mathcal{S}(\mathbf{R}\times\mathbf{R}^n)$ with support where $\{\,|t| < c\}$. Then, by taking the derivative and using Remark A.1 and (A.3), we obtain for any $|t| < c$ that

$$\begin{aligned} \partial_t M_t(u) &= 2\operatorname{Re}\langle Bu, \partial_t u\rangle + \langle(\partial_t B)u, u\rangle \\ &= \langle(\partial_t B)u, u\rangle + 2\operatorname{Re}\langle Bu, iPu\rangle + 2\langle Bu, ABu\rangle \\ &\quad + 2\operatorname{Re}\langle Bu, -iRu\rangle \\ &\ge 2\operatorname{Re}\langle Bu, iPu\rangle + (2-\gamma)\langle Bu, ABu\rangle - C\|u\|^2 \end{aligned} \tag{A.19}$$

since $2\operatorname{Re}\langle Bu, -iRu\rangle = 2\operatorname{Re}\langle -iRu, Bu\rangle = 2\operatorname{Im}\langle Ru, Bu\rangle$. We have

$$2\operatorname{Re}\langle Bu, iPu\rangle = 2\operatorname{Re}\langle iPu, Bu\rangle = -2\operatorname{Im}\langle Pu, Bu\rangle, \tag{A.20}$$

thus, integrating (A.19) with respect to t, we obtain

$$(2-\gamma)\int \langle ABu, Bu\rangle\,dt \le \int 2\operatorname{Im}\langle Pu, Bu\rangle + C\|u\|^2\,dt. \tag{A.21}$$

This proves (A.15) and the lemma. □

Appendix B. Wick operators

In this appendix we shall define the Wick operators we are going to use. We assume that $f \in S(H^{-1}, G)$ where G is σ temperate, $G/G^\sigma \le H^2 \le 1$ and $G = Hg/h$, where $\sup g/g^\sigma = h^2 \le 1$.

Choose an partition of unity $\{\phi_\nu(w)\}_\nu$ and $\{\psi_\nu(w)\}_\nu \in S(1,g)$ such that $\psi_\nu \ge 0$ and $\phi_\nu \ge 0$, $\sum_\nu \phi_\nu^2 = 1$, $\psi_\nu \equiv 1$ on $\operatorname{supp}\phi_\nu$ and ψ_ν has compact support where $g \cong g_\nu = g_{w_\nu}$ and $h \cong h_\nu = h(w_\nu)$. For each g_ν there exists a unique symplectic intermediate metric $g_\nu^\sharp$, such that

$$g_\nu \le g_\nu^\sharp = g_\nu^{\sharp\,\sigma} \le g_\nu^\sigma. \tag{B.1}$$

We also find $h_\nu^{-1} g_\nu \le g_\nu^\sharp \le h_\nu g_\nu^\sigma$. Also, $G \cong H g_\nu / h_\nu$ in $\operatorname{supp}\psi_\nu$.

Following Lerner [7], we define the local Wick quantization: for $f \in L^\infty(T^*\mathbf{R}^n)$ we let

$$f^{Wick_\nu}(x, D_x) = \int_{T^*\mathbf{R}^n} f(w)\Sigma^w_{\nu,w}(x, D_x)\, dw \tag{B.2}$$

where $\Sigma_{\nu,w_0}(w) = 2^n \exp(-2\pi g_\nu^\sharp(w - w_0))$. By Proposition 4.2 in [7], we have

$$f \ge 0 \implies f^{Wick_\nu}(x, D_x) \ge 0 \qquad \text{on } L^2(\mathbf{R}^n) \tag{B.3,}$$

$(f^{Wick_\nu})^* = (\overline{f})^{Wick_\nu}$, and

$$\|f^{Wick_\nu}(x, D_x)\|_{\mathcal{L}(L^2(\mathbf{R}^n))} \le \|f\|_{L^\infty(\mathbf{R}^n)}. \tag{B.4}$$

If $f \in S(\Lambda, \Lambda^{-1} g_\nu^\sharp)$ for bounded $\Lambda \ge 1$, we find

$$f^{Wick_\nu} = f^w + r^{Wick_\nu}, \tag{B.5}$$

where r is uniformly bounded (see Proposition 4.3 in [7], where Λ need not be constant). For more general symbols we make the following definition:

Definition B.1. *For f measurable satisfying $|f| \le Ch^{-N}$, we define*

$$f^{Wick} = \sum_\nu \phi_\nu^w f_\nu^{Wick_\nu} \phi_\nu^w \tag{B.6}$$

where $f_\nu = \psi_\nu f$.

Here we use the symbols $\{\phi_\nu\}_\nu$ and $\{\psi_\nu\}_\nu \in S(1,g)$ above. Since $|f_\nu| \le Ch^{-N}(w_\nu)$ we find $f_\nu \in L^\infty(\mathbf{R}^n)$, so this is a well-defined quantization.

Proposition B.2. *We find that f^{Wick} maps $\mathcal{S}(\mathbf{R}^n) \mapsto L^2(\mathbf{R}^n)$. The quantization has the following properties: $\|f^{Wick}(x, D_x)\|_{\mathcal{L}(L^2(\mathbf{R}^n))} \le C\|f\|_{L^\infty(\mathbf{R}^n)}$, $(f^{Wick})^* = (\overline{f})^{Wick}$ on $\mathcal{S}$, and*

$$f \ge 0 \implies f^{Wick} \ge 0 \qquad \text{on } L^2. \tag{B.7}$$

For $f \in S(H^{-1}, G)$ we find $f^{Wick} = f^w + R$ where $\|R\|_{\mathcal{L}(L^2(\mathbf{R}^n))} \le C$.

Proof. If $u, v \in \mathcal{S}(\mathbf{R}^n)$, we find from (B.4) that

$$|\langle f^{Wick}u, v\rangle| \le C\sum_\nu h^{-N}(w_\nu)|\langle \phi_\nu^w u, \phi_\nu^w v\rangle| \le C\|\Phi u\|\|v\| \tag{B.8}$$

where $\Phi(w) = \{h^{-N}(w_\nu)\phi_\nu(w)\}_\nu \in S(h^{-N}, g)$. This gives $\|f^{Wick}u\| \le C\|\Phi u\| \le C'$ when $u \in \mathcal{S}$.

We have that $f \ge 0$ implies $f_\nu^{Wick_\nu} \ge 0$ in L^2 by (B.3), thus we find $f^{Wick} \ge 0$ in L^2. Now, if $f \in S(H^{-1}, G)$ we find that

$$f_\nu \in S(H^{-1}, Hg_\nu/h_\nu) \subseteq S(H^{-1}, Hg_\nu^\sharp). \tag{B.9}$$

Thus, we find $f_\nu^{Wick_\nu} = f_\nu^w + r_\nu^{Wick_\nu}$ by (B.5), where r_ν is uniformly bounded. We obtain that $\|r_\nu^{Wick_\nu}\| \le C$ uniformly in ν. This gives $f^{Wick} = \sum_\nu \phi_\nu^w f_\nu^w \phi_\nu^w + R$, where $R = \sum_\nu \phi_\nu^w r_\nu^{Wick_\nu}\phi_\nu^w$ is bounded in L^2. Since $\sum_\nu \phi_\nu^w f_\nu^w \phi_\nu^w \cong f^w$ modulo $\operatorname{Op} S(H, G)$, we obtain the proposition. □

Remark B.3. It is easy to see that if $f \in C^1$ and $|f| + |\partial_t f| \le Ch^{-N}$, then $\partial_t f^{Wick} = (\partial_t f)^{Wick}$.

References

[1] R. Beals and C. Fefferman, On local solvability of linear partial differential equations, *Ann. of Math.* **97** (1973), 482–498.

[2] N. Dencker, The solvability of non L^2 solvable operators, Journees “Équations aux dérivées Partielles” Jean-de-Monts 1996.

[3] L. Hörmander, *The analysis of linear partial differential operators, I-IV,* Springer Verlag, Berlin, Heidelberg, New York, Tokyo, 1983–1985.

[4] ——— On the solvability of pseudodifferential equations, *Structure of solutions of differential equations*, 183–213, World Scientific, Singapore, New Jersey, London, Hong Kong, editors M. Morimoto and T. Kawai, 1996.

[5] N. Lerner, Sufficiency of condition (Ψ) for local solvability in two dimensions, *Ann. of Math.* **128** (1988), 243–258.

[6] ——— Nonsolvability in L^2 for a first order operator satisfying condition (Ψ), *Ann. of Math.* **139** (1994), 363–393.

[7] ——— Energy methods via coherent states and advanced pseudodifferential calculus, preprint.

[8] A. Menikoff, On local solvability of pseudo-differential equations, *Proc. Amer. Math. Soc.* **43** (1974), 149–154.

[9] L. Nirenberg and F. Treves, On local solvability of linear partial differential equations. I: Necessary conditions, *Comm. Pure Appl. Math.* **23** (1970), 1–38; II: Sufficient conditions *Comm. Pure Appl. Math.* **23** (1970), 459–509; Correction, *Comm. Pure Appl. Math.* **24** (1971), 279–288.

[10] J.-M. Trépreau, Sur la résolubilité analytique microlocale des opérateurs pseudodifférentiels de type principal, Ph.D. thesis, Université de Reims, 1984.

Department of Mathematics
University of Lund, Box 118
S-221 00 Lund, Sweden
e-mail address: dencker@maths.lth.se

On the uniqueness of the Cauchy problem under partial analyticity assumptions

Lars Hörmander

1. Introduction

A few years ago Robbiano [7] proved a striking and surprising uniqueness theorem for hyperbolic differential operators in $\mathbf{R}^{n+1}$ of the form

$$P = D_t^2 - A(x, D_x), \tag{1.1}$$

where A is a positive elliptic second order operator in an open set $\Omega \subset \mathbf{R}^n$ with real principal symbol $a(x,\xi)$. By a quantitatively more precise form of his result given in Hörmander [5] there is unique continuation of solutions of the equation $Pu = 0$ across a timelike surface with conormal (τ, ξ) at a point (t, x) provided that

$$27\tau^2/23 - a(x,\xi) < 0. \tag{1.2}$$

It is of course classical that there is unique continuation across spacelike surfaces, that is, when $\tau^2 - a(x,\xi) > 0$. If the coefficients of A are analytic, then Holmgren's theorem proves uniqueness when $\tau^2 - a(x,\xi) < 0$, which is a weaker condition than (1.2), so it was natural to ask if this would be true without the analyticity assumption. An affirmative answer was given recently by Tataru [8] as a special case of a more general result combining features of Holmgren's uniqueness theorem and of the uniqueness theorems involving convexity conditions based on Carleman estimates (see Hörmander [3, Chapter VIII] and [4, Chapter XXVIII]).

Tataru considered a differential operator $P(x, D)$ of arbitrary order in $\mathbf{R}^n$ such that with a splitting $x = (x', x'')$ of the coordinates all coefficients are real analytic with respect to x' and the principal symbol $p(x,\xi)$ satisfies one of the following conditions:

(i) the coefficients of $p(x,\xi)$ are real and independent of x';
(ii) the coefficients of $p(x,\xi)$ are entire analytic functions of x' of order 2, and $p(x,\xi) \neq 0$ when $\xi \in \mathbf{R}^n \setminus \{0\}$ and $\xi' = 0$.

Then there is unique continuation across a surface for which the pseudoconvexity condition in [4, Chapter XXVIII] is satisfied just for cotangent vectors with $\xi' = 0$.

In this paper we shall improve on these results. In particular we shall prove that the hypothesis (i) can be relaxed to assuming that the

Typeset by $\mathcal{A}_{\mathcal{M}}\mathcal{S}$-TEX

coefficients of $p(x,(0,\xi''))$ are real and independent of x'. This is an invariant condition which is not the case for (i). In fact, the conormal bundle of a leaf $\Sigma_c = \{(x',c)\}$ in the foliation is naturally locally flat since it is isomorphic to the pullback of the corresponding cotangent space at c of the space $\mathbf{R}^{n''}$ of leaves whereas the restriction of the full cotangent bundle to a leaf does not have a natural (local) trivialisation. We shall also weaken the very restrictive global analyticity assumption in (ii) to just local analyticity in the x' variables. Similar results will be proved under a principal normality condition.

The proofs in [8] depend on two new ideas. The traditional Carleman method for proving uniqueness of the Cauchy problem is based on the remark that if $u \in L^2_{\text{comp}}(\mathbf{R}^n)$, $\varphi \in C(\mathbf{R}^n)$, and

$$\int |e^{\tau\varphi}u|^2\,dx \le C, \quad \tau > \tau_0, \tag{1.3}$$

then $u = 0$ in $\{x \in \mathbf{R}^n; \varphi(x) > 0\}$. Tataru observed that the same conclusion holds if

$$\int |e^{-Q(D)/\tau}e^{\tau\varphi}u|^2\,dx \le C, \quad \tau > \tau_0, \tag{1.4}$$

where Q is a positive definite quadratic polynomial in $D' = -i\partial/\partial x'$, so that $e^{-Q(D)/\tau}$ is a highly regularizing operator in the x' variables. We shall give a general discussion of this point in Section 2. Estimates of the form (1.3) have usually been based on Carleman estimates such as

$$\int |e^{\tau\varphi}u|^2\,dx \le C\int |e^{\tau\varphi}P(x,D)u|^2\,dx, \quad u \in C_0^\infty(\Omega),\ \tau > \tau_0, \tag{1.5}$$

where Ω is a neighborhood of the point x^0 where uniqueness is to be proved. The second main point in [8] is the observation that if $\varphi(x^0) = 0$ and

$$\begin{aligned}\int |e^{-Q(D)/\tau}e^{\tau\varphi}u|^2\,dx &\le C\int |e^{-Q(D)/\tau}e^{\tau\varphi}P(x,D)u|^2\,dx \\ &+ Ce^{-2\tau c}\int |u(x)|^2\,dx, \quad u \in C_0^\infty(\Omega),\ \tau > \tau_0,\end{aligned} \tag{1.6}$$

where $c > 0$, then (1.4) follows with φ replaced by $\varphi + c$ if $P(x,D)u = 0$ when $\varphi + c > 0$, hence $u = 0$ when $\varphi > -c$. Admitting the last term in (1.6) is essential, for introducing $e^{-Q(D)/\tau}e^{\tau\varphi}u$ as a new unknown one is forced to consider the inverse of the operator $e^{-Q(D)/\tau}$ which has very bad continuity properties. To make it manageable and to gain control of x' derivatives one has to introduce cutoffs leading to errors which are controlled by the last term in (1.6). We shall explain this in greater detail in Section 2.

Section 3 is devoted to technical lemmas required for the proof of estimates of the form (1.6). To warm up we then prove in Section 4 case (i) of Tataru's result. At first we assume that all coefficients of $P(x, D)$ are independent of x', for only a very small part of the lemmas in Section 3 is needed in that case, but the result already contains the desired uniqueness theorem for the operator (1.1). Section 5 proves a uniqueness theorem corresponding to case (ii) of Tataru's result but with just a local analyticity assumption on the leading coefficients too. In Section 6 we discuss operators such that only $p(x, (0, \xi''))$ satisfies condition (i). Section 7 is devoted to an extension to operators of principal type. Thus the main theorem in Section 7 contains that in Section 6 which in turn contains that in Section 4. We hope that this successive increase in generality will explain the reason for the technicalities in the proofs, and we shall not repeat the arguments which carry over from a simpler to a more general case.

Two months after this paper was completed I was informed by L. Robbiano and C. Zuily that they had also proved the uniqueness theorems which appear here in Sections 4, 5, 6, and that they hoped to prove a result like that in Section 7 here. This was done in the preprint [9] which they sent me a month later. Their methods differ by relying strongly on Sjöstrand's FBI transforms where Gaussian mollifiers are built in whereas such mollifiers are used directly in this paper.

2. Modification of the Carleman method

Following [8, Section 4] we shall now prove a general result on the conclusions which can be drawn from weakened versions of (1.3) such as (1.4).

Proposition 2.1. *Let $u \in \mathcal{E}'^k(\mathbf{R}^n)$, and let φ be a real valued function in $C^k(\Omega)$ where Ω is a neighborhood of* $\operatorname{supp} u$. *For $\tau > 0$ let A_τ be continuous (bounded) functions in $\mathbf{R}^n$ such that $A_\tau(\xi) \to 1$ uniformly on every compact set when $\tau \to +\infty$. If*

$$\|A_\tau(D)e^{\tau\varphi}u\|_{L^2} \le C \quad \text{when } \tau > \tau_0, \tag{2.1}$$

it follows that $\varphi \le 0$ in $\operatorname{supp} u$.

Proof. Without restriction we may assume that $\varphi \in C_0^k(\mathbf{R}^n)$. Choose $f \in \mathcal{S}(\mathbf{R}^n)$ with Fourier transform $\hat{f} \in C_0^\infty(\mathbf{R}^n)$, and set $h = \varphi_*(fu)$, that is,

$$\langle h, \chi\rangle = \langle fu, \varphi^*\chi\rangle, \quad \chi \in C^\infty(\mathbf{R}), \tag{2.2}$$

where $\varphi^*\chi = \chi \circ \varphi$. Then $h \in \mathcal{E}'(\mathbf{R})$ and $\operatorname{supp} h \subset \{\varphi(x); x \in \operatorname{supp} u\}$. Since

$$\hat{h}(\tau) = \langle fu, e^{-i\tau\varphi}\rangle = \langle u, e^{-i\tau\varphi} f\rangle = \langle e^{-i\tau\varphi}u, f\rangle, \quad \tau \in \mathbf{C},$$

we have

$$|\hat{h}(\tau)| = |\langle u, e^{-i\tau\varphi} f\rangle| \le C_f(1+|\tau|)^k \le C_f 2^k |\tau + i|^k, \quad \tau \in \mathbf{R}.$$

For sufficiently large $\tau > 0$ we obtain using (2.1) and the fact that $\hat{f} \in C_0^\infty$

$$\begin{aligned}|\hat{h}(i\tau)| = |\langle f, e^{\tau\varphi}u\rangle| &= |\langle A_\tau(-D)^{-1}f, A_\tau(D)e^{\tau\varphi}u\rangle| \\ &\le 2\|f\|_{L^2}\|A_\tau(D)e^{\tau\varphi}u\|_{L^2} \le 2C\|f\|_{L^2}.\end{aligned}$$

Since $\hat{h}$ is an entire function of exponential type it follows from the Phragmén-Lindelöf theorem first that $\hat{h}(\tau)/(\tau+i)^k$ is bounded when $\operatorname{Im}\tau > 0$, and then that

$$|\hat{h}(\tau)| \le C_f 2^k |\tau + i|^k, \quad \operatorname{Im}\tau > 0.$$

By the Paley-Wiener theorem this implies that $\operatorname{supp} h \subset (-\infty, 0]$. If $\chi \in C_0^\infty((0,\infty))$ we have therefore proved that

$$0 = \langle fu, \varphi^*\chi\rangle = \langle (\varphi^*\chi)u, f\rangle.$$

Since f is any function in a dense subset of $\mathcal{S}$ it follows that $(\varphi^*\chi)u = 0$, which means that $u = 0$ when $\varphi > 0$. The proposition is proved.

As in [8] we shall in what follows take $A_\tau(D) = e^{-Q(D)/\tau}$ for some positive semi-definite quadratic form Q. This is convenient since the Fourier transform of a Gaussian is an explicit Gaussian, and it is reminiscent of the proofs in [5] and [7]. They were based on an application of $e^{-D_t^2/\tau}$ to a cutoff of a solution of the equation $(D_t^2 - A(x, D_x))u = 0$, followed by an application of standard Carleman estimates to the function obtained when t varies on a parallel of the imaginary axis. However, many other choices of A_τ could be made.

The standard approach to Carleman estimates of the form (1.5) starts from the observation that

$$e^{\tau\varphi}P(x,D)u = P(x,D,\tau)(e^{\tau\varphi}u),$$

where

$$(2.3) \qquad P(x,D,\tau) = e^{\tau\varphi}P(x,D)e^{-\tau\varphi} = P(x, D + i\tau\varphi'(x)),$$

since $e^{\tau\varphi}D_j e^{-\tau\varphi} = D_j + i\tau\partial_j\varphi$. Estimates such as (1.5) are therefore equivalent to estimates for the operator $P(x,D,\tau)$ which no longer involve a weight function. Note that if $p(x,\xi)$ is the principal symbol of $P(x,D)$, of order m, then the symbol of $P(x,D,\tau)$ is $p(x,\xi+i\tau\varphi'(x))$ apart from terms of order $< m$ in (ξ,τ).

To prove estimates of the form (1.6), where Q is a positive semi-definite quadratic form, it is similarly natural to consider

$$e^{-Q(D)/\tau}e^{\tau\varphi}P(x,D)u = e^{-Q(D)/\tau}P(x,D,\tau)(e^{\tau\varphi}u).$$

The operator $e^{-Q(D)/\tau}$ commutes with differentiations but commuting it through the coefficients of $P(x,D,\tau)$ is more delicate. If $v \in \mathcal{S}(\mathbf{R}^n)$ then $e^{-Q(D)/\tau}(x_j v)$ has the Fourier transform

$$e^{-Q(\xi)/\tau}i\partial\hat{v}(\xi)/\partial\xi_j = i\partial(e^{-Q(\xi)/\tau}\hat{v}(\xi))/\partial\xi_j + \tfrac{i}{\tau}Q_j(\xi)e^{-Q(\xi)/\tau}\hat{v}(\xi),$$

where $Q_j(\xi) = \partial Q(\xi)/\partial\xi_j$, which means that

$$e^{-Q(D)/\tau}(x_j v) = x_j e^{-Q(D)/\tau}v + \tfrac{i}{\tau}Q_j(D)e^{-Q(D)/\tau}v.$$

More generally, if f is a polynomial it follows that

$$e^{-Q(D)/\tau}(f(x)v) = f(x + \tfrac{i}{\tau}Q'(D))e^{-Q(D)/\tau}v.$$

Note that

$$[x_j + \tfrac{i}{\tau}Q_j(D), x_k + \tfrac{i}{\tau}Q_k(D)] = \tfrac{1}{\tau}Q_{jk} - \tfrac{1}{\tau}Q_{kj} = 0,$$

so the expression is well defined. We therefore get

$$(2.4) \quad e^{-Q(D)/\tau}e^{\tau\varphi}P(x,D)u = P(x + iQ'(D)/\tau, D, \tau)(e^{-Q(D)/\tau}e^{\tau\varphi}u),$$

if the coefficients of $P(x,D,\tau)$ are polynomials. The interpretation of the right-hand side is that a term $c(x)D^\alpha\tau^j$ in $P(x,D,\tau)$ is replaced by $c(x + iQ'(D)/\tau)D^\alpha\tau^j$. This makes sense if the coefficients $c(x)$ are polynomial along parallels of $\mathcal{A} = \{Q'(\xi); \xi \in \mathbf{R}^n\}$ which is a linear subspace of $\mathbf{R}^n$. In particular, the operator is defined if $P(x, D)$ is translation invariant in the directions of $\mathcal{A}$ and φ is a polynomial, for then the coefficients of $P(x, D, \tau) = P(x, D + i\tau\varphi'(x))$ are invariant for such translations apart from factors involving derivatives of φ. If φ is a quadratic polynomial, which may be assumed in the proof of uniqueness theorems, then $\varphi'(x + iQ'(D)/\tau) = \varphi'(x) + i\varphi''Q'(D)/\tau$, where φ'' is a constant symmetric matrix, so the operator on the right-hand side of (2.4) becomes

$$P(x, D - \varphi''Q'(D) + i\tau\varphi'(x)) \tag{2.5}$$

which is a small perturbation of $P(x, D, \tau) = P(x, D + i\tau\varphi'(x))$ if the coefficients of Q are small. We shall discuss in Section 4 how this simple formula leads to the improvement in [8] of the results of [5] and [7] mentioned in the introduction.

The operator in the right-hand side of (2.4) can be given a sense if the coefficients of P are entire analytic in the direction $\mathcal{A}$, with suitable properties. This was used in [8]. However, we just want to assume local analyticity properties. That will require that the operators $x + iQ'(D)/\tau$ are cut off in a suitable manner so that the error committed can be absorbed in the last exponentially small term in an estimate such as (1.6). These matters will be discussed in Section 3 before we return in Sections 4 – 6 to proving uniqueness theorems along the lines indicated above.

3. Technical lemmas

In this section we shall examine the effect of the Gaussian regularizer discussed in Section 2. To simplify notation we shall take $Q(D) = \varepsilon|D|^2/2$ where ε is a small positive number. Later on we shall apply the operator $e^{-\varepsilon|D|^2/2\tau}$ only in a group of the variables in $\mathbf{R}^n$. The L^2 norm will just be denoted by $\|\cdot\|$, and we shall write

$$\|u\|_{(\sigma,\tau)} = \Big((2\pi)^{-n}\int|\hat{u}(\xi)|^2(|\xi|^2+\tau^2)^\sigma\,d\xi\Big)^{\frac{1}{2}}, \quad u \in H_{(\sigma)}(\mathbf{R}^n),$$

where $H_{(\sigma)}$ is the Sobolev space consisting of temperate distributions u with $\hat{u} \in L^2_{\text{loc}}$ for which the right-hand side is finite.

The reason why the Gaussian regularization is useful in the proof of estimates of the form (1.6) is that it allows one to estimate a derivative by a small multiple of τ, with an exponentially small error. To make this precise we fix a cutoff function $\chi \in C_0^\infty(\mathbf{R}^n)$ such that $0 \le \chi \le 1$, $\chi(x) = 1$ when $|x| < 1$ and $\chi(x) = 0$ when $|x| > 2$, and we set $\chi_\kappa = \chi(\cdot/\kappa)$, $\tilde{\chi}_\kappa = 1 - \chi_\kappa$ where $\kappa > 0$.

Lemma 3.1. *For arbitrary $\varepsilon > 0$, $\kappa > 0$, $\tau > 0$, we have*

$$(3.1)\qquad \|\chi_\kappa(\varepsilon D/\tau)(\varepsilon D/\tau)W\| \le 2\kappa\|W\|, \quad W \in \mathcal{S}(\mathbf{R}^n),$$

and for fixed $\sigma \in \mathbf{R}$, $\varepsilon > 0$, $\kappa > 0$, we have for $\tau > \tau_0(\sigma, \varepsilon, \kappa)$

$$(3.2)\quad \|\tilde{\chi}_\kappa(\varepsilon D/\tau)(\varepsilon D/\tau)e^{-\varepsilon|D|^2/2\tau}w\| \le e^{-\tau\kappa^2/4\varepsilon}\|w\|_{(\sigma,\tau)}, \quad w \in \mathcal{S}(\mathbf{R}^n).$$

Proof. (3.1) is trivial since $|\varepsilon\xi/\tau| \le 2\kappa$ when $\chi_\kappa(\varepsilon\xi/\tau) \ne 0$. By Parseval's formula (3.2) follows if we prove that

$$\tilde{\chi}(\varepsilon\xi/\kappa\tau)|\varepsilon\xi/\tau|(|\xi|^2 + \tau^2)^{-\frac{1}{2}\sigma} \le \exp(\varepsilon|\xi|^2/2\tau - \tau\kappa^2/4\varepsilon).$$

Since $\tilde{\chi}(\varepsilon\xi/\kappa\tau) \ne 0$ implies $|\varepsilon\xi| \ge \kappa\tau$, we have then

$$\varepsilon|\xi|^2/2\tau - \tau\kappa^2/4\varepsilon \ge \kappa|\xi|(\tfrac{1}{2} - \tfrac{1}{4}) = \kappa|\xi|/4 \ge \kappa^2\tau/4\varepsilon,$$

and the estimate follows because the left-hand side is $O(|\xi|^{|\sigma|+1}\tau^{|\sigma|-1})$.

Combining (3.1) and (3.2) we obtain when $\tau > \tau_0(\sigma, \varepsilon, \kappa)$

$$(3.3)\quad \|De^{-\varepsilon|D|^2/2\tau}w\| \le \tau\big(2\kappa\varepsilon^{-1}\|e^{-\varepsilon|D|^2/2\tau}w\| + e^{-\tau\kappa^2/4\varepsilon}\varepsilon^{-1}\|w\|_{(\sigma,\tau)}\big), \quad w \in \mathcal{S}(\mathbf{R}^n).$$

When κ/ε is small this confirms that a differentiation to the left of the regularizer can be estimated by a small multiple of τ times the regularizer apart from an exponentially small error. This remains true for higher order derivatives,

$$\begin{aligned}\||D|^k e^{-\varepsilon|D|^2/2\tau}w\| &\le (2\kappa\tau/\varepsilon)\||D|^{k-1}e^{-\varepsilon|D|^2/2\tau}w\| \\ &+ (\tau/\varepsilon)e^{-\tau\kappa^2/4\varepsilon}\||D|^{k-1}w\|_{(\sigma,\tau)} \le (2\kappa\tau/\varepsilon)^k\|e^{-\varepsilon|D|^2/2\tau}w\| \\ &+ (\tau/\varepsilon)e^{-\tau\kappa^2/4\varepsilon}(\||D|^{k-1}w\|_{(\sigma,\tau)} + \dots + (2\kappa\tau/\varepsilon)^{k-1}\|w\|_{(\sigma,\tau)}) \\ &\le (2\kappa\tau/\varepsilon)^k\|e^{-\varepsilon|D|^2/2\tau}w\| + e^{-\tau\kappa^2/4\varepsilon}\|w\|_{(\sigma+k+1,\tau)}\end{aligned}$$

for large τ. Replacing σ by $\sigma - k - 1$ we conclude that for $\tau > \tau_0(\sigma, \varepsilon, \kappa, k)$

$$(3.3)'\quad \||D|^k e^{-\varepsilon|D|^2/2\tau}w\| \le (2\kappa\tau/\varepsilon)^k\|e^{-\varepsilon|D|^2/2\tau}w\| + e^{-\tau\kappa^2/4\varepsilon}\|w\|_{(\sigma,\tau)}, \quad w \in \mathcal{S}(\mathbf{R}^n).$$

Before examining how localized the operator $e^{-\varepsilon|D|^2/2\tau}$ is in the space variables we must prove some simple and well-known facts on the norms $\|\cdot\|_{(\sigma,\tau)}$.

Lemma 3.2. *For every $\sigma \in \mathbf{Z}$ there is a constant C_σ such that*

$$\|fu\|_{(\sigma,\tau)} \le C_\sigma \|u\|_{(\sigma,\tau)} \max_{|\alpha|\le|\sigma|} \tau^{-|\alpha|} \sup_B |D^\alpha f|, \quad u \in H_{(\sigma)}(\mathbf{R}^n),$$

when $B = \mathbf{R}^n$ or B is any ball in $\mathbf{R}^n$ with radius $\ge 1/\tau$ containing $\operatorname{supp} u$*, and $D^\alpha f \in L^\infty(B)$ when $|\alpha| \le |\sigma|$.*

Proof. Let us first assume that $B = \mathbf{R}^n$. By duality it suffices to prove the estimate when σ is an integer ≥ 0, and then it follows at once from the fact that for $|\alpha| \le \sigma$

$$\begin{aligned}\tau^{\sigma-|\alpha|}\|D^\alpha(fu)\| &\le \tau^{\sigma-|\alpha|} \sum_{\beta+\gamma=\alpha} \tbinom{\alpha}{\beta} \|D^\beta u\| \sup|D^\gamma f| \\ &\le \tau^{\sigma-|\alpha|} \sum_{\beta+\gamma=\alpha} \tbinom{\alpha}{\beta} \tau^{|\beta|-\sigma} \|u\|_{(\sigma,\tau)} \sup|D^\gamma f| \\ &\le \|u\|_{(\sigma,\tau)} 2^{|\alpha|} \max_{|\gamma|\le|\alpha|} \tau^{-|\gamma|} \sup|D^\gamma f|.\end{aligned}$$

If B is a ball with finite radius, the statement follows if we combine the preceding special case with the extension of f given in the following lemma.

Lemma 3.3. *Let σ be an integer ≥ 0, and let $1 \le p \le \infty$. If $D^\alpha f \in L^p(B)$ for $|\alpha| \le \sigma$, where B is a ball with radius $\ge 1/\tau$, then there is an extension $\tilde f$ of f to $\mathbf{R}^n$ such that*

$$\max_{|\alpha|\le\sigma} \tau^{\sigma-|\alpha|}\|D^\alpha \tilde f\|_{L^p(\mathbf{R}^n)} \le C_\sigma \max_{|\alpha|\le\sigma} \tau^{\sigma-|\alpha|}\|D^\alpha f\|_{L^p(B)}. \tag{3.4}$$

Proof. Introducing τx as a new variable instead of x shows that we may assume that $\tau = 1$, and that $B = B_R = \{x \in \mathbf{R}^n; |x| < R\}$ for some $R \ge 1$. We can then define $\tilde f = 0$ in $\complement B_{R+1}$ and define $\tilde f$ in $B_{R+1} \setminus B_R$ using the Seeley extension,

$$\tilde f((R+r)\omega) = \chi(r)\sum_0^{\sigma-1} a_j f((R+\lambda_j r)\omega), \quad 0 \le r < 1, \ |\omega| = 1,$$

where the numbers $\lambda_j \in (-1,0)$ are all different, $\sum_0^{\sigma-1} a_j \lambda_j^\nu = 1$ for $0 \le \nu < \sigma$, and $\chi \in C_0^\infty((-\frac12, \frac12))$ is equal to 1 in $(-\frac14, \frac14)$.

We can now prove an analogue of Lemma 3.1 for localization in space. As in the preceeding proof we shall write $B_R = \{x \in \mathbf{R}^n; |x| < R\}$.

Lemma 3.4. *For arbitrary $\kappa > 0$ we have*

$$\|\chi_\kappa(x)xW\| \le 2\kappa\|W\|, \quad W \in \mathcal{S}(\mathbf{R}^n), \tag{3.5}$$

and for fixed $\sigma \in \mathbf{R}$, $\varepsilon > 0$, $\kappa > 0$, we have for $\tau > \tau_0(\sigma, \varepsilon, \kappa)$

$$\|\tilde\chi_\kappa(x)xe^{-\varepsilon|D|^2/2\tau}w\| \le e^{-\tau\kappa^2/4\varepsilon}\|w\|_{(\sigma,\tau)}, \quad w \in C_0^\infty(B_{\kappa/4}). \tag{3.6}$$

Proof. The estimate (3.5) is trivial since $|\chi_\kappa(x)x| \le 2\kappa$. To prove (3.6) we may assume that σ is an integer ≤ 0, and we note that the kernel k of $e^{-\varepsilon|D|^2/2\tau}$ is given explicitly by

$$k(x,y) = (\tau/2\pi\varepsilon)^{n/2}e^{-\tau|x-y|^2/2\varepsilon}. \tag{3.7}$$

Using Lemma 3.3 we can estimate $e^{-\varepsilon|D|^2/2\tau}w(x)$ by $\|w\|_{(\sigma,\tau)}$ times the norm of $k(x,\cdot)$ in $B_{\kappa/4}$ corresponding to the right-hand side of (3.4), with $p = 2$ and σ replaced by $|\sigma|$. If we apply D_y to $k(x,y)$, we get a factor $\tau(x-y)/\varepsilon$, and differentiations of higher order give a product of such factors with possibly an even number of them replaced by $\sqrt{\tau/\varepsilon}$, which altogether can introduce in the estimate of the derivatives of order $\le |\sigma|$ a factor of the form $(\tau|x-y|/\varepsilon + \sqrt{\tau/\varepsilon})^{|\sigma|}$. When $|x| > \kappa$ the square of the L^2 norm of $k(x,\cdot)$ over $B_{\kappa/4}$ can be estimated by

$$\Big(\frac{\tau}{2\pi\varepsilon}\Big)^n\int_{|y|<\kappa/4} e^{-\tau(|x|-|y|)^2/\varepsilon}\,dy \le (\tau/2\pi\varepsilon)^n e^{-9\tau|x|^2/16\varepsilon}(\kappa/4)^n m(B_1),$$

which implies that

$$|e^{-\varepsilon|D|^2/2\tau}w(x)| \\ \le C_\sigma(\tau\kappa/\varepsilon)^{\frac12 n}e^{-9\tau|x|^2/32\varepsilon}(\tau|x|/\varepsilon + \sqrt{\tau/\varepsilon} + \tau)^{|\sigma|}\|w\|_{(\sigma,\tau)}.$$

The square of the L^2 norm when $|x| > \kappa$ of the product by $|x|$ can be estimated by

$$C'_\sigma(\tau\kappa/\varepsilon)^n e^{-9\tau\kappa^2/16\varepsilon}(\tau/\varepsilon)^{|\sigma|-1-n/2}(1+\sqrt{\tau/\varepsilon})^{2|\sigma|+n}\|w\|^2_{(\sigma,\tau)},$$

and since $9/16 > 1/2$ the lemma is proved.

Lemmas 3.1 and 3.4 suffice for the discussion of partially translation invariant operators in Theorem 4.1, but the more general results in Theorem 4.9 and Sections 5, 6, 7 require a more detailed study of the commutation of multiplication operators through the Gaussian regularizer. The identity

$$(x + i\varepsilon D/\tau)e^{-\varepsilon|D|^2/2\tau}w = e^{-\varepsilon|D|^2/2\tau}(xw)$$

proved in Section 2 is not useful as it stands if one wants to iterate and then use a power series expansion to replace x by a function $c(x)$ which is just analytic in a neighborhood of the origin. With the notation in Lemmas 3.1 and 3.4 we shall therefore consider the truncated operator

$$X_\kappa(x,D) = \chi(x/\kappa)x + i\chi(\varepsilon D/\kappa\tau)\varepsilon D/\tau, \tag{3.8}$$

where κ is a small positive parameter, and estimate the remainder

$$\widetilde{X}_\kappa(x,D) = x + i\varepsilon D/\tau - X_\kappa(x,D) = \tilde{\chi}(x/\kappa)x + i\tilde{\chi}(\varepsilon D/\kappa\tau)\varepsilon D/\tau. \tag{3.9}$$

We have estimated the terms in $X_\kappa(x,D)$ and in $\widetilde{X}_\kappa(x,D)$ in Lemmas 3.1 and 3.4. In particular, (3.1), (3.5) and (3.2), (3.6) give

$$\|X_\kappa(x,D)W\| \le 4\kappa\|W\|, \quad W \in \mathcal{S}(\mathbf{R}^n), \tag{3.10}$$

$$\|\widetilde{X}_\kappa(x,D)e^{-\varepsilon|D|^2/2\tau}w\| \le 2e^{-\tau\kappa^2/4\varepsilon}\|w\|_{(\sigma,\tau)}, \quad w \in C_0^\infty(B_{\kappa/4}). \tag{3.11}$$

Write

$$(x + i\varepsilon D/\tau)^\alpha e^{-\varepsilon|D|^2/2\tau} = X_\kappa(x,D)^\alpha e^{-\varepsilon|D|^2/2\tau} + R_{\kappa\alpha}.$$

Then $R_{\kappa\alpha}$ is a component of $\widetilde{X}_\kappa(x,D)$ if $|\alpha| = 1$, hence estimated in (3.11) then. We shall next give a similar estimate for $|\alpha| > 1$.

If $x^\alpha = x_j x^\beta$, $|\beta| \ge 1$, then

$$e^{-\varepsilon|D|^2/2\tau}x^\alpha w = (X_{\kappa j}(x,D) + \widetilde{X}_{\kappa j}(x,D))e^{-\varepsilon|D|^2/2\tau}x^\beta w$$

$$= X_{\kappa j}(x,D)(X_\kappa(x,D)^\beta e^{-\varepsilon|D|^2/2\tau} + R_{\kappa\beta})w + \widetilde{X}_{\kappa j}(x,D)e^{-\varepsilon|D|^2/2\tau}x^\beta w.$$

Thus

$$R_{\kappa\alpha}w = X_{\kappa j}(x,D)R_{\kappa\beta}w + \widetilde{X}_{\kappa j}(x,D)e^{-\varepsilon|D|^2/2\tau}x^\beta w,$$

and by (3.10), (3.11) we obtain

$$\|R_{\kappa\alpha}w\| \le 4\kappa\|R_{\kappa\beta}w\| + 2e^{-\tau\kappa^2/4\varepsilon}\|x^\beta w\|_{(\sigma,\tau)}, \quad w \in C_0^\infty(B_{\kappa/4}),$$

if τ is large enough. If $M_\nu = \max_{|\alpha|=\nu}\|R_{\kappa\alpha}w\|$ it follows when $\kappa\tau \ge 4$ by applying Lemma 3.2 in the second term that

$$M_{\nu+1} \le 4\kappa M_\nu + C'_\sigma(\nu+1)^{|\sigma|+1}e^{-\tau\kappa^2/4\varepsilon}(\kappa/4)^\nu A, \quad A = \|w\|_{(\sigma,\tau)},$$

which by (3.11) remains true for $\nu = 0$ with $M_0 = 0$. Thus we have

$$M_{\nu+1}/(4\kappa)^{\nu+1} \le M_\nu/(4\kappa)^\nu + C'_\sigma(\nu+1)^{|\sigma|+1}e^{-\tau\kappa^2/4\varepsilon}16^{-\nu}A/4\kappa,$$

and adding the inequalities gives with $C''_\sigma = C'_\sigma\sum_0^\infty(\nu+1)^{|\sigma|+1}16^{-\nu}$

$$M_\nu \le C''_\sigma(4\kappa)^{\nu-1}e^{-\tau\kappa^2/4\varepsilon}A.$$

We have therefore proved

Lemma 3.5. *Let $\varepsilon, \sigma, \kappa$ be fixed. Then we have for $w \in C_0^\infty(B_{\kappa/4})$ and large τ, with $X_\kappa(x,D)$ defined by (3.8)*

$$(3.12)\quad \|e^{-\varepsilon|D|^2/2\tau} x^\alpha w - X_\kappa(x,D)^\alpha e^{-\varepsilon|D|^2/2\tau} w\| \le C_\sigma (4\kappa)^{|\alpha|-1} e^{-\tau\kappa^2/4\varepsilon} \|w\|_{(\sigma,\tau)}.$$

Remark. One can of course take $C_\sigma = 4\kappa$, for $C_{\sigma-1}/\tau \le 4\kappa$ when τ is large enough. The order of the factors in $X_\kappa(x,D)^\alpha$ is arbitrary.

Corollary 3.6. *If c is an analytic function in the polydisc*

$$\Delta = \{z \in \mathbf{C}^n; |z_j| < 5\kappa,\ j = 1, \dots, n\}, \quad \textit{and} \quad M = 5^n \sup_{z\in\Delta} |c(z)|,$$

then $c(X_\kappa(x,D))$ can be defined by the power series expansion of c, and for large τ

$$(3.13)\quad \|e^{-\varepsilon|D|^2/2\tau}(cw) - c(X_\kappa(x,D)) e^{-\varepsilon|D|^2/2\tau} w\| \le M e^{-\tau\kappa^2/4\varepsilon} \|w\|_{(\sigma,\tau)},\quad w \in C_0^\infty(B_{\kappa/4}),$$

$$(3.14)\qquad \|c(X_\kappa(x,D))W\| \le M\|W\|, \quad W \in \mathcal{S}(\mathbf{R}^n).$$

Proof. The power series expansion

$$c(z) = \sum_\alpha c_\alpha z^\alpha$$

converges absolutely in Δ, and $(5\kappa)^{|\alpha|}|c_\alpha| \le M/5^n$ by the Cauchy inequalities. Hence $\sum_\alpha (4\kappa)^{|\alpha|}|c_\alpha| \le M$. By (3.10) this proves that $c(X_\kappa(x,D)) = \sum c_\alpha X_\kappa(x,D)^\alpha$ converges in the L^2 operator norm and that the norm of the sum is $\le M$. The sum of the error terms in (3.12) is also convergent, which yields (3.13) if $C_\sigma = 4\kappa$ as in the remark above.

Corollary 3.6 allows us to move an analytic factor $c(x)$ from the right-hand side of the Gaussian regularizer to the left-hand side as $c(X_\kappa(x,D))$, provided that κ is small enough. However, we need to know that $c(X_\kappa(x,D))$ is approximately equal to a pseudodifferential operator with a symbol which can be calculated. To do so we shall compare $X_\kappa(x,D)^\alpha$ with the operator $\mathrm{Op}(X_\kappa(x,\xi)^\alpha)$ having the symbol

$$X_\kappa(x,\xi)^\alpha = (\chi(x/\kappa)x + i\chi(\varepsilon\xi/\kappa\tau)\varepsilon\xi/\tau)^\alpha.$$

For fixed α this can be done using pseudo-differential calculus, but we must also examine how the estimates depend on α. In the product $X_\kappa(x,D)^\alpha$ we have to move the terms involving D to the right. It suffices to estimate the commutator in one such step together with the norms of the other factors to get the necessary bounds, for we are now dealing with operators which will stand to the left of the regularizer and only need to gain a factor $(|D|^2+\tau^2)^{-\frac{1}{2}}$ or just τ^{-1}.

Set

$$a_\kappa(\xi)=\chi(\varepsilon\xi/\kappa\tau)\varepsilon\xi/\tau, \quad b_\kappa(x)=\chi(x/\kappa)x,$$

thus $X_\kappa(x,\xi)=b_\kappa(x)+ia_\kappa(\xi)$. Then we have for all α,β

$$(3.15)\quad |a_\kappa^{(\alpha)}(\xi)|\le C_\alpha\kappa(|\xi|+\kappa\tau/\varepsilon)^{-|\alpha|}, \quad |b_{\kappa(\beta)}(x)|\le C_\beta\kappa(|x|+\kappa)^{-|\beta|}.$$

The second estimate follows from the first by taking $\varepsilon=\tau=1$. To prove the first we observe that $\varepsilon|\xi|\le 2\kappa\tau$ in $\operatorname{supp} a_\kappa$, hence

$$\begin{aligned}|a_\kappa^{(\alpha)}(\xi)|&\le 2\kappa\sup|\chi^{(\alpha)}|(\varepsilon/\kappa\tau)^{|\alpha|}+|\alpha|\sup_{|\beta|=|\alpha|-1}|\chi^{(\beta)}|(\varepsilon/\kappa\tau)^{|\alpha|-1}\varepsilon/\tau\\&\le C'_\alpha\kappa(\varepsilon/\kappa\tau)^{|\alpha|}\le 3^{|\alpha|}C'_\alpha\kappa(|\xi|+\kappa\tau/\varepsilon)^{-|\alpha|},\end{aligned}$$

which proves the claim. We can therefore regard $a_\kappa(\xi)$ and $b_\kappa(x)$ as symbols of weight κ with respect to the metric

$$(3.16)\qquad g=\frac{|dx|^2}{(|x|+\kappa)^2}+\frac{|d\xi|^2}{(|\xi|+\kappa\tau/\varepsilon)^2}.$$

(See [4, Chapter XVIII] for the definition of symbols with respect to weights and metrics.) Note that

$$(|x|+\kappa)(|\xi|+\kappa\tau/\varepsilon)\ge\kappa(|\xi|+\kappa\tau/\varepsilon)\ge\kappa^2\tau/\varepsilon\ge 1, \quad \text{if } \tau>\varepsilon/\kappa^2.$$

Hence the weight of the symbol of a commutator $[a_{\kappa j}(D),b_{\kappa k}(x)]$ is at most $C\kappa^2/(\kappa(\kappa\tau/\varepsilon+|\xi|))\le C\varepsilon/\tau$. The operator norm in L^2 is therefore $O(\varepsilon/\tau)$, and the operator norm of the product to the right by D is $O(\kappa)$.

A monomial $X_\kappa(x,D)^\alpha=(b_\kappa(x)+ia_\kappa(D))^\alpha$ is a sum of $2^{|\alpha|}$ terms each of which is a product of $|\alpha|$ factors $a_{\kappa j}(D)$ and $b_{\kappa k}(x)$. If every factor $a_{\kappa j}(D)$ is moved to the right, we obtain the desired pseudodifferential operator $\operatorname{Op}(b_\kappa(x)+ia_\kappa(\xi))^\alpha$. However, we must estimate the commutator terms which occur in this process. We can start with moving to the right the factor $a_{\kappa j}(D)$ farthest to the right which has a right hand neighbor $b_{\kappa k}$. When a commutator appears we stop. Moving the $a_{\kappa j}$'s to the right will yield fewer than $|\alpha|^2$ terms of the form

$$(3.17)\qquad (a_\kappa, b_\kappa)^{\alpha'}[a_{\kappa j}, b_{\kappa k}]b_\kappa^{\alpha''}a_\kappa^{\alpha'''},\quad |\alpha'|+|\alpha''|+|\alpha'''| = |\alpha|-2,$$

in addition to the desired term with all $a_{\kappa j}$ to the right. As operator in L^2 the commutator has norm $\le C\varepsilon/\tau$, and the norm of the other factors is $\le 2\kappa$, so the norm of such a term is $\le (C\varepsilon/\tau)(2\kappa)^{|\alpha|-2}$. Summing up, the norm in L^2 of $\mathrm{Op}\,(X_\kappa(x,\xi)^\alpha) - (\mathrm{Op}\,X_\kappa(x,\xi))^\alpha$ is $\le 2^{|\alpha|}|\alpha|^2(C\varepsilon/\tau)(2\kappa)^{|\alpha|-2}$. We have therefore proved

Lemma 3.7. *For the difference* $S_{\kappa,\alpha} = (\mathrm{Op}\,X_\kappa(x,\xi))^\alpha - \mathrm{Op}\,(X_\kappa(x,\xi)^\alpha)$ *we have if* $\varepsilon \le \kappa^2\tau$

$$\tau\|S_{\kappa,\alpha}W\| \le C|\alpha|^2(4\kappa)^{|\alpha|-2}\varepsilon\|W\|,\quad W\in\mathcal{S}(\mathbf{R}^n).$$

The powers of $|\alpha|$ here do not matter, and we get immediately a result on commutation with functions which are analytic in the polydisc Δ of Corollary 3.6:

Corollary 3.8. *Let* c *and* M *be as in Corollary 3.6. Then for* $\tau > \tau_0(\sigma,\varepsilon,\kappa)$

$$(3.18)\qquad \begin{aligned}&\|e^{-\varepsilon|D|^2/2\tau}(cw) - \tilde c(x,D)e^{-\varepsilon|D|^2/2\tau}w\|\\ &\qquad\le M(e^{-\tau\kappa^2/4\varepsilon}\|w\|_{(\sigma,\tau)} + C(\varepsilon/\kappa^2\tau)\|e^{-\varepsilon|D|^2/2\tau}w\|),\end{aligned}$$

if $w\in C_0^\infty(B_{\kappa/4})$. *Here the symbol* $\tilde c$ *is defined by*

$$(3.19)\qquad \begin{aligned}&\tilde c(x,\xi) = c(X_\kappa(x,\xi))\in S(1,g),\\ &\qquad g = \frac{|dx|^2}{(|x|+\kappa)^2} + \frac{|d\xi|^2}{(|\xi|+\kappa\tau/\varepsilon)^2},\end{aligned}$$

thus

$$(3.20)\qquad \begin{aligned}&|\tilde c^{(\alpha)}_{(\beta)}(x,\xi)| \le C_{\alpha\beta}(|x|+\kappa)^{-|\beta|}(|\xi|+\kappa\tau/\varepsilon)^{-|\alpha|}M_0,\\ &\qquad M_0 = \sup_{z\in\Delta}|c(z)|.\end{aligned}$$

For the proof of (3.20) we just have to note that in a term in the derivative $\partial^{\alpha+\beta}c(X_\kappa(x,\xi))/\partial x^\beta\partial\xi^\alpha$ where c has been differentiated ν times the derivative of c can be estimated by $\nu!M_0/\kappa^\nu$ while we have ν factors which are derivatives of a component of a_κ or b_κ contributing a factor κ each according to (3.15).

Remark 3.9. If $|\alpha|+|\beta|\neq 0$ we may replace M_0 in (3.20) by

$$M_1=\sup_{z\in\Delta}|c(z)-c(0)|, \tag{3.21}$$

for replacing c by $c-c(0)$ does not change the derivatives of $\tilde{c}$. If $|\alpha||\beta|\neq 0$ we may replace M_0 by

$$M_2=\sup_{z\in\Delta}|c(z)-c(0)-\langle z,c'(0)\rangle|, \tag{3.22}$$

for subtracting a linear function from c means subtracting from $\tilde{c}$ the sum of a function of x and a function of ξ, which disappears when one differentiates with respect to both x and ξ.

In Sections 5 – 7 we shall also have to estimate the difference

$$\tilde{c}(x,\xi)-\tilde{c}(x,0)=c(b_\kappa(x)+ia_\kappa(\xi))-c(b_\kappa(x)).$$

From (3.21), (3.22) it follows that

$$\begin{aligned}|\partial_\xi^\alpha\partial_x^\beta(\tilde{c}(x,\xi)-\tilde{c}(x,0))|&\\ \le C_{\alpha\beta}(|x|+\kappa)^{-|\beta|}(|\xi|+\kappa\tau/\varepsilon)^{-|\alpha|}\cdot&\begin{cases}M_1 & \text{if } \beta=0,\\ M_2 & \text{if } \beta\neq 0.\end{cases}\end{aligned} \tag{3.23}$$

The case where $\beta=0$, $\alpha\neq 0$ is an immediate consequence of (3.20), since $\tilde{c}(x,0)$ drops out, and so is the case where $|\alpha||\beta|\neq 0$. Also the case where $\alpha=0$ and $\beta\neq 0$ follows from (3.20) by Taylor's formula, for

$$\begin{aligned}|\partial_x^\beta(\tilde{c}(x,\xi)-\tilde{c}(x,0))|&\le C_\beta|\xi|(|x|+\kappa)^{-|\beta|}(\kappa\tau/\varepsilon)^{-1}M_2\\ &\le 2C_\beta(|x|+\kappa)^{-|\beta|}M_2,\quad \text{if } |\xi|\le 2\kappa\tau/\varepsilon,\end{aligned}$$

and $\tilde{c}(x,\xi)-\tilde{c}(x,0)=0$ when $|\xi|\ge 2\kappa\tau/\varepsilon$. The case $\alpha=\beta=0$ follows in the same way using Taylor's formula.

4. Operators with real and partially translation invariant principal symbol

In this section we shall just prove one of the uniqueness theorems of Tataru [8] as an introduction to the improvements which will be given in Sections 5 – 7. We shall consider a differential operator $P(x,D)$ in an open set $\Omega\subset\mathbf{R}^n$ with real principal symbol, and assume at first that the whole operator is invariant under translations in the directions of a linear subspace. The coordinates can be chosen so that with the notation

$$x' = (x_1, \dots, x_\nu), \quad x'' = (x_{\nu+1}, \dots, x_n),$$

the subspace is defined by $x'' = 0$, that is, the coefficients of $P(x, D)$ are functions of x'' defined in an open neighborhood Ω'' of $0 \in \mathbf{R}^{n-\nu}$. Thus the Gaussian regularizer $e^{-\varepsilon|D'|^2/2\tau}$ operates in the x' variables for fixed x'', so we can apply the results of Section 3 regarding x'' as a parameter.

Theorem 4.1. *Let $P(x, D)$ be a differential operator of order m with C^∞ coefficients in a neighborhood Ω of $0 \in \mathbf{R}^n$. Assume that the coefficients are independent of $x' = (x_1, \dots, x_\nu)$ and that those of the principal symbol $p(x, \xi)$ are real valued. Let $\psi \in C^2(\Omega)$ be real valued, $\psi'(0) \neq 0$, and assume that the level surface at 0 is conormally strictly pseudoconvex at 0 in the sense that, with $\{\cdot, \cdot\}$ denoting a Poisson bracket,*

$$(4.1) \quad \{p, \{p, \psi\}\}(0, \xi) > 0 \quad \text{if } 0 \neq \xi = (0, \xi'') \in \mathbf{R}^n \text{ and } p(0, \xi) = \langle p'_\xi(0, \xi), \psi'(0)\rangle = 0,$$

$$(4.2) \quad \{\overline{p(x, \xi + i\tau\psi'(x))}, p(x, \xi + i\tau\psi'(x))\}/2i\tau > 0 \quad \text{if } x = 0,\ \xi = (0, \xi'') \in \mathbf{R}^n, \tau > 0 \text{ and } p(0, \xi + i\tau\psi'(0)) = \langle p'_\xi(0, \xi + i\tau\psi'(0)), \psi'(0)\rangle = 0.$$

If $u \in H^{\text{loc}}_{(m-1)}(\Omega)$ satisfies the differential equation $P(x, D)u = 0$ and $\psi \le \psi(0)$ in supp u, *it follows that* $0 \notin$ supp u.

At the end of this section we shall prove that the result remains valid when the coefficients of the lower order terms are just analytic in x'. In Section 6 we shall prove that it suffices to assume that the coefficients of $p((x', x''), (0, \xi''))$ are real and independent of x'. This is an invariant condition which the stronger hypothesis in Theorem 4.1 is not.

To facilitate references in the following sections, the proof of Theorem 4.1 will be given as a series of lemmas. The first lemma concerns the improvement of the hypothesis made possible by substituting for ψ a convex increasing function of ψ such as $e^{\lambda\psi}$ with a large positive λ. (See also [4, Proposition 28.3.3].)

Lemma 4.2. *If $\varphi = e^{\lambda\psi}$ then we have for $\tau \neq 0$ and $\tau_1 = \tau\lambda\varphi(x)$*

$$(4.3)\quad \{\overline{p(x,\xi+i\tau\varphi'(x))}, p(x,\xi+i\tau\varphi'(x))\}/2i\tau$$
$$= \lambda\varphi(x)\big(\{\overline{p(x,\xi+i\tau_1\psi'(x))}, p(x,\xi+i\tau_1\psi'(x))\}/2i\tau_1 +$$
$$\lambda|\langle p'_\xi(x,\xi+i\tau_1\psi'(x)),\psi'(x)\rangle|^2\big).$$

Proof. The left-hand side can be written more explicitly in the form

$$(4.4)\quad \sum_{j,k=1}^{n} \frac{\partial^2\varphi(x)}{\partial x_j\partial x_k} p^{(j)}(x,\xi+i\tau\varphi'(x))\overline{p^{(k)}(x,\xi+i\tau\varphi'(x))}$$
$$+ \operatorname{Re}\sum_{j=1}^{n} \overline{p^{(j)}(x,\xi+i\tau\varphi'(x))}p_{(j)}(x,\xi+i\tau\varphi'(x))/i\tau.$$

Here and in what follows we use the notation $p^{(j)}(x,\xi) = \partial p(x,\xi)/\partial\xi_j$ and $p_{(j)}(x,\xi) = \partial p(x,\xi)/\partial x_j$ as in [4]. Since

$$\tau\varphi'(x) = \tau\lambda\varphi(x)\psi'(x) = \tau_1\psi'(x),$$
$$\varphi''(x) = \lambda\varphi(x)\psi''(x) + \lambda^2\varphi(x)\psi'(x)\otimes\psi'(x),$$

the identity (4.3) follows from (4.4).

Since $p(x,\xi)$ is real valued, the Poisson bracket on the left-hand side of (4.3) vanishes identically when $\tau = 0$, so the left-hand side of (4.3) is a homogeneous polynomial in (ξ,τ) of degree $2m-2$, which reduces to $\{p,\{p,\varphi\}\}(x,\xi)$ when $\tau = 0$ as is easily seen from (4.4). Thus (4.3) implies

$$(4.3)'\quad \{p,\{p,\varphi\}\}(x,\xi)$$
$$= \lambda\varphi(x)\big(\{p,\{p,\psi\}\}(x,\xi) + \lambda|\langle p'_\xi(x,\xi),\psi'(x)\rangle|^2\big).$$

Now it follows from the hypotheses (4.1) and (4.2) that we can choose $\lambda > 0$ so large that (4.3) is positive when $x = 0$ for all $(\xi,\tau)\in\mathbf{R}^{n+1}$ with $\xi' = 0$, $|\xi|^2+\tau^2 = 1$ and $p(x,\xi+i\tau\varphi'(x)) = 0$. The continuity in $\varphi''(0)$ which is obvious from (4.4) proves that this remains true if we replace φ by the quadratic polynomial

$$\varphi_\delta(x) = \sum_{|\alpha|\le 2} (\partial^\alpha\varphi)(0)x^\alpha/\alpha! - \delta|x|^2$$

provided that δ is small enough. Then we have $\varphi_\delta(x) < \varphi_\delta(0)$ when $x \in \Omega_1 \cap \operatorname{supp} u \setminus \{0\}$ if Ω_1 is a sufficiently small neighborhood of the origin. Changing notation, we have proved

Lemma 4.3. *When the hypotheses of Theorem 4.1 are fulfilled there is a real valued quadratic polynomial φ and a neighborhood Ω_1 of $0 \in \mathbf{R}^n$ such that $\varphi(x) < \varphi(0)$ when $x \in \Omega_1 \cap \operatorname{supp} u \setminus \{0\}$ and*

$$(4.5)\quad \{p, \{p, \varphi\}\}(x, \xi) > 0 \quad \text{if } x = 0,\ 0 \neq \xi \in \mathbf{R}^n,\ \xi' = 0,\ p(x,\xi) = 0;$$

$$(4.6)\quad \{\overline{p(x, \xi + i\tau\varphi'(x))}, p(x, \xi + i\tau\varphi'(x))\}/2i\tau > 0, \quad \text{if } x = 0,\ \xi \in \mathbf{R}^n,\ \xi' = 0,\ \tau > 0,\ p(x, \xi + i\tau\varphi'(x)) = 0.$$

In what follows we shall write $p(x, \xi, \tau) = p(x, \xi + i\tau\varphi'(x))$.

Lemma 4.4. *From* (4.5) *and* (4.6) *it follows if Ω_1 is a sufficiently small neighborhood of* 0 *that there are constants C_j such that*

$$(4.7)\quad (|\xi|^2 + \tau^2)^{m-1} \leq C_1 \{\overline{p(x,\xi,\tau)}, p(x,\xi,\tau)\}/2i\tau + C_2 |p(x,\xi,\tau)|^2/(|\xi|^2+\tau^2) + C_3 |\xi'|^{2(m-1)}, \quad x \in \overline{\Omega}_1,\ \xi \in \mathbf{R}^n.$$

Proof. The terms in the right-hand side are homogeneous of degree $2m-2$ in (ξ, τ) and continuous when $|\xi|^2 + \tau^2 = 1$, since $p(x, \xi, 0)$ is real. By hypothesis we can choose C_1 so that (4.7) is valid with strict inequality when $|\xi|^2 + \tau^2 = 1$, $p(x, \xi, \tau) = 0$, $x = 0$ and $\xi' = 0$, hence in a neighborhood. This proves that (4.7) is valid for sufficiently large C_2 and C_3 when Ω_1 is small enough.

Lemma 4.5. *The estimate* (4.7) *implies that*

$$(4.8)\quad \tfrac{1}{2}\|V\|^2_{(m-1,\tau)} \leq C_1 \|p(x,D,\tau)V\|^2/\tau + C_3 \||D'|^{m-1}V\|^2, \quad V \in C_0^\infty(K),$$

if K is a compact neighborhood of 0 *contained in Ω_1 and τ is sufficiently large.*

Proof. The proof of (4.8) follows from the fact that

$$\|p(x,D,\tau)V\|^2 \geq \|p(x,D,\tau)V\|^2 - \|p(x,D,\tau)^*V\|^2 = (F(x,D,\tau)V, V)$$

where the symbol of F is equal to

$$(4.9)\qquad \{\overline{p(x,\xi,\tau)}, p(x,\xi,\tau)\}/i$$

apart from terms in $S((|\xi| + \tau)^{2m-2}, g_\tau)$ where g_τ is the metric $|dx|^2 + |d\xi|^2/(|\xi|^2 + \tau^2)$. (See [4, Chap. XVIII] for this notation.) Hence it follows from the elementary Gårding inequality and (4.7), with $p(x,\xi,\tau)$ cut off outside a neighborhood of K, that

$$\tfrac{1}{2}\|V\|^2_{(m-1,\tau)} \le C_1(F(x,D,\tau)V,V)/2\tau + C_2\|p(x,D,\tau)V\|^2_{(-1,\tau)} + C_3\||D'|^{m-1}V\|^2,$$

when $V \in C_0^\infty(K)$, and this proves (4.8).

Proof of Theorem 4.1. The estimate (4.8) would become a conventional Carleman estimate if V is replaced by $e^{\tau\varphi}u$. To prove estimates of the form (1.6) with Ω replaced by a smaller neighborhood of $0 \in \mathbf{R}^n$ we recall that by (2.5), with $Q(D) = \varepsilon|D'|^2/2$, we have

$$e^{-\varepsilon|D'|^2/2\tau}e^{\tau\varphi}P(x,D)u = P(x,D-\varepsilon\varphi''(D',0)+i\tau\varphi'(x))v, \tag{4.10}$$

where $v = e^{-\varepsilon|D'|^2/2\tau}e^{\tau\varphi}u$. If $u \in C_0^\infty(B_{\kappa/4})$, $B_\delta = \{x \in \mathbf{R}^n; |x| \le \delta\}$, then $\operatorname{supp} v \subset \{x \in \mathbf{R}^n; |x''| \le \kappa/4\}$ which is not a compact set. However, if we choose $\chi \in C_0^\infty(\mathbf{R}^\nu)$ as in Section 3 with $\chi(x') = 1$ when $|x'| < 1$ and $\chi(x') = 0$ when $|x'| > 2$, then the support of $V(x) = \chi(x'/2\kappa)v(x)$ is contained in

$$K_\kappa = \{(x',x'') \in \mathbf{R}^n; |x'| \le 4\kappa,\ |x''| \le \kappa/4\}.$$

For small κ we can therefore apply (4.8) to V. From (3.6), with $\sigma = -1$, and w replaced by a derivative of $e^{\tau\varphi}u$ of order $< m$, we obtain in view of (4.10) for large τ

$$\|P(x,D-\varepsilon\varphi''(D',0)+i\tau\varphi'(x))V - \chi(x'/2\kappa)e^{-\varepsilon|D'|^2/2\tau}e^{\tau\varphi}P(x,D)u\| \le Ce^{-\tau\kappa^2/4\varepsilon}\|e^{\tau\varphi}u\|_{(m-1,\tau)}. \tag{4.11}$$

In fact, the commutator $[P(x,D-\varepsilon\varphi''(D',0)+i\tau\varphi'(x)),\chi(x'/2\kappa)]$ is of order $\le m-1$ in (D,τ), and $2\kappa \le |x'| \le 4\kappa$ in the support of the coefficients, which implies that $\tilde\chi_\kappa(x') = 1$.

The estimate (4.8) remains valid with the constant $\frac{1}{3}$ instead of $\frac{1}{2}$ if $p(x,\xi,\tau)$ is replaced by a small perturbation which remains real when $\tau = 0$, for the estimate (4.7) remains valid with the left-hand side multiplied by $\frac{2}{3}$. If ε is sufficiently small we may therefore replace $p(x,\xi,\tau)$ by $p(x,\xi-\varepsilon\varphi''(\xi',0)+i\tau\varphi'(x))$. Hence

$$\tfrac{1}{2}\|V\|_{(m-1,\tau)} \le \sqrt{C_1}\tau^{-\frac{1}{2}}\|P(x,D-\varepsilon\varphi''(D',0)+i\tau\varphi'(x))V\| + \sqrt{C_3}\||D'|^{m-1}V\|, \quad V \in C_0^\infty(K_\kappa), \tag{4.12}$$

when τ is large enough, for terms in $P(x, D-\varepsilon\varphi''(D',0)+i\tau\varphi'(x))V$ of order $< m$ in (D,τ) can be estimated by a constant times $\|V\|_{(m-1,\tau)}$ so their contribution to the right-hand side is less than $(\sqrt{\frac{1}{3}}-\frac{1}{2})\|V\|_{(m-1,\tau)}$ for large τ.

Combining (4.12) with (4.11) and its analogue

$$\||D'|^{m-1}V\| \le \||D'|^{m-1}v\| + Ce^{-\tau\kappa^2/4\varepsilon}\|e^{\tau\varphi}u\|_{(m-2,\tau)}, \tag{4.11$'$}$$

we now obtain with v as in (4.10)

$$\begin{aligned}\tfrac{1}{2}\|V\|_{(m-1,\tau)} \le \sqrt{C_1}\tau^{-\frac{1}{2}}&\|e^{-\varepsilon|D'|^2/2\tau}e^{\tau\varphi}P(x,D)u\| \\ &+ \sqrt{C_3}\||D'|^{m-1}v\| + C_4e^{-\tau\kappa^2/4\varepsilon}\|e^{\tau\varphi}u\|_{(m-1,\tau)}.\end{aligned}$$

Now Lemma 3.1, or rather (3.3)$'$, gives

$$\||D'|^{m-1}v\| \le (2\kappa\tau/\varepsilon)^{m-1}\|v\| + Ce^{-\tau\kappa^2/4\varepsilon}\|e^{\tau\varphi}u\|.$$

When κ is so small that $\sqrt{C_3}(2\kappa/\varepsilon)^{m-1} < \frac{1}{4}$, we obtain

$$\begin{aligned}\tfrac{1}{2}\|V\|_{(m-1,\tau)} \le \sqrt{C_1}\tau^{-\frac{1}{2}}\|e^{-\varepsilon|D'|^2/2\tau}e^{\tau\varphi}P(x,D)u\| + \tfrac{1}{4}\tau^{m-1}\|v\| \\ + C_5e^{-\tau\kappa^2/4\varepsilon}\|e^{\tau\varphi}u\|_{(m-1,\tau)}.\end{aligned}$$

Since $v - V = \tilde{\chi}(x'/2\kappa)e^{-\varepsilon|D'|^2/2\tau}e^{\tau\varphi}u$, it follows from (3.6) that for large τ

$$\|v\|_{(m-1,\tau)} \le \|V\|_{(m-1,\tau)} + Ce^{-\tau\kappa^2/4\varepsilon}\|e^{\tau\varphi}u\|_{(m-1,\tau)},$$

so we have proved that for small κ and large τ

$$\begin{aligned}\tfrac{1}{4}\|e^{-\varepsilon|D'|^2/2\tau}&e^{\tau\varphi}u\|_{(m-1,\tau)} \\ &\le \sqrt{C_1}\tau^{-\frac{1}{2}}\|e^{-\varepsilon|D'|^2/2\tau}e^{\tau\varphi}P(x,D)u\| \\ &\quad + C_6e^{-\tau\kappa^2/4\varepsilon}\|e^{\tau\varphi}u\|_{(m-1,\tau)}, \quad u \in C_0^\infty(B_{\kappa/4}).\end{aligned} \tag{4.13}$$

By approximation the estimate extends to all u in $H_{(m-1)}$ with $P(x,D)u$ in L^2 and $\operatorname{supp} u \subset B_{\kappa/4}$.

Now let u be the function in Theorem 4.1, let $\varphi(0) = 0$, and choose a cutoff function $h \in C_0^\infty(B_{\kappa/4})$ which is equal to 1 in a neighborhood of 0. Since $\varphi \le 0$ in $\operatorname{supp}(hu)$ we have for large τ

$$C_6e^{-\tau\kappa^2/4\varepsilon}\|e^{\tau\varphi}(hu)\|_{(m-1,\tau)} \le C_7\tau^{m-1}e^{-\tau\kappa^2/4\varepsilon},$$

We have $P(x,D)(hu) \in L^2$ and the support is contained in $\operatorname{supp} u \cap \operatorname{supp} h'$, hence $\varphi(x) \le -c < 0$ there. This implies that the right-hand side of (4.13) with u replaced by hu is $O(e^{-c'\tau})$ as $\tau \to +\infty$ if $c' = \min(\kappa^2/5\varepsilon, c)$, so it follows that

$$\|e^{-\varepsilon|D'|^2/2\tau} e^{\tau(\varphi+c')}(hu)\|_{(m-1,\tau)} \quad \text{is bounded as } \tau \to +\infty.$$

By Proposition 2.1 it follows that $\varphi + c' \le 0$ in $\operatorname{supp}(hu)$. Since $\varphi(0) = 0$ and $h(0) = 1$, we conclude that $0 \notin \operatorname{supp} u$. The proof is complete.

Remark 4.6. If the level surface of ψ is characteristic at 0, that is, $p(0, \psi'(0)) = 0$, then the condition (4.2) for $\xi = 0$ means by (4.4) that

$$\sum_{j,k=1}^{n} \frac{\partial^2 \psi}{\partial x_j \partial x_k} p^{(j)}(x, \psi'(x)) \overline{p^{(k)}(x, \psi'(x))} \\ + \operatorname{Re} \sum_{j=1}^{n} p_{(j)}(x, \psi'(x)) \overline{p^{(j)}(x, \psi'(x))} > 0$$

when $x = 0$. If we write $\eta = \psi'(0)$ then

$$\begin{aligned} \operatorname{Re}\{\bar p, \{p, \psi\}\}(0,\eta) &= \sum_{j,k=1}^{n} \frac{\partial^2 \psi(0)}{\partial x_j \partial x_k} p^{(j)}(0,\eta) \overline{p^{(k)}(0,\eta)} \\ &\quad + \operatorname{Re} \sum_{j,k=1}^{n} \left(\overline{p^{(k)}(0,\eta)} p^{(j)}_{(k)}(0,\eta) - \overline{p_{(k)}(0,\eta)} p^{(jk)}(0,\eta)\right)\eta_j \\ &= \sum_{j,k=1}^{n} \frac{\partial^2 \psi(0)}{\partial x_j \partial x_k} p^{(j)}(0,\eta) \overline{p^{(k)}(0,\eta)} + \operatorname{Re} \sum_{k=1}^{n} p_{(k)}(0,\eta) \overline{p^{(k)}(0,\eta)}, \end{aligned}$$

where we have used the homogeneity of p to calculate a sum with respect to j. The condition (4.2) for $\xi = 0$ which occurs in the characteristic case is therefore identical to the condition

$$(4.1)' \qquad \operatorname{Re}\{\bar p, \{p, \psi\}\}(0,\eta) > 0, \quad \text{if } \eta = \psi'(0),\ p(0,\eta) = 0.$$

When p has real coefficients it reduces to (4.1), but with later results in mind we have allowed complex coefficients in this calculation. Unless $\partial\psi(0)/\partial x_j = 0$ for $j = 1, \dots, \nu$, it is not a special case of (4.1) as it is in the uniqueness theorems of [3, 4].

We shall now compare Theorem 4.1 with the results of [5, 7] by making the special case where P is of second order more explicit. As just observed, condition (4.2) with $\xi = 0$ means (for any order) that if $p(0, \psi'(0)) = 0$ then $\{p, \{p, \psi\}\}(0, \psi'(0)) > 0$. Geometrically this means that if the level surface of ψ is characteristic at 0 then it must be strictly convex in the direction of the corresponding tangential bicharacteristic. This condition occurs also in the analytic case although it can then be considerably weakened by means of results on propagation of analytic singularities. Apart from that, condition (4.2) is void in the second order case, for roots of the equation $p(0, \xi + \tau\psi'(0)) = 0$ occur in complex conjugate pairs so they must be simple if they are not real. The conditions in Theorem 4.1 are therefore reduced to the strict convexity condition $\{p, \{p, \psi\}\}(0, \xi) > 0$ if $0 \neq \xi$, $p(0, \xi) = 0$, $\{p, \psi\}(0, \xi) = 0$ and $\xi' = 0$ or $\xi = \psi'(0)$.

The convexity conditions are satisfied for all ψ with $\psi'(0) = \nu_0$ if and only if $p(0, \nu_0) \neq 0$ and $p(0, \xi) \neq 0$ when $0 \neq \xi \in \mathcal{A}^\perp$ and $\pi_0(\xi, \nu_0) = 0$, where π_0 is the symmetric bilinear form on T_0^* defined by $p(0, \xi)$, and $\mathcal{A}^\perp \subset T_0^*$ is the orthogonal space of the space $\{(x', 0)\}$ of translation invariance. If $\mathcal{A}^\perp \setminus \{0\}$ does not intersect the characteristic set then this is true for arbitrary ν_0 with $p(0, \nu_0) \neq 0$. If $p(0, \xi)$ has Lorentz signature $1, n-1$ we shall otherwise distinguish two cases:

(i) If $\mathcal{A}^\perp \setminus \{0\}$ does not intersect $\{\xi; p(0, \xi) > 0\}$ but only the closure, then the intersection with the closure is generated by a covector $\xi_0 \neq 0$ with $p(0, \xi_0) = 0$, and $\mathcal{A}^\perp$ is contained in the π_0 orthogonal space of ξ_0. The conditions on ν_0 are then $p(0, \nu_0) \neq 0$ and $\pi_0(\xi_0, \nu_0) \neq 0$. The orthogonal space $\mathcal{A}$ contains no timelike vectors but an isotropic vector t with $\pi_0(\xi_0, \cdot) = \langle t, \cdot \rangle$.

(ii) If $\mathcal{A}^\perp$ intersects the timelike cone $\{\xi; p(0, \xi) > 0\}$ of covectors, then ν_0 is not π_0 orthogonal to any $\xi \in \mathcal{A}^\perp \setminus \{0\}$ with $p(0, \xi) = 0$ if ν_0 belongs to the sum of the cone of timelike vectors and the π_0 orthogonal space $\mathcal{A}^\flat$ of $\mathcal{A}^\perp$. This is the space corresponding to $\mathcal{A}$ under the identification of tangent and cotangent vectors defined by π_0.

Thus we have proved:

Corollary 4.7. *If P satisfies the conditions in Theorem 4.1 and is of second order with principal symbol p of Lorentz signature, then there is unique continuation across any non-characteristic surface at* 0 *if the space $\mathcal{A} = \{(x', 0)\}$ of translation invariance at* 0 *contains a timelike vector. If not, but $\mathcal{A}$ contains an isotropic vector $t \neq 0$, then there is unique continuation across all noncharacteristic surfaces for which t is not a tangent at* 0. *If $\mathcal{A} \setminus \{0\}$ only contains spacelike vectors, then there is unique continuation across all noncharacteristic surfaces with conormal which is congruent to a timelike covector modulo $\mathcal{A}^\flat$, the* *subspace of T_0^* corresponding to $\mathcal{A}$ under the identification of tangent and cotangent vectors defined by $p(0,\xi)$.*

Remark. In Theorem 4.1 it would suffice to assume that the coefficients of the principal part are in C^1 while those in the lower order terms are just in L^∞. In fact, instead of appealing to the Gårding inequality we could use the lengthier arguments in [3, Chapter VIII]. Also L^q classes with lower values of q can be allowed if $|\alpha| < m-1$ (see [8]). In Corollary 4.7 we could replace C^1 by Lipschitz since there are no convexity conditions any longer.

There is of course an analogous result for other signatures. It becomes interesting when $p(0,\xi)$ or $-p(0,\xi)$ restricted to $\mathcal{A}^\perp$ has Lorentz signature, but we leave the statement and proof for the reader. Corollary 4.7 is already a vast improvement of the results of [5, 7].

Theorem 4.1 remains valid if the coefficients of the lower order terms are just analytic in the x' variables rather than independent of them. For the proof we need a lemma:

Lemma 4.8. *Let $R(x,D)$ be a differential operator of order $m-1$ with coefficients which can be extended to bounded functions in $\{(z',x'') \in \mathbf{C}^\nu \times \mathbf{R}^{n-\nu}; |z'| < r, |x''| < r\}$ which are analytic with respect to z' for fixed x''. Then it follows that for $u \in C_0^\infty(B_{\kappa/4})$, $5\kappa \le r$,*

$$\begin{aligned}(4.14)\quad &\|e^{-\varepsilon|D'|^2/2\tau}e^{\tau\varphi}R(x,D)u\| \\ &\le C\|e^{-\varepsilon|D'|^2/2\tau}e^{\tau\varphi}u\|_{(m-1,\tau)} + Ce^{-\tau\kappa^2/4\varepsilon}\|e^{\tau\varphi}u\|_{(m-1,\tau)}.\end{aligned}$$

Proof. Writing $R(x,D) = \sum_{|\alpha|\le m-1} c_\alpha D^\alpha$ we note that

$$\begin{aligned}e^{-\varepsilon|D'|^2/2\tau}e^{\tau\varphi}R(x,D)u &= \sum_{|\alpha|\le m-1} e^{-\varepsilon|D'|^2/2\tau}c_\alpha(x)(D+i\tau\varphi')^\alpha e^{\tau\varphi}u \\ &= \sum_{|\alpha|+\nu\le m-1} e^{-\varepsilon|D'|^2/2\tau}c_{\alpha,\nu}(x)D^\alpha\tau^\nu e^{\tau\varphi}u,\end{aligned}$$

where the coefficients $c_{\alpha,\nu}$ have the same analyticity properties as c_α. If (3.13) and (3.14) are applied in the x' variables we obtain for $u \in C_0^\infty(B_{\kappa/4})$, $5\kappa \le r$,

$$\begin{aligned}\|e^{-\varepsilon|D'|^2/2\tau}c_{\alpha,\nu}D^\alpha\tau^\nu e^{\tau\varphi}u\| \le C\|e^{-\varepsilon|D'|^2/2\tau}&D^\alpha\tau^\nu e^{\tau\varphi}u\| \\ &+ Ce^{-\tau\kappa^2/4\varepsilon}\|D^\alpha\tau^\nu e^{\tau\varphi}u\|,\end{aligned}$$

which proves (4.14).

Theorem 4.9. *Theorem 4.1 remains valid if the terms of lower order in $P(x, D)$ are not independent of x' but just analytic in the x' variables.*

Proof. It follows from (4.14) that (4.13) remains valid for large τ with a slight change of constants if $P(x, D)$ is replaced by $P(x, D) + R(x, D)$.

5. Transversally elliptic operators

In Section 4 we mainly studied differential operators $P(x, D)$ in $\mathbf{R}^n$ of order m, such that the coefficients were independent of $x' = (x_1, \dots, x_\nu)$. In the end we saw, in Theorem 4.9, that the coefficients of the terms of order $< m$ could be allowed to depend analytically on x'. In this section we shall examine operators where the coefficients of the principal symbol $p(x, \xi)$ are also allowed to depend analytically on x'. However, we shall avoid the difficulties which occur at real characteristics by assuming that $p(x, \xi) \neq 0$ when $0 \neq \xi \in \mathbf{R}^n$ and $\xi' = 0$. Under this hypothesis Tataru proved uniqueness when the coefficients of p are entire analytic functions of order 2 in the x' variables. We shall only assume local analyticity in the following theorem:

Theorem 5.1. *Let $P(x, D)$ be a differential operator in a neighborhood Ω of $0 \in \mathbf{R}^n$ of order m with C^∞ coefficients in the principal symbol $p(x, \xi)$, bounded coefficients in the lower order terms, and all coefficients analytic with respect to $x' = (x_1, \dots, x_\nu)$ in a neighborhood of 0. Thus they are defined in $\{(z', x'') \in \mathbf{C}^\nu \times \mathbf{R}^{n-\nu}; |z'| < r, |x''| < r\}$ for some $r > 0$. Assume that $p(x, \xi) \neq 0$ when $0 \neq \xi \in \mathbf{R}^n$ and $\xi' = 0$. Let $\psi \in C^2(\Omega)$ be real valued and assume that the pseudoconvexity condition (4.2) is fulfilled. If $u \in H^{\text{loc}}_{(m)}(\Omega)$ and $\psi \leq \psi(0)$ in $\operatorname{supp} u$, it follows that $0 \notin \operatorname{supp} u$.*

Using Lemma 4.2 as in the proof of Lemma 4.3, we obtain an analogue of Lemma 4.3:

Lemma 5.2. *When the hypotheses of Theorem 5.1 are fulfilled there is a real valued quadratic polynomial φ and a neighborhood Ω_1 of $0 \in \mathbf{R}^n$ such that $\varphi(x) < \varphi(0)$ when $x \in \Omega_1 \cap \operatorname{supp} u \setminus \{0\}$ and*

$$\{\overline{p(x, \xi + i\tau\varphi'(x)), p(x, \xi + i\tau\varphi'(x))}\}/2i\tau > 0, \quad \text{if } x = 0, \tag{5.1}$$
$$\xi \in \mathbf{R}^n,\ \xi' = 0,\ \tau > 0,\ p(x, \xi + i\tau\varphi'(x)) = 0.$$

In what follows we shall write $p(x, \xi, \tau) = p(x, \xi + i\tau\varphi'(x))$.

Lemma 5.3. *From (5.1) it follows if Ω_1 is a sufficiently small neighborhood of 0 that there are constants C_j such that*

$$(|\xi|^2 + \tau^2)^m \leq C_1\tau\{\overline{p(x, \xi, \tau)}, p(x, \xi, \tau)\}/2i + C_2|p(x, \xi, \tau)|^2 + C_3|\xi'|^{2m}, \tag{5.2}$$

when $x \in \overline{\Omega}_1$, $\xi \in \mathbf{R}^n$ *and* $\tau \geq 0$.

Proof. The terms in the right-hand side are homogeneous of degree $2m$ in (ξ, τ) and continuous when $|\xi|^2 + \tau^2 = 1$. By hypothesis we can choose C_1 so that the estimate is valid with strict inequality when $|\xi|^2 + \tau^2 = 1$, $p(x, \xi, \tau) = 0$, $x = 0$, $\xi' = 0$ and $\tau \geq 0$, for this implies $\tau > 0$. Hence the inequality is valid also in a neighborhood, so it follows for $|\xi|^2 + \tau^2 = 1$, $\tau \geq 0$, $x \in \overline{\Omega}_1$ for sufficiently large C_2 and C_3 and sufficiently small Ω_1.

As in Lemma 4.5 we obtain an a priori estimate from (5.2):

Lemma 5.4. *The estimate* (5.2) *implies that*

$$\begin{aligned}(5.3)\quad \tfrac{1}{2}\|V\|^2_{(m,\tau)} \leq \tfrac{1}{2}C_1\tau([p(x,D,\tau)^*, p(x,D,\tau)]V, V) \\ + C_2\|p(x,D,\tau)V\|^2 + C_3\||D'|^m V\|^2, \quad V \in C_0^\infty(K),\end{aligned}$$

if K *is a compact neighborhood of* 0 *contained in* Ω_1 *and* τ *is sufficiently large.*

Proof. The right-hand side can be written $(F(x, D, \tau)V, V)$ where the symbol of F belongs to $S((|\xi|+\tau)^{2m}, g_\tau)$ and is equal to the right-hand side of (5.2) apart from terms which are in $S((|\xi|+\tau)^{2m-1}, g_\tau)$. Here g_τ is the metric $|dx|^2 + |d\xi|^2/(|\xi|^2 + \tau^2)$. Hence (5.3) follows from the elementary Gårding inequality and (5.2).

As in Section 4 we must examine

$$e^{-\varepsilon|D'|^2/2\tau} e^{\tau\varphi} p(x, D)u = e^{-\varepsilon|D'|^2/2\tau} p(x, D + i\tau\varphi'(x))(e^{\tau\varphi}u)$$

where φ is a quadratic polynomial. With the notation $p(x, \xi, \tau) = p(x, \xi + i\tau\varphi'(x))$ in Lemmas 5.3, 5.4 the main term in the right-hand side is $e^{-\varepsilon|D'|^2/2\tau}p(x, D, \tau)(e^{\tau\varphi}u)$. If $u \in H_{(m)}$ has support sufficiently close to 0, then Corollary 3.8 shows that this differs by an acceptably small error from $\mathcal{P}_\kappa(x, D, \tau)e^{-\varepsilon|D'|^2/2\tau}e^{\tau\varphi}u$ where

$$\begin{aligned}(5.4)\quad \mathcal{P}_\kappa(x, \xi, \tau) = p(X_\kappa(x, \xi), \xi, \tau), \\ X_\kappa(x, \xi) = (\chi(x'/\kappa)x' + i\chi(\varepsilon\xi'/\kappa\tau)\varepsilon\xi'/\tau, x''),\end{aligned}$$

is well defined for sufficiently small κ since the coefficients of p are analytic with respect to x'. Our next aim is to prove that $p(x, D, \tau)$ can be replaced by $\mathcal{P}_\kappa(x, D, \tau)$ in (5.3) if $u \in C_0^\infty(B_\kappa)$ and κ, ε are small enough.

Lemma 5.5. *If ε and κ are sufficiently small, $\kappa \le \varepsilon$, then*

$$(5.5)\quad \tfrac{1}{4}\|V\|^2_{(m,\tau)} \le \tfrac{1}{2}C_1\tau([\mathcal{P}_\kappa(x,D,\tau)^*, \mathcal{P}_\kappa(x,D,\tau)]V,V) \\ + C_2\|\mathcal{P}_\kappa(x,D,\tau)V\|^2 + C_3\||D'|^m V\|^2, \quad V \in C_0^\infty(B_\kappa),$$

when τ is sufficiently large.

Proof. The estimate (5.3) remains valid if $p(x,\xi,\tau)$ is replaced by

$$(5.6)\qquad p_\kappa(x,\xi,\tau) = p((\chi(x'/\kappa)x', x''), \xi, \tau)$$

provided that $V \in C_0^\infty(B_\kappa)$ where κ is so small that $B_\kappa \subset K$, for $p(x,\xi,\tau) = p_\kappa(x,\xi,\tau)$ when $x \in B_\kappa$. If $p(x,\xi,\tau) = \sum_{|\alpha|+\nu=m} c_{\alpha,\nu}\xi^\alpha\tau^\nu$ then

$$\mathcal{T}_\kappa(x,\xi,\tau) = \mathcal{P}_\kappa(x,\xi,\tau) - p_\kappa(x,\xi,\tau) = \sum_{|\alpha|+\nu=m} (\tilde{c}_{\alpha,\nu}(x,\xi) - \tilde{c}_{\alpha,\nu}(x,0))\xi^\alpha\tau^\nu$$

with the notation (3.19) applied in the x' variables. Hence it follows from Remark 3.9, applied in the x' variables, that $\mathcal{T}_\kappa(x,\xi,\tau)$ is bounded in $S(\kappa(|\xi|+\tau)^m, G)$ for small ε, κ, when $|x''| < 2\kappa$. Here G is the metric

$$(5.7)\qquad G = \frac{|dx'|^2}{\kappa^2} + |dx''|^2 + \frac{|d\xi'|^2}{|\xi'|^2 + \kappa^2\tau^2/\varepsilon^2} + \frac{|d\xi''|^2}{|\xi|^2 + \tau^2},$$

and we have used that $M_1 = O(\kappa)$ in (3.23). Note that if G^σ is the dual of G with respect to the symplectic form then

$$G/G^\sigma \le \left(\kappa^2(|\xi'|^2 + \kappa^2\tau^2/\varepsilon^2)\right)^{-1} \le (\varepsilon/\tau\kappa^2)^2 \le 1$$

for large τ. Hence the calculus of pseudo-differential operators gives for large τ

$$(5.8)\qquad \|\mathcal{T}_\kappa(x,D,\tau)V\| \le C\kappa\|V\|_{(m,\tau)}, \quad V \in C_0^\infty(B_\kappa).$$

To handle the commutator term in (5.3) we write

$$(5.9)\quad [p_\kappa(x,D,\tau)^*, p_\kappa(x,D,\tau)] - [\mathcal{P}_\kappa(x,D,\tau)^*, \mathcal{P}_\kappa(x,D,\tau)] \\ = -[\mathcal{T}_\kappa(x,D,\tau)^*, \mathcal{T}_\kappa(x,D,\tau)] - [\mathcal{T}_\kappa(x,D,\tau)^*, p_\kappa(x,D,\tau)] \\ - [p_\kappa(x,D,\tau)^*, \mathcal{T}_\kappa(x,D,\tau)].$$

Apart from

$$(5.10)\quad -\left(\{\overline{\mathcal{T}_\kappa(x,\xi,\tau)}, \mathcal{T}_\kappa(x,\xi,\tau)\} + \{\overline{\mathcal{T}_\kappa(x,\xi,\tau)}, p_\kappa(x,\xi,\tau)\} \\ + \{\overline{p_\kappa(x,\xi,\tau)}, \mathcal{T}_\kappa(x,\xi,\tau)\}\right)/i$$

the symbol of the right-hand side of (5.9) is in $S(\tau^{-2}(|\xi|+\tau)^{2m},G)$ (with bounds depending on ε,κ). The first term in (5.10) is bounded in $S((\varepsilon/\tau)(|\xi|+\tau)^{2m},G)$ for $\kappa^2(\varepsilon/\tau\kappa^2)=\varepsilon/\tau$. Now it follows from (3.23) that $\partial\mathcal{T}_\kappa(x,\xi,\tau)/\partial x$ is bounded in $S(\kappa(|\xi|+\tau)^m,G)$, for $M_2=O(\kappa^2)$ and $\kappa^2/(|x'|+\kappa)\le\kappa$, and $\partial\mathcal{T}_\kappa(x,\xi,\tau)/\partial\xi$ is bounded in $S((\varepsilon/\tau)(|\xi|+\tau)^m,G)$ since $M_1=O(\kappa)$ and $\kappa/(|\xi'|+\kappa\tau/\varepsilon)\le\varepsilon/\tau$. Thus (5.10) is bounded in $S((\varepsilon+\kappa)\tau^{-1}(|\xi|+\tau)^{2m},G)$, and it follows that for large τ

$$(5.11)\quad ([p_\kappa(x,D,\tau)^*,p_\kappa(x,D,\tau)]V,V) \le ([\mathcal{P}_\kappa(x,D,\tau)^*,\mathcal{P}(x,D,\tau)]V,V) + C(\varepsilon+\kappa)\tau^{-1}\|V\|^2_{(m,\tau)},\quad V\in C_0^\infty(B_\kappa).$$

When $\varepsilon+\kappa$ is sufficiently small it follows from (5.3), (5.8) and (5.11) that (5.5) is valid.

Proof of Theorem 5.1. From (5.5) it follows at once that

$$(5.5)'\quad \tfrac12\|V\|_{(m,\tau)}\le\sqrt{C_1\tau}\|\mathcal{P}_\kappa(x,D,\tau)V\| + \sqrt{C_3}\||D'|^mV\|,\quad V\in C_0^\infty(B_\kappa).$$

We shall now apply Corollary 3.8 to return to the operator $p(x,D,\tau)$. By (3.18) we have for $w\in C_0^\infty(B_{\kappa/4})$

$$(5.12)\quad \|e^{-\varepsilon|D'|^2/2\tau}p(x,D,\tau)w-\mathcal{P}_\kappa(x,D,\tau)e^{-\varepsilon|D'|^2/2\tau}w\| \le C(e^{-\tau\kappa^2/4\varepsilon}\|w\|_{(m,\tau)}+(\varepsilon/\kappa^2\tau)\|e^{-\varepsilon|D'|^2/2\tau}w\|_{(m,\tau)}).$$

Set $v=e^{-\varepsilon|D'|^2/2\tau}w$, and let $V=\chi(4x'/\kappa)v$. The cutoff takes place well outside the support of w if $w\in C_0^\infty(B_{\kappa/32})$, so it follows from (3.6) as in (4.11) that for large τ

$$(5.13)\quad \|\mathcal{P}_\kappa(x,D,\tau)V-\chi(4x'/\kappa)\mathcal{P}_\kappa(x,D,\tau)v\| \le Ce^{-\tau\kappa^2/256\varepsilon}\|w\|_{(m-1,\tau)},\quad w\in C_0^\infty(B_{\kappa/32}).$$

From (5.5)′ and (5.12) it follows now that

$$\tfrac12\|V\|_{(m,\tau)}\le\sqrt{C_1\tau}\|e^{-\varepsilon|D'|^2/2\tau}p(x,D,\tau)w\|+\sqrt{C_3}\||D'|^mV\| + C_4e^{-\tau\kappa^2/256\varepsilon}\|w\|_{(m,\tau)}+C_5(\varepsilon/\kappa^2\tau)\|v\|_{(m,\tau)}.$$

By (3.6) again the estimate remains valid with V replaced by v in both sides if C_4 is replaced by some larger constant. For large τ we can then cancel the last term and obtain

$$\tfrac14\|v\|_{(m,\tau)}\le\sqrt{C_1\tau}\|e^{-\varepsilon|D'|^2/2\tau}p(x,D,\tau)w\|+\sqrt{C_3}\||D'|^mv\| + C_4'e^{-\tau\kappa^2/256\varepsilon}\|w\|_{(m,\tau)}.$$

Now $(3.3)'$ gives

$$\text{(5.14)} \qquad |||D'|^m v\| \le (2\kappa\tau/\varepsilon)^m \|v\| + Ce^{-\tau\kappa^2/256\varepsilon}\|w\|.$$

When κ is so small that $\sqrt{C_3}(2\kappa/\varepsilon)^m < 1/20$ it follows that

$$\tfrac{1}{5}\|v\|_{(m,\tau)} \le \sqrt{C_1\tau}\|e^{-\varepsilon|D'|^2/2\tau}p(x,D,\tau)w\| + C_4''e^{-\tau\kappa^2/256\varepsilon}\|w\|_{(m,\tau)},$$

and if we put $w = e^{\tau\varphi}u$ this means that

$$\tfrac{1}{5}\|e^{-\varepsilon|D'|^2/2\tau}e^{\tau\varphi}u\|_{(m,\tau)} \le \sqrt{C_1\tau}\|e^{-\varepsilon|D'|^2/2\tau}p(x,D,\tau)e^{\tau\varphi}u\| \\ + C_4''e^{-\tau\kappa^2/256\varepsilon}\|e^{\tau\varphi}u\|_{(m,\tau)}, \quad u \in C_0^\infty(B_{\kappa/32}).$$

Now $e^{\tau\varphi}P(x,D)u - p(x,D,\tau)e^{\tau\varphi}u = e^{\tau\varphi}R(x,D,\tau)u$ where $R(x,D,\tau)$ is of order $m-1$ with respect to (D,τ). It is clear that (4.14) remains valid for such τ dependent R, so we obtain when τ is large enough

$$\text{(5.15)} \quad \tfrac{1}{6}\|e^{-\varepsilon|D'|^2/2\tau}e^{\tau\varphi}u\|_{(m,\tau)} \le \sqrt{C_1\tau}\|e^{-\varepsilon|D'|^2/2\tau}e^{\tau\varphi}P(x,D)u\| \\ + C_4'''e^{-\tau\kappa^2/256\varepsilon}\|e^{\tau\varphi}u\|_{(m,\tau)}, \quad u \in C_0^\infty(B_{\kappa/32}).$$

This estimate remains valid when $u \in H_{(m)}$ and $\operatorname{supp} u \subset B_{\kappa/32}$. It is a perfect substitute for (4.13), so the end of the proof of the theorem is just like that of the proof of Theorem 4.1. We leave the repetition of the argument for the reader.

Remark. It is easy to reduce the assumption $u \in H_{(m)}^{\text{loc}}$ to $u \in \mathcal{D}'$. In fact, using the norms $\|\cdot\|_{(\sigma,\tau)}$ with respect to the x' variables instead of the L^2 norm, we can replace $\|e^{\tau\varphi}u\|_{(m,\tau)}$ by $\|(\tau^2+|D'|^2)^{\sigma/2}e^{\tau\varphi}u\|_{(m,\tau)}$ in (5.15). If u is the function in Theorem 5.1 and $h \in C_0^\infty(\Omega_1)$ then $(1+|D'|^2)^{\sigma/2}(1+|D|^2)^{m/2}(hu) \in L^2$ if σ is a large negative number since $P(x,D)$ is partially elliptic. This implies that we have the bound $\|(\tau^2+|D'|^2)^{\sigma/2}e^{\tau\varphi}(hu)\|_{(m,\tau)} = O(\tau^{m+|\sigma|})$ as $\tau \to +\infty$, and the proof of uniqueness works as before.

6. Operators with real and partially translation invariant transversal principal symbol

As pointed out in the introduction, the conormal bundle of a leaf Σ in a foliation has a natural (local) trivialization as the pullback of the cotangent space at Σ of the space of leaves. In this section we shall again consider differential operators $P(x,D)$ in $\mathbf{R}^n$ with coefficients analytic with respect to $x' = (x_1,\dots,x_\nu)$ in a neighborhood of the origin and principal symbol $p(x,\xi)$ of order m. We shall assume that $p(x',x'',0,\xi'')$ is real and independent of x'. By the preceding observation this hypothesis is invariant under changes of variables preserving the analyticity assumption.

Theorem 6.1. *Let $P(x,D)$ be a differential operator in a neighborhood Ω of $0 \in \mathbf{R}^n$ of order m with C^∞ coefficients in the principal symbol $p(x,\xi)$, bounded coefficients in the lower order terms, and all coefficients analytic with respect to $x' = (x_1,\dots,x_\nu)$ in a neighborhood of 0. Thus they are defined in $\{(z',x'') \in \mathbf{C}^\nu \times \mathbf{R}^{n-\nu}; |z'| < r, |x''| < r\}$ for some $r > 0$. Assume that $p(x',x'',0,\xi'')$ is real and independent of x', and that $\psi \in C^2(\Omega)$ is a real valued function satisfying the pseudo-convexity conditions* (4.2) *and*

$$\text{(6.1)}\quad \operatorname{Re}\{\bar p,\{p,\psi\}\}(0,\xi) > 0 \quad \text{if } 0 \neq \xi = (0,\xi'') \in \mathbf{R}^n \text{ and } p(0,\xi) = \langle p'_\xi(0,\xi),\psi'(0)\rangle = 0.$$

If $u \in H^{\text{loc}}_{(m-1)}(\Omega)$ satisfies the differential equation $P(x,D)u = 0$ and $\psi \le \psi(0)$ in $\operatorname{supp} u$, *it follows that* $0 \notin \operatorname{supp} u$.

Note that we have not assumed that all coefficients of $p(x,\xi)$ are real. However, since by assumption

$$\text{(6.2)}\qquad p(x,(0,\xi'')) = p^0(x'',\xi'')$$

is independent of x' and has real coefficients, we have

$$\text{(6.3)}\qquad \{\overline{p(x,\xi)}, p(x,\xi)\} = 0, \quad \text{if } \xi' = 0.$$

In fact, if we write $p(x,\xi) = p^0(x'',\xi'') + r^0(x,\xi)$ then the Poisson bracket $\{\overline{p^0(x'',\xi'')}, p^0(x'',\xi'')\}$ vanishes since p^0 is real valued, and $\{\overline{p^0(x'',\xi'')}, r^0(x,\xi)\}$ as well as $\{\overline{r^0(x,\xi)}, r^0(x,\xi)\}$ vanish when $\xi' = 0$ since $r^0(x,\xi)$ and its derivatives with respect to x and ξ'' vanish when $\xi' = 0$. From (6.3) it follows that

$$\{\overline{p(x,\xi+i\tau\varphi'(x))}, p(x,\xi+i\tau\varphi'(x)\}/2i\tau$$

is a polynomial of degree $2m-2$ in (ξ'',τ) when $\xi' = 0$, so the proof of Lemma 4.3 gives

Lemma 6.2. *When the hypotheses of Theorem 6.1 are fulfilled there is a real quadratic polynomial φ and a neighborhood Ω_1 of $0 \in \mathbf{R}^n$ such that $\varphi(x) < \varphi(0)$ when $x \in \Omega_1 \cap \operatorname{supp} u \setminus \{0\}$ and*

$$\text{(6.4)}\quad \operatorname{Re}\{\bar p,\{p,\varphi\}\}(x,\xi) > 0, \quad \text{if } x = 0,\ 0 \neq \xi \in \mathbf{R}^n,\ \xi' = 0,\ p(x,\xi) = 0;$$

$$\text{(6.5)}\quad \{\overline{p(x,\xi+i\tau\varphi'(x))}, p(x,\xi+i\tau\varphi'(x))\}/2i\tau > 0, \quad \text{if } x = 0,\ \xi \in \mathbf{R}^n,\ \xi' = 0,\ \tau > 0,\ p(x,\xi+i\tau\varphi'(x)) = 0.$$

In the following analogue of Lemma 4.4 we use the notation $p(x,\xi,\tau)$ for $p(x,\xi+i\tau\varphi'(x))$.

Lemma 6.3. *From* (6.4) *and* (6.5) *it follows if* Ω_1 *is a sufficiently small neighborhood of* 0 *that there are constants* C_j *such that*

$$(6.6)\quad (|\xi|^2+\tau^2)^{m-1} \le C_1\{\overline{p(x,\xi,\tau)}, p(x,\xi,\tau)\}/2i\tau + C_2|p(x,\xi,\tau)|^2/(|\xi|^2+\tau^2) + C_3(|\xi|^2+\tau^2)^{(m-1)}|\xi'|/\tau, \quad \text{if } x\in\overline{\Omega}_1,\ \xi\in\mathbf{R}^n,\ \tau>0.$$

Proof. The terms in the right-hand side are homogeneous of degree $2m-2$ in (ξ,τ) and continuous when $|\xi|^2+\tau^2=1$ and $\xi'=0$, by (6.3). By hypothesis we can choose C_1 and Ω_1 so that the estimate is valid with strict inequality when $|\xi|^2+\tau^2=1$, $p(x,\xi,\tau)=0$ and $\xi'=0$. Write $q(x,\xi,\tau)=\{\overline{p(x,\xi,\tau)}, p(x,\xi,\tau)\}$. Since $q(x,(0,\xi''),\tau)$ is divisible by τ we conclude that for sufficiently large C_2 and C_4 and sufficiently small Ω_1

$$(|\xi|^2+\tau^2)^{m-1} \le C_1 q(x,(0,\xi''),\tau)/2i\tau + C_2|p(x,\xi,\tau)|^2/(|\xi|^2+\tau^2) + C_4|\xi'|^{2(m-1)},$$

when $x\in\overline{\Omega}_1$. Now

$$q(x,(0,\xi''),\tau)/2i\tau \le q(x,\xi,\tau)/2i\tau + C_5(|\xi|^2+\tau^2)^{m-1}|\xi'|/\tau,$$

and since $|\xi'|^{2(m-1)} \le (|\xi'|^2+\tau^2)^{m-1}|\xi'|/\tau$, we obtain (6.6) with $C_3 = C_4+C_5$.

As in Lemma 5.4 we can obtain an a priori estimate from (6.6). However, it is convenient to weaken (6.6) first to

$$(6.6)'\quad \tfrac{3}{4}(|\xi|^2+\tau^2)^{m-1} \le C_1\{\overline{p(x,\xi,\tau)}, p(x,\xi,\tau)\}/2i\tau + C_2|p(x,\xi,\tau)|^2/(|\xi|^2+\tau^2) + C_3^2(|\xi|^2+\tau^2)^{m-1}|\xi'|^2/\tau^2, \quad \text{if } x\in\overline{\Omega}_1,\ \xi\in\mathbf{R}^n,\ \tau>0.$$

which follows from (6.6) since $C_3|\xi'|/\tau \le \frac{1}{4} + C_3^2|\xi'|^2/\tau^2$.

Lemma 6.4. *The estimate* (6.6)′ *implies that*

$$(6.7)\quad \tfrac{1}{2}\|V\|^2_{(m-1,\tau)} \le \tfrac{1}{2}C_1([p(x,D,\tau)^*, p(x,D,\tau)]V,V)/\tau + C_2\|p(x,D,\tau)V\|^2_{(-1,\tau)} + C_3^2\||D'|V\|^2_{(m-1,\tau)}/\tau^2, \quad V\in C_0^\infty(K),$$

if K *is a compact neighborhood of* 0 *contained in* Ω_1 *and* τ *is sufficiently large.*

Proof. We can cut off the coefficients of p so that they are in C_0^∞ but are not changed in a neighborhood of K. The right-hand side of (6.7) can be written $(F(x,D,\tau)V,V)$ where the symbol of F is in $S((|\xi|^2+\tau^2)^m/\tau^2, g_\tau)$ and is equal to the right-hand side of (6.6)′ apart from terms in $S((|\xi|^2+\tau^2)^{m-1}/\tau, g_\tau)$. Here g_τ is the metric $|dx|^2+|d\xi|^2/(|\xi|^2+\tau^2)$. By the Fefferman-Phong inequality [4, Theorem 18.6.8] and (6.6)′ it follows that

$$(F(x,D,\tau)V,V)-\tfrac{3}{4}\|V\|^2_{(m-1,\tau)} \ge -C\tau^{-1}\|V\|^2_{(m-1,\tau)},\quad V\in C_0^\infty(K),$$

which proves (6.7).

Recall the definition of $\mathcal{P}_\kappa$ in (5.4) and the definition of p_κ in (5.6). Next we shall prove that as in Lemma 5.4 it is possible to replace $p(x,\xi,\tau)$ in (6.7) by $\mathcal{P}_\kappa(x,\xi,\tau)$.

Lemma 6.5. *If $\kappa<\varepsilon$ and ε is sufficiently small, then*

$$(6.8)\quad \tfrac{1}{3}\|V\|^2_{(m-1,\tau)} \le \tfrac{1}{2}C_1([\mathcal{P}_\kappa(x,D,\tau)^*,\mathcal{P}_\kappa(x,D,\tau)]V,V)/\tau \\ +C_2\|\mathcal{P}_\kappa(x,D,\tau)V\|^2_{(-1,\tau)}+C_4\||D'|V\|^2_{(m-1,\tau)}/\tau^2,\quad V\in C_0^\infty(B_\kappa),$$

if $C_4>C_3^2$ and τ is sufficiently large.

Proof. We argue as in the proof of Lemma 5.5, noting first that (6.7) remains valid if p is replaced by p_κ and $V\in C_0^\infty(B_\kappa)$. As there we let $\mathcal{T}_\kappa=\mathcal{P}_\kappa-p_\kappa$. The proof of (5.8) gives with no essential change that

$$(6.9)\qquad \|\mathcal{T}_\kappa(x,D,\tau)V\|_{(-1,\tau)}\le C\kappa\|V\|_{(m-1,\tau)},\quad V\in C_0^\infty(B_\kappa).$$

However, estimating the commutator term in (6.7) requires much more care. Set $r(x,\xi,\tau)=p(x,\xi,\tau)-p^0(x'',\xi'')$, which is thus of degree $<m$ with respect to ξ'', and define corresponding r_κ and $\mathcal{R}_\kappa$ as we defined p_κ and $\mathcal{P}_\kappa$ in (5.6) and (5.4). Then we have

$$\mathcal{T}_\kappa(x,\xi,\tau)=\mathcal{P}_\kappa(x,\xi,\tau)-p_\kappa(x,\xi,\tau)=\mathcal{R}_\kappa(x,\xi,\tau)-r_\kappa(x,\xi,\tau),$$

and we conclude as in Section 5 that $\mathcal{T}_\kappa(x,\xi,\tau)$ is bounded for small ε and κ in $S(\kappa(|\xi'|+\tau)(|\xi|+\tau)^{m-1},G)$. This implies that the symbol of $[\mathcal{T}_\kappa(x,D,\tau)^*,\mathcal{T}_\kappa(x,D,\tau)]$ is bounded for small ε and τ in the symbol class $S((\varepsilon/\tau)(|\xi'|^2+\tau^2)(|\xi|^2+\tau^2)^{m-1},G)$, for $\kappa^2(\varepsilon/\tau\kappa^2)=\varepsilon/\tau$. Hence

$$(6.10)\quad |([\mathcal{T}_\kappa(x,D,\tau)^*,\mathcal{T}_\kappa(x,D,\tau)]V,V)|/\tau \\ \le C\varepsilon(\||D'|V\|^2_{(m-1,\tau)}/\tau^2+\|V\|^2_{(m-1,\tau)}),\quad V\in C_0^\infty(B_\kappa).$$

The last two terms in (5.9) are adjoint, so it suffices to examine the commutator $[p_\kappa(x,D,\tau)^*,\mathcal{T}_\kappa(x,D,\tau)]$. The symbol of the commutator

$[p^0(x'',D'')^*, \mathcal{T}_\kappa(x,D,\tau)]$ is bounded in $S(\kappa(|\xi'|+\tau)(|\xi|+\tau)^{2m-2},G)$ since p^0 only depends on (x'',ξ''). It is clear that the symbol of $r_\kappa(x,D,\tau)^*$ is bounded in $S((|\xi'|+\tau)(|\xi|+\tau)^{m-1},G)$, so the symbol of $[r_\kappa(x,D,\tau)^*, \mathcal{T}_\kappa(x,D,\tau)]$ differs from $\{\overline{r_\kappa(x,\xi,\tau)}, \mathcal{T}_\kappa(x,\xi,\tau)\}/i$ by a symbol in $S((|\xi|^2+\tau^2)^{m-1},G)$ (with bounds depending on ε and κ). As in the proof of Lemma 5.5 we see that $\partial\mathcal{T}_\kappa(x,\xi,\tau)/\partial x$ is bounded in $S(\kappa(|\xi'|+\tau)(|\xi|+\tau)^{m-1},G)$ and that the symbol of $\partial\mathcal{T}_\kappa(x,\xi,\tau)/\partial\xi$ is bounded in $S((\varepsilon/\tau)(|\xi'|+\tau)(|\xi|+\tau)^{m-1},G)$. Hence the Poisson bracket $\{\overline{r_\kappa(x,\xi,\tau)}, \mathcal{T}_\kappa(x,\xi,\tau)\}$ is bounded in $S((\varepsilon+\kappa)\tau^{-1}(|\xi'|^2+\tau^2)(|\xi|^2+\tau^2)^{m-1},G)$. Combined with (6.10) this proves that for large τ

$$\begin{aligned}(6.11)\quad &([p_\kappa(x,D,\tau)^*, p_\kappa(x,D,\tau)]V,V)/\tau\\ &\le ([\mathcal{P}_\kappa(x,D,\tau)^*, \mathcal{P}_\kappa(x,D,\tau)]V,V)/\tau\\ &+ C(\varepsilon+\kappa)(\||D'|V\|^2_{(m-1,\tau)}/\tau^2 + \|V\|^2_{(m-1,\tau)}),\quad V\in C_0^\infty(B_\kappa).\end{aligned}$$

Now (6.8) follows from (6.7), (6.9) and (6.11).

Remark 6.6. The proof of (6.11) did not use the hypothesis that p_0 is real valued, so we shall be able to exploit this estimate in Section 7 also.

Proof of Theorem 6.1. It follows from (6.8) that for large τ

$$\begin{aligned}(6.12)\quad \tfrac{1}{3}\|V\|^2_{(m-1,\tau)} &\le C_1\tau^{-1}\|\mathcal{P}_\kappa(x,D,\tau)V\|^2\\ &+ C_4\||D'|V\|_{(m-1,\tau)}/\tau^2,\quad V\in C_0^\infty(B_\kappa).\end{aligned}$$

(Note that if we had simplified (6.7) in this way, the passage to (6.12) would not have been possible.) We can now continue as in Section 5. Let $w\in C_0^\infty(B_{\kappa/32})$ and set $v=e^{-\varepsilon|D'|^2/2\tau}w$, $V=\chi(x'/2\kappa)v$. Then (5.13) holds, and (5.12) can be improved to

$$\begin{aligned}(6.13)\quad &\|e^{-\varepsilon|D'|^2/2\tau}p(x,D,\tau)w - \mathcal{P}_\kappa(x,D,\tau)e^{-\varepsilon|D'|^2/2\tau}w\|\\ &\le C(e^{-\tau\kappa^2/256\varepsilon}\|w\|_{(m-1,\tau)} + (\varepsilon/\kappa^2)\|e^{-\varepsilon|D'|^2/2\tau}w\|_{(m-1,\tau)}),\\ &\qquad\text{if } w\in C_0^\infty(B_{\kappa/32}),\end{aligned}$$

for large τ. Here we have used that the coefficients of $p_0(x'',D'')$, which contains the terms in $p(x,D,\tau)$ of order m in D'', are independent of x' so that they commute with the Gaussian regularizer, and that

$$\begin{aligned}\|e^{-\varepsilon|D'|^2/2\tau}|D'|w\|_{(m-1,\tau)} \le C\tau(&\|e^{-\varepsilon|D'|^2/2\tau}w\|_{(m-1,\tau)}\\ &+ e^{-\tau\kappa^2/4\varepsilon}\varepsilon^{-1}\|w\|_{(m-1,\tau)})\end{aligned}$$

by (3.3) applied to the x' derivatives of w. For sufficiently large τ we conclude that

$$\begin{aligned}\tfrac{1}{2}\|V\|_{(m-1,\tau)} \le \sqrt{C_1/\tau}\|e^{-\varepsilon|D'|^2/2\tau}p(x,D,\tau)w\| \\ + \sqrt{C_4}\||D'|(|D|^2+\tau^2)^{\frac{1}{2}(m-1)}V\|/\tau \\ + C_5 e^{-\tau\kappa^2/256\varepsilon}\|w\|_{(m-1,\tau)} + C_6(\varepsilon/\kappa^2\sqrt{\tau})\|v\|_{(m-1,\tau)}.\end{aligned}$$

By (3.6) we may replace V by v in both sides if the constants are modified appropriately. For large τ we can then cancel the last term and obtain

$$\begin{aligned}\tfrac{1}{4}\|v\|_{(m-1,\tau)} \le \sqrt{C_1/\tau}\|e^{-\varepsilon|D'|^2/2\tau}p(x,D,\tau)w\| \\ + \sqrt{2C_4}\||D'|(|D|^2+\tau^2)^{\frac{1}{2}(m-1)}v\|/\tau + C_5' e^{-\tau\kappa^2/256\varepsilon}\|w\|_{(m-1,\tau)}.\end{aligned}$$

Using $(3.3)'$ and taking κ so small that $\sqrt{2C_4}(2\kappa/\varepsilon) < 1/20$, we obtain

$$\begin{aligned}\tfrac{1}{5}\|v\|_{(m-1,\tau)} \le \sqrt{C_1/\tau}\|e^{-\varepsilon|D'|^2/2\tau}p(x,D,\tau)w\| \\ + C_5'' e^{-\tau\kappa^2/256\varepsilon}\|w\|_{(m-1,\tau)}.\end{aligned}$$

If we put $w = e^{\tau\varphi}u$ this means that

$$\begin{aligned}\tfrac{1}{5}\|e^{-\varepsilon|D'|^2/2\tau}e^{\tau\varphi}u\|_{(m-1,\tau)} \le \sqrt{C_1/\tau}\|e^{-\varepsilon|D'|^2/2\tau}p(x,D,\tau)e^{\tau\varphi}u\| \\ + C_5'' e^{-\tau\kappa^2/256\varepsilon}\|e^{\tau\varphi}u\|_{(m-1,\tau)}, \quad u \in C_0^\infty(B_{\kappa/32}).\end{aligned}$$

As in the proof of Theorem 4.9, using Lemma 4.8 we may replace $p(x,D,\tau)e^{\tau\varphi}$ by $e^{\tau\varphi}P(x,D)$ in the right-hand side, if the constant $\frac{1}{5}$ is replaced by $\frac{1}{6}$ and τ is large enough. The estimate obtained,

$$\begin{aligned}\tfrac{1}{6}\|e^{-\varepsilon|D'|^2/2\tau}e^{\tau\varphi}u\|_{(m-1,\tau)} \le \sqrt{C_1/\tau}\|e^{-\varepsilon|D'|^2/2\tau}e^{\tau\varphi}P(x,D)u\| \\ + C_5''' e^{-\tau\kappa^2/256\varepsilon}\|e^{\tau\varphi}u\|_{(m-1,\tau)}, \quad u \in C_0^\infty(B_{\kappa/32}),\end{aligned}$$

remains valid when $u \in H_{(m-1)}$ and $\operatorname{supp} u \subset B_{\kappa/32}$. Again this is a perfect substitute for (4.13) so the end of the proof is a repetition of that of Theorem 4.1, which we leave for the reader.

7. Operators with partially translation invariant transversally principally normal symbol

Carleman estimates and uniqueness theorems were proved in [3, Chapter VIII] first for elliptic operators, then for operators with real principal part, and finally for principally normal operators. Extension of the first two cases to the partially analytic case have been given in Sections 5 and 6, and we shall now discuss an extension of the principally normal case.

In [3] a differential operator with principal symbol $p(x,\xi)$ was called principally normal if

$$\{\overline{p(x,\xi)}, p(x,\xi)\}/i = \mathrm{Re}(p(x,\xi)a(x,\xi))$$

where $a(x,\xi)$ is a homogeneous polynomial in ξ of degree $m-1$ with C^1 coefficients. It was pointed out by Lerner [6] that using pseudodifferential operators, which were not yet available when [3] was written, one can weaken the condition to existence of a smooth homogeneous function $a(x,\xi)$. Thus elliptic operators as well as operators with real principal symbol are principally normal in this sense. In [4] the Fefferman-Phong estimate made it possible to reduce the condition further to

$$|\{\overline{p(x,\xi)}, p(x,\xi)\}| \le C|p(x,\xi)||\xi|^{m-1}, \tag{7.1}$$

which is the definition we shall adopt here. (The importance of the condition is underlined by examples given by Alinhac [1] and Colombini-Del Santo [2].) We shall of course only assume that this condition is fulfilled when $\xi' = 0$, that is,

$$|\{\overline{p(x,\xi)}, p(x,\xi)\}| \le C_K|p(x,\xi)||\xi|^{m-1}, \quad \text{if } x \in K,\ \xi \in \mathbf{R}^n,\ \xi' = 0. \tag{7.2}$$

However, we shall need an additional hypothesis in order to be able to control the terms which occur in the symbolic calculus. The following proposition gives another indication of what this condition should be:

Proposition 7.1. *Assume that*

$$|\{\overline{q(x,\xi)}, q(x,\xi)\}| \le C_q|q(x,\xi)||\xi|^{m-1}, \quad x \in K,\ \xi' = 0, \tag{7.3}$$

for every smooth homogeneous symbol q of degree m such that $q(x,\xi) = p(x,\xi)$ when $\xi' = 0$. Then it follows that

$$|\{\overline{p(x,\xi)}, p(x,\xi)\}| \le C|p(x,\xi)||\xi|^{m-1}, \quad \sum_1^{\nu} |\partial p(x,\xi)/\partial x_j| \le C|p(x,\xi)|;\ x \in K,\ \xi' = 0. \tag{7.4}$$

Conversely, (7.3) *follows from* (7.4).

Proof. The first part of (7.4) is just (7.3) when $q = p$. The general q occurring in (7.3) can be written in the form

$$q(x,\xi) = p(x,\xi) + \sum_1^\nu q_j(x,\xi)\xi_j,$$

where q_j is a homogeneous polynomial of degree $m-1$. When $\xi' = 0$ we have

$$\{\overline{q(x,\xi)}, q(x,\xi)\} = \{\overline{p(x,\xi)}, p(x,\xi)\} + 2i \operatorname{Im} \sum_1^\nu \overline{q_j(x,\xi)} p_{(j)}(x,\xi).$$

If (7.3) is valid and we take $q_j(x,\xi) = c_j \xi''^\alpha$, $|\alpha| = m-1$, $\max |c_j| = 1$, it follows that the second part of (7.4) must be valid. Conversely, (7.3) follows immediately from (7.4).

Remark. If $\partial p(x,\xi)/\partial x' = 0$ for $\xi' = 0$ then the proof shows that $\{\overline{q(x,\xi)}, q(x,\xi)\} = \{\overline{p(x,\xi)}, p(x,\xi)\}$. In a special case this was already observed in (6.3).

Proposition 7.1 shows that (7.4) is a natural, obviously invariant, version of the principal normality condition associated with the analytic foliation by the planes where x'' is constant. Perhaps a more convincing reason to believe in the relevance of (7.4) is that, with the notation (5.4), $\{\overline{\mathcal{P}_\kappa(x,\xi,\tau)}, \mathcal{P}_\kappa(x,\xi,\tau)\}$ is not just equal to $\{\overline{p(x,\xi,\tau)}, p(x,\xi,\tau)\}$ with x replaced by $X_\kappa(x,\xi)$, for there is an additional term equal to

$$-2i \sum_{j=1}^\nu |p_{(j)}(x' + i\varepsilon\xi'/\tau, x'', \xi)|^2 \varepsilon/\tau$$

when $|x'| < \kappa$ and $|\xi'| < \kappa\tau/\varepsilon$. To handle this contribution it seems plausible that a condition such as (7.4) is needed. However, using only (7.4) we have not been able to find sufficiently good estimates of the difference $\mathcal{P}_\kappa - p_\kappa$ as in Sections 5 and 6. In the following theorem we shall therefore, as in Section 6, make the much stronger assumption that the sum in (7.4) is identically 0. Thus our goal is only to prove:

Theorem 7.2. *Let $P(x,D)$ be a differential operator in a neighborhood Ω of $0 \in \mathbf{R}^n$, of order m, with C^∞ coefficients in the principal symbol $p(x,\xi)$, bounded coefficients in the lower order terms, and all coefficients analytic with respect to $x' = (x_1, \dots, x_\nu)$ in a neighborhood of 0. Thus they are defined in $\{(z', x'') \in \mathbf{C}^\nu \times \mathbf{R}^{n-\nu}; |z'| < r, |x''| < r\}$*

for some $r > 0$. Assume that $p(x', x'', 0, \xi'')$ is independent of x', that the transversal principal normality condition (7.2) *is fulfilled in a neighborhood K of the origin, and that $\psi \in C^2(\Omega)$ is real valued and satisfies the pseudoconvexity conditions* (6.1) *and* (4.2). *If $u \in H^{\rm loc}_{(m-1)}(\Omega)$ satisfies the differential equation $P(x, D)u = 0$ and $\psi \le \psi(0)$ in* $\operatorname{supp} u$, *it follows that* $0 \notin \operatorname{supp} u$.

Note that it follows from Proposition 7.1 that the transversal principal normality condition means precisely that $p_0(x'', \xi'') = p(x, (0, \xi''))$ is principally normal in the sense of (7.1) (with $n - \nu$ dimensions).

Lemma 7.3. *When the hypotheses of Theorem 7.2 are fulfilled, there is a compact neighborhood $K_1 \subset K$ of 0 such that*

$$(7.5) \quad (|\xi|^2 + \tau^2)^{m-1} \le C\{\overline{p(x, \xi + i\tau\psi'(x))}, p(x, \xi + i\tau\psi'(x))\}/2i\tau,$$
$$\text{if } p(x, \xi + i\tau\psi'(x)) = \langle p'_\xi(x, \xi + i\tau\psi'(x)), \psi'(x)\rangle = 0, \ x \in K_1, \ \xi' = 0,$$

with the right-hand side interpreted as $C \operatorname{Re}\{\overline{p(x, \xi)}, \{p(x, \xi), \psi(x)\}\}$ when $\tau = 0$. Moreover, there is some $\delta > 0$ such that (7.5) *remains valid with ψ replaced by another function $\tilde\psi$, at any point $x \in K_1$ where $|\tilde\psi'(x) - \psi'(x)| \le \delta$, $|\tilde\psi''(x) - \psi''(x)| \le \delta$.*

Proof. In the formula (cf. (4.4))

$$(7.6) \quad \{\overline{p(x, \xi + i\tau\psi'(x))}, p(x, \xi + i\tau\psi'(x))\}/2i\tau$$
$$= \sum_{j,k=1}^{n} \frac{\partial^2 \psi(x)}{\partial x_j \partial x_k} p^{(j)}(x, \xi + i\tau\psi'(x))\overline{p^{(k)}(x, \xi + i\tau\psi'(x))}$$
$$+ \operatorname{Re} \sum_{j=1}^{n} \overline{p^{(j)}(x, \xi + i\tau\psi'(x))} p_{(j)}(x, \xi + i\tau\psi'(x))/i\tau$$

the last term is singular when $\tau \to 0$, for

$$2 \operatorname{Re} \sum_1^n \overline{p^{(j)}(x, \xi + i\tau N)} p_{(j)}(x, \xi + i\tau N)/i\tau$$
$$= \{\overline{p(x, \xi)}, p(x, \xi)\}/i\tau + 2 \operatorname{Re} \sum_1^n N_k \{\overline{p(x, \xi)}, p^{(k)}(x, \xi)\}$$
$$+ O(\tau |N|^2 |\xi + i\tau N|^{2m-3}),$$

by Taylor's formula. If $\xi' = 0$ we can estimate the first term by (7.2) and then use Taylor's formula again to prove the estimate

$$(7.7)\quad |\operatorname{Re}\sum_1^n \overline{p^{(j)}(x,\xi+i\tau N)}, p_{(j)}(x,\xi+i\tau N)/i\tau - \operatorname{Re}\sum_1^n N_k\{\overline{p(x,\xi)}, p^{(k)}(x,\xi)\}| \le C'\big(|p(x,\xi+i\tau N)|\,|\xi+i\tau N|^{m-1}/\tau+|\langle p'_\xi(x,\xi+i\tau N),N\rangle||\xi+i\tau N|^{m-1} + \tau|N|^2|\xi+i\tau N|^{2m-3}\big),\quad x\in K_1,\ \xi'=0.$$

This is inequality (28.2.9) in [4], where it was proved for all ξ under the full hypothesis (7.1). The proof is the same; in fact, (7.7) only uses (7.1) for the same ξ.

It follows from (7.7) that the left-hand side of (7.6) is a continuous function of x, ξ, τ, $\psi'(x)$ and $\psi''(x)$ when $p(x,\xi+i\tau\psi'(x))=0$ and $\langle p'_\xi(x,\xi+i\tau\psi'(x)),\psi'(x)\rangle=0$, if the definition is extended to $\tau=0$ as

$$(7.8)\quad \sum_{j,k=1}^n \frac{\partial^2\psi(x)}{\partial x_j\partial x_k}\overline{p^{(j)}(x,\xi)}p^{(k)}(x,\xi) + \operatorname{Re}\sum_1^n\{\overline{p(x,\xi)},p^{(k)}(x,\xi)\}\frac{\partial\psi(x)}{\partial x_k} = \operatorname{Re}\{\overline{p(x,\xi)},\{p(x,\xi),\psi(x)\}\}.$$

The lemma is an immediate consequence.

Lemma 7.4. *When the hypotheses of Theorem 7.2 are fulfilled there is a real quadratic polynomial φ and a compact neighborhood K_2 of $0\in\mathbf{R}^n$ such that $\varphi(x)<\varphi(0)$ when $x\in K_2\cap\operatorname{supp}u\setminus\{0\}$ and*

$$(7.9)\quad \tau(|\xi|^2+\tau^2)^{m-1}\le C_1\{\overline{p(x,\xi+i\tau\varphi'(x))},p(x,\xi+i\tau\varphi'(x))\}/i + C_2|p(x,\xi+i\tau\varphi'(x))|^2/\tau + C_3|\xi'|(|\xi|^2+\tau^2)^{m-1},\quad x\in K_2,\ \tau>0.$$

Proof. The proof is close to that of [4, Prop. 28.3.3]. The first step is to observe that (7.5) with $\tau=0$ implies

$$\tfrac12|\xi|^{2m-2}\le C\Big(\sum_{j,k=1}^n\frac{\partial^2\psi(x)}{\partial x_j\partial x_k}p^{(j)}(x,\xi)\overline{p^{(k)}(x,\xi)} + \operatorname{Re}\sum_{j=1}^n\{\overline{p(x,\xi)},p^{(k)}(x,\xi)\}\frac{\partial\psi(x)}{\partial x_k}\Big) + C_2|p(x,\xi)||\xi|^{m-2} + C_3|\langle p'_\xi(x,\xi),\psi'(x)\rangle||\xi|^{m-1},\quad \text{if } x\in K_1,\ \xi'=0.$$

It suffices to note that this follows from (7.5) for reasons of continuity in a conic neighborhood of any point $\xi \neq 0$ with $\xi' = 0$ where the last two terms vanish, and that it is true for sufficiently large C_2, C_3 in a conic neighborhood of any point where one of them is not equal to zero. Now we apply (7.7) to the second sum and replace ξ by $\xi + i\tau\psi'(x)$ in the other terms which gives an error $O(\tau|\xi|^{2m-3})$. Hence we obtain if $\tau/|\xi|$ is sufficiently small and $\xi' = 0$, with new constants C_2, C_3, and $E(\xi,\tau) = (|\xi|^2 + \tau^2)^{m-1}$

$$
\begin{aligned}
\tfrac{1}{3}E(\xi,\tau) \leq C\Big(\sum_{j,k=1}^{n} \frac{\partial^2\psi(x)}{\partial x_j \partial x_k} p^{(j)}(x,\xi+i\tau\psi'(x))\overline{p^{(k)}(x,\xi+i\tau\psi'(x))} \\
+ \operatorname{Re}\sum_{j=1}^{n} \overline{p^{(j)}(x,\xi+i\tau\psi'(x))}p_{(j)}(x,\xi+i\tau\psi'(x))/i\tau\Big) \\
+ (C_2|p(x,\xi+i\tau\psi'(x)|/\tau + C_3|\langle p'_\xi(x,\xi+i\tau\psi'(x)),\psi'(x)\rangle|)\sqrt{E(\xi,\tau)}.
\end{aligned}
$$

To remove the restriction $\xi' = 0$ we note that the difference between the terms in this estimate for a general $\xi = (\xi', \xi'')$ and the terms obtained when ξ' is replaced by 0 is $O(|\xi'|(|\xi|+\tau)^{2m-3})$ except in the terms with a τ in the denominator where the difference is $O(|\xi'|(|\xi|+\tau)^{2m-2}/\tau)$. Hence we have for $x \in K_1$ and $\xi \in \mathbf{R}^n$, if $0 < \tau < \gamma|\xi|$, say,

$$
\begin{aligned}
\tfrac{1}{3}E(\xi,\tau) \leq C\Big(\sum_{j,k=1}^{n} \frac{\partial^2\psi(x)}{\partial x_j \partial x_k} p^{(j)}(x,\xi+i\tau\psi'(x))\overline{p^{(k)}(x,\xi+i\tau\psi'(x))} \\
+ \operatorname{Re}\sum_{j=1}^{n} \overline{p^{(j)}(x,\xi+i\tau\psi'(x))}p_{(j)}(x,\xi+i\tau\psi'(x))/i\tau\Big) \\
+ \big(C_2|p(x,\xi+i\tau\psi'(x))|/\tau + C_3|\langle p'_\xi(x,\xi+i\tau\psi'(x)),\psi'(x)\rangle|\big)\sqrt{E(\xi,\tau)} \\
+ C_4|\xi'|E(\xi,\tau)/\tau.
\end{aligned}
$$

Since

$$
\begin{aligned}
\big(C_2|p(x,\xi+i\tau\psi'(x))|/\tau + C_3|\langle p'_\xi(x,\xi+i\tau\psi'(x)),\psi'(x)\rangle|\big)\sqrt{E(\xi,\tau)} \\
\leq 6C_2^2|p(x,\xi+i\tau\psi'(x))|^2/\tau^2 \\
+ 6C_3^2|\langle p'_\xi(x,\xi+i\tau\psi'(x)),\psi'(x)\rangle|^2 + \tfrac{1}{12}E(\xi,\tau),
\end{aligned}
$$

it follows that for $x \in K_1$ and $0 < \tau < \gamma|\xi|$

$$
\begin{aligned}
(7.10)\quad & \tfrac{1}{4}(|\xi|^2+\tau^2)^{m-1} \\
&\le C\Big(\sum_{j,k=1}^{n} \frac{\partial^2\psi(x)}{\partial x_j \partial x_k} p^{(j)}(x,\xi+i\tau\psi'(x))\overline{p^{(k)}(x,\xi+i\tau\psi'(x))} \\
&\qquad + \lambda|\langle p'_\xi(x,\xi+i\tau\psi'(x)),\psi'(x)\rangle|^2 \\
&\qquad + \operatorname{Re}\sum_{j=1}^{n} \overline{p^{(j)}(x,\xi+i\tau\psi'(x))}p_{(j)}(x,\xi+i\tau\psi'(x))/i\tau\Big) \\
&\qquad + 6C_2^2|p(x,\xi+i\tau\psi'(x))|^2/\tau^2 + C_4|\xi'|(|\xi|^2+\tau^2)^{m-1}/\tau
\end{aligned}
$$

if $\lambda \ge 6C_3^2/C$. Next assume that $\tau \ge \gamma|\xi|$. Then it follows from (7.5) that the estimate (7.10) is valid even with $\lambda = C_2 = C_4 = 0$ in a conic neighborhood of every point $(x,\xi,\tau) \in K_1 \times \mathbf{R}^{n+1}$ with $\xi' = 0$, $p(x,\xi+i\tau\psi'(x)) = 0$ and $\langle p'_\xi(x,\xi+i\tau\psi'(x)),\psi'(x)\rangle = 0$. In a conic neighborhood of any other point with $\tau \ge \gamma|\xi|$ the estimate is obviously valid for large λ, C_2, C_4 so (7.10) is proved for suitable λ, C_2, C_4 and K_1.

Let $\psi(0) = 0$ and set $\varphi = e^{\lambda\psi}$, thus $\varphi(0) = 1$. Then $\varphi \le \varphi(0)$ in $\operatorname{supp} u$, and $\varphi' = \lambda\varphi\psi'$, $\varphi'' = \lambda\varphi\psi'' + \lambda^2\varphi\psi'\otimes\psi'$. If we multiply (7.10) by $\lambda\varphi$ and replace τ by $\tau\lambda\varphi$, it follows that

$$
\begin{aligned}
(7.11)\quad & \tfrac{1}{4}\lambda\varphi(|\xi|^2+(\tau\lambda\varphi)^2)^{m-1} \\
&\le C\Big(\sum_{j,k=1}^{n} \frac{\partial^2\varphi(x)}{\partial x_j \partial x_k} p^{(j)}(x,\xi+i\tau\varphi'(x))\overline{p^{(k)}(x,\xi+i\tau\varphi'(x))} \\
&\qquad + \operatorname{Re}\sum_{j=1}^{n} \overline{p^{(j)}(x,\xi+i\tau\varphi'(x))}p_{(j)}(x,\xi+i\tau\varphi'(x))/i\tau\Big) \\
&\quad + 6C_2^2|p(x,\xi+i\tau\varphi'(x))|^2/(\tau^2\lambda\varphi) + C_4|\xi'|(|\xi|^2+(\tau\lambda\varphi)^2)^{m-1}/\tau.
\end{aligned}
$$

As in the proof of Theorem 4.1 we have to replace φ by the quadratic polynomial

$$\tilde\varphi = \sum_{|\alpha|\le 2} \partial^\alpha\varphi(0)x^\alpha/\alpha! - \tilde\delta|x|^2$$

for some $\tilde\delta > 0$. We can write $\tilde\varphi = e^{\lambda\tilde\psi}$ where $\tilde\psi = \lambda^{-1}\log\tilde\varphi$ for small $\tilde\delta$ is so close to ψ that $|\tilde\psi' - \psi'| \le \delta$, $|\tilde\psi'' - \psi''| \le \delta$ in a neighborhood $K_2 \subset K_1$ of 0, where δ is the number in Lemma 7.3. The preceding argument is then applicable to $\tilde\psi$ in K_2, so the estimate (7.11) is valid for $x \in K_2$ if φ is replaced by $\tilde\varphi$. Changing notation we have now found a quadratic polynomial φ and a compact neighborhood K_2 of 0 such that $\varphi(x) < \varphi(0)$ if $x \in K_2 \cap \operatorname{supp} u \setminus \{0\}$ and (7.9) is valid. This completes the proof of the lemma.

From now on we shall use the notation $p(x,\xi,\tau) = p(x,\xi + i\tau\varphi'(x))$. Just as we weakened (6.6) to (6.6)′ we weaken (7.9) to

$$(7.9)' \quad \tfrac{3}{4}\tau(|\xi|^2+\tau^2)^{m-1} \le C_1\{\overline{p(x,\xi,\tau)}, p(x,\xi,\tau)\}/i \\ + C_2|p(x,\xi,\tau)|^2/\tau + C_3^2|\xi'|^2(|\xi|^2+\tau^2)^{m-1}/\tau, \quad x \in K_2,\ \tau > 0.$$

Lemma 7.5. *The estimate* (7.9)′ *implies that*

$$(7.12) \quad \tfrac{1}{2}\tau\|V\|^2_{(m-1,\tau)} \le C_1([p(x,D,\tau)^*, p(x,D,\tau)]V, V) \\ + 3C_2\|p(x,D,\tau)V\|^2/\tau + 2C_3^2\||D'|V\|^2_{(m-1,\tau)}/\tau, \quad V \in C_0^\infty(K),$$

if K is a compact neighborhood of 0 *contained in the interior of K_2, and τ is sufficiently large.*

Proof. We can cut off the coefficients of p so that they are in C_0^∞ but are not changed in a neighborhood of K. Set

$$(7.13) \quad F(x,D,\tau) = C_1[p(x,D,\tau)^*, p(x,D,\tau)] \\ + C_2(p(x,D,\tau)^*p(x,D,\tau) + p(x,D,\tau)p(x,D,\tau)^*)/\tau \\ + C_3^2|D'|^2(|D|^2+\tau^2)^{m-1}/\tau - \tfrac{3}{4}\tau(|D|^2+\tau^2)^{m-1}.$$

Then $F \in S((|\xi|^2+\tau^2)^m/\tau, g_\tau)$ where $g_\tau = |dx|^2 + |d\xi|^2/(\tau^2+|\xi|^2)$. To calculate the symbol it is convenient to use the Weyl symbol $a(x,\xi,\tau)$ of $p(x,D,\tau)$, for the Weyl symbol of $p(x,D,\tau)^*$ is then $\overline{a(x,\xi,\tau)}$ and the Weyl symbol of $p(x,D,\tau)^*p(x,D,\tau) + p(x,D,\tau)p(x,D,\tau)^*$ is $2|a|^2$ apart from terms in $S(|\xi|^2+\tau^2)^{m-1}, g_\tau)$, since $\{\bar a, a\} + \{a, \bar a\} = 0$. Apart from terms in $S((|\xi|^2+\tau^2)^{m-1}, g_\tau)$, the symbol of the commutator in (7.13) is $\{\overline{p(x,\xi,\tau)}, p(x,\xi,\tau)\}/i$. Since

$$|p(x,\xi,\tau)|^2 \le 2|a(x,\xi,\tau)|^2 + C(|\xi|^2+\tau^2)^{m-1},$$

it follows from (7.9)′ that the Weyl symbol of $F(x,D,\tau)$ is bounded below by a symbol in $S((|\xi|^2+\tau^2)^{m-1}, g_\tau)$, so the Fefferman-Phong estimate gives when $V \in C_0^\infty(B_\kappa)$

$$(\tfrac{3}{4}\tau - C)\|V\|^2_{(m-1,\tau)} \le (C_1 - C_2/\tau)([p(x,D,\tau)^*, p(x,D,\tau)]V, V) \\ + 2C_2\|p(x,D,\tau)V\|^2/\tau + C_3^2\||D'|V\|^2_{(m-1,\tau)}/\tau.$$

If we divide by $1 - C_2/C_1\tau$, the lemma follows.

Recalling that $\mathcal{P}_\kappa$ and p_κ are defined in (5.4) and (5.6), we shall now replace p by $\mathcal{P}_\kappa$ in (7.12):

Lemma 7.6. *If $\kappa < \varepsilon$ and ε is sufficiently small, then*

$$(7.14)\quad \tfrac{1}{3}\tau\|V\|^2_{(m-1,\tau)} \le C_1([\mathcal{P}_\kappa(x,D,\tau)^*, \mathcal{P}_\kappa(x,D,\tau)]V,V) \\ + 4C_2\|\mathcal{P}_\kappa(x,D,\tau)V\|^2/\tau + C_4\||D'|V\|^2_{(m-1,\tau)}/\tau, \quad V \in C_0^\infty(B_\kappa),$$

if $C_4 > 2C_3^2$ and τ is sufficiently large.

Proof. When $V \in C_0^\infty(B_\kappa)$ the estimate (7.12) is not changed if p is replaced by p_κ. With the notation used in the proof of Lemma 6.5 we recall that

$$\mathcal{T}_\kappa(x,\xi,\tau) = \mathcal{P}_\kappa(x,\xi,\tau) - p_\kappa(x,\xi,\tau) = \mathcal{R}_\kappa(x,\xi,\tau) - r_\kappa(x,\xi,\tau)$$

is bounded in $S(\kappa(|\xi'|+\tau)(|\xi|+\tau)^{m-1}, G)$ for small ε and κ. Hence

$$\|\mathcal{T}_\kappa(x,D,\tau)V\|^2 \le C\kappa^2\|(|D'|+\tau)V\|^2_{(m-1,\tau)} \\ \le 2C\kappa^2(\||D'|V\|^2_{(m-1,\tau)} + \tau^2\|V\|^2_{(m-1,\tau)}),$$

which implies that

$$\|p_\kappa(x,D,\tau)V\|^2 \le \tfrac{4}{3}\|\mathcal{P}_\kappa(x,D,\tau)V\|^2 + 4\|\mathcal{T}_\kappa(x,D,\tau)V\|^2 \\ \le \tfrac{4}{3}\|\mathcal{P}_\kappa(x,D,\tau)V\|^2 + 8C\kappa^2(\||D'|V\|^2_{(m-1,\tau)} + \tau^2\|V\|^2_{(m-1,\tau)}).$$

As observed in Remark 6.6 the proof of (6.11) is valid under the hypotheses in Theorem 7.2, so (7.14) follows from (7.12).

Proof of Theorem 7.2. It follows from (7.14) that for large τ

$$\tfrac{1}{3}\tau\|V\|^2_{(m-1,\tau)} \le 2C_1\|\mathcal{P}_\kappa(x,D,\tau)V\|^2 + C_4\||D'|V\|^2_{(m-1,\tau)}/\tau, \\ \text{if } V \in C_0^\infty(B_\kappa).$$

This is precisely the estimate (6.12) (with C_1 replaced by $2C_1$), so the rest of the proof of Theorem 6.1 is applicable without any change.

References

[1] S. Alinhac, Non-unicité du problème de Cauchy, *Ann. of Math.* **117** (1983), 77–108.

[2] F. Colombini and D. Del Santo, Condition (P) is not sufficient for uniqueness in the Cauchy problem, *Comm. in Partial Differential Equations* **20** (1995), 2113–2128.

[3] L. Hörmander, *Linear Partial Differential Operators*, Springer Verlag, 1963.

[4] ——— *The Analysis of Linear Partial Differential Operators*, Springer Verlag, 1983–1985.

[5] ——— A uniqueness theorem for second order hyperbolic differential equations, *Comm. Partial Differential Equations* **17** (1992), 699–714.

[6] N. Lerner, Unicité de Cauchy pour des opérateurs de type principalement normaux, *J. Math. Pure Appl.* **64** (1985), 1–11.

[7] L. Robbiano, Théorème d'unicité adapté au contrôle des solutions des problèmes hyperboliques, *Comm. Partial Differential Equations* **16** (1991), 789–800.

[8] D. Tataru, Unique continuation for solutions to PDE's; between Hörmander's theorem and Holmgren's theorem, *Comm. Partial Differential Equations* **20** (1995), 855–884.

[9] L. Robbiano and C. Zuily, Uniqueness in the Cauchy problem for operators with partially holomorphic coefficients, Preprint, 48 pp.

Department of Mathematics
University of Lund
Box 118, S-221 00 Lund, Sweden
email: lvh@maths.lth.se

Nonlinear wave diffraction

John K. Hunter

1. Introduction

The high-frequency asymptotics of solutions of wave equations are described by ray theory in which the wave energy propagates along a set of curves in space-time, called rays. Two examples of ray theories are the geometrical optics theory of light and semi-classical quantum mechanics. When applicable, ray theory is extremely powerful and it often provide surprisingly accurate quantitative results.

The simplest ray theories do not take account of wave diffraction. Even for high-frequency waves, diffraction is important in many circumstances, such as the focusing of waves at caustics, the spreading of beams and wavepackets, the diffraction of waves into shadow zones behind obstacles, the tunelling of waves through shadow zones, and the scattering of waves by nonuniformities in the medium.

General asymptotic expansions which are valid at caustics were derived by Ludwig [12]. J. B. Keller [10] developed a geometrical theory of diffraction (GTD) for linear waves to describe the diffraction of waves around obstacles, and other effects. He introduced new sets of diffracted rays together with various canonical diffraction problems whose solution provides initial data for the diffracted rays. Microlocal analysis provides another point of view on these diffraction problems ([1] [5] [16]).

For nonlinear waves, one usually finds that there is a critical scaling of the wave amplitude relative to the wavelength at which nonlinearity becomes significant. The nature of the nonlinear effects varies from equation to equation, and depends crucially on whether the wave is dispersive or nondispersive [19]. Here, we consider nondispersive waves modelled by hyperbolic conservation laws, such as the compressible Euler equations, and problems involving shock diffraction. This class of nonlinear waves has important physical applications and there is a large body of experimental data. For example, [18] contains a beautiful collection of photographs illustrating various aspects of shock diffraction including the transition from regular to Mach reflection (fig. 235-236), the diffraction of a shock over a step (fig. 242), and shock focusing (fig. 244). All of these problems can be analyzed by the use of nonlinear ray theory provided that the shocks are weak. The diffraction of strong shocks is poorly understood from a mathematical point of view, despite the great success of numerical computations.

A nonlinear ray theory for sound waves was first developed by Whitham, who called it the "nonlinearization technique". Whitham introduced a nonlinear correction to the linearized ray location [19]. A systematic formal derivation of nonlinear ray theory was given by Choquet-Bruhat [2]. The equivalence between Whitham's nonlinearization technique and Choquet-Bruhat's expansion is shown in [9]. There are also a number of rigorous justifications of nonlinear ray theories.

Straightforward nonlinear ray theories are invalid when diffraction becomes important. One would like to develop a nonlinear geometrical theory of diffraction (NGTD) analogous to the the linear theory. Even the simplest canonical equations which combine nonlinearity and diffraction turn out to be very hard to analyze. This fact indicates that the asymptotics capture the fundamental qualitative features of nonlinear wave diffraction, enabling one to study them in the simplest setting.

In the next section, we briefly summarize the basic two-scale nonlinear ray theory which applies when diffraction is negligible. For hyperbolic conservation laws, the main result is an inviscid Burgers equation for the wave amplitude,

$$u_t + \left(\frac{1}{2}u^2\right)_x = 0. \tag{1.1}$$

We then describe the simplest three-scale expansion which combines the effects of diffraction and nonlinearity. This expansion leads to a two-dimensional generalization of the inviscid Burgers equation,

$$\begin{aligned} &u_t + \left(\frac{1}{2}u^2\right)_x + v_y = 0, \\ &u_y - v_x = 0. \end{aligned} \tag{1.2}$$

This equation was originally derived in transonic aerodynamics [17] where it is called the unsteady transonic small disturbance equation [3], but it arises in many other physical contexts as well.

2. Nonlinear ray theories

Suppose that $u : \mathbf{R}^{d+1} \to \mathbf{R}^m$ satisfies a strictly hyperbolic system of conservation laws,

$$f^\alpha(u)_{x^\alpha} = 0. \tag{2.1}$$

Here $x = (x^0, \ldots, x^d) \in \mathbf{R}^{d+1}$ is a space-time variable and we sum over repeated indices. We look for a high-frequency asymptotic solution of

(2.1) of the form [2]

$$u = \varepsilon u_1(\theta, x) + \varepsilon^2 u_2(\theta, x) + O(\varepsilon^3), \tag{2.2}$$

as $\varepsilon \to 0+$, where the fast phase variable θ is evaluated at

$$\theta = \frac{\varphi(x)}{\varepsilon}.$$

The function $\varphi : \mathbf{R}^{d+1} \to \mathbf{R}$ is the phase of the wave. Using (2.2) in (2.1), Taylor expanding the result with respect to ε, and equating coefficients of ε^0 and ε^1 to zero, we obtain that

$$\varphi_\alpha A^\alpha u_{1\theta} = 0, \tag{2.3}$$

$$\varphi_\alpha A^\alpha u_{2\theta} + A^\alpha u_{1x^\alpha} + \frac{1}{2}\varphi_\alpha D^2 f^\alpha(0) \cdot (u_1, u_1)_\theta = 0. \tag{2.4}$$

Here, D denotes the derivative with respect to u and

$$\varphi_\alpha = \varphi_{x^\alpha}, \qquad A^\alpha = Df^\alpha(0).$$

Equation (2.3) has a nontrivial solution provided that φ satisfies the linearized eikonal equation

$$\det(\varphi_\alpha A^\alpha) = 0.$$

A solution for u_1 is then

$$u_1(\theta, x) = a(\theta, x) r(x), \tag{2.5}$$

where a is an arbitrary scalar amplitude and r is a right nullvector such that

$$\varphi_\alpha A^\alpha r = 0.$$

We denote a left nullvector of this matrix by ℓ. Taking the inner product of (2.4) with ℓ and using (2.5) in the result, we find that a satisfies

$$a_s + \left(\frac{1}{2}\mathcal{G}a^2\right)_\theta + \mathcal{H}a = 0. \tag{2.6}$$

In (2.6), ∂_s is a derivative along the rays associated with φ,

$$\partial_s = \ell \cdot A^\alpha r \, \partial_{x^\alpha},$$

and the coefficients $\mathcal{G}(x)$ and $\mathcal{H}(x)$ are given by

$$\mathcal{G} = \ell \cdot \varphi_\alpha D^2 f^\alpha(0) \cdot (r, r), \qquad \mathcal{H} = \ell \cdot A^\alpha r_{x^\alpha}.$$

If $\mathcal{G}$ and $\mathcal{H}$ are smooth and $\mathcal{G} \neq 0$, then a suitable change of variables transforms (2.6) into the Burgers equation (1.1) in which $x = \theta$ is the phase variable.

To derive a diffractive version of equation (1.1), we look for an asymptotic solution of (2.1) which depends on three different scales [6],

$$u = \varepsilon u_1(\theta, \eta, x) + \varepsilon^{3/2} u_2(\theta, \eta, x) + \varepsilon^2 u_3(\theta, \eta, x) + O(\varepsilon^{5/2}), \qquad (2.7)$$

where the fast variable θ and the intermediate variable η are evaluated at

$$\theta = \frac{\varphi(x)}{\varepsilon}, \qquad \eta = \frac{\psi(x)}{\varepsilon^{1/2}}.$$

The function ψ must satisfy (2.11) below, but is otherwise arbitrary.

Using (2.7) in (2.1) and expanding the result, we find that

$$\varphi_\alpha A^\alpha u_{1\theta} = 0, \qquad (2.8)$$

$$\varphi_\alpha A^\alpha u_{2\theta} + \psi_\alpha A^\alpha u_{1\eta} = 0, \qquad (2.9)$$

$$\varphi_\alpha A^\alpha u_{3\theta} + \psi_\alpha A^\alpha u_{2\eta} + A^\alpha u_{1x^\alpha} + \frac{1}{2}\varphi_\alpha D^2 f^\alpha(0) \cdot (u_1, u_1)_\theta = 0, \qquad (2.10)$$

where $\psi_\alpha = \psi_{x^\alpha}$. A solution of (2.8) for u_1 is

$$u_1(\theta, \eta, x) = a(\theta, \eta, x) r(x).$$

Taking the inner product of (2.9) with the left null vector ℓ, we get

$$\ell \cdot A^\alpha r \, \psi_\alpha = 0. \qquad (2.11)$$

This condition implies that ψ is constant along the rays associated with φ. A solution of (2.9) for u_2 is then given by

$$u_2 = b(\theta, \eta, x) s(x),$$

where the scalar amplitude b satisfies

$$a_\eta - b_\theta = 0, \qquad (2.12)$$

and the vector s is a solution of

$$\varphi_\alpha A^\alpha s + \psi_\alpha A^\alpha r = 0.$$

Taking the inner product of equation (2.10) with ℓ, we obtain

$$a_s + \left(\frac{1}{2}\mathcal{G} a^2\right)_\theta + \mathcal{H} a + \mathcal{N} b_\eta = 0, \qquad (2.13)$$

where ∂_s, $\mathcal{G}$, and $\mathcal{H}$ are the same as in (2.6) and

$$\mathcal{N} = \ell \cdot A^{\alpha} s \psi_{\alpha}.$$

Equations (2.12) and (2.13) are a system of equations for a and b. If the coefficients are smooth functions of x and $\mathcal{G}$ is nonzero, then we can transform this system into a two-dimensional Burgers equation of the form

$$\begin{aligned} &u_t + \left(\frac{1}{2}u^2\right)_x + \nu(t)v_y = 0, \\ &u_y - v_x = 0. \end{aligned} \tag{2.14}$$

In the case of plane waves, with

$$\varphi(x) = \kappa_\alpha x^\alpha \qquad \psi(x) = \lambda_\alpha x^\alpha,$$

the coefficient ν is constant and, provided that ν is nonzero, we can normalize it to one which gives (1.2).

3. An initial-boundary value problem

The two-dimensional Burgers equation (1.2) is strictly hyperbolic but t and x are characteristic directions. In this section we show that a characteristic inital boundary value problem for (1.2) in the left-half space $x < 0$ is is well-posed forward in time. The resulting IBVP consists of (1.2) in the space-time quadrant $\{(x, y, t) : x \in \mathbf{R}^-, y \in \mathbf{R}, t \in \mathbf{R}^+\}$ together with the intitial and boundary conditions $u(x, y, 0) = f(x, y)$ and $v(0, y, t) = g(y, t)$. To obtain a well-posed evolution equation backward in time, (1.2) must be solved in the right-half space $x > 0$.

The nonlinear term does not play a large role in determining appropriate auxilliary conditions for (1.2). First, we consider the linearized equation where well-posedness can be established in a standard way by use of the Fourier-Laplace transform or by energy estimates.

We consider the following IBVP for the linearized two-dimensional Burgers equation,

$$\begin{aligned} &u_t - v_y = 0, \qquad (x, y, t) \in \mathbf{R}^+ \times \mathbf{R} \times \mathbf{R}^+, \\ &u_y - v_x = 0, \\ &u(x, y, 0) = 0, \qquad (x, y) \in \mathbf{R}^+ \times \mathbf{R}, \\ &v(0, y, t) = g(y, t), \qquad (y, t) \in \mathbf{R} \times \mathbf{R}^+. \end{aligned} \tag{3.1}$$

Here we have changed $u \to -u$ and $x \to -x$ in comparison with (1.2) so that the right-half space $x > 0$ for (3.1) corresponds to the left-half space $x < 0$ for (1.2). We also assume zero initial data without loss of generality. The following theorem implies that (3.1) is well-posed [11].

Theorem 3.1 *Suppose that $g(y,t)$ is smooth and there are constants $C, \alpha \in \mathbf{R}$ such that*

$$\int_{\mathbf{R}} |g(y,t)|^2\, dy \le Ce^{\alpha t}. \tag{3.2}$$

Then for any $T > 0$ there is a constant K_T such that the solution of (3.1) satisfies

$$\int_0^T \int_{\mathbf{R}} \int_{\mathbf{R}^+} |u(x,y,t)|^2\, dxdydt \le K_T \int_0^T \int_{\mathbf{R}} |g(y,t)|^2\, dydt. \tag{3.3}$$

Proof. Fourier transforming (3.1) in y and Laplace transforming in t we get that

$$\begin{aligned} s\hat{u} - i\ell\hat{v} &= 0, \\ i\ell\hat{u} - \hat{v}_x &= 0, \\ \hat{v}(0,\ell,s) &= \hat{g}(\ell,s), \end{aligned} \tag{3.4}$$

where

$$\begin{aligned} \hat{u}(x,\ell,s) &= \mathcal{F}_y\mathcal{L}_t[u(x,y,t)], \\ \hat{v}(x,\ell,s) &= \mathcal{F}_y\mathcal{L}_t[v(x,y,t)]. \end{aligned}$$

The solution of (3.4) is given by

$$\hat{u} = \frac{i\ell}{s}\hat{g}(\ell,s)e^{-\ell^2 x/s}, \qquad \hat{v} = \hat{g}(\ell,s)e^{-\ell^2 x/s}. \tag{3.5}$$

Use of Parseval's theorem to estimate the space-time norm of u implies that, for any $T > 0$ and $\eta > 0$,

$$\begin{aligned} \|u\|_T^2 &= \int_0^T \int_{\mathbf{R}} \int_{\mathbf{R}^+} |u(x,y,t)|^2\, dxdydt \\ &\le e^{2\eta T} \int_{\mathbf{R}^+} \int_{\mathbf{R}} \int_{\mathbf{R}^+} e^{-2\eta t} |u(x,y,t)|^2\, dxdydt \\ &= \frac{e^{2\eta T}}{(2\pi)^2} \int_{\mathbf{R}} \int_{\mathbf{R}} \int_{\mathbf{R}^+} |\hat{u}(x,\ell,\eta + i\tau)|^2\, dxd\ell d\tau. \end{aligned}$$

Use of (3.5) followed by Parseval's theorem implies that

$$\begin{aligned} \int_{\mathbf{R}} \int_{\mathbf{R}} \int_{\mathbf{R}^+} |\hat{u}(x,\ell,\eta + i\tau)|^2\, dxd\ell d\tau &= \frac{1}{2\eta} \int_{\mathbf{R}} \int_{\mathbf{R}} |\hat{g}(\ell,\eta + i\tau)|^2\, d\ell d\tau \\ &= \frac{1}{2\eta} \int_{\mathbf{R}^+} \int_{\mathbf{R}} e^{-2\eta t} |g(y,t)|^2\, dydt \end{aligned}$$

Combining these equations, we conclude that

$$\|u\|_T^2 \le \frac{1}{(2\pi)^2}\frac{e^{2\eta T}}{2\eta}\int_{\mathbf{R}^+}\int_{\mathbf{R}} e^{-2\eta t}|g(y,t)|^2\,dydt.$$

By a standard argument, the value of g for $t > T$ does not affect the solution for $t \le T$, so we can assume $g(y,t) = 0$ for $t > T$. It then follows that

$$\|u\|_T^2 \le K_T \int_0^T \int_{\mathbf{R}} |g(y,t)|^2\,dydt,$$

where

$$K_T = \frac{1}{(2\pi)^2}\inf_{\eta>0}\frac{e^{2\eta T}}{2\eta} = \frac{Te}{(2\pi)^2}.$$

This proves the theorem. □

Next we derive an energy estimate which proves that the linearized two-dimensional Burgers equation is well-posed and that the solution at a given point depends only on the initial data in the intersection of the backwards characteristic cone with $t = 0$. We consider the IBVP

$$\begin{aligned}
&u_t + cu_x + v_y = 0, \qquad (x,y,t) \in \mathbf{R}\times\mathbf{R}\times\mathbf{R}^+\\
&u_y - v_x = 0,\\
&u(x,y,0) = f(x,y),\\
&v(x,y,t) = 0, \quad \text{for } x > Ct\,.
\end{aligned} \tag{3.6}$$

Here, the coefficient $c(x,y,t)$ is an arbitrary smooth, uniformly bounded function. We use the same x-direction as in (1.2), and suppose that $x \in \mathbf{R}$. We impose a boundary condition on the right by requiring that $v = 0$ for $x > Ct$ with $C = \sup c$. To ensure that the initial data is compatible with this boundary condition, we assume that $f(x,y) = 0$ when $x > 0$.

The characteristic surfaces of (3.6) are given by $\varphi(x,y,t) = $ constant where φ satisfies the eikonal equation

$$\varphi_x\varphi_t + c\varphi_x^2 + \varphi_y^2 = 0. \tag{3.7}$$

We consider a truncated backward characteristic cone Ω_T for $0 \le t \le T$,

$$\Omega_T = \{(x,y,t) : \varphi(x,y,t) > 0, \quad 0 < t < T\}.$$

Since time is characteristic, this cone is unbounded. The surface of the cone is a paraboloid opening to the right. For example, when c is constant, we can take

$$\varphi = x - x_0 - c(t - t_0) + \frac{(y-y_0)^2}{4(t-t_0)}$$

for any $t_0 > T$. We denote the bottom, top, and side of the cone by

$$\begin{aligned} \Sigma_0 &= \{(x,y,0) : \varphi(x,y,0) > 0\}, \\ \Sigma_T &= \{(x,y,T) : \varphi(x,y,T) > 0\}, \\ \Lambda_T &= \{(x,y,t) : \varphi(x,y,t) = 0, \quad 0 < t < T\}. \end{aligned}$$

Theorem 3.2 *Any smooth solution of* (3.6) *satisfies the energy estimate*

$$\int_{\Sigma_T} |u(x,y,T)|^2 \, dxdy \le K_T \int_{\Sigma_0} |u(x,y,0)|^2 \, dxdy, \tag{3.8}$$

where $K_T = \exp(C_T T)$ *with* $C_T = \|c_x\|_{L^\infty(\Omega_T)}$.

Proof. We denote an outward space-time normal of Λ_T by

$$\mathbf{n} = -\varphi_x^{-1}(\varphi_t, \nabla\varphi) = \left(c + \nu^2, -1, \nu\right), \tag{3.9}$$

where $\nu = -\varphi_y/\varphi_x$. Multiplying the first equation in (3.6) by u, the second equation by v, and adding the results, we obtain

$$\left(u^2\right)_t + \left(cu^2 - v^2\right)_x + (2uv)_y = c_x u^2.$$

Integrating this equation over Ω_T, then using the divergence theorem and equation (3.9), we get

$$\int_{\Sigma_T} u^2 \, dxdy + \int_{\Lambda_T} (\nu u + v)^2 \|\mathbf{n}\|^{-1} dS = \int_{\Sigma_0} u^2 \, dxdy + \int_{\Omega_T} c_x u^2 \, dxdydt. \tag{3.10}$$

The integrals extend over a bounded region since the characteristic cone opens to the right and u and v are zero for sufficiently large positive values of x. We denote the energy in the spatial cross-section of the characteristic cone at time t by

$$E(t) = \int_{\Sigma_t} u^2 \, dxdy.$$

Use of (3.10) with $T = t$ implies that

$$E(t) \le E(0) + f(t),$$

where

$$f(t) = \int_{\Omega_t} c_x u^2 \, dxdydt'.$$

Since $\Omega_t \subset \Omega_T$ for $0 \le t \le T$, we have

$$\begin{aligned} |f(t)| &\le \sup_{\Omega_T} |c_x| \int_{\Omega_t} u^2 \, dx dy dt' \\ &\le C_T \int_0^t E(t')\, dt'. \end{aligned}$$

It follows that

$$E(t) \le E(0) + C_T \int_0^t E(t')\, dt'.$$

An application of Gronwall's inequality gives (3.8). □

The energy estimate (3.8) can be used to prove short time existence of smooth solutions of the quasilinear system

$$\begin{aligned} & u_t + \left(\frac{1}{2}u^2\right)_x + v_y = 0, \\ & u_y - v_x = 0, \\ & u(x,y,0) = f(x,y), \\ & v(x,y,t) = 0 \quad \text{for } x > \xi(y,t). \end{aligned} \tag{3.11}$$

by a standard iteration [13]. This argument establishes the short-time existence of smooth (C^1) solutions u if $f \in H^s(\mathbf{R}^2)$ for some $s > 2$ and the support of f is contained in $\mathbf{R}^- \times \mathbf{R}$. The existence time depends only on the H^s-norm of f. One peculiarity of the characteristic IBVP is that v need not be smooth unless $s > 3$. There is a loss of one y-derivative in reconstructing v from u by means of $v_x = u_y$ and there are no direct a priori estimates on v.

Smooth solutions of (3.11) may break down in finite time due to the formation of shocks. Global existence of weak solutions, even for small data, is an open question.

4. Regularized Burgers equation

The fact that time is a characteristic variable means that standard numerical schemes for the solution of hyperbolic conservation laws cannot be applied directly to (1.2). It is surprisingly difficult to construct a stable numerical scheme to solve the characteristic initial-boundary value problem (3.11). One obvious strategy is to solve the second equation for v in terms of u,

$$v(x,y,t) = \partial_x^{-1} u_y(x,y,t) = \int_{+\infty}^{x} u_y(x',y,t)dx',$$

and then evolve u in time according to

$$u_t + \left(\frac{1}{2}u^2\right)_x + \partial_x^{-1} u_{yy} = 0.$$

Naive discretizations of this equation are invariably unstable to numerical oscillations which vary rapidly in y.

To remove this instability, we introduce the following regularized two-dimensional Burgers equation,

$$\begin{aligned} u_t + \left(\frac{1}{2}u^2\right)_x + v_y &= 0, \\ u_y &= v_x + \varepsilon v_{yy}, \end{aligned} \tag{4.1}$$

where the "viscosity" ε is a small positive parameter. This regularization is not completely smoothing since weak solutions of (4.1) may contain jumps in u.

Equation (4.1) is in conservation form. Admissible weak solutions of (4.1) satisfy an entropy inequality (with equality if u is smooth),

$$\left(\frac{1}{2}u^2\right)_t + \left(\frac{1}{3}u^3 - \frac{1}{2}v^2\right)_x + \left(v\left[u - \varepsilon v_y\right]\right)_y \leq -\varepsilon v_y^2.$$

The limit as $\varepsilon \to 0+$ of a strongly convergent sequence of solutions of the regularized two-dimensional Burgers equation is therefore an entropy solution of (1.2).

The viscosity ε in (4.1) does not correspond to the usual physical viscosity. If one applies the diffractive asymptotic expansion to the compressible Navier-Stokes equations, with an appropriately scaled viscosity, one obtains the following viscous two-dimensional Burgers equation,

$$\begin{aligned} u_t + \left(\frac{1}{2}u^2\right)_x + v_y &= \mu u_{xx}, \\ u_y &= v_x. \end{aligned}$$

The physical viscosity μ in this equation damps x-waves but leaves y-waves undamped, exactly the reverse of the viscosity in (4.1). A small physical viscosity does not stabilize numerical schemes for (1.2). The viscous equation (4.1) can be derived from a primitive system of hyperbolic conservation laws if one assumes a highly anisotropic viscosity matrix with much larger y-viscosity than x-viscosity.

The system (4.1) can be rewritten as a Burgers equation for u with a nonlocal source term,

$$u_t + \left(\frac{1}{2}u^2\right)_x + \mathcal{L}_\varepsilon[u] = 0, \tag{4.2}$$

where

$$\mathcal{L}_\varepsilon = \left(\partial_x + \varepsilon\partial_y^2\right)^{-1}\partial_y^2.$$

The operator $\mathcal{L}_\varepsilon$ is bounded when $\varepsilon > 0$ but when $\varepsilon = 0$ it reduces to the unbounded operator $\partial_x^{-1}\partial_y^2$.

Linearizing the regularized two-dimensional Burgers equation (4.1) about $u = 0$ and eliminating v, we obtain the equation

$$u_{xt} + \varepsilon u_{yyt} + u_{yy} = 0. \tag{4.3}$$

The Fourier solutions of (4.3) are

$$u(x, y, t) = \hat{u}\exp(ikx + i\ell y - i\omega t),$$

where the frequency ω satisfies the linearized dispersion relation

$$\omega = \frac{\ell^2}{k + i\varepsilon\ell^2} = \frac{k\ell^2}{k^2 + \varepsilon^2\ell^4} - i\frac{\varepsilon\ell^4}{k^2 + \varepsilon^2\ell^4}. \tag{4.4}$$

For $\varepsilon = 0$, this equation reduces to the linearized dispersion relation of (1.2) $\omega = \ell^2/k$. For $\varepsilon > 0$, non-physical waves propagating in the y-direction ($k = 0$) are strongly damped, but waves propagating in the x-direction ($\ell = 0$) are undamped.

An interesting special case of the nonlinear equation is that of oblique plane waves. The resulting nonlocal, one-dimensional conservation law is similar to a relaxing gas equation, with diffraction playing the role of relaxation. We consider solutions of (4.1) of the form

$$u = u(z, t), \qquad v = v(z, t), \qquad z = kx + \ell y.$$

After an integration, equation (4.1) becomes

$$\begin{aligned} &u_t + \left(\frac{k}{2}u^2\right)_z + \ell v_z = 0, \\ &\varepsilon k v_z + v = \ell u_z. \end{aligned} \tag{4.5}$$

This equation can be written as a one-dimensional Burgers equation with a lower order nonlocal damping term,

$$u_t + \left(\frac{k}{2}u^2\right)_z + \frac{1}{\varepsilon}\int_{-\infty}^{z} \exp\left(-\frac{k}{\varepsilon\ell^2}(z - \xi)\right) u_\xi(\xi, t)\, d\xi = 0.$$

Expanding the integral on the right hand side of this equation as $\varepsilon \to 0+$ for fixed $k > 0$, we obtain a viscous Burgers equation,

$$u_t + \left(\frac{\ell^2}{k}u + \frac{k}{2}u^2\right)_z = \frac{\varepsilon\ell^4}{k^2}u_{zz}.$$

The linearized dispersion relation of this equation is consistent with the small ε limit of the linearized dispersion relation (4.4) and it shows the direction-dependent viscosity of (4.1).

To study the profile of oblique shocks we look for traveling wave solutions of (4.5). We set $k = 1$ and $\ell = -a$ and let

$$u = u(s), \qquad v = v(s), \qquad s = \frac{z - ct}{\varepsilon} = \frac{x - ay - ct}{\varepsilon}.$$

Without loss of generality, we normalize the shock strength to one, so that

$$\begin{aligned} u(s) &\to 0 \qquad \text{as } s \to +\infty, \\ u(s) &\to 1 \qquad \text{as } s \to -\infty. \end{aligned}$$

The shock speed c is then given by

$$c = \frac{1}{2} + a^2.$$

After a little algebra, we obtain the following ordinary differential equation for u:

$$(u - c)u' = \frac{1}{2}u(1 - u). \tag{4.6}$$

There are two cases, depending on whether the shock is supersonic ($c > 1$) or subsonic ($c < 1$) with respect to the state $u = 1$ behind the shock. If $c > 1$, corresponding to $a > 1/\sqrt{2}$, then (4.6) has a smooth solution which decreases monotonically from $u = 1$ to $u = 0$. Thus, the regularization in (4.1) gives smooth profiles for supersonic shocks.

If $c < 1$, corresponding to $a < 1/\sqrt{2}$, then the singular point $u = c$ of (4.6) lies between zero and one. In this case, there is a shock at the rear of the profile across which u jumps down from $u = 1$ to $u = 2a^2 < c$. The speed of this shock is equal to the traveling wave speed. The shock is followed by a smooth wave in which u decreases monotonically from $u = 2a^2$ to 0. This precursor wave can be seen in front of some of the oblique shocks in the numerical solutions shown below. In the limiting case of normal shocks, $a = 0$, the precursor wave disappears and the shock profile is completely sharp.

In the sonic case, $c = 1$ and $a = 1/\sqrt{2}$, the shock profile ends in a corner. The analytical solution in this case is particularly simple, namely

$$u(s) = \begin{cases} 1 & \text{if } s \le 0 \\ \exp(-s/2) & \text{if } s > 0 \end{cases}.$$

5. Shock focusing and shock reflection

In this section we present some numerical solutions of (3.11) which describe shock reflection and shock focusing. Our numerical scheme is based on the regularized equation (4.1). We use a first order artificial viscosity $\varepsilon = c\Delta x$, where c is a constant (typically, $c = 1$) and Δx is the x-grid spacing. Given $u(x, y, t)$ at time t, we obtain v by solving the one-dimensional heat equation

$$v_x + \varepsilon v_{yy} = u_y$$

backward in x. We use a backward implicit scheme since it damps grid scale oscillations. We then use the equation

$$u_t + \left(\frac{1}{2}u^2\right) + v_y = 0$$

to update u in time. This procedure splits the solution of the original two-dimensional problem into two one-dimensional problems, one in (x, y) and one in (x, t).

Numerical solutions of the two-dimensional Burgers equation for shock reflection and focusing were computed by Tabak and Rosales [15] using a completely different scheme. They made a coordinate change $(x, t) \rightarrow (x', t')$ to nocharacteristic, primed coordinates $x' = x - t$, $t' = x + t$. In the primed coordinates, the two-dimensional Burgers equation is a nonlinear wave equation which can be solved using standard numerical methods for hyperbolic conservation laws. There is good overall agreement between the numerical solutions shown here and the ones shown in [15].

The first problem arises in describing the transition from regular to Mach reflection for weak shocks reflecting off thin wedges ([7] [8] [15]). The solution is symmetric in y with

$$u(x, -y, t) = u(x, y, t), \qquad v(x, -y, t) = -v(x, y, t).$$

The initial data depends on a parameter $a \in \mathbf{R}^+$ and when $y > 0$ it is given by

$$u(x, y, 0) = \begin{cases} 0 & \text{if } x > ay \\ 1 & \text{if } x < ay \end{cases},$$

$$v(x, y, 0) = \begin{cases} 0 & \text{if } x > ay \\ -a & \text{if } x < ay \end{cases}.$$

A numerical solution for $a = 0.5$ is shown in Figures 1–3. The problem is self-similar so that u and v depend only on x/t and y/t. A very clear Mach reflection emerges. Equation (1.2) does not allow pure triple points in which three plane shocks meet at a point, so the structure of the solution near the apparent triple point in the numerical solution is an interesting question. Proposed explanations have included a continuous reflected wave [4], a local subsonic singularity in v [15], and a supersonic expansion fan [8], but there is no current agreement on the answer.

The second problem is a generalization of the first. The initial data consist of a constant strength, nonplanar shock located at $x = s(y)$ with

$$u(x,y,0) = \begin{cases} 0 & \text{if } x > s(y) \\ u_0 & \text{if } x < s(y) \end{cases},$$

$$v(x,y,0) = \begin{cases} 0 & \text{if } x > s(y) \\ -u_0 s'(y) & \text{if } x < s(y) \end{cases}.$$

For the focusing problem, we took $s(y)$ to be a plane wavefront with a Gaussian bump of width 0.4 in y and depth 0.25 in x. A transition from a linear fish-tail pattern to a nonlinear Mach stem pattern is observed experimentally as the shock strength is increased [14]. The transition occurs because the shock intensifies and accelerates as it focuses. If the initial shock strength is large enough, the acceleration of the central part of the shock prevents the shock from crossing over itself and forming a fish-tail. In Figures 4–5, we show a numerical solution for u at $t = 1.0$ when $u_0 = 0.2$ which is in the nonlinear, Mach stem regime. The solution is computed in a reference frame moving with the planar part of the shock and the spatial axes are labelled by grid point number rather than by x and y. Figures 6–7 show a numerical solution for u when $u_0 = 0.05$ which is in the linear, fish-tail regime. In this case, nonlinear effects delay the crossing of the shock although they do not prevent it. As a result, the fish-tail is significantly smaller than it would be for a purely linear wave.

Acknowledgements. This work was partially supported by the NSF.

References

[1] Beals, M., *Propagation and Interaction of Singularities in Nonlinear Hyperbolic Problems*, Birkhäuser, Boston, 1989.

[2] Choquet-Bruhat, Y., Ondes asymptotiques et approchées pour systèmes non linéaires d'équations aux dérivées partielles non linéaires, *J. Math. Pure et Appl.* **48** (1969), 117–158.

[3] Cole, J. D., and Cook L. P., *Transonic Aerodynamics*, Elsevier, New York, 1986.

[4] Colella, P., and Henderson, R., The von Neumann paradox for the diffraction of weak shock waves, *J. Fluid Mech.* **213** (1990), 71–94.

[5] Hörmander, L., *The Analysis of Linear Partial Differential Operators*, Vol. I–IV, Springer-Verlag, Berlin, 1983.

[6] Hunter, J. K., Transverse diffraction of nonlinear waves and singular rays, *SIAM J. Appl. Math* **48** (1988), 1–37.

[7] Hunter, J. K., Nonlinear geometrical optics, in *Multidimensional Hyperbolic Problems and Computations*, Vol. 29, IMA Volumes in Mathematics and Its Applications, ed. A. J. Majda and J. Glimm, Springer-Verlag, New York (1991), 179–197.

[8] Hunter, J. K., and Brio, M., Irregular reflection of weak shocks I., submitted to *J. Fluid Mech.*

[9] Hunter, J. K., and Keller, J. B., Weakly nonlinear high frequency waves, *Comm. Pure Appl. Math.* **36** (1983), 547–569.

[10] Keller, J. B., Rays waves and asymptotics, *Bulletin of the AMS* **84** (1978), 727–750.

[11] Kreiss, H.-O., and Lorenz, J., *Initial-Boundary Value Problems and the Navier-Stokes Equations*, Academic Press, 1989.

[12] Ludwig, D., Uniform asymptotic expansions at a caustic, *Comm. Pure Appl. Math.* **19** (1966), 215–250.

[13] Majda, A. J., *Compressible Fluid Flow and Systems of Conservation Laws in Several Space Variables*, Springer-Verlag, 1984.

[14] Sturtevant, B., and Kulkarny, V. A., The focusing of weak shock waves, *J. Fluid Mech.* **73** (1983), 651–671.

[15] Tabak, E., and Rosales, R. R., Weak shock focusing and the von Neumann paradox of oblique shock reflection, *Phys. Fluids* **6** (1994), 1874–1892.

[16] Taylor, M., *Pseudodifferential Operators and Nonlinear Partial Differential Equations*, Progress in Mathematics, Vol. 100, Birkhäuser, Boston, 1991.

[17] Timman, R., Unsteady motion in transonic flow, in *Symposium Transonicum*, ed. K. Oswatitsch, Springer-Verlag, Aachen (1962), 394–401.

[18] Van Dyke, M., *An Album of Fluid Motion*, Parabolic Press, Stanford, 1982.

[19] Whitham, G. B., *Linear and Nonlinear Waves*, Wiley, New York, 1974.

University of California at Davis

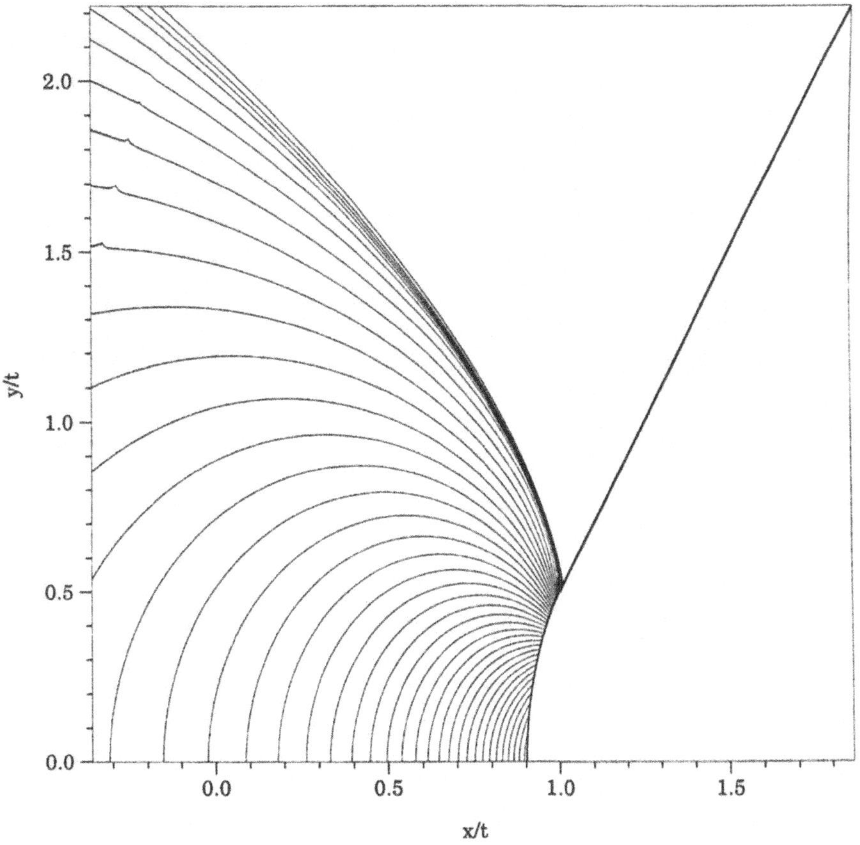

Figure 1: Shock reflection: contour plot of u for $a = 0.5$.

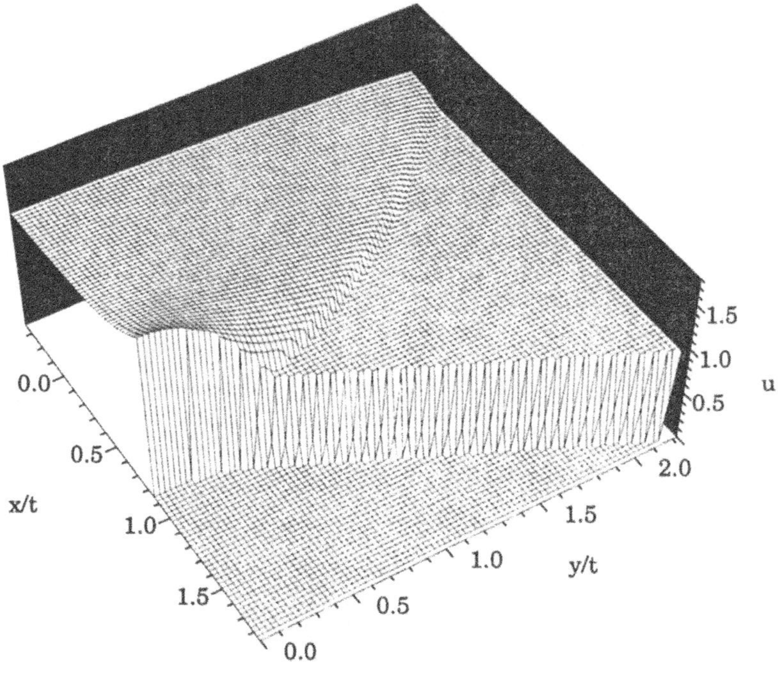

Figure 2: Shock reflection: surface plot of u for $a = 0.5$.

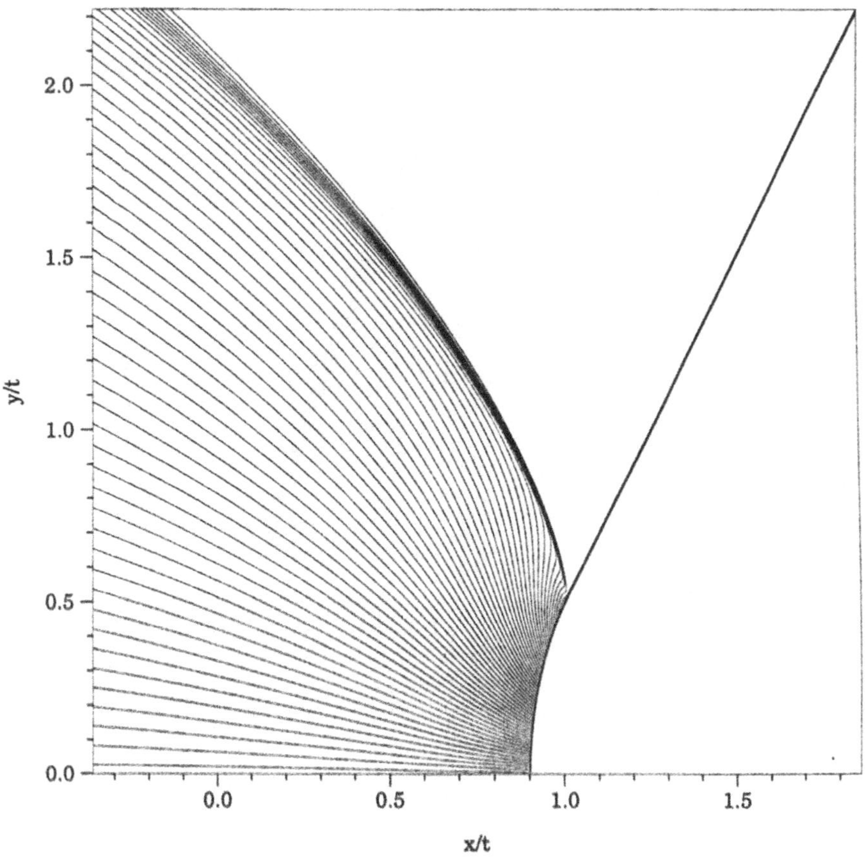

Figure 3: Shock reflection: contour plot of v for $a = 0.5$.

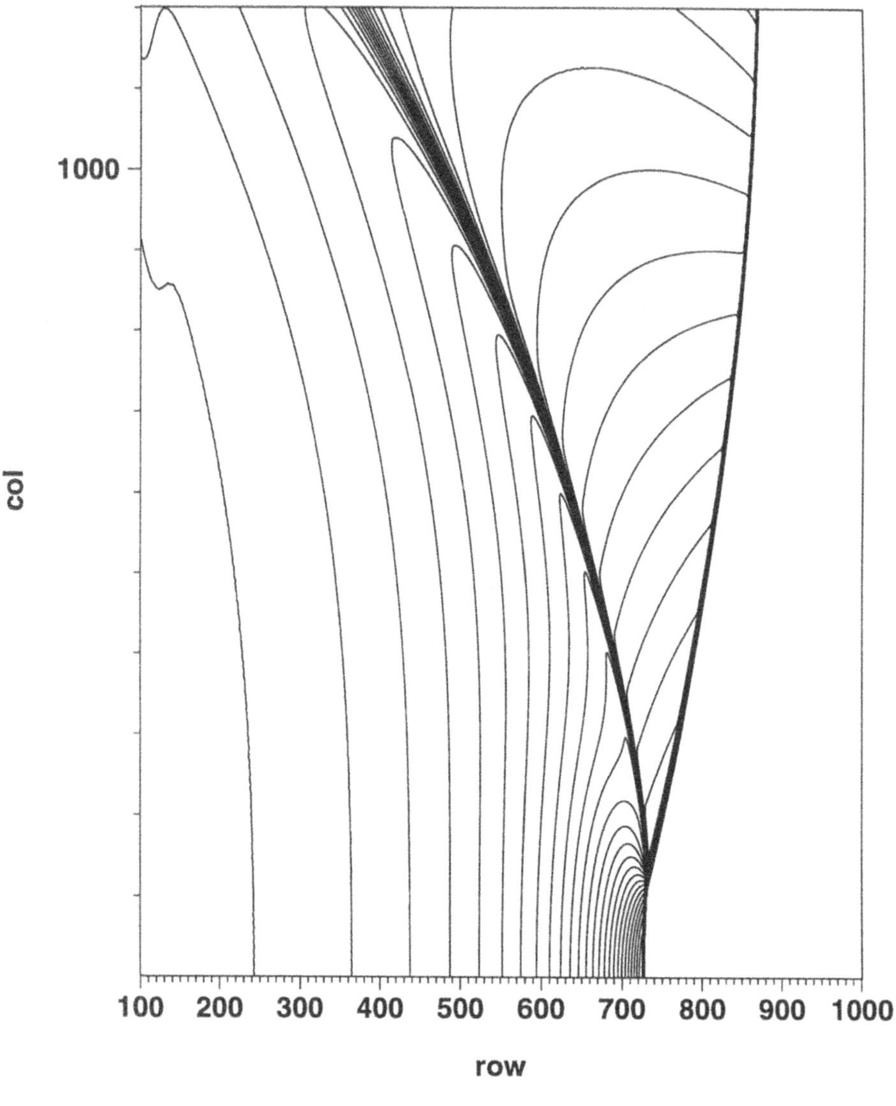

Figure 4: Shock focusing: contour plot of u at $t = 1.0$ for $u_0 = 0.2$.

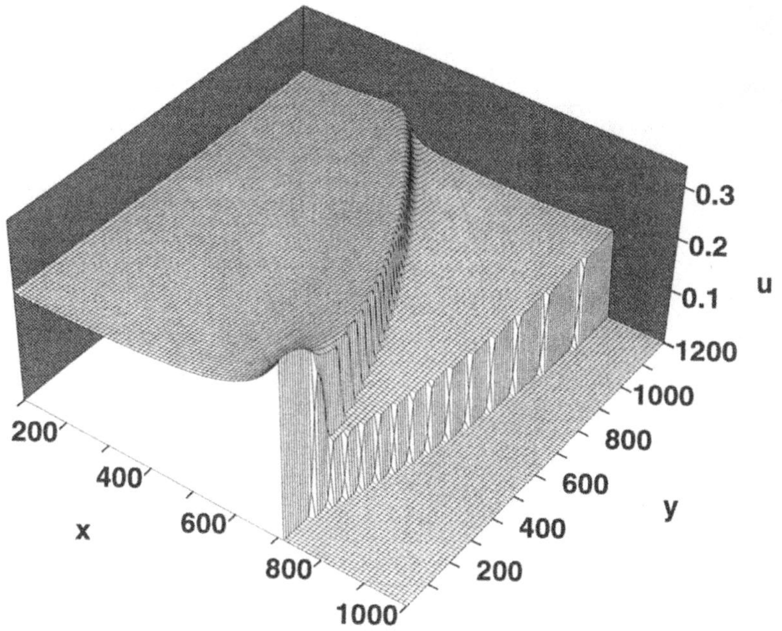

Figure 5: Shock focusing: surface plot of u at $t = 1.0$ for $u_0 = 0.2$.

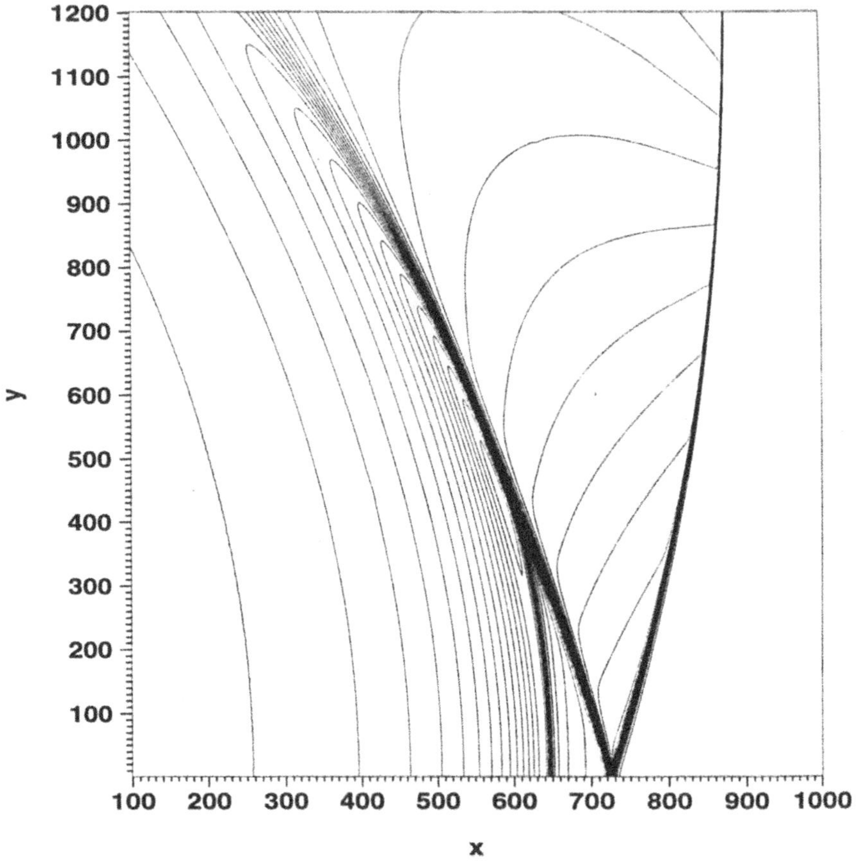

Figure 6: Shock focusing: contour plot of u at $t = 1.0$ for $u_0 = 0.05$.

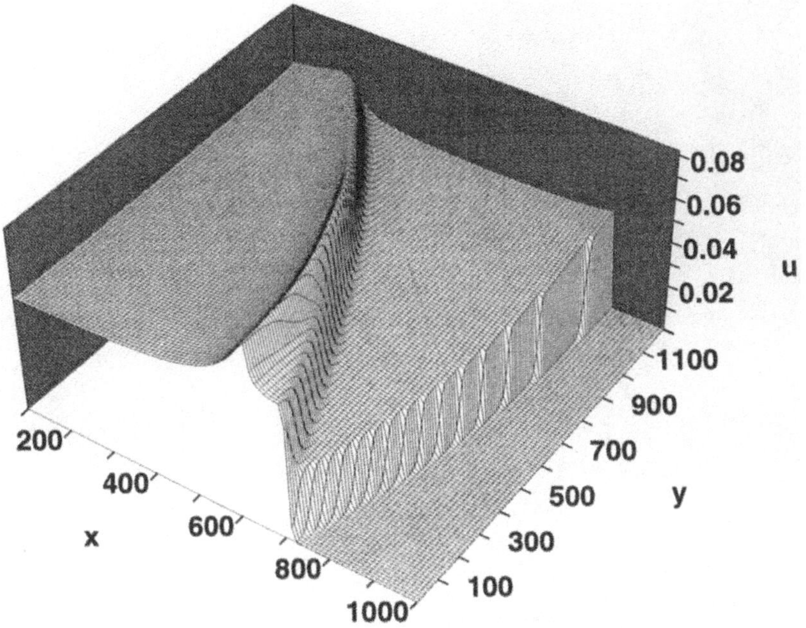

Figure 7: Shock focusing: surface plot of u at $t = 1.0$ for $u_0 = 0.05$.

Caustics for dissipative semilinear oscillations

Jean-Luc Joly, Guy Métivier, and Jeffrey Rauch

Introduction

Consider in $\mathbb{R}^{1+d}$ the semilinear wave equation

$$\Box u + f(\partial u) = 0 \tag{1.1}$$

where $\partial u := (\partial_t u, \partial_x u)$ and f is a smooth function from $\mathbb{R}^{1+d}$ into $\mathbb{R}^{1+d}$. Consider oscillatory Cauchy data

$$\begin{cases} u^\varepsilon_{|t=0} &= \underline{u}_0(x) + \varepsilon\, \mathbf{U}_0(x, \psi(x)/\varepsilon), \\ \partial_t u^\varepsilon_{|t=0} &= \underline{u}_1(x) + \mathbf{U}_1(x, \psi(x)/\varepsilon), \end{cases} \tag{1.2}$$

where ψ is a smooth function with nonvanishing differential on $\omega \subset \mathbb{R}^d$, $\mathbf{U}_0(x,\theta)$ and $\mathbf{U}_1(x,\theta)$ are smooth, 2π-periodic in θ with mean 0 and compactly supported in $\omega \times \mathbb{T}$ and $\underline{u}_0, \underline{u}_1 \in C_0^\infty(\mathbb{R}^d)$. In this problem, the scales of the wavelength and of the amplitude of the oscillations are chosen so that the nonlinear effects are expected to appear in time $O(1)$.

The behaviour of u^ε for small times is given in [JR]. There is $T > 0$ such that for all $\varepsilon \in]0, \varepsilon_0]$ the solution u^ε exists on $[0, T]$ and satisfies

$$u^\varepsilon = \underline{u} + \varepsilon\, \mathbf{U}_+(t, x, \varphi_+(t,x)/\varepsilon) + \varepsilon\, \mathbf{U}_-(t, x, \varphi_-(t,x)/\varepsilon) + \varepsilon^2 \dots, \tag{1.3}$$

where $\varphi_\pm$ are the solutions of the eikonal equation for $\Box$ with data ψ at $t = 0$ and the profiles $\mathbf{U}_\pm$ satisfy transport equations, coupled to a nonlinear wave equation for $\underline{u}$.

In general, $\varphi_\pm$ develop singularities in finite time. The singular locus is the caustic set $\mathcal{C}$. The question we discuss in this article is the behavior of the solutions for large time, in particular after the formation of caustics. Equation (1.1) serves only as a model for this discussion and the assumptions on the data could be considerably weakened (see [JMR 5]).

However, this question does not make sense for general equations. In the linear case, amplitudes tend to infinity as one approaches $\mathcal{C}$ and the description (1.3) breaks down. In the nonlinear case, the large amplitudes can be amplified by nonlinearities and therefore strongly nonlinear phenomena can occur. For example, consider in space dimension $d = 3$, the equation

$$\Box u^\varepsilon \; - \; |\partial_t u^\varepsilon|^2 \, \partial_t u^\varepsilon \; = \; 0, \tag{1.4}$$

and initial data (1.2) with $\psi(x) = |x|$, $\underline{u}_0 = \underline{u}_1 = 0$, $\mathbf{U}_1 = \partial_\theta \mathbf{U}_0$, $\mathbf{U}_0$ odd in θ and $\mathbf{U}_0 = 0$ near $x = 0$. The caustic set is $C = \{x = 0\}$. Call T the smallest t such that a ray starting from a point in the support of $\mathbf{U}_0$ reaches C at time t. Then, the solutions u^ε blow up *before* T. More precisely, there is $T' < T$, such that (1.4) does not have a family of solutions bounded in $L^3([0, T'] \times \mathbb{R}^d)$ (see [JMR 2] [JMR 4]). Of course, one does not expect that general equations (1.1) have global solutions. The important point in the example above is that the blow up always occurs before the first time of focusing T, even for arbitrary small data $\mathbf{U}_0$.

In this example, the blow up is created by the principal oscillations themselves and focusing effects due to the phases $\varphi_\pm$ created by ψ. This is called *direct focusing* in [JMR 1]. But nonlinear interactions make the problem much harder. Focusing and blow up can be created by phases not present in the principal term of the expansion, but which are generated after several interactions. This phenomenon is explored in detail in [JMR 1], where it is called *hidden focusing.*

In the opposite direction, there are equations (1.1) which always have global solutions. This happens when f is globally Lipschitzean or when it is dissipative. A typical example of a dissipative equation is

$$\Box u \; + \; |\partial_t u|^{p-1} \, \partial_t u \; = \; 0. \tag{1.5}$$

Since the Cauchy data (1.2) define bounded families in $H^1(\mathbb{R}^d)$ and $L^2(\mathbb{R}^d)$, there are unique solutions u^ε which are bounded in $C^0([0, +\infty[; H^1(\mathbb{R}^d))$ with $\partial_t u^\varepsilon$ bounded in $C^0([0, +\infty[; L^2(\mathbb{R}^d)) \cap L^{p+1}([0, +\infty[\times \mathbb{R}^d)$, see [L], [LS], [Str].

For linear equations, the substitute to (1.3) is well known. The phases are replaced by Lagrangian manifolds $\Lambda_\pm$ which are globally defined and smooth and the oscillations $\mathbf{U}_\pm(\,.\,, \varphi_\pm/\varepsilon)$ are replaced by Lagrangian distributions associated with $\Lambda_\pm$ (see e.g. [Du], [DH], [Hö 1],

[Hö 2]). This representation by oscillatory integrals shows that after $\mathcal{C}$ more phases are required to describe the asymptotics of u^ε. Our goal is to show that the solutions of nonlinear problems can still be globally described by similar Lagrangian distributions. The case of a semilinear equation with globally Lipschitzean nonlinearity f is worked out in [JMR 3]. In this case, one can work entirely within an L^2 or H^1 framework which is well adapted both to oscillatory integrals and Lipschitzean nonlinearities. The case of dissipative equations is studied in [JMR 5]. In this article, we present the results of [JMR 5] on the example of equation (1.5). The solutions u^ε and their time derivative ∂u^ε are approximated in L^{p+1} by similar oscillatory integral.

When working with nonlinearities f which are not globally Lipschitzean, as in (1.5), one is forced to leave the L^2 framework. In particular, a key new element in [JMR 5] is to prove sharp uniform estimates in L^p for oscillatory integrals. The analysis reveals a critical exponent p_c associated to each point of the caustic set where rays focus. For generic caustic points, this exponent is 3. For radial focusing in $\mathbb{R}^d$, $d \geq 2$, it is $1 + 2/(d-1)$. When $p \geq p_c$ the dissipative mechanism is strong and oscillations are absorbed at caustics: there are no oscillations past the focusing point. This extends [JMR 2] to general caustics. When $p < p_c$, oscillations cross caustics and amplitudes can be computed after focusing. In contrast to the linear case where they remain smooth, the symbols of the oscillatory integrals can be singular at points $\rho \in \Lambda_\pm$ which project over $\mathcal{C}$, even if the projection is regular at ρ itself, but then singular at another point ρ' which has the same projection.

For complete proofs we refer to [JMR 5].

2. Oscillatory integrals

For the linear wave equation, the Lagrangian distributions to consider are given by integrals of the form

$$(2.1)\qquad (2\pi\varepsilon)^{-d} \iint e^{i\Phi_\pm(t,x,y,\xi)/\varepsilon}\, a(t,y)\, dy\, d\xi\,,$$

with

$$(2.2)\qquad \Phi_\pm(t,x,y,\xi) := \pm t\,|\xi| + (x-y)\cdot\xi + \psi(y)\,.$$

These integrals are associated to Lagrangian manifolds, $\Lambda_\pm \subset T^*\mathbb{R}^{1+d}$. $\Lambda_\pm$ is the set of points (t, x, τ, ξ) such that there exists y such that

$$x = y \mp t \frac{\xi}{|\xi|}, \quad \xi = d\psi(y), \quad \tau = \pm|\xi|. \tag{2.3}$$

Remark that (2.3) provides a global parametrization of the Lagrangians with (t, y) as parameters. In particular, we consider $a(t, y)$ as the *symbol* of the integral (2.1).

In the nonlinear case one must take care of harmonics and consider symbols depending on the extra variable $\theta \in \mathbb{T} := \mathbb{R}/2\pi\mathbb{Z}$. This leads to periodic symbols with mean 0

$$\mathcal{A}(t, y, \theta) = \sum_{n \neq 0} a_n(t, y)\, e^{i\; n\; \theta} \tag{2.4}$$

and oscillatory integrals

$$I^\varepsilon_\pm(\mathcal{A}) = \sum_{n \neq 0} \frac{|n|^d}{(2\pi\varepsilon)^d} \iint e^{i\; n\; \Phi_\pm(t,x,y,\xi)/\varepsilon}\, a_n(t, y)\, dy\, d\xi. \tag{2.5}$$

The integral is well defined in $C^0([0, +\infty[; L^2(\mathbb{R}^d))$ as soon as $\mathcal{A}(.,., \psi(.)/\varepsilon)$ belongs to the same space. This is the case when

$$\begin{cases} \mathcal{A} \in C^0([0, +\infty[; L^2(\mathbb{R}^d \times \mathbb{T})), \quad \partial_\theta \mathcal{A} \in C^0([0, +\infty[; L^2(\mathbb{R}^d \times \mathbb{T})), \\ \mathcal{A} \text{ is supported in } [0, +\infty[\times \omega \times \mathbb{T}. \end{cases} \tag{2.6}$$

Introduce $g(y) := d\psi(y)/|d\psi(y)|$. The rays are the lines $t \to q_\pm(t, y) := y \mp t\, g(y)$. With (t, y) as parameters on $\Lambda_\pm$, the projection $(t, x, \tau, \xi) \to (t, x)$ from $\Lambda_\pm$ to $\mathbb{R}^{1+d}$ is the mapping

$$\jmath_\pm \;:\; (t, y) \;\to\; (t, q_\pm(t, y)). \tag{2.7}$$

Introduce the set $\mathcal{S}_\pm$ where $\jmath_\pm$ is not a local diffeomorphism:

$$\mathcal{S}_\pm := \{ (t, y) \;:\; \det \partial_y q_\pm(t, y) = 0 \}. \tag{2.8}$$

The caustic set is $\mathcal{C} := \mathcal{C}_+ \cup \mathcal{C}_-$, where $\mathcal{C}_\pm := \jmath_\pm(\mathcal{S}_\pm)$. $\mathcal{C}_\pm$, is the set of points (t, x) where at least one of the critical point of $(y, \xi) \to \Phi_\pm(t, x, y, \xi)$ is degenerate. Thus, outside the caustic set, the classical stationary phase expansion yields, for smooth and compactly supported symbols $\mathcal{A}$, a phase-amplitude expansion

$$I^\varepsilon_\pm(\mathcal{A}) = J^\varepsilon_\pm(\mathcal{A}) \; + \; O(\varepsilon)\,, \tag{2.9}$$

where

(2.10)

$$J^\varepsilon_\pm(\mathcal{A})\,(t,x) \;:=\; \sum_{\{y \,|\, q_\pm(t,y)=x\}} \frac{1}{\Delta_\pm(t,y)}\,\big(\mathcal{H}^{m_\pm(t,y)}\mathcal{A}\big)\,(t,y,\psi(y)/\varepsilon)\,,$$

with $m_\pm(t,y) \;:=\; \frac{1}{2}$ sign $\partial^2_{(y,\xi)}\Phi_\pm(t,x,y,d\psi(y))$ evaluated at $x = q_\pm(t,y)$ and $\Delta_\pm(t,y) := |\det d_y\, q_\pm(t,y)\,|^{1/2}$. $\mathcal{H}$ denotes the Hilbert transform acting on Fourier series by

$$\mathcal{H}\Big(\sum_{n\neq 0} a_n\; e^{in\theta}\Big) \;:=\; \sum_{n\neq 0} i^{\mathrm{sign} n}\; a_n\; e^{in\theta}\,.$$

Locally, there is a regular labelling $y_{\pm,k}$ of points y such that $y \mp tg(y) = x$. Then

$$\varphi_{\pm,k}(t,x) \;:=\; \psi(y_{\pm,k}) \tag{2.11}$$

are solutions of the eikonal equation, and $J^\varepsilon_\pm(A)$ is of the form

$$J^\varepsilon_\pm(A) \;=\; \sum_k \mathbf{A}_{\pm,k}(t,x,\varphi_{\pm,k}(t,x)/\varepsilon)\,, \tag{2.12}$$

with

$$\mathbf{A}_{\pm,k}(t,x,\theta) \;:=\; \frac{1}{\Delta_\pm(t,y_{\pm,k})}\,\big(\mathcal{H}^{m(t,y_{\pm,k})}\,A\big)(t,y_{\pm,k},\theta)\,, \tag{2.13}$$

This formula links the profile $\mathcal{A}$ on the Lagrangian and the profile $\mathbf{A}$ on the base.

Assumptions 2.1. *i)* (Finiteness) *There is an integer ℓ_*, such that all $(t,x)\notin C_\pm$, the number of points $y\in\omega$ such that $x = q_\pm(t,y)$, is less than or equal to ℓ_*.*

ii) (Non resonance) *For all $(t,x)\notin C$, the phases $\varphi_{\pm,k}$ defined in* (2.11) *satisfy the following property: for all real numbers $\{\alpha_{+,k},\alpha_{-,j}\}$*

such that at least two of them are not equal to zero, the function $\varphi := \sum \alpha_{+,k}\,\varphi_{+,k} + \sum \alpha_{-,j}\,\varphi_{-,j}$ *satisfies*

$$(2.14) \qquad (\partial_t\varphi)^2 - |\partial_x\varphi|^2 \neq 0 \quad \text{a.e. on a neighborhood of } (t,x)\,.$$

iii) ψ *is real analytic on* ω.

Examples and Remarks.

1. Examples of phases which satisfy these assumptions are
   a) $\psi(y) = |y|$, away from the origin
   b) $\psi(y) = a_1\,y_1^2 + a_2\,y_2^2$, on $\mathbb{R}^2\backslash 0$
   c) $\psi(y) = y_d + y_1^2 + \ldots + y_{d-1}^2$ in $\mathbb{R}^d$.
2. The third assumption can be relaxed. However, for C^∞ phases ψ, technical assumptions, which are generically satisfied, are required. See [JMR 5] for details.
3. Much more important is the nonresonance condition. This assumption is *much weaker than coherence.* Coherence would require that the phases φ in (2.14) are *nowhere* characteristic. Moreover, the coherence assumption *is not satisfied* in the examples b) or c) above, while the nonresonance condition (2.14) is satisfied in these examples.

We refer the reader to [JMR 3] [JMR 5] for further examples and discussions.

3. L^q estimates

A trigonometric polynomial is a finite sum (2.4) $\mathcal{A}$, with coefficients $a_n \in C_0^\infty([0,+\infty[\times\omega)$. The space of trigonometric polynomials is denoted by $\mathcal{P}$.

Using the nonresonance assumption, the local L^q norm of $J_\pm^\varepsilon(\mathcal{A})$ is asymptotically estimated to be the L^q norm of the profiles $\mathbf{A}_{\pm,j}$ in (2.12). Since the Jacobian of $\jmath_\pm$ is $\Delta_\pm^2$, the L^q norm of $\mathbf{A}_{\pm,j}$ is equal to a weighted L^q norm of $\mathcal{A}$. This suggests the following definition.

Definition 3.1. *For* $q \in]1,+\infty[$, $\mathcal{L}_\pm^q$ *denotes the space of* L^q *integrable functions on* $[0,+\infty[\times\omega \times \mathbb{T}$ *with respect to the measure* $\Delta_\pm(t,y)^{2-q}\,dt\,dy\,d\theta$.

For $0 < s < 1$, $\mathcal{L}_\pm^{s,q}$ *denotes the space of functions* $\mathcal{U} \in \mathcal{L}_\pm^q$ *such that*

$$\int_0^1 \int_{[0,T]\times\omega\times\mathbb{T}} \Delta_\pm(t,y)^{2-q} \left| \frac{\mathcal{U}_k(t,y,\theta+h) - \mathcal{U}_k(t,y,\theta)}{h^s} \right|^q dt\, dy\, d\theta\, \frac{dh}{h} < +\infty .$$

Proposition 3.2. *i) For all $q \in]1,+\infty[$, there is a constant $C > 0$ such that for all trigonometric polynomials $\mathcal{A}$ which vanish on a neighborhood of $\mathcal{S}_\pm$ and all $T \geq 0$,*

$$\begin{aligned} \frac{1}{C}\, \|\mathcal{A}\|_{\mathcal{L}^q_\pm([0,T]\times\omega\times\mathbb{T})} &\leq \liminf_{\varepsilon\to 0}\, \|J^\varepsilon_\pm(\mathcal{A})\|_{L^q([0,T]\times\mathbb{R}^d)} \\ \leq \limsup_{\varepsilon\to 0}\, \|J^\varepsilon_\pm(\mathcal{A})\|_{L^q([0,T]\times\mathbb{R}^d)} &\leq C\, \|\mathcal{A}\|_{\mathcal{L}^q_\pm([0,T]\times\omega\times\mathbb{T})} . \end{aligned} \tag{3.1}$$

ii) For $1/q < s < 1$ and each $\varepsilon \in]0,1]$, $J^\varepsilon_\pm$ extends to a bounded linear operator from $\mathcal{L}^{s,q}_\pm$ to $L^q([0,+\infty[\times\mathbb{R}^d)$ and

$$\sup_{\varepsilon\in]0,1]} \|J^\varepsilon_\pm(\mathcal{A})\|_{L^q([0,T]\times\mathbb{R}^d)} \leq C\, \|\mathcal{A}\|_{\mathcal{L}^{s,q}_\pm([0,T]\times\omega\times\mathbb{T})} . \tag{3.2}$$

The method sketched in §4 is by duality and requires nonsmooth profiles. Estimate (3.1) implies that $J^\varepsilon_\pm$ can be extended *asymptotically* as a map from $\mathcal{L}^q_\pm$ to $L^q([0,+\infty[\times\mathbb{R}^d)$, so that the substitution $\theta = \psi(y)$ in $\mathcal{A}(t,y,\theta)$, which has no direct meaning for $\mathcal{A} \in \mathcal{L}^q_k$, makes sense *asymptotically.* More precisely, one shows that the trigonometric polynomials which vanish near $\mathcal{S}_\pm$ are dense in $\mathcal{L}^q_\pm$. With (3.1), this implies the following result.

Proposition 3.3. *For all $\mathcal{U} \in \mathcal{L}^q_\pm$ there exists a bounded family u^ε in $L^q([0,+\infty[\times\mathbb{R}^d)$ such that for all $\delta > 0$, there are $\varepsilon_\delta > 0$ and a trigonometric polynomial $\mathcal{U}_\delta$ which vanishes near $\mathcal{S}_\pm$, such that*

$$\|\mathcal{U} - \mathcal{U}_\delta\|_{\mathcal{L}^q_\pm} \leq \delta , \tag{3.3}$$

$$\forall \varepsilon \in]0,\varepsilon_\delta] , \quad \|u^\varepsilon - J^\varepsilon_\pm(\mathcal{U}_\delta)\|_{L^q([0,+\infty[\times\mathbb{R}^d)} \leq \delta . \tag{3.4}$$

When u^ε satisfies (3.3) (3.4), we write $u^\varepsilon \sim \tilde{J}^\varepsilon_\pm(\mathcal{U})$ in L^q. If $u^\varepsilon \sim \tilde{J}^\varepsilon_\pm(\mathcal{U})$ and $v^\varepsilon \sim \tilde{J}^\varepsilon_\pm(\mathcal{U})$ in L^q, then $u^\varepsilon - v^\varepsilon$ converges strongly to 0 in $L^q([0,+\infty[\times\mathbb{R}^d)$. More generally, we write $u^\varepsilon \sim v^\varepsilon$ in L^q, when u^ε and v^ε are bounded families in $L^q([0,+\infty[\times\mathbb{R}^d)$ such that $u^\varepsilon - v^\varepsilon$ converges strongly to 0 in $L^q([0,+\infty[\times\mathbb{R}^d)$. The notations are consistent: when $\mathcal{U} \in \mathcal{L}^{s,q}_\pm$ and $s > 1/q$, $J^\varepsilon_\pm(\mathcal{U})$ is defined for each ε, and

$$J^\varepsilon_k(\mathcal{U}) \sim \tilde{J}^\varepsilon_\pm(\mathcal{U}) \quad \text{in} \quad L^q([0,+\infty[\times\mathbb{R}^d) \tag{3.5}$$

in the sense defined above.

There is a similar treatment for functions on $\mathbb{R}^d$, i.e. for initial data. Given $\mathcal{U}_0 \in L^2(\omega \times \mathbb{T})$, one defines bounded families in $L^2(\mathbb{R}^d)$, denoted $u^\varepsilon_0 \sim \tilde{J}^\varepsilon_0(\mathcal{U}_0)$, and such that

$$J^\varepsilon_0(\mathcal{U}_0)\,(y) \;:=\; \mathcal{U}_0(y, \psi(y)/\varepsilon)\,, \tag{3.6}$$

when $\mathcal{U}_0$ is smooth, or more generally whenever the substitution $\theta = \psi(y)$ makes sense, for example when $\mathcal{U}_0 \in L^2(\omega\times\mathbb{T})$ and $\partial_\theta\mathcal{U}_0 \in L^2(\omega\times\mathbb{T})$.

When $u^\varepsilon \sim \tilde{J}^\varepsilon_\pm(\mathcal{U})$, the profile $\mathcal{U}$ is unique. More precisely, one has

Proposition 3.4. *Suppose that $\mathcal{U} \in \mathcal{L}^q_\sigma$, $\mathcal{V} \in \mathcal{L}^{q'}_{\sigma'}$, $u^\varepsilon \sim \tilde{J}^\varepsilon_\sigma(\mathcal{U})$ in L^q, and $v^\varepsilon \sim \tilde{J}^\varepsilon_{\sigma'}(\mathcal{V})$ in $L^{q'}$ with $\sigma \in \{\pm\}$ $\sigma' \in \{\pm\}$, $1 < q < +\infty$ and $1/q + 1/q' = 1$.*

Then, u^ε converges weakly to zero in L^q and

$$\int u^\varepsilon\, \overline{v^\varepsilon}\; dt\, dx \;\to\; \delta_{\sigma,\sigma'} \int \mathcal{U}\, \overline{\mathcal{V}}\; dt\, dy\, d\theta \quad as \quad \varepsilon \;\to\; 0\,. \tag{3.7}$$

Here and below, $d\theta$ denotes the invariant measure on $\mathbb{T} := \mathbb{R}/2\pi\mathbb{Z}$ of total mass equal to one. In (3.7), $\delta_{\sigma,\sigma'}$ denotes the Kronecker's symbol.

A key point in the analysis is to obtain sharp L^q estimates for oscillatory integrals $I^\varepsilon_\pm$. The boundedness in L^2 is clear, as well as the blow up in L^∞ near caustics. The question is to determine the set of q such that $I^\varepsilon_k(\mathcal{A})$ is bounded in L^q. The analysis reveals a critical exponent, which surprisingly does not depend on the full complexity of the caustic but only on the multiplicity $\mu_\pm$.

Definition 3.5. *i) The multiplicity of $\mathcal{S}_\pm$ at a point (t,y) is the algebraic multiplicity of $1/t$ as an eigenvalue of $\pm g'(y)$. We denote it by $\mu_\pm(t,y)$.*

ii) For $q \in [2,+\infty]$, $\mathcal{S}_\pm^-(q)$ denotes the set of points $(t,y) \in \mathcal{S}_k$ such that the eigenvalue of g' which is equal to $1/t$ at y has constant multiplicity on a neighborhood of y and $\mu_\pm(t,y) < 2/(q-2)$. The remaining set is denoted by $\mathcal{S}_\pm^+(q) := \mathcal{S}_\pm \backslash \mathcal{S}_\pm^-(q)$.

Introduce ω_0, the set of points $y \in \omega$ such that the eigenvalues of g' have constant multiplicity near y. The real analyticity of ψ implies that ω_0 is a dense open subset of full measure in ω. Points in $\mathcal{S}_\pm^-(q)$ are thought of as *subcritical* (for L^q) while points in $\mathcal{S}_\pm^+$ are *supercritical.* Note that for $q = 2$, every point is subcritical. In addition, note that the multiplicities μ always satisfy

$$1 \leq \mu_\pm(t,y) \leq d-1\,. \tag{3.8}$$

Theorem 3.6. *Let $q \geq 2$, and $\mathcal{A} \in \mathcal{P}$. Suppose that the coefficients a_n vanish on a neighborhood of the points $(t,y) \in \mathcal{S}_\pm^+(q)$. Then,*

$$\sup_{\varepsilon \in]0,1]} \| I_\pm^\varepsilon(\mathcal{A}) \|_{L^q(\mathbf{R}^{1+d})} < +\infty\,. \tag{3.9}$$

Moreover,

$$I_\pm^\varepsilon(\mathcal{A}) \sim J_\pm^\varepsilon(\mathcal{A}) \quad in \quad L^q\,. \tag{3.10}$$

The estimates (3.9) are not and do not follow from the sharp L^p estimates on Fourier integral operators which can be found for example in [Ste] and [So] (see also the references therein).

Remark 3.7. Suppose that $(\underline{t},\underline{y}) \in \mathcal{S}_\pm$ with $\underline{y} \in \omega_0$. Then on a neighborhood G of $(\underline{t},\underline{y})$, $\mathcal{S}_\pm$ is a smooth manifold of codimension one where $\Delta_\pm$ vanishes exactly at the order $\mu_\pm(\underline{t},\underline{y})/2$. If $\mathcal{A} \in \mathcal{P}$ is supported in $G \times \mathbb{T}$ and does not vanish at $(\underline{t},\underline{y})$, $\mathcal{A} \in \mathcal{L}_\pm^q$ if and only if Δ^{2-q} is integrable near $(\underline{t},\underline{y})$, that is if and only if $\mu_\pm(1-q/2) < 1$, or equivalently, $(\underline{t},\underline{y}) \in \mathcal{S}_\pm^-(q)$.

Similarly, if (3.9) holds, then cutting off a small neighborhood of $\mathcal{C}$ of arbitrary small measure, and using (2.9) outside $\mathcal{C}$ and (3.1), one can show that $\mathcal{A}$ must belong to $\mathcal{L}^q_\pm$, and therefore must vanish at supercritical points. This proves that the limit $\mu_k < 2/(q-2)$ is sharp.

Note that L^∞ bounds for integrals like (2.1) with symbols which vanish on $\mathcal{S}_\pm$ can be found in [CDMM] (see also references in [Ste]).

4. The main result

For linear equations, the substitute for (1.3) after the formation of caustics is

$$u^\varepsilon \sim \underline{u} + \varepsilon I^\varepsilon_+(\mathcal{U}_+) + \varepsilon I^\varepsilon_-(\mathcal{U}_-)\,. \tag{4.1}$$

(see e.g. [Du], [Hö 2]). We show that a similar result holds for a dissipative semilinear equation. Consider Cauchy data (1.2). Since they are bounded in $H^1(\mathbb{R}^d)$ and $L^2(\mathbb{R}^d)$ respectively, the problem (1.5), (1.2) has a unique solution u^ε. Moreover, the family u^ε is bounded in $C^0([0,+\infty[;H^1(\mathbb{R}^d))$ and $\partial_t u^\varepsilon$ is bounded in $C^0([0,+\infty[;L^2(\mathbb{R}^d)) \cap L^{p+1}([0,+\infty[\times\mathbb{R}^d)$. We study the asymptotic behavior of the gradient of u^ε.

An abbreviated version of the main result is the following.

Theorem 4.1. *There is a unique $\underline{u} \in C^0([0,+\infty[;H^1(\mathbb{R}^d))$ with $\partial_t\underline{u} \in C^0([0,+\infty[;L^2(\mathbb{R}^d)) \cap L^{p+1}([0,+\infty[\times\mathbb{R}^d)$, such that for all $T < +\infty$,*

$$u^\varepsilon \to \underline{u} \quad \textit{strongly in} \quad L^2([0,T]\times\mathbb{R}^d)\,. \tag{4.2}$$

There are unique $\mathcal{V}_\pm \in \mathcal{L}^{p+1}_\pm \cap C^0([0,+\infty[;L^2(\omega\times\mathbb{T}))$ with $\mathcal{V}_\pm \in \mathcal{L}^{s,p+1}_\pm$ for all $s < 2/(p+1)$ and $\partial_\theta\mathcal{V}_\pm \in C^0([0,+\infty[;L^2(\omega\times\mathbb{T}))$, such that for all $T < +\infty$,

(4.3)
$$\begin{cases} \partial_t u^\varepsilon \sim \partial_t\underline{u} + J^\varepsilon_+(\mathcal{V}_+) + J^\varepsilon_-(\mathcal{V}_-) & \text{in} \quad L^{p+1}([0,T[\times\mathbb{R}^d)\,, \\ \partial_{x_j} u^\varepsilon \sim \partial_{x_j}\underline{u} + J^\varepsilon_+(\mathcal{V}_{+,j}) + J^\varepsilon_-(\mathcal{V}_{-,j}) & \text{in} \quad L^2([0,T[\times\mathbb{R}^d)\,, \end{cases}$$

with

$$\mathcal{V}_{\pm,j} = \pm\frac{\partial_{y_j}\psi}{|d\psi|}\,\mathcal{V}_\pm\,. \tag{4.4}$$

Extracting subsequences, one can introduce $\underline{u}$, the weak limit of u^ε in H^1. Then, locally, $u^\varepsilon \to \underline{u}$, strongly in L^2. Moreover, since $v^\varepsilon := \partial_t u^\varepsilon$ is bounded in $L^{p+1}([0,+\infty[\times\mathbb{R}^d)$, $v^\varepsilon \to \underline{v} := \partial_t\underline{u}$ weakly in L^{p+1}. Finally, since $f^\varepsilon := f(\partial_t u^\varepsilon)$ is bounded in $L^{1+1/p}$, one can assume that $f^\varepsilon \to \underline{f}$ weakly in $L^{1+1/p}$.

Next, we determine the profiles $\mathcal{V}_\pm$. If (4.3) holds, then Proposition 3.4 shows that for all $\mathcal{A} \in \mathcal{P}$ which vanish on a neighborhood of $\mathcal{S}_\pm$, one has

$$\int v^\varepsilon\ \overline{J^\varepsilon_\pm(\mathcal{A})}\ dt\,dx \ \to\ \int \mathcal{V}_\pm\ \overline{\mathcal{A}}\ dt\,dx\,d\theta\,. \tag{4.5}$$

Conversely, one can extract a subsequence such that the left hand side converges for all $\mathcal{A}$. Using Proposition 3.2, the limit determines a unique profile $\mathcal{V}_\pm$ such that (4.5) holds. This is an extension of the multiscale analysis introduced for instance in [N], [A], [E], [ES], [JMR 2], with oscillatory test functions replaced by oscillatory integrals. Similarly, one defines profiles $\mathcal{F}_\pm$ and $\mathcal{V}_{\pm,j}$ by the property that for all $\mathcal{A} \in \mathcal{P}$ which vanish on a neighborhood of $\mathcal{S}_\pm$, one has

$$\int f^\varepsilon\ \overline{J^\varepsilon_\pm(\mathcal{A})}\ dt\,dx \longrightarrow \int \mathcal{F}_\pm\ \overline{\mathcal{A}}\ dt\,dx\,d\theta\,. \tag{4.6}$$

$$\int \partial_{x_j} u^\varepsilon\ \overline{J^\varepsilon_\pm(\mathcal{A})}\ dt\,dx \longrightarrow \int \mathcal{V}_{\pm,j}\ \overline{\mathcal{A}}\ dt\,dx\,d\theta\,. \tag{4.7}$$

Proposition 4.2. *i) There is a subsequence u^ε and profiles $\mathcal{V}_\pm \in \mathcal{L}^{p+1}_\pm$, $\mathcal{F}_\pm \in \mathcal{L}^{1+1/p}_\pm$ and $\mathcal{V}_{\pm,j} \in \mathcal{L}^2_\pm$ such that* (4.5) (4.6) *and* (4.7) *are satisfied for all $\mathcal{A} \in \mathcal{P}$ which vanish on a neighborhood of $\mathcal{S}_\pm$.*

ii) The relations (4.4) *are satisfied.*

iii) Introduce $G_\pm(p+1) := ([0,+\infty[\times\omega)\backslash\mathcal{S}^+_\pm(p+1)$. Then, $\mathcal{F}_\pm$ is locally integrable on $G_\pm(p+1)\times\mathbb{T}$ and one has, in the sense of distributions,

$$2\,\partial_t\mathcal{V}_\pm\ +\ \mathcal{F}_\pm\ =\ 0\quad on\ G_\pm(p+1)\times\mathbb{T}\,,\qquad \mathcal{V}_{\pm|t=0}\ =\ \mathcal{V}^0_\pm \tag{4.8}$$

with $\mathcal{V}^0_\pm(x,\theta) := \frac{1}{2}\left(\mathbf{U}_1(x,\theta) \pm |d\psi(x)|\,\partial_\theta\mathbf{U}_0(x,\theta)\right)$.

iv) The weak limits satisfy

$$\Box\underline{u}+\underline{f}\ =\ 0\quad \text{on } [0,+\infty[\times\mathbb{R}^d\,,\qquad \underline{u}_{|t=0}\ =\ \underline{u}_0\,,\qquad \partial_t\underline{u}_{|t=0}\ =\ \underline{u}_1\,. \tag{4.9}$$

Part *i)* follows from Proposition 3.2. The relations $\partial_{x_j} v^\varepsilon = \partial_t(\partial_{x_j} u_j^\varepsilon)$ imply (4.4). To prove *iii)*, note that the weight $\Delta_\pm^{1-p}$ is locally integrable near subcritical points, that is on $G_\pm(p+1)$. The stationary phase expansion (2.9) and (4.6) imply that

$$\int f^\varepsilon\, \overline{I_\pm^\varepsilon(\mathcal{A})}\; dt\, dx \longrightarrow \int \mathcal{F}_\pm\, \overline{\mathcal{A}}\; dt\, dx\, d\theta \tag{4.10}$$

for all $\mathcal{A} \in \mathcal{P}$ which vanish near $\mathcal{S}_\pm$. Theorem 3.6 implies that (4.10) extends to all $\mathcal{A} \in \mathcal{P}$ which vanish only on neighborhoods of the set of supercritical points, $\mathcal{S}_\pm(p+1)$. The standard calculus of oscillatory integrals implies that for all $\mathcal{A} \in \mathcal{P}$, $\Box I_\pm^\varepsilon(\mathcal{A}) = \partial_t I_\pm^\varepsilon(\mathcal{B}) + O(\varepsilon)$, with $\mathcal{B} = 2\partial_t\mathcal{A} + O(\varepsilon)$. Using (4.10) for test profiles $\mathcal{A}$ which vanish near supercritical points, this implies (4.8).

Propositions 3.2 and 3.3 shows that (4.5) (resp. (4.6)) can be extended to nonsmooth profiles $\mathcal{A} \in \mathcal{L}_\pm^{1+1/p}$ (resp. $\mathcal{A} \in \mathcal{L}_\pm^{p+1}$). Taking $a^\varepsilon \sim \tilde{J}_\pm^\varepsilon(\mathcal{A})$ in $L^{1+1/p}$ (resp in L^{p+1}), one has

$$\begin{gathered}\int v^\varepsilon\, \overline{a^\varepsilon}\; dt\, dx \longrightarrow \int \mathcal{V}_\pm\, \overline{\mathcal{A}}\; dt\, dx\, d\theta \\ \left(\text{resp.} \int f^\varepsilon\, \overline{a^\varepsilon}\; dt\, dx \longrightarrow \int \mathcal{F}_\pm\, \overline{\mathcal{A}}\; dt\, dx\, d\theta\ \right).\end{gathered} \tag{4.11}$$

We say that $\mathcal{V}_\pm$ or $\mathcal{F}_\pm$ defined by (4.5) or (4.6), are the *weak profiles* associated to the subsequence v^ε of f^ε.

Next we construct an approximate solution.

Proposition 4.3. *There is a bounded family u_1^ε in $C^0([0,+\infty[;H^1(\mathbb{R}^d))$ with $v_1^\varepsilon := \partial_t u_1^\varepsilon$ bounded in $C^0([0,+\infty[;L^2(\mathbb{R}^d)) \cap L^{p+1}([0,+\infty[\times\mathbb{R}^d)$, such that for all $T < +\infty$,*

$$u_1^\varepsilon \to \underline{u} \quad \textit{strongly in} \quad L^2([0,T]\times\mathbb{R}^d), \tag{4.12}$$

(4.13)
$$\begin{cases} \partial_t u_1^\varepsilon \sim \partial_t \underline{u} + \tilde{J}_+^\varepsilon(\mathcal{V}_+) + \tilde{J}_-^\varepsilon(\mathcal{V}_-) & \textit{in} \quad L^{p+1}([0,T[\times\mathbb{R}^d), \\ \partial_{x_j} u_1^\varepsilon \sim \partial_{x_j}\underline{u} + \tilde{J}_+^\varepsilon(\mathcal{V}_{+,j}) + \tilde{J}_-^\varepsilon(\mathcal{V}_{-,j}) & \textit{in} \quad L^2([0,T[\times\mathbb{R}^d) \end{cases},$$

$$\Box u_1^\varepsilon \sim -\underline{f} - \tilde{J}_+^\varepsilon(\mathcal{F}_+) + \tilde{J}_-^\varepsilon(\mathcal{F}_-) \quad \text{in} \quad L^{1+1/p}, \tag{4.14}$$

$$(4.15) \qquad \partial_{t,x} u^\varepsilon_{1|t=0} \sim \partial_{t,x} u^\varepsilon{}_{|t=0} \quad in\ L^2(\mathbb{R}^d)\,.$$

The construction uses the ideas of Proposition 3.3. To show that one can choose a family u^ε_1 which satisfies (4.14), one uses the equations (4.8) (4.9). Note that the profiles are not yet known to be smooth, and we need here the extended definition $\tilde{J}^\varepsilon$.

To prove (4.3), we show that $\delta u^\varepsilon := u^\varepsilon - u^\varepsilon_1$ and $\delta v^\varepsilon := v^\varepsilon - v^\varepsilon_1 = \partial_t \delta u^\varepsilon$ satisfy, for all $T > 0$,

$$(4.16) \quad \sup_{0\le t\le T} \|\,\delta u^\varepsilon(t)\,\|_{H^1(\mathbb{R}^d)} \to 0\,, \quad \|\,\delta v^\varepsilon(t)\,\|_{L^{p+1}([0,T]\times\mathbb{R}^d)} \to 0,.$$

The proof uses the energy inequality

(4.17)

$$\|\,\delta u^\varepsilon(t)\,\|^2_{H^1(\mathbb{R}^d)} + \|\,\delta v^\varepsilon(t)\,\|^2_{L^2(\mathbb{R}^d)} + c\,\|\,\delta v^\varepsilon\,\|^{1+p}_{L^{1+p}([0,t]\times\mathbb{R}^d)}$$
$$\le \|\,\delta u^\varepsilon(0)\,\|^2_{H^1(\mathbb{R}^d)} - 2\,\mathrm{Re}\int_{[0,t]\times\mathbb{R}^d} (\Box u^\varepsilon_1 + f(v^\varepsilon_1))\cdot\overline{\delta v^\varepsilon_1}\,dt\,dx\,,$$

This estimate uses that $f(\lambda) := |\lambda|^{p-1}\lambda$ is strictly monotone, that is:

$$\mathrm{Re}\big(f(\lambda) - f(\mu)\big)\,\overline{(\lambda-\mu)} \ \ge\ c\,|\lambda-\mu|^{p+1}\,.$$

To prove (4.16), it remains to prove that the integral on the right hand side of (4.17) tends to 0 as $\varepsilon \to 0$. An important step is to analyse $f(v^\varepsilon_1)$.

Proposition 4.4. *One has in* $L^{1+1/p}$,

(4.18)

$$f(v^\varepsilon_1) \sim \underline{\mathcal{E}}(\underline{v},\mathcal{V}_+,\mathcal{V}_-) + \tilde{J}^\varepsilon_+(\mathcal{E}_+(\underline{v},\mathcal{U}_+,\mathcal{V}_-) + \tilde{J}^\varepsilon_-(\mathcal{E}_-(\underline{v},\mathcal{U}_+,\mathcal{V}_-) + h^\varepsilon\,,$$

where $\underline{\mathcal{E}}$ *and* $\mathcal{E}_\pm$ *are nonlinear operators acting from* $L^{p+1}\times\mathcal{L}^{p+1}_+\times\mathcal{L}^{p+1}_-$ *into* $L^{1+1/p}$ *and* $\mathcal{L}^{1+1/p}_\pm$ *respectively. The family* h^ε *is bounded in* $L^{1+1/p}$ *and satisfies the following property:*

For all bounded families w^ε *in* $C^0([0,+\infty[;H^1(\mathbb{R}^d))$ *such that* $\partial_t w^\varepsilon_1$ *is bounded in* $C^0([0,+\infty[;L^2(\mathbb{R}^d))\cap L^{p+1}([0,+\infty[\times\mathbb{R}^d)$ *and* $\Box w^\varepsilon$ *is bounded in* $L^{1+1/p}([0,+\infty[\times\mathbb{R}^d)$*, one has for all* $T>0$,

$$(4.19) \qquad \int_{[0,T]\times\mathbb{R}^d} h_\varepsilon\,\partial_t w^\varepsilon \to 0 \quad \text{as} \quad \varepsilon\to 0\,.$$

We sketch the construction, assuming that the profiles $\mathcal{V}_\pm$ are smooth and vanish near $\mathcal{S}_\pm$. The actual proof would follow by regularization.

Suppose that Ω is a small domain in $[0,+\infty[\times\mathbb{R}^d$ which does not intersect $\mathcal{C}$. Then, using notation as in (2.11) (2.12),

$$v_1^\varepsilon \sim \underline{v} + \sum \mathbf{V}_{\pm,k}(\,\cdot\,,\varphi_{\pm,k}/\varepsilon)\,. \tag{4.20}$$

Thus,

$$f(v_1^\varepsilon) \sim \mathbf{F}(t,x,\varphi_*(t,x)/\varepsilon)\,, \tag{4.21}$$

where

$$\mathbf{F}(t,x,\theta_*) := f\Big(\underline{v}(t,x)+\sum \mathbf{V}_{\pm,k}(t,x,\theta_{\pm,k})\Big)\,.$$

We have used the notation, $\theta_* := \{\theta_{\pm,k}\}$, $\varphi_* := \{\varphi_{\pm,k}\}$.

The idea is now to extract from (4.20) the characteristic oscillations. Introduce first the total average

$$\underline{\mathcal{E}}(\underline{v},\mathcal{V}_+,\mathcal{V}_-)\,(t,x) := \int \mathbf{F}(t,x,\theta_*)\,d\theta_*\,. \tag{4.22}$$

Next, for (t,y) such that $x := y - tg(y) \notin \mathcal{C}_+$, label the $y_{\pm,k}$ so that $y = y_{+,1}$. Introduce the average of $\mathbf{F}$ with respect to the variables θ_* except the first one $\theta_{+,1}$:

$$\mathbf{F}_{+,1}(\underline{v},\mathcal{V}_+,\mathcal{V}_-)\,(t,y,\theta_{\pm,1}) := \int \mathbf{F}(t,x,\theta_*)\,d\theta_{+,2}\dots d\theta_{-,1}\dots\,. \tag{4.23}$$

Its oscillatory part is

$$\mathbf{F}^{osc}_{+,1}(t,y,\theta) := \mathbf{F}_{+,1}(\underline{v},\mathcal{V}_+,\mathcal{V}_-)\,(t,y,\theta) - \underline{\mathcal{E}}(\underline{v},\mathcal{V}_+,\mathcal{V}_-)(t,x)\,. \tag{4.24}$$

To lift it to the Lagrangian, introduce

$$\mathcal{E}_+(\underline{v},\mathcal{V}_+,\mathcal{V}_-)\,(t,y,\theta) := \Delta_+(t,y)\,\big(\mathcal{H}^{-m(t,y)}\,\mathbf{F}^{osc}_{+,1}\big)\,(t,y,\theta)\,. \tag{4.25}$$

The definition of $\mathcal{E}_-(\underline{v},cv_+,cv_-)$ is similar. (4.22) and (4.25) define $\underline{\mathcal{E}}$ and $\mathcal{E}_\pm$ outside the caustic set $\mathcal{C}$, thus almost everywhere.

Expanding $\mathbf{F}(t,x,\theta_*)$ in Fourier series in θ_* and using the nonresonance assumption, one shows that

$$h^\varepsilon \sim f(v_1^\varepsilon) - \underline{\mathcal{E}}(\underline{v},\mathcal{V}_+,\mathcal{V}_-) - \tilde{J}_+^\varepsilon(\mathcal{E}_+(\underline{v},\mathcal{U}_+,\mathcal{V}_-)) - \tilde{J}_-^\varepsilon(\mathcal{E}_-(\underline{v},\mathcal{U}_+,\mathcal{V}_-))$$

satisfies the following property: h^ε is bounded in $L^{1+1/p}$ and for all $\delta > 0$ there is a finite sum

$$h_1^\varepsilon(t,x) = \sum h_\alpha(t,x)\, e^{i\varphi_\alpha(t,x)/\varepsilon} \tag{4.26}$$

where $(\partial_t\varphi_\alpha)^2 - |\partial_x\varphi_\alpha|^2 \neq 0$ almost everywhere, and

$$\limsup_{\varepsilon\to 0} \|h^\varepsilon - h_1^\varepsilon\|_{L^{1+1/p}} \leq \delta\,. \tag{4.27}$$

This implies (4.19).

Propositions 4.3 and 4.4 imply that

$$g^\varepsilon := \Box u_1^\varepsilon + f(v_1^\varepsilon) \sim \underline{g} + \tilde{J}_+^\varepsilon(\mathcal{G}_+) + \tilde{J}_-^\varepsilon(\mathcal{G}_-) + h^\varepsilon\,, \tag{4.28}$$

with $\underline{g} := \underline{f} + \underline{\mathcal{E}}(\underline{u},\mathcal{V}_+,\mathcal{V}_-)$ and $\mathcal{G}_\pm := \mathcal{F}_\pm + \mathcal{E}_\pm(\underline{u},\mathcal{V}_+,\mathcal{V}_-)$. Moreover, δu^ε is bounded in $C^0([0,+\infty[;H^1(\mathbb{R}^d))$, $\delta v^\varepsilon = \partial_t\delta u^\varepsilon$ is bounded in L^{p+1} and $\Box\delta u^\varepsilon$ is bounded in $L^{1+1/p}$. Thus

$$\lim_{\varepsilon=0}\int_{[0,T]\times\mathbb{R}^d} h^\varepsilon\,\overline{\delta v^\varepsilon} = 0\,.$$

By definition, $\partial_t\underline{u}$ and $\mathcal{V}_\pm$ are the weak limit and the profiles of the subsequence v^ε. Proposition 3.4 and (4.13) show that they are also the weak limit and the profiles of v_1^ε. Therefore, the weak limit and the profiles of δv^ε vanish. Using (4.11), this implies that

$$\int_{[0,t]\times\mathbb{R}^d} \left(\underline{g} + \tilde{J}_+^\varepsilon(\mathcal{G}_+) + \tilde{J}_-^\varepsilon(\mathcal{G}_-)\right)\overline{\delta v^\varepsilon} \to 0\,.$$

Thus the right hand side of (4.17) tends to zero. Moreover, one can show that the convergence is uniform for $t \leq T$. Thus, $u^\varepsilon - u_1^\varepsilon \to 0$ in $C^0([0,+\infty[;H^1(\mathbb{R}^d))$, and $\partial_t(u^\varepsilon - u_1^\varepsilon) \to 0$ in $C^0([0,+\infty[;L^2(\mathbb{R}^d)) \cap L^{p+1}([0,+\infty[\times\mathbb{R}^d)$. Therefore,

Proposition 4.5. *For the extracted subsequence, one has for all* $T < +\infty$,

$$u^\varepsilon \to \underline{u} \quad \textit{strongly in} \quad L^2([0,T] \times \mathbb{R}^d), \tag{4.29}$$

(4.30)
$$\begin{cases} \partial_t u^\varepsilon \sim \partial_t \underline{u} + \tilde{J}_+^\varepsilon(\mathcal{V}_+) + \tilde{J}_-^\varepsilon(\mathcal{V}_-) & \textit{in} \quad L^{p+1}([0,T[\times\mathbb{R}^d), \\ \partial_{x_j} u^\varepsilon \sim \partial_{x_j} \underline{u} + \tilde{J}_+^\varepsilon(\mathcal{V}_{+,j}) + \tilde{J}_-^\varepsilon(\mathcal{V}_{-,j}) & \textit{in} \quad L^2([0,T[\times\mathbb{R}^d), \end{cases}$$

5. The profile equations and absorption of oscillations

The next step is to prove that the weak limits $(\underline{u}, \underline{f})$ and the profiles $(\mathcal{V}_\pm, \mathcal{F}_\pm)$ satisfy

$$\underline{f} = \underline{\mathcal{E}}(\underline{u}, \mathcal{V}_+, \mathcal{V}_-), \quad \mathcal{F}_\pm = \mathcal{E}_\pm(\underline{u}, \mathcal{V}_+, \mathcal{V}_-). \tag{5.1}$$

Knowing (4.28), Proposition 4.4 applies to u^ε. In particular, $f(\partial_t u^\varepsilon)$ satisfies (4.18). Approximating h^ε by finite sums (4.26), one shows that h^ε converges weakly to zero in $L^{1+1/p}$ and

$$\int h^\varepsilon \, \overline{J_\pm^\varepsilon(\mathcal{A})} \, dt \, dx \to 0 \quad \text{as } \varepsilon \to 0, \tag{5.2}$$

for all $\mathcal{A} \in \mathcal{P}$ which vanish on a neighborhood of $\mathcal{S}_\pm$. Taking the weak limit of the right hand side of (4.18) implies that $\underline{f} = \underline{\mathcal{E}}(\underline{u}, \mathcal{V}_+, \mathcal{V}_-)$. Multiplying (4.18) by $J_\pm^\varepsilon(\mathcal{A})$ and passing to the limit yields $\mathcal{F}_\pm = \mathcal{E}_\pm(\underline{u}, \mathcal{V}_+, \mathcal{V}_-)$.

With Proposition 4.2, this implies the following result.

Proposition 5.1. *The limit* $\underline{u}$ *and the profiles* $\mathcal{V}_\pm$ *satisfy*

$$\begin{cases} \Box \underline{u} + \underline{\mathcal{E}}(\underline{u}, \mathcal{V}_+, \mathcal{V}_-) = 0 & \textit{on } [0, +\infty[\times\mathbb{R}^d, \\ \underline{u}_{|t=0} = \underline{u}_0, & \partial_t \underline{u}_{|t=0} = \underline{u}_1 \end{cases} \tag{5.3}$$

$$\begin{cases} 2\,\partial_t \mathcal{V}_\pm + \mathcal{E}_\pm(\underline{u}, \mathcal{V}_+, \mathcal{V}_-) = 0 & \textit{on } G_\pm(p+1) \\ \mathcal{V}_{\pm|t=0} = \mathcal{V}_\pm^0. \end{cases} \tag{5.4}$$

The system (5.3), (5.4) inherits dissipativity from the original one. It satisfies

Proposition 5.2. *The system* (5.3), (5.4) *has a unique solution* $(\underline{u}, \mathcal{V}_+, \mathcal{V}_-)$ *such that*

$$(5.5)\qquad \begin{cases} \underline{u} \in C^0([0,+\infty[; H^1(\mathbb{R}^d))\,, \\ \partial_t \underline{u} \in C^0([0,+\infty[; L^2(\mathbb{R}^d)) \cap L^{1+p}([0,+\infty[\times\mathbb{R}^d)\,, \\ \mathcal{V}_\pm \in \mathcal{L}_\pm^{p+1}\,. \end{cases}$$

Moreover, it satisfies

$$(5.6)\qquad \begin{cases} \mathcal{V}_\pm \in C^0([0,+\infty[\, L^2(\omega \times \mathbb{T}))\,, \\ \partial_\theta \mathcal{V}_\pm \in L^\infty([0,+\infty[\, L^2(\omega\times\mathbb{T}))\,, \\ \mathcal{V}_\pm \in \mathcal{L}^{s,p+1} \quad \textit{for all } s < 2/(p+1)\,. \end{cases}$$

The existence is a consequence of (5.1) and Proposition 4.2. Uniqueness follows from monotonicity of the operator $\mathcal{E} := (\underline{\mathcal{E}}, \mathcal{E}_+, \mathcal{E}_-)$ as a function of $(\partial_t \underline{u}, \mathcal{V}_+, \mathcal{V}_-)$ (see [JMR 5]). If $(\underline{u}, \mathcal{V}_\pm)$ and $(\underline{u}', \mathcal{V}'_\pm)$ are two solutions of the equations in (5.3) (5.4), $(\delta\underline{u}, \delta\mathcal{V}_\pm) := (\underline{u} - \underline{u}', \mathcal{V}_\pm - \mathcal{V}'_\pm)$ satisfies (5.7)

$$\begin{aligned} &\|\delta\underline{u}(t)\|^2_{H^1(\mathbb{R}^d)} + \|\partial_t\delta\underline{u}\|^2_{L^2(\mathbb{R}^d)} + c\,\|\partial_t\delta\underline{u}\|^{1+p}_{L^{1+p}([0,t]\times\mathbb{R}^d)} \\ &\quad + \|\delta\mathcal{V}_+(t)\|^2_{L^2(\omega\times\mathbb{T})} + \|\delta\mathcal{V}_-(t)\|^2_{L^2(\omega\times\mathbb{T})} \\ &\quad + c\,\|\delta\mathcal{V}_+\|^{p+1}_{\mathcal{L}_+^{p+1}} + c\,\|\delta\mathcal{V}_-\|^{p+1}_{\mathcal{L}_-^{p+1}} \\ &\le \|\delta\underline{u}(0)\|^2_{H^1(\mathbb{R}^d)} + \|\delta\mathcal{V}_+(0)\|^2_{L^2(\omega\times\mathbb{T})} + \|\delta\mathcal{V}_-(0)\|^2_{L^2(\omega\times\mathbb{T})}\,. \end{aligned}$$

A tricky point in the proof of (5.7) is that the equation (5.4) does not hold everywhere. Note that $\mathcal{E}_\pm(\underline{u}, \mathcal{V}_+, \mathcal{V}_-)$ belongs to the weighted space $\mathcal{L}_\pm^{1+1/p}$ and is not always locally integrable near supercritical points. This is why the equations (5.4) are settled only on $G_\pm(p+1)$. However, near supercritical points, the property $\mathcal{V}_\pm \in \mathcal{L}_\pm^{p+1}$ is strong and implies some vanishing of $\mathcal{V}_\pm$ on $\mathcal{S}_\pm^+(p+1)$. This is why the equation (5.4) away from $\mathcal{S}_\pm^+(p+1)$ plus the information $\mathcal{V}_\pm \in \mathcal{L}_\pm^{p+1}$ near $\mathcal{S}_\pm^+(p+1)$ are sufficient to prove (5.7), which implies uniqueness.

The equations in (5.3)–(5.4) are invariant by translation in θ. Thus one can use (5.7) to compare $\mathcal{V}_\pm(t,y,\theta)$ to $\mathcal{V}_\pm(t,y,\theta+h)$. Together with the smoothness of the initial data, this implies (5.6).

End of the proof of Theorem 4.1.

The uniqueness of $(\underline{u}, \mathcal{V}_\pm)$ implies that the full sequence u^ε satisfies (4.29)–(4.30). Choosing $s \in]1/(p+1), 2/(p+1)[$, (5.6) and Proposition

3.2 allow us to define $J^\varepsilon_\pm(\mathcal{V}_\pm)$ for each fixed ε and one has the stronger form (4.3) of the asymptotics.

Remarks 5.3. 1. (4.3) provides a phase-amplitude representation of ∇u^ε outside $\mathcal{C}$, see for example (4.20). The propagation equation for the amplitude $\mathbf{V}_{\pm,j}$ follows from the equation for $\mathcal{V}_\pm$. The $\mathbf{V}_{\pm,j}$ satisfy transmission conditions which include the usual phase shift. These conditions reflect the continuity of the profiles $\mathcal{V}_\pm$.

2. The nonlinear interactions $\mathcal{E}_\pm$ are singular above the caustic set. This is due to the presence of the factors Δ in (2.13) and (4.25). We refer the reader to [JMR 3] and [JMR 5] for examples of nonsmooth tranfer of energy at caustics and creation of singularities for the profiles $\mathcal{V}_\pm$. This phenomenon is truly nonlinear.

Near supercritical points, *strong* dissipation and *absorption* occur: the transport equation (5.4) is satisfied along the rays before the singular point, $\mathcal{V}_\pm$ tends to zero at this point, and $\mathcal{V}_\pm$ vanishes after it. This follows from the following estimate.

Proposition 5.4. *Introduce*

$$\sigma_\pm(t,y) \;:=\; \int_{\mathbf{T}} |\,\mathcal{V}_\pm(t,y,\theta)\,|^2 \, d\theta\,. \tag{5.8}$$

Then, $\sigma_\pm \in C^0([0,+\infty[;L^1(\mathbb{R}^d))$, $\partial_t\sigma_\pm$ *and* $\Delta_\pm(t,y)^{1-p}\,\sigma_\pm(t,y)^{(1+p)/2}$ *belong to* $L^1([0,+\infty[\times\mathbb{R}^d)$ *and there is a constant* $c>0$ *such that*

$$\partial_t\sigma_\pm(t,y) \;+\; c\,\Delta_\pm(t,y)^{1-p}\,\sigma_\pm(t,y)^{(1+p)/2} \;\leq\; 0\,. \tag{5.9}$$

$\sigma_\pm$ is the density of energy along the ray $t \to (t,y)$ in the Lagrangian. The inequality (5.9) follows from the dissipativity of the original equation (1.5). For the linear equation $\Box u = 0$, the analogue would be $\partial_t\sigma_\pm = 0$ which expresses the asymptotic conservation of energy along the rays. Note that (5.9) holds everywhere, including neighborhoods of supercritical points. This is in contrast to the profile equation (5.4) which holds only on the complement of the supercritical points.

Definition 5.5. *For* $y \in \omega$, $T^{abs}_\pm(y)$ *denotes the smallest value of* $t>0$ *such that* $(t,y) \in S^+_\pm(p+1)$ *the set of* $p+1$*-supercritical point in* $S_\pm$, *i.e. satisfying* $\mu_\pm(t,y) \geq 2/(p-1)$.

$1/T_\pm^{abs}(y)$ is the largest positive eigenvalue of $\pm g'(y)$ of multiplicity larger than or equal to $2/(p-1)$. $T_\pm^{abs}(y)$ is infinite if there is no such eigenvalue. $T_\pm^{abs}$ is C^∞ on ω_0 where the eigenvalues of $g'(y)$ are smooth. Near a supercritical point $(\underline{t}, \underline{y}) \in \mathcal{S}_\pm^+(p+1)$ with $\underline{y} \in \omega_0$, $\Delta_\pm^{1-p}$ is not integrable, and the analysis of (5.9) shows that $\sigma_\pm(t,y) \to 0$ as $t \to \underline{t}$. This implies the following result.

Theorem 5.6. *The profiles $\mathcal{V}_\pm$ vanish for $t > T_\pm^{abs}(y)$, and the equation* (5.4) *holds in L^1_{loc} on the domains $t < T_\pm^{abs}(y)$*

Examples 5.7. 1. If $p \geq 3$, all points in $\mathcal{S}_k$ are supercritical, since then $2/(p-1) \leq 1 \leq \mu_\pm(t,y)$. Thus $T_\pm^{abs}(y)$ is the first time of focusing along the ray.

2. In the extreme opposite case, when $p < 1 + \frac{2}{d-1}$, all points in $\mathcal{S}_\pm$ are subcritical since then (3.8) implies that $\frac{2}{p-1} > d-1 \geq \mu_\pm(t,y)$. In this case $T_\pm^{abs}$ is infinite and equation (5.4) is satisfied everywhere.

3. For radial focusing, i.e. an initial phase of the form $\psi(\xi) := \chi(|\xi|)$, g' has only one positive eigenvalue equal to $1/|y|$, of multiplicity $d-1$. The causic set is $\mathcal{C} = \{x = 0\}$. Absorption occurs if and only if $p \geq 1 + 2/(d-1)$. The if part was proved in [JMR 2].

4. In space dimension $d = 2$, the multiplicity μ is always equal to 1 (see (3.8)). The critical exponent is $p = 3$. If $p < 3$, $T_\pm^{abs}(y)$ is infinite. Conversely if $p \geq 3$, absorption occurs at the first crossing of caustics. Examples are the familiar phases

$$
(5.10) \qquad \begin{cases} \psi_1(y_1, y_2) &= y_2 + y_1^2 \,, \\ \psi_2(y_1, y_2) &= a_1 y_1^2 + a_2 y_2^2 \quad \text{with} \quad 0 < a_1 < a_2 \,. \end{cases}
$$

For ψ_1 the caustic set is the manifold $\mathcal{C}$ with equation $x_1^2 = (t^{2/3} - 1)^3$, which has a cusp singularity at $x_1 = 0, t = 1$, corresponding to the rays from $y_1 = 0$.

5. More generally, $\mu_\pm(t,y) = 1$ for all points $(t,y) \in \mathcal{S}_\pm$ when all the nonvanishing eigenvalues of $g'(y)$ are simple. This happens when all the rays meet the caustic at generic points (folds).

6. Consider

$$
(5.11) \qquad \psi_3(y) := y_d + \frac{1}{2}\,(y_1^2 + \cdots + y_{d-1}^2) =: y_d + \frac{1}{2}\,|y'|^2 \,.
$$

For $y' \neq 0$, there are two distinct positive eigenvalues: $1/\delta$ of multiplicity $d-2$ and $1/\delta^3$ of multiplicity 1, where $\delta^2 := 1 + |y'|^2$. For

$y' = 0$, these eigenvalues are equal. Therefore, $(t = 1, y' = 0)$ is a crossing point for two different eigenvalues, but it is the only point where multiplicities are not locally constant. Nevertheless Assumption 2.1 is satisfied.

There are two focusing times, δ and δ^3. Accordingly, the caustic set has two components. The first one is the plane $\mathcal{C}_1 = \{x' = 0\}$. The second is the manifold $\mathcal{C}_2$ with equation $|x'|^2 = (t^{2/3} - 1)^3$, which has a singular locus at $x' = 0, t = 1$, corresponding to the rays from $y' = 0$. Since $\delta < \delta^3$, the rays cross $\mathcal{C}_1$ before $\mathcal{C}_2$.

If $p \geq 1 + 2/(d-2)$, then $\mathcal{C}_1$ is supercritical. Oscillations are absorbed at $\mathcal{C}_1$ and never reach $\mathcal{C}_2$. If $p < 1 + 2/(d-2)$, then both $\mathcal{C}_1$ and $\mathcal{C}_2$ are subcritical and oscillations cross $\mathcal{C}_1$ and $\mathcal{C}_2$.

References

[A] G. Allaire, Homogeneization and two-scale analysis, *S.I.A.M. J. Math. Anal.* **23** (1992), 1482–1518.

[CDMM] M. Cowling, S. Disney, G. Mauceri and D. Müller, Damping oscilllatory integrals, *Invent. Math.* **101** (1990), 237–260.

[Du] J.J. Duistermaat, Oscillatory integrals, Lagrangian immersions and unfolding of singularities, *Comm. Pure Appl. Math.* **27** (1974), 207–281.

[DH] J.J. Duistermaat and L. Hörmander, Fourier integral operators II, *Acta. Math.* **128** (1972), 183–269.

[E] W. E, Homogeneization of linear and nonlinear transport equations, *Comm. Pure Appl. Math* **45** (1992), 301–326.

[ES] W. E, D. Serre, Correctors for the homogeneization of conservation laws with oscillatory forcing terms, *Asymptotic Analysis* **5** (1992), 311–316.

[Ha] A. Haraux, *Semilinear hyperbolic problems in bounded domains*, Mathematical Reports, vol. 3, Part 1, Harwood Academic Publishers, 1987.

[Hö 1] L. Hörmander, Fourier integral operators I, *Acta. Math.* **127**(1971), 79–183.

[Hö 2] L. Hörmander, *The Analysis of linear partial differential operators*, Springer-Verlag, 1991.

[HMR] J. Hunter, A. Majda and R. Rosales, Resonantly interacting weakly nonlinear hyperbolic waves II: several space variables, *Stud. Appl. Math.* **75** (1986), 187–226.

[JMR 1] J.-L. Joly, G. Métivier and J. Rauch, Coherent and focusing multidimensional nonlinear geometric optics,*Annales Scientifique de l'École Normale Supérieure* **28** (1995), 51–113.

[JMR 2] J.-L. Joly, G. Métivier and J. Rauch, Focusing at a point and absorption of nonlinear oscillations, *Trans. Amer. Math. Soc.* **347** (10) (1995), 3921–3969.

[JMR 3] J.-L. Joly, G. Métivier and J. Rauch, Nonlinear oscillations beyond caustics, *Comm. Pure Appl. Math.*, to appear.

[JMR 4] J.-L. Joly, G. Métivier and J. Rauch, *Several Recent Results in Nonlinear Geometric Optics, Partial Differential Equations and Mathematical Physics*, The Danish-Swedish Analysis Seminar, Birkhauser, 1995, 181–207.

[JMR 5] J.-L. Joly, G. Métivier and J. Rauch, L^p Estimates for oscillatory integrals and caustics for dissipative semilinear oscillations, preprint 1996.

[JR] J.L. Joly and J. Rauch, Justification of multidimensional single phase semilinear geometric optics, *Trans. Amer. Math. Soc.* **330** (1992), 599–625.

[La] P. Lax, Asymptotic solutions of oscillatory initial value problems, *Duke Math. Journal* **24** (1957), 627–645.

[L] J.-L. Lions, *Quelques méthodes de résolution de problèmes aux limites non linéaires*, Dunod, Gauthier-Villars, Paris, 1969.

[LS] J.-L. Lions and W. Strauss, Some nonlinear evolution equations, *Bull. Soc. Math. France* **93** (1965), 43–96.

[Lu] D. Ludwig, Uniform asymptotic expansions at a caustic, *Comm. Pure Appl. Math.* **13** (1966), 85–114.

[McLPT] D.W. McLaughlin, G. Papanicolaou and L. Tartar, Weak limits of semi-linear hyperbolic systems with oscillating data, in *Macroscopic Modelling of Turbulent Flow*, Lecture Notes in Physics **230** (1985) 277–298.

[N] G. Nguetseng, A general convergence result for a functional related to the theory of homogeneization, *Siam J. Math. Anal.* **20** (1989), 608–623.

[Ste] E. M. Stein, *Harmonic analysis, real variable methods, orthogonality and oscillatory integrals*, Princeton University Press, 1993.

[Str] W. Strauss, *The energy method in nonlinear partial differential equations. Notas de matematica 47*, Instituto de Matematica Pura e Applicata, Rio de Janeiro, 1969.

[So] C. D. Sogge, *Fourier integrals in classical analysis*, Cambridge Tracts in Mathematics, 105, Cambridge University Press, 1993

Jean-Luc Joly
MAB, Université Bordeaux I
33405 Talence Cedex, FRANCE

Guy Métivier
IRMAR, Université Rennes I
35042 Rennes Cedex, FRANCE

Jeffrey Rauch
Department of Mathematics
The University of Michigan
Ann Arbor MI, 48109, USA

Geometric optics and the bottom of the spectrum

Richard B. Melrose

Abstract. The continuous spectrum, near its infimum, of the Laplacian for certain classes of 'doubly warped' complete Riemannian metrics on the interiors of compact manifolds with boundary is examined using geometric optics. General conditions on the geometry of an algebra of vector fields leading to a calculus of pseudodifferential operators are described and, once verified in this particular case, the associated notion of wavefront set can be used to show the existence of an absolute scattering matrix. This operator is conjectured to be a Fourier integral operator; the confirmation of this conjecture is discussed in the special case of a 'double-zero' metric where the fibrations are trivial.

Introduction

In this note the structure of the bottom part, which is to say near 0, of the spectrum of the Laplacian (on functions) is discussed, for a complete metric on the interior of a compact manifold with boundary exhibiting a 'doubly warped' structure near the boundary. Namely if X is the compact manifold with boundary and ∂X is its boundary, then it is assumed that there is a two-level fibration of the boundary

$$\partial X \xrightarrow{\phi_1} Y_1 \xrightarrow{\phi_2} Y_2, \ \psi = \phi_2 \circ \phi_1.$$

Here both $\phi_1 : \partial X \longrightarrow Y_1$ and $\phi_2 : Y_1 \longrightarrow Y_2$ are fibrations of compact manifolds. By taking a product decomposition of X near its boundary these fibrations can be extended to a neighbourhood of the boundary; denote such a product extension by

$$[0,\delta) \times \partial X \xrightarrow{\tilde{\phi}_1} [0,\delta) \times Y_1 \xrightarrow{\tilde{\phi}_2} [0,\delta) \times Y_2, \ \tilde{\psi} = \tilde{\phi}_2 \circ \tilde{\phi}_1.$$

Consider a metric on the interior of X which, for some such choice of extensions takes the form

$$g = \frac{\tilde{\psi}^* \tilde{h}_2}{x^4} + \frac{\tilde{\phi}_1^* \tilde{h}_1}{x^2} + \tilde{h}_0, \text{ near } \partial X. \tag{1}$$

Typeset by $\mathcal{A}_{\mathcal{M}}\mathcal{S}$-TEX

Here, $\tilde{h}_2$, $\tilde{h}_1$ and $\tilde{h}_0$ are, respectively, metrics on $[0,\delta)\times Y_2$, $[0,\delta)\times Y_1$ and X and it is also assumed that, near ∂X, $x=\tilde{\psi}^*\tilde{x}$ where $\tilde{x}$ is the distance from the boundary in the $\tilde{h}_2$ metric. In fact it is enough that $\tilde{h}_1$ and $\tilde{h}_0$ be smooth symmetric 2-cotensors which restrict to metrics on the fibres of ϕ_2 and on the fibres ψ respectively. Under these assumptions near the boundary g, further assumed to be a Riemannian metric on the interior of X, is necessarily complete.

The Laplacian associated to g, Δ, is an essentially self-adjoint operator on the domain $\dot{C}^\infty(X)\subset C^\infty(X)$ consisting of the smooth functions vanishing to all orders at the boundary. The general structure of the spectrum of such an operator can be described as follows.

Theorem 1. *For any metric of the type* (1), *with* $\dim Y_1+\dim Y_2>0$, *the essential spectrum of the Laplacian on functions is absolutely continuous of infinite multiplicity on* $[0,\infty)$.

In this theorem I do not exclude the possibility of embedded discrete spectrum. In all likelihood it does not occur at all if the fibres of ϕ_1 have dimension zero and in general there should be no embedded spectrum near 0 assuming $\dim Y_1+\dim Y_2>0$. If it occurs, the eigenfunctions corresponding to such discrete spectrum are in the space $\dot{C}^\infty(X)$.

Although this gives a reasonably full description of the spectrum of Δ, a more detailed description of the generalized eigenfunctions is possible through the approach of 'geometric' scattering theory. This can be described in a variety of ways. One rather transparent approach is through a generalized (or degenerate) boundary problem.

Theorem 2. *For a metric of type* (1), *if* $\lambda\in\mathbb{R}$, $0<\lambda^2<A^2$, *where* A^2 *is the infimum of the positive eigenvalues of the Laplacian on the fibres of* ϕ_1, *then for each* $f\in C^\infty(Y_1)$ *there is a unique generalized eigenfunction with eigenvalue* λ^2 *of the form*

(2)

$$u=P_\lambda(f)=e^{i\lambda/x}x^q u_+ + e^{-i\lambda/x}x^q u_-,\ u_\pm\in C^\infty(X),\ u_+\restriction\partial X=\phi_1^* f,$$
$$q=\frac{1}{2}\dim Y_1+\dim Y_2,$$

orthogonal in L^2 *to any eigenfunctions in* $\dot{C}^\infty(X)$ *with eigenvalue* λ^2.

The proof of this result is outlined below, with particular emphasis on the newer aspects.

The map whose existence is asserted by this proposition is the *Poisson operator* for the generalized boundary value problem at 'wave number' λ. Since the Laplacian is real, the uniqueness part of this result shows that

$$P_{-\lambda}f=\overline{P_\lambda\overline{f}}.$$

For this reason it is enough to suppose throughout that $\lambda > 0$. Furthermore the *absolute scattering operator* can be defined by

$$A_\lambda : C^\infty(Y_1) \longrightarrow C^\infty(Y_1),\ A_\lambda f = f_-,\ u_- \upharpoonright \partial X = \phi_1^* f_-,$$

where $u_\pm$ are as in (2); this operator is necessarily invertible. Its microlocal properties are described here in some cases and the structure in the general case is conjectured. Let me give (rather *ad hoc*) nomenclature for some of the special cases of the metrics in (1) and hence discuss some results previously established. First, in case ϕ_1 is the identity and Y_2 is a point, so $\tilde{h}_2 = d\tilde{x}^2$ and $\tilde{h}_0$ is absent, we arrive at the notion of a scattering metric considered in [8] and [10].

Theorem 3. [10] *In the case of a scattering metric (ϕ_1 has trivial fibres, ϕ_2 has one fibre) A_λ is a Fourier integral operator associated with geodesic flow at time π for the induced metric on the boundary.*

The case in which $\dim Y_2 = 0$ but ϕ_1 can have non-trivial fibres is described in [8] and in more detail in [3] where such metrics are called 'fibred cusp' metrics. The same result is shown there in this case:

Theorem 4. [3] *For a fibred cusp metric the scattering operator, for $0 < \lambda^2 < A^2$, is a Fourier integral operator on $C^\infty(Y_1)$ associated with geodesic flow at time π, for the metric induced by $\tilde{h}_1$ on Y_1.*

In the special case where ϕ_1 and ϕ_2 are isomorphisms, so it can be assumed that $Y_2 = \partial X$, the metric can be described as a 'double-zero' metric (because the associated Lie algebra of vector fields, discussed below, consists of the vector fields which vanish to second order at the boundary). This is the main new case discussed here. The integration of this case with the 'fibred cusp' case can be expected to allow the full analysis of the general doubly warped metric in (1).

In the case of a double-zero metric the construction of a parametrix for the Poisson operator is outlined here. Carried to its proper conclusion this yields

Theorem 5. *If h is a non-singular Riemann metric on a compact manifold with boundary X and $x \in C^\infty(X)$ is equal to the distance from the boundary up to a quadratic term (i.e. modulo $x^2C^\infty(X)$ near the boundary) then the absolute scattering operator defined in* (2) *is a pseudodifferential operator.*

Formally combining this result and that from [10] one can give a conjecture in the general case. The identity wavefront relation for the scattering operator implied by Theorem 5 can be understood in terms of the limiting behaviour of complete geodesics on X which tend uniformly to the boundary. This limiting geodesic relation can be analyzed in terms of the rescaled symplectic geometry which arises from the (formal part of) the algebra of pseudodifferential operators discussed below. In this way the various boundary variables in (1) can be seen to move at 'different speeds' according to the associated powers of x. The variables in the fibres of ϕ_1 are 'fast' and the problem is global in these variables. The variables in the fibres of ϕ_2 are 'finite speed' while those in Y_2 are 'slow'. The limiting geodesic relation for a doubly warped metric descends to a relation on T^*Y_1 in which motion is along the fibres, with the distance scaled from π to 0 according to the length of the projection of the initial covector into the fibre normal. Using this formal reasoning (described also in [7]) one can expect the following.

Conjecture. *If g is a doubly warped Riemann metric on a compact manifold with boundary X and $0 < \lambda^2 < A^2$, the absolute scattering operator A_λ, defined in (2), is a Fourier integral operator associated to the limiting geodesic relation at infinity.*

The initial step in the proof of the existence of generalized eigenfunctions as in (2) is the construction of a suitably tailored algebra of pseudodifferential operators, with its associated wavefront set. I shall use this opportunity to outline what I believe is the 'appropriate' structure to give such an algebra and also a related 'large' calculus; this is done in §1. The doubly warped metric is then shown, in §2, to correspond to a Lie algebra of vector fields satisfying these general conditions. In the particular case of double-zero metrics the calculus so defined, and its associated wavefront set, are used in §3 to show the existence of generalized eigenfunctions; the general case should follow by combining the methods here with those of [3]. In §4 a more constructive approach is taken, as in [10], which leads to a a parametrix for the Poisson operator and hence a proof of Theorem 5. The same approach in the general setting should yield the Conjecture.

The author thanks the organizers of the meeting at Cortona for the opportunity to speak there and would also like to acknowledge discussions with Rafe Mazzeo, who also kindly commented on the manuscript, and Andras Vasy, who suggested an alternative proof of a slightly weaker form of Theorem 5 and the Conjecture which restored the author's faith!

§1: Boundary structures

To construct an algebra of pseudodifferential operators appropriate to the analytic questions of interest here a general procedure of 'microlocalization' will be invoked. This is based on the differential analysis of manifolds with corners and I shall therefore describe some general definitions and results. These should eventually appear in [9]; in the meantime they can be extracted from various places: [1], [4], [5], [6], [7].

Manifolds with corners

An intrinsic definition of a manifold with corners is as a (connected) Hausdorff, paracompact topological manifold with boundary X with a ring of functions $C^\infty(X)$ which can be embedded as a submanifold with corners of a smooth manifold of the same dimension. That is, there is a continuous map $i : X \longrightarrow \widetilde{X}$ into a C^∞ manifold $\widetilde{X}$, such that i is a homeomorphism onto its image, $C^\infty(X) = i^* C^\infty(\widetilde{X})$ and there are k non-negative functions $\rho_i \in C^\infty(\widetilde{X})$, $i = 1, \dots, k$ such that

$$i(X) = \{p \in \widetilde{X}; \rho_i(p) \geq 0,\ i = 1, \dots, k\} \text{ and}$$
$$\forall\, p \in \widetilde{X},\ \{d\rho_i(p); \rho_i(p) = 0\} \text{ is an independent set.}$$

This is the same as a coordinate-covering definition, with C^∞ transitions, based on the local model space $[0,\infty)^r \times \mathbb{R}^{n-r}$ (for fixed n but variable r) except that it ensures that the boundary hypersurfaces, which are the components of the sets $i^{-1}(\{\rho_i = 0\})$, are embedded. These components are each manifolds with corners with the C^∞ structure given by restriction of $C^\infty(X)$. The set of these boundary hypersurfaces is denoted $M_1(X)$; each of these necessarily has a defining function $\rho_H \in C^\infty(X)$, $H \in M_1(X)$, such that

$$\rho_H \geq 0,\ H = \{\rho_H = 0\} \text{ and } d\rho_H(p) \neq 0 \;\forall\, p \in H.$$

Here $df(p) \in T_p^* X$ is an element of the intrinsic cotangent bundle which is identified with $T^*_{i(X)}\widetilde{X}$ by i, i.e. it is the pull-back under i of $T^*\widetilde{X}$. Similar definitions apply to any of the usual natural bundles over a manifold without boundary.

p-submanifolds

There are various notions of submanifold of a manifold with corners, the most restrictive of which is a *p-submanifold* where the 'p' stands for product. Notice that a manifold with corners always has a product decomposition near each point p, in the sense that p has a neighbourhood O_p which is diffeomorphic to a product $[0,1)^k \times (-1,1)^{n-k}$. A closed subset $Y \subset X$ is a p-submanifold if each point $p \in Y$ has such a neighbourhood O_p with a diffeomorphism which also (i.e. simultaneously) reduces $Y \cap O_p$ to a product $\{0\}^r \times [0,1)^{k-r} \times \{0\}^s \times (-1,1)^{n-k-s}$. It follows that Y itself is a manifold with corners.

b-maps

A $\mathcal{C}^\infty$ map $f : X \longrightarrow Y$ between manifolds with corners is defined by the usual condition that $f^*\mathcal{C}^\infty(Y) \subset \mathcal{C}^\infty(X)$. It is useful to deal with a restricted class of smooth maps which 'respect the boundaries'. Namely consider the notion of an *interior b-map* defined by the requirement in terms of boundary defining functions that

$$f^*\rho'_G = a_G \prod_{H \in M_1(X)} \rho_H^{e(H,G)} = a_G \rho^{e(G)},$$
$$0 < a_G \in \mathcal{C}^\infty(X), \ \forall\ G \in M_1(Y).$$

Since f is assumed smooth, the *boundary exponents* $e(H,G)$ are non-negative integers for each $H \in M_1(X)$ and $G \in M_1(X)$. Note that $e(H,G)$ is positive if and only if $f(H) \subset G$. The obvious multi-index notation is used here to write the powers of boundary defining functions with $e(G) : M_1(X) \longrightarrow \mathbb{N}_0$ the 'multi-index'.

A general b-map is an interior b-map into one of the boundary faces of the range space.

Tangent vector fields

The smooth vector fields on X, meaning the smooth sections of TX, form a Lie algebra. A more significant Lie subalgebra is the space, $\mathcal{V}_\mathrm{b}(X)$, of smooth vector fields tangent to each boundary hypersurface. This has many properties similar to the whole Lie algebra. In particular it is the space of sections of a smooth vector bundle over X,

$$\mathcal{V}_\mathrm{b}(X) = \mathcal{C}^\infty(X; {}^bTX).$$

This new bundle is bundle-isomorphic to TX but not in any natural way. There is a natural $\mathcal{C}^\infty$ vector bundle map

$$\iota_\mathrm{b} : {}^bTX \longrightarrow TX$$

which is a diffeomorphism over the interior and has fibre rank exactly k at any boundary point of codimension k, i.e. any point which lies in precisely k boundary hypersurfaces. The null space of ι_p at such a point is a vector space, ${}^bN_pX$, which is naturally the product of k oriented lines, each corresponding to one of the hypersurfaces passing through p.

b-differential

By continuity from the usual differential in the interior, an interior b-map, as defined above, has both a push-forward and a pull-back b-differential (dual to each other)

$$^bf_* : {}^bT_pX \longrightarrow {}^bT_{f(p)}Y \text{ and } {}^bf^* : {}^bT^*_{f(p)}Y \longrightarrow {}^bT^*_pX.$$

It follows automatically that at any boundary point

$$^bf_* : {}^bN_pX \longrightarrow {}^bN_{f(p)}Y.$$

Using these maps, two surjectivity properties for an interior b-map can be identified. Namely f is said to be a *b-submersion* if it is surjective and has surjective b-differential at each point. On the other hand it is said to be *b-normal* if at each boundary point the restricted b-differential on the b-normal spaces is surjective. If it is both b-normal and is a b-submersion it is said to be a *b-fibration.*

Each of the four classes just defined, interior b-maps, b-submersions, b-normal maps and b-fibrations is closed under composition. Some of their analytic properties are described below.

Generalized blow-down maps

Another class of b-maps that typically arise from blow-up constructions has a weaker surjectivity property than b-submersions. Namely a b-map between manifolds with corners of the same dimension is said to be a *generalized blow-down map* if it restricts to a diffeomorphism of the interiors and its pull-back differential has the property

$$\mathcal{C}^\infty(X) \cdot {}^bf^*\mathcal{C}^\infty(Y; {}^bT^*Y) \supset \rho^a \mathcal{C}^\infty(X; {}^bT^*X) \text{ for some multi-index } a.$$

This implies in particular that if $\nu_b(Y)$ is a positive b-density on Y then pull-back of densities induces a map of the form

$$f^*\nu_b(Y) = a\rho^w \nu_b(X),\ w : M_1(X) \longrightarrow \mathbb{N}_0,\ 0 < a \in \mathcal{C}^\infty(X).$$

It further follows that a generalized blow-down map always induces an isomorphism on functions and densities

$$f^* : \dot{\mathcal{C}}^\infty(Y) \longleftrightarrow \dot{\mathcal{C}}^\infty(X),\ f_* : \dot{\mathcal{C}}^\infty(X; \Omega) \longleftrightarrow \dot{\mathcal{C}}^\infty(Y, \Omega).$$

This in turn implies that it induces an isomorphism from the space of extendible distributional sections of any bundle

$$f^* : \mathcal{C}^{-\infty}(Y; G) \longleftrightarrow \mathcal{C}^{-\infty}(X; f^*G),\ \mathcal{C}^{-\infty}(Y; G) = \left(\dot{\mathcal{C}}^\infty(Y; G' \otimes \Omega)\right)'.$$

Thus a generalized blow-down map can be thought of as a 'distributional isomorphism' in the sense that it changes the structure near the boundary but only to finite order. Notice that the composite of two generalized blow-down maps is again a generalized blow-down map.

The generalized blow-down maps considered here are constructed from radial blow-down maps. Part of the construction shows that the blow down map for blow up of a boundary p-submanifold, i.e. a p-submanifold of a proper boundary face, is always a generalized blow-down map in this sense.

Lifting

If $f : X \longrightarrow Y$ is a generalized blow-down map let $K \subset Y$ be the 'singular set' which is the image of the set near which it is not locally a diffeomorphism. Then the lift of a subset $C \subset Y$ is defined provided *either* C is the closure of its intersection with the complement of K i.e. $C = \mathrm{Clos}(C \cap (Y \setminus K))$, when by definition it is the set

$$f^*(C) = \mathrm{Clos}(f^{-1}(C \cap (Y \setminus K))$$

or in case $C \subset K$ when it is taken to be

$$f^*(C) = f^{-1}(C).$$

This definition reflects the commutation properties of blow ups.

Boundary structure

By a boundary structure (this is a special case of the more general notion of a boundary-fibration structure) on a compact manifold with corners X, I shall mean a Lie subalgebra $\mathcal{W} \subset \mathcal{V}_b(X)$ which can be 'resolved' in a sense made precise in terms of the notions introduced above. These conditions are just formal properties of the products X^2 and X^3 in their relation to X in the boundaryless case.

(i) There must be two manifolds with corners $X^2_{\mathcal{W}}$ and $X^3_{\mathcal{W}}$ and generalized blow-down maps

$$\beta^2 : X^2_{\mathcal{W}} \longrightarrow X^2, \ \beta^3 : X^3_{\mathcal{W}} \longrightarrow X^3$$

such that the symmetry groups $X^2 \ni (x, y) \longrightarrow (y, x) \in X^2$ and the similar exchange maps on X^3 lift to diffeomorphisms of $X^2_{\mathcal{W}}$ and $X^3_{\mathcal{W}}$ respectively.

(ii) Under β^2 the diagonal $\Delta = \{(x, x) \in X^2; x \in X\}$, lifts to a p-submanifold $\Delta_{\mathcal{W}} \subset X^2_{\mathcal{W}}$ such that β^2 restricts to a diffeomorphism of $\Delta_{\mathcal{W}}$ to Δ.

(iii) The two composite maps

$$\pi_{O,\mathcal{W}} = \pi_O \circ \beta^2_{\mathcal{W}},\ O = R, L$$

are b-fibrations, where π_R and π_L are the right and left projections from X^2 to X

(iv) The elements of $\mathcal{W}$, lifted to X^2 as vector fields on either the left or right factor, lift under β^2 into $\mathcal{V}_{\mathrm{b}}(X^2_{\mathcal{W}})$ to give a space of vector fields transversal to $\Delta_{\mathcal{W}}$.

(v) Let π_O for $O = S, C, F$ be the three projections $\pi_O : X^3 \longrightarrow X^2$ corresponding, respectively, to dropping the left, the central and the right factor of X. Then there are required to be three b-fibrations each giving a commutative diagramme

$$\begin{array}{ccc} X^3_{\mathcal{W}} & \xrightarrow{\ \pi_{O,\mathcal{W}}\ } & X^2_{\mathcal{W}} \\ {\scriptstyle \beta^3}\big\downarrow & & \big\downarrow{\scriptstyle \beta^2} \\ X^3 & \xrightarrow[\ \pi_O\ ]{} & X^2 \end{array} \qquad O = F, C, L,$$

such that under $\pi_{O,\mathcal{W}}$ the submanifold $\Delta_{\mathcal{W}} \subset X^2_{\mathcal{W}}$ lifts to a p-submanifold of $\Delta_{O,\mathcal{W}} \subset X^3_{\mathcal{W}}$ to which the other two $\pi_{O',\mathcal{W}}$, $\{S, C, F\} \ni O' \neq O$ restrict to diffeomorphisms.

Given the symmetry condition it suffices to check the last two conditions for one case. Over the interiors of X, X^2 and X^3 all these conditions are trivial except that they ensure that $\mathcal{W}$ must be unconstrained there. At the boundary the existence of such a resolution is distinctly non-trivial. Even in the simplest of cases, such as for $\mathcal{V}_{\mathrm{b}}(X)$ described next, such a resolution need not be unique.

One property which follows from these conditions is that $\mathcal{W} = \mathcal{C}^\infty(X; {}^{\mathcal{W}}TX)$ for a vector bundle ${}^{\mathcal{W}}TX$ over X.

Notation

Essentially by definition a boundary structure arises from some sort of 'uniformity' structure near the boundary. The Lie algebra of vector fields is generally denoted $\mathcal{V}_{\mathrm{n}}(X)$ for some abbreviated name n. Then the spaces and maps are denoted X^2_{n}, β^2_{n} etc.; note that the 'name' refers to the *resolution* rather than just the space of vector fields.

Basic example, $\mathcal{V}_{\mathrm{b}}(X)$

For any compact manifold with corners the space of all tangent vector fields $\mathcal{V}_{\mathrm{b}}(X)$ is a boundary structure on X. This is not however completely trivial! In particular it is certainly not the case that the double space X_b^2 is the product X^2, rather it is obtained from this product by blowing up all the codimension two boundary faces $H \times H$ where $H \in M_1(X)$. The triple product X_b^3 is similarly defined from X^3 by first blowing up all the triple products, $H^3 \subset X^3$ and then all the double products of the form $X \times H^2$, $H \times X \times H$ or $H^2 \times X$.

Even in the case of a manifold with boundary this resolution is not always unique. If the boundary has two components, H_1, and $H_2 \in M_1(X)$ then one can also define an 'overblown' resolution with X_{ob}^2 defined by blowing up $H_1 \times H_2$ and $H_2 \times H_1$ in addition to H_1^2 and H_2^2.

Conormal distributions at a p-submanifold

Conormal distributions with respect to an interior p-submanifold (one that meets the interior) can be defined just as in the original case of an embedded submanifold of a manifold without boundary as considered by Hörmander in [2]. Indeed it is always possible to 'double' a manifold with corners successively across each boundary hypersurface in such a way that a given p-submanifold extends to a closed embedded submanifold of an extension $\widetilde{X}$. Then conormal distributions with respect to an interior p-submanifold $Y \subset X$ are just the restrictions of conormal distributions for the extension $\widetilde{Y} \subset \widetilde{X}$. The resulting spaces are independent of the choice of extension and their properties can be deduced from the boundaryless case. The space of conormal distributions with respect to a p-submanifold can then be denoted $I^m(X;Y)$. The 1-step polyhomogeneous subspace $I^m_{\text{phg}}(X;Y)$ is similarly well defined, as are the corresponding spaces of sections of any vector bundle. These spaces are always $\mathcal{C}^\infty(X)$ modules with principal symbol maps as in the boundaryless case.

Conormal coefficients

As well as the natural space, $\mathcal{C}^\infty(X)$, of smooth functions on a manifold with corners it is frequently useful to consider larger spaces of distributions conormal with respect to the boundary. Since the boundary may consist of a number of intersecting hypersurfaces these distributions are somewhat more complicated than in the case of a p-submanifold. However, the 'spaces with bounds' are easily defined. Let ν_b be a smooth positive b-density on X, assumed compact for convenience. Thus if

$$\rho_{\text{tot}} = \prod_{H \in M_1(X)} \rho_H$$

is a 'total boundary defining function' then $0 < \rho_{\text{tot}}\nu_b \in \mathcal{C}^\infty(X;\Omega)$ is a smooth positive measure. Now, define

$$L^2_{\mathrm{b}}(X) = \left\{ f \in L^2_{\mathrm{loc}}(X^\circ); \int_X |f|^2 \nu_{\mathrm{b}} < \infty \right\}$$

and more generally if $\alpha : M_1(X) \longrightarrow \mathbb{R}$ is a 'multi-index' set

$$\rho^\alpha L^2_{\mathrm{b}}(X) = \left\{ f \in L^2_{\mathrm{loc}}(X^\circ); \int_X |\rho^{-\alpha} f|^2 \nu_{\mathrm{b}} < \infty \right\}$$
$$\text{where } \rho^\alpha = \prod_{H \in M_1(X)} \rho_H^{\alpha(H)}.$$

Since $L^2_{\mathrm{loc}}(X^\circ)$ is a well-defined subspace of the space of distributions on the interior, derivatives can be interpreted distributionally and then

$$\rho^\alpha H^\infty_{\mathrm{b}}(X) = \left\{ f \in \rho^\alpha L^2_{\mathrm{b}}(X); V_1 \dots V_l f \in \rho^\alpha L^2_{\mathrm{b}}(X),\ \forall\, V_i \in \mathcal{V}_{\mathrm{b}}(X),\ \forall\, l \right\}.$$

These are the L^2-based conormal distributions (or functions since they are necessarily C^∞ in X°.) There are similar L^∞-based conormal functions, obtained by replacing $L^2_{\mathrm{b}}(X)$ by $L^\infty(X)$ throughout. These are more naturally thought of as symbols so we set

$$S^\alpha(X) = \left\{ f \in \rho^{-\alpha} L^\infty(X); V_1 \dots V_l f \in \rho^{-\alpha} L^\infty(X),\ \forall\, V_i \in \mathcal{V}_{\mathrm{b}}(X),\ \forall\, l \right\}.$$

A form of the Sobolev embedding theorem then gives the 'Sandwich Theorem'

$$S^{-\alpha-\epsilon}(X) \subset \rho^\alpha H^\infty_{\mathrm{b}}(X) \subset S^{-\alpha}(X)\ \forall\, \alpha$$

provided ϵ is a positive multi-index, i.e. all entries are strictly positive.

Polyhomogeneity

As well as these conormal distributions defined by bounds there are polyhomogeneous space. By definition an *index set* is a subset

$$E \subset \mathbb{C} \times \mathbb{N}_0,\ \mathbb{N}_0 = \{0, 1, 2, \dots\}$$

satisfying the following four conditions

$$E \text{ is discrete}$$
$$(z_l, k_l) \in E,\ l \in \mathbb{N},\ |z_l| + k_l \to \infty \Longrightarrow \mathrm{Re}(z_l) \to \infty$$
$$(z, k) \in E,\ l \in \mathbb{N}_0,\ 0 \le l \le k \Longrightarrow (z, l) \in E$$
$$(z, k) \in E,\ r \in \mathbb{N} \Longrightarrow (z + r, k) \in E.$$

An *index family* $\mathfrak{E}$ for X is an assignment of an index set $\mathfrak{E}(H)$ to each $H \in M_1(X)$.

If $\mathfrak{E}$ is an index family for X then the space of polyhomogeneous conormal distributions, $\mathcal{A}^{\mathfrak{E}}(X)$, at the boundary with powers determined by $\mathfrak{E}$ can be defined so that it contains the functions

$$\prod_{H \in M_1(X)} \rho_H^{z(H)} \log^{k(H)} \rho_H,$$

provided $(z(H), k(H)) \in \mathfrak{E}(H)$ and in addition has appropriate completeness properties. These spaces are modules over $C^\infty(X)$ so one can also define

$$\mathcal{A}^{\mathfrak{E}} I^m(X;Y) = \mathcal{A}^{\mathfrak{E}}(X) \hat{\otimes}_{C^\infty(X)} I^m(X;Y)$$

as appropriate completions of the tensor products. A similar argument allows the corresponding spaces of sections of any vector bundle over X to be defined.

Kernels and operators

The space of pseudodifferential operators on a compact manifold without boundary, acting on half-densities, is identified with the corresponding space of kernels

$$\Psi^m(X;\Omega^{\frac{1}{2}}) = I^m(X^2, \Delta; \Omega^{\frac{1}{2}}).$$

In the case of a compact manifold with corners an appropriate form of the Schwartz kernel theorem identifies all continuous linear operators

$$A : \dot{C}^\infty(X;\Omega^{\frac{1}{2}}) \longrightarrow C^{-\infty}(X;\Omega^{\frac{1}{2}})$$

with their kernels which are arbitrary elements of $C^{-\infty}(X^2;\Omega^{\frac{1}{2}})$. The assumption that $\beta^2 : X^2_{\mathcal{W}} \longrightarrow X^2$ is a generalized blow-down map implies that the identification over the interior extends by continuity to give an isomorphism

$$(\beta^2)^* : C^{-\infty}(X^2;\Omega^{\frac{1}{2}}) \longleftrightarrow C^{-\infty}(X^2_{\mathcal{W}};\Omega^{\frac{1}{2}}).$$

Thus a space of operators on X can be defined by specifying the kernels as half-densities on $X^2_{\mathcal{W}}$.

The identity

The relation between $X^2_{\mathcal{W}}$ and X demanded of a resolution implies that β^2 induces a diffeomorphism from the lifted diagonal to the diagonal of X^2. From this, and the assumption that it is a generalized blow-down map, it follows that the kernel of the identity operator, which is a non-vanishing Dirac section of $\Omega^{\frac{1}{2}}$ for the diagonal, lifts to a distribution

$$\mathrm{Id} \in \mathcal{C}^{-\infty}(X^2_{\mathcal{W}}; \Omega^{\frac{1}{2}}),\ \mathrm{Id} = \rho^{\kappa} A,\ \ A \in I^0(X^2_{\mathcal{W}}, \Delta_{\mathcal{W}}; \Omega^{\frac{1}{2}}), A > 0.$$

That is, apart from a boundary weight ρ^{κ} it too is a positive Dirac section of the half-density bundle for the lifted diagonal. The index κ is only determined for the boundary hypersurfaces of $X^2_{\mathcal{W}}$ which meet the lifted diagonal (they are in one to one correspondence with the boundary hypersurfaces of X).

$\mathcal{W}$-pseudodifferential operators

Using the assumption that the Lie algebra $\mathcal{W}$ is transversal to the lifted diagonal it follows that the algebra of differential operators on half-densities generated by $\mathcal{W}$ and $\mathcal{C}^{\infty}(X)$ is identified with the space

$$\left\{A \in I^{\mathbb{N}_0}(X^2_{\mathcal{W}}, \Delta_{\mathcal{W}}; \rho^{\kappa}\Omega^{\frac{1}{2}}); \mathrm{supp}(A) \subset \Delta_{\mathcal{W}}\right\}.$$

That is, the space of $\mathcal{W}$-differential operators is identified with the space of all smooth Dirac sections over the lifted diagonal of the blown-up space $X^2_{\mathcal{W}}$.

This identification of the $\mathcal{W}$-differential operators makes it natural to define the 'small' calculus of $\mathcal{W}$-pseudodifferential operators in terms of kernels by

$$\Psi^m_{\mathcal{W}}(X; \Omega^{\frac{1}{2}}) =$$
$$\left\{A \in I^m(X^2_{\mathcal{W}}, \Delta_{\mathcal{W}}; \rho^{\kappa}\Omega^{\frac{1}{2}}); A \equiv 0 \text{ at } H \in M_1(X^2_{\mathcal{W}}) \text{ if } H \cap \Delta_{\mathcal{W}} = \emptyset\right\}.$$

Larger spaces of $\mathcal{W}$-pseudodifferential operators can be defined for any choice of index set $\mathfrak{E}$ for $X^2_{\mathcal{W}}$ by

$$\Psi^{m,\mathfrak{E}}_{\mathcal{W}}(X; \Omega^{\frac{1}{2}}) = \mathcal{A}^{\mathfrak{E}} I^m(X^2_{\mathcal{W}}, \Delta_{\mathcal{W}}; \rho^{\kappa}\Omega^{\frac{1}{2}}).$$

If $\mathfrak{E}$ satisfies the compatibility condition that $\mathfrak{E}(H) \subset \mathbb{N}_0 \times \{0\}$ if $H \cap \Delta_{\mathcal{W}} \neq \emptyset$, then $\mathcal{C}^{\infty}(X^2_{\mathcal{W}}) \subset \mathcal{A}^{\mathfrak{E}}(X^2_{\mathcal{W}})$ and then

$$\mathrm{Diff}^m_{\mathcal{W}}(X; \Omega^{\frac{1}{2}}) \subset \Psi^m_{\mathcal{W}}(X; \Omega^{\frac{1}{2}}).$$

These definitions can be extended to operators acting on sections of any vector bundle(s).

Pull-back

Conormal functions can be pulled back under general interior b-maps provided they have a transversality property with respect to the p-submanifold involved (if there is one). Since it suffices for present purposes, attention can be restricted here to b-fibrations. Then each boundary hypersurface $H \in M_1(X)$ is mapped onto either the whole of Y or onto a boundary hypersurface $G = F(H) \in M_1(Y)$. If $\mathfrak{E}$ is an index family for Y define $\mathfrak{F} = f^{\#}\mathfrak{E}$ by setting $\mathfrak{F}(H) = \mathbb{N} \times \{0\}$ (the index set of $\mathcal{C}^\infty$) in the first case and

$$\mathfrak{F}(H) = \{(z', k); z' = e(H, G)z,\ (z, k) \in \mathfrak{E}(G)\}$$

in the second case.

Proposition. *Suppose $f : X \longrightarrow Y$ is a b-fibration of compact manifolds with corners and that $B \subset Y$ is an interior p-submanifold. Then $f^*(B) \subset X$ is a p-submanifold and pull back defines linear maps*

$$f^* : \mathcal{A}^{\mathfrak{E}} I^m_{\text{phg}}(Y, B; G) \longrightarrow \mathcal{A}^{\mathfrak{F}} I^{m'}_{\text{phg}}(X, f^*B; f^*G),\ \mathfrak{F} = f^{\#}\mathfrak{E}\,.$$

Here f^*G is the pull-back of the bundle G and $m' - m$ depends on the relative dimensions.

This result follows from a direct characterization of the polyhomogeneous conormal functions in terms of expansions.

Push-forward

There is a similar result for push-forward. For an index set E write $E > 0$ to mean that $(z, k) \in E \Longrightarrow \operatorname{Re} z > 0$. For a b-fibration $f : X \longrightarrow Y$ and an index set $\mathfrak{E}$ on X define an index set $\mathfrak{F} = f_{\#}\mathfrak{E}$ as follows. For each $G \in M_1(Y)$ set

$$\mathfrak{F}(G) = \{(z, k); \exists\ (z_i, k_i) \in \mathfrak{E}(H_i),\ F(H_i) \subset G,\\ z = \sum_i \frac{z_i}{e(H_i, G)},\ k + 1 \le \sum_i (k_i + 1)\}.$$

Proposition. *Suppose $f : X \longrightarrow Y$ is a b-fibration of compact manifolds with corners and $B \subset X$ is an interior p-submanifold to which f is (b-)normal. Then $f(B) \subset Y$ is an interior p-submanifold and, provided $\mathfrak{E}(H) > 0$ for all $H \in M_1(X)$ mapped onto Y, push-forward defines linear maps*

$$f_* : \mathcal{A}^{\mathfrak{E}} I^m_{\text{phg}}(X, B; f^*G \otimes \Omega_b) \longrightarrow \mathcal{A}^{\mathfrak{F}} I^{m'}_{\text{phg}}(X, f(B); G \otimes \Omega_b),\ \mathfrak{F} = f_{\#}\mathfrak{E};$$

here $m' - m$ depends on the relative dimensions.

This has further extensions to the case, which occurs in the composition of pseudodifferential operators, where B is replaced by two transversally intersecting submanifolds, each of which is mapped onto Y but the intersection of the two is mapped onto a proper p-submanifold. A similar result then holds for spaces of conormal distributions defined with respect to the intersecting pair. The important point for the present discussion is that these contain the products of conormal distributions associated to the two submanifolds; again this is just as in the boundaryless case.

Operators

As discussed above the kernels of these $\mathcal{W}$-pseudodifferential operators are defined on $X^2_{\mathcal{W}}$ and hence equivalently as extendible distributions on X^2. They therefore define operators from $\dot{\mathcal{C}}^\infty(X)$ to $\mathcal{C}^{-\infty}(X)$. For the small calculus

$$A \in \Psi^m_{\mathcal{W}}(X) \Longrightarrow A : \rho^a \mathcal{C}^\infty(X) \longrightarrow \rho^a \mathcal{C}^\infty(X)$$

for any multi-index a. Since the calculus is closed under passage to adjoints these operators act on extendible distributions. Elements of the large calculus determined by a choice of index family for $X^2_{\mathcal{W}}$ may not act on any fixed space, depending on the index families, but do map appropriately 'integrable' polyhomogeneous spaces to polyhomogeneous spaces with transformed index families.

Symbol and normal morphisms

The symbol of an element of the small calculus is the symbol of its kernel, as a conormal distribution. This is an equivalence class of functions (or in general sections of a bundle) on the conormal bundle to the lifted diagonal. The hypotheses above identify this bundle with the structure bundle ${}^{\mathcal{W}}T^*X$.

Elements of the small calculus act on $\rho^a\mathcal{C}^\infty(X)$ and hence by restriction define operators on $\mathcal{C}^\infty(H)$ for each $H \in M_1(X)$. These boundary actions are determined by the restriction of the kernels to the corresponding boundary hypersurface of $X^2_{\mathcal{W}}$. There restricted kernels in fact contain more information and can be viewed as the kernels of $\mathcal{W}_H$-pseudodifferential operators (with additional invariance properties) for a boundary(-fibration) structure on ${}^bNH$. Thus even to handle operators associated to boundary structures one needs the more general theory including the fibration part of boundary-fibration structures. For the cases discussed below these normal operators can be handled directly.

Composition

Using the push-forward and pull-back theorems just described, and extensions as in the boundaryless case it follows that the small calculus is always an algebra containing the $\mathcal{W}$-differential operators and the large calculus is a two-sided module over the small calculus. Elements of the larger calculus can only be composed under appropriate conditions on the index families but once these are satisfied the composite is in the calculus.

Sobolev spaces

There is a natural L^2 space, $L^2_{\mathcal{W}}(X)$, defined by a density given locally by the wedge product of a basis of ${}^{\mathcal{W}}T^*X$. Using this basic space and the action of the small calculus on distributions, Sobolev spaces can be defined in terms of weighting by powers of boundary defining functions

$$H^{a,m}_{\mathcal{W}}(X) = \{u \in C^{-\infty}(X); Au \in \rho^a L^2_{\mathcal{W}}(X) \ \forall \ A \in \Psi^{-m}_{\mathcal{W}}(X)\}.$$

Blow up

Essentially by definition the spaces $X^2_{\mathcal{W}}$ and $X^3_{\mathcal{W}}$ are defined by 'blow-up'. The radial blow up of a p-submanifold Y of a manifold with corners X is a well-defined procedure giving a new manifold with corners $[X;Y]$ with natural blow-down map $\beta : [X;Y] \longrightarrow X$. This new manifold has an additional boundary hypersurface compared to X which replaces Y. Polar coordinates in the normal variables to Y in X lift to local coordinates near this 'front face'. If Y is a boundary p-submanifold then β is a generalized blow-down map in the sense described above.

Amongst the important properties of blow up are two commutation properties. If $Y_1 \subset Y_2$ are two p-submanifolds then under blow-up of either of them the other lifts to be a p-submanifold and denoting the lifts by the same letter the two blow ups are naturally isomorphic:

$$[X;Y_1;Y_2] \equiv [X;Y_1;Y_2], \ Y_1 \subset Y_2 \text{ p-submanifolds.}$$

Here the naturality is that the isomorphism gives a commutative diagram with the blow-down maps. The same result is true if Y_1 and Y_2 meet transversally, but not in general.

§2: The doubly warped calculus

The 'elliptic' part of the analysis of the Laplacian associated to a metric as in (1) consists in the construction of an algebra of pseudodifferential operators on X which contains the resolvent away from the spectrum. In this case it is the 'small calculus' discussed above for the appropriate Lie algebra of smooth vector fields which describes the degeneration of the metric in the precise sense that the Laplacian is a (symbolically) elliptic operator in the corresponding enveloping algebra.

Theorem 6. *The space $\mathcal{V}_{\mathrm{dw}}(X)$ consisting of those smooth vector fields on X which have bounded length with respect to the metric g in* (1), *is a Lie algebra which is a boundary structure in the sense of the previous section.*

Proof. The space $\mathcal{V}_{\mathrm{dw}}(X)$ is easily described. Local coordinates near any boundary point of the form x, z, y, t can be introduced with x being the distance function $\tilde{x}$ lifted from $[0,\delta) \times Y_2$, the t's are coordinates lifted from Y_2, the y's coordinates lifted from the fibre of ϕ_2 in Y_1 through p and the z's are coordinates on the fibre of ϕ_1 through p. A smooth vector field on X is an element of $\mathcal{V}_{\mathrm{dw}}(X)$ if and only if expressed in terms of these coordinates it is a smooth linear combination of the basis vector fields

$$x^2\frac{\partial}{\partial x},\ x^2\frac{\partial}{\partial t},\ x\frac{\partial}{\partial y} \text{ and } \frac{\partial}{\partial z}.$$

Thus the theorem just states that this Lie algebra of vector fields has a resolution in the sense described above. Thus the spaces X^2_{dw} and X^3_{dw} and appropriate maps must be constructed. These spaces will be constructed from X^2 and X^3 respectively by iterated radial blow up.

Let us consider first X^2_{dw}; for simplicity assume that ∂X is connected and that the fibrations have connected fibres. The first stage in its construction is the blow up of the fibre diagonal at the corner of X^2 corresponding to the total fibration ψ. In the corner $(\partial X)^2 \subset X^2$ of X^2 consider the submanifold which is the ψ-fibre diagonal, namely

$$B_2 = \{(p,q) \in (\partial X)^2; \psi(p) = \psi(q)\}.$$

This is a closed embedded submanifold of the corner (which is a manifold without boundary); in terms of local coordinates as discussed above, x, y, t, z and x', y', t', z' in the two factors of X it is $y = y'$, $x = x' = 0$. As such it is a p-submanifold of X^2 so it can be blown up radially to give the new manifold with corners $[X^2; B_2]$ and associated blow-down map $\beta : [X^2; B_2] \longrightarrow X^2$.

The second stage of the blow up involves the 'lift' of the fibre diagonal associated to $\tilde{\phi}_1$. The fibre diagonal, defined near the boundary,

$$B_1' = \{(p,q); \tilde{\phi}_1(p) = \tilde{\phi}_1(q)\}$$

lifts under the first blow up to a p-submanifold near the boundary and B_1 is by definition the intersection of this lift with the boundary. It is again a boundary p-submanifold. Then the stretched double space is given by iterated blow up

$$X^2_{\mathrm{dw}} = [X^2; B_2; B_1].$$

The total blow down map is therefore a generalized blow down map. The conditions in the first four conditions can also be checked.

The construction of X^3_{dw} is similar but more involved since it has four stages. In X^3 there are three analogues of B_2, namely the corresponding fibre diagonals in the three pairs of factors; these can be denoted $B_{2,O}$, $O = S, C, F$. These intersect in the triple fibre diagonal $B_{2,T}$. All are boundary p-submanifolds. The first step is to blow up $B_{2,T}$. Under this blow up the three fibre diagonals lift to disjoint p-submanifolds which can therefore be blown up independently.

The submanifold B_1 similarly has three analogues denoted, $B_{1,O}$, for $O = S, C, F$, in the blown up version of X^3 produced so far. They are all p-submanifolds intersecting in pairs in the same p-submanifold $B_{1,T}$. The remaining two stages in the construction of X^3_{dw} are the blow up of $B_{1,T}$ after which the remaining $B_{1,O}$ lift to be disjoint and are then blown up. This defines X^3_{dw}.

The main step in checking the properties required above of a resolution of the Lie algebra $\mathcal{V}_{\mathrm{dw}}(X)$ is to show the existence of the stretched projections $\pi_{O,\mathrm{dw}}$, for $O = S, C, F$. This is done using the commutation properties of blow up outlined above as follows for $O = F$ with $\tilde{B}_1 = B_{1,C} \cup B_{1,S}$ and $\tilde{B}_2 = B_{2,C} \cup B_{2,S}$:

$$\begin{aligned}
X^3_{\mathrm{dw}} &= [X^3; B_{2,T}; B_{2,F}; \tilde{B}_2; B_{1,T}; B_{1,F}; \tilde{B}_1] \\
&\xrightarrow{\beta(\tilde{B}_1)} [X^3; B_{2,T}; B_{2,F}; \tilde{B}_2; B_{1,T}; B_{1,F}], \\
&\equiv [X^3; B_{2,F}; B_{2,T}; \tilde{B}_2; B_{1,F}; B_{1,T}],\ B_{2,T} \subset B_{2,F},\ B_{1,T} \subset B_{1,F}, \\
&\xrightarrow{\beta(B_{1,T})} [X^3; B_{2,F}; B_{2,T}; \tilde{B}_2; B_{1,F}] \\
&\equiv [X^3; B_{2,F}; B_{2,T}; B_{1,F}; \tilde{B}_2],\ B_{1,F} \subset \tilde{B}_2 \\
&\xrightarrow{\beta(\tilde{B}_2)} [X^3; B_{2,F}; B_{2,T}; B_{1,F}] \equiv [X^3; B_{2,F}; B_{1,F}; B_{2,T}]],\ B_{2,T} \subset B_{1,F} \\
&\xrightarrow{\beta(B_{2,T})} [X^3; B_{2,F}; B_{1,F}] \equiv [X \times X^2; X \times B_2; X \times B_1] \equiv X \times X^2_{\mathrm{dw}} \\
&\xrightarrow{\pi} X^2_{\mathrm{dw}}.
\end{aligned}$$

It is then relatively straightforward to check the desired properties of these maps. □

The calculus of operators given by the results outlined in the previous section then gives a description of the resolvent.

Proposition. *The Laplacian of the metric g in (1) is an element of $\mathrm{Diff}^2_{\mathrm{dw}}(X)$ for the corresponding doubly warped structure and the resolvent $(\Delta - \sigma)^{-1} \in \Psi^{-2}_{\mathrm{dw}}(X)$ is a holomorphic function of $\sigma \in \mathbb{C} \setminus [0, \infty)$.*

Proof. This result can be shown constructively, essentially as the same result obtained in the boundaryless case using the normal operator and the usual symbol map. □

§3: Double zero metrics

Next I shall briefly describe the proof of the existence of generalized eigenfunctions as in (2), restricting the discussion to the special case of a double-zero metric, where the two fibrations maps ϕ_i are isomorphisms and can be forgotten. This represents well the changes needed to the proof of the corresponding fact in the scattering calculus ([6]) and the fibred-cusp calculus ([8], [3]). In terms of the coordinates introduced above there are no z or y variables; the metric becomes

$$g = \frac{h}{x^4}$$

where h is a non-degenerate metric on X and x is a boundary defining function equal to the h-distance to second order.

In the special case that x is the distance for h,

$$\Delta = -(x^2\partial_x)^2 + (2n-4)x^3\partial_x + x^4\Delta_{h(x)} + x^2 W, \; W \in \mathcal{V}_{\mathrm{dz}}(X) \tag{3}$$

where $\Delta_{h(x)}$ is the tangential Laplacian of the x-dependent metric. In the general case the same decomposition is valid up to a term in $x^2\,\mathrm{Diff}^2_{\mathrm{dz}}(X)$ (which annihilates 1). An easy inductive argument then shows that there are formal solutions as in (2). More precisely, for any $f \in C^\infty(\partial X)$ there is a function $\tilde{f} \in C^\infty(X)$, unique up to $\dot{C}^\infty(X)$, i.e. in Taylor series at the boundary, such that

$$(\Delta - \lambda^2)\tilde{u}_f = e_f \in \dot{C}^\infty(X), \; \tilde{u}_f = e^{i\lambda/x}x^{n-1}\tilde{f}, \; 0 \neq \lambda \in \mathbb{R}, \; n = \dim X.$$

The actual eigenfunction in (2) is then constructed as

$$u_f = \tilde{u}_f - \lim_{\epsilon \downarrow 0}(\Delta - (\lambda - i\epsilon)^2)^{-1} e_f. \tag{4}$$

Even if there is discrete spectrum this limit exists.

Thus the proof of the existence of u_f reduces to an appropriate form of the limiting absorption principle, the examination of the limit of the resolvent on the real axis. This is not really an 'elliptic' problem but rather has more similarity to hyperbolic equations. To explain this I shall first discuss some properties of the calculus $\Psi^*_{\mathrm{dz}}(X)$ which is defined as a special case of the discussion above.

The double-zero calculus has a global quantization map. Consider the vector bundle ${}^{\mathrm{dz}}T^*X$ which is the dual of the 'structure bundle' ${}^{\mathrm{dz}}TX$ of which $\mathcal{V}_{\mathrm{dz}}(X) = \mathcal{C}^\infty(X; {}^{\mathrm{dz}}TX)$ is the space of all sections. Radial compactification of the fibres of this bundle defines a compact manifold with corners which I shall denote ${}^{\mathrm{dz}}\overline{T}^*X$. Since X is a manifold with boundary (assumed connected for notational simplicity) ${}^{\mathrm{dz}}\overline{T}^*X$ has two boundary hypersurfaces (assuming that $\dim X > 1$ as we shall). Let ρ_∂ and ρ_σ be defining functions for these two boundary hypersurfaces. Then for any integers k, l the spaces $\rho_\partial^k \rho_\sigma^l \mathcal{C}^\infty({}^{\mathrm{dz}}\overline{T}^*X)$ are special case of the polyhomogeneous conormal distributions discussed above. There is then a global quantization map for all k, l which is itself independent of k, l

$$q : \rho_\partial^k \rho_\sigma^{-l} \mathcal{C}^\infty({}^{\mathrm{dz}}\overline{T}^*X) \longrightarrow \rho^k \Psi^l_{\mathrm{dz}}(X).$$

The map q has null space contained in $\dot{\mathcal{C}}^\infty({}^{\mathrm{dz}}\overline{T}^*X)$ and range a complement to $\rho^\infty \Psi^{-\infty}_{\mathrm{dz}}(X)$. In consequence it induces a 'full symbol isomorphism'

$$\tilde{q} : \rho_\partial^k \rho_\sigma^{-l} \mathcal{C}^\infty({}^{\mathrm{dz}}\overline{T}^*X) / \dot{\mathcal{C}}^\infty({}^{\mathrm{dz}}\overline{T}^*X) \longrightarrow \rho^k \Psi^l_{\mathrm{dz}}(X) / \rho^\infty \Psi^{-\infty}_{\mathrm{dz}}(X)$$

and thus induces a 'star product' on the domain space, the space of Laurent series at the boundary, i.e. an associative product given to any order by bilinear differential operators

$$a \star b \equiv \sum_{j=0}^{\infty} P_j(a, b), \ \ P_0(a, b) = ab.$$

The microlocality of this product arises here from the limiting commutativity of the Lie algebra $\mathcal{W}$. The principal symbol map, defined as the leading part of the inverse of q, is independent of all choices and gives a short exact sequence for each k, l

$$0 \longrightarrow \rho_\partial^{k+1} \Psi^{l-1}_{\mathrm{dz}}(X) \hookrightarrow \rho_\partial^k \Psi^l_{\mathrm{dz}}(X) \xrightarrow{\sigma_{l,k}} \rho_\partial^k \rho^{-l} \mathcal{C}^\infty({}^{\mathrm{dz}}\overline{T}^*X) / \rho_\partial^{k+1} \rho_\sigma^{-l+1} \mathcal{C}^\infty({}^{\mathrm{dz}}\overline{T}^*X) \longrightarrow 0.$$

This identification (even though it is by no means unique) shows that there is an associated notion of wavefront set. Namely for an element of the algebra the wavefront set is the support (in the boundary) of the associated Laurent series:

$$\mathrm{WF}'_{\mathrm{dz}}(A) = \mathrm{supp}([a]) \subset C_{\mathrm{dz}}(X) = \partial^{\mathrm{dz}}\overline{T}^*X,$$
$$A \in \rho^k \Psi^l_{\mathrm{dz}}(X),\ \tilde{q}([a]) = [A].$$

This wavefront set makes the calculus microlocal in the sense that, $\mathrm{WF}'_{\mathrm{dz}}(A) \subset \mathrm{WF}'_{\mathrm{dz}}(A) \cap \mathrm{WF}'_{\mathrm{dz}}(B)$. Then for extendable distributions on X one can set, as in the boundaryless case,

$$\mathrm{WF}_{\mathrm{dz}}(u) = \bigcap \left\{ \mathrm{WF}'_{\mathrm{dz}}(A); A \in \rho^k \Psi^l_{\mathrm{dz}}(X),\ Au \in \dot{C}^\infty(X) \right\}$$

with the wavefront set independent of k, l.

This wavefront set has the usual properties, for example

$$u \in C^{-\infty}(X),\ \mathrm{WF}_{\mathrm{dz}}(u) = \emptyset \Longrightarrow u \in \dot{C}^\infty(X)$$

and if the elliptic set $\mathrm{Ell}_{k,l}(A)$ of $A \in \rho^k \Psi^l_{\mathrm{dz}}(X)$ is defined to be the set where $\sigma_{k,l}(A)$ is invertible then

$$u \in C^{-\infty}(X),\ Au \in \dot{C}^\infty(X) \longrightarrow \mathrm{WF}_{\mathrm{dz}}(u) \cap \mathrm{Ell}_{k,l}(A) = \emptyset.$$

Returning to the question of eigenfunctions, suppose $u \in C^{-\infty}(X)$ is a tempered eigenfunction of the Laplacian, so

$$(\Delta - \lambda^2)u = 0.$$

Then it follows immediately that

$$\mathrm{WF}_{\mathrm{dz}}(u) \subset \{\sigma_{0,2}(\Delta) = \lambda^2\} \subset C_{\mathrm{dz}}X.$$

The symbol of the Laplacian is represented in $\rho_\sigma^{-2} C^\infty({}^{\mathrm{dz}}\overline{T}^*X)$ by the square of the metric length function on ${}^{\mathrm{dz}}T^*X$. Thus the wavefront set of an eigenfunction u must be contained in the characteristic set of $\Delta - \lambda^2$, which is the appropriate sphere bundle over the boundary

$$\mathrm{WF}_{\mathrm{dz}}(u) \subset \{|\Xi|^2 = \lambda^2\} \subset {}^{\mathrm{dz}}T^*_{\partial X}X \longrightarrow C_{\mathrm{dz}}X.$$

To analyze the limit in (4) a uniform (in ϵ) version of this wavefront set, of finite order relative to weighted Sobolev spaces, can be used. Thus, if $u_\epsilon \in C^{-\infty}(X)$ is a family of distributions depending on $\epsilon \in$

P, a parameter space, then a point $Z \in C_{\mathrm{dz}}X$ is not in the uniform wavefront set $\mathrm{WF}^{k,l}_{\mathrm{dz},P}$ relative to the finite order Sobolev space $H^{k,l}_{\mathrm{dz}}(X)$ if it is bounded in some such Sobolev space and there is an operator $A \in \rho^{-l}\Psi^k_{\mathrm{dz}}(X)$ which is elliptic at Z and such that Au_ϵ is bounded in $L^2_{\mathrm{dz}}(X)$.

The starting point is the self-adjointness of Δ from which it follows that u_ϵ, the solution of (4), i.e.

$$(5) \qquad (\Delta - (\lambda - i\epsilon)^2)u_\epsilon = g \in \dot{C}^\infty(X),\ \epsilon > 0,\ 0 \neq \lambda \in \mathbb{R},$$

has ϵu_ϵ bounded in $L^2_{\mathrm{dz}}(X)$ for $\epsilon \in (0,1]$.

Consider a family v_ϵ which is bounded in $H^{k',l'}_{\mathrm{dz}}(X)$ for $\epsilon \in (0,1] = P$ and some k', l' and suppose that

$$(6) \qquad (\Delta - (\lambda - i\epsilon)^2)v_\epsilon = f_\epsilon \text{ is bounded in } \dot{C}^\infty(X).$$

Directly from the definition of the uniform wavefront set it follows that

$$\mathrm{WF}^{k,l}_{\mathrm{dz},P}(v_\epsilon) \subset \{|\Xi| = \lambda^2\}\ \forall\ k,l,\ P = (0,1].$$

Using commutation arguments as in [6] improved microlocal bounds can also be deduced from (6). First there are no singularities in the 'incoming' subset.

Proposition. *For a family v_ϵ bounded in $H^{k',l'}_{\mathrm{dz}}(X)$ for some k', l' satisfying* (6) *and any $k, l \in \mathbb{R}$*

$$\mathrm{WF}^{k,l}_{\mathrm{dz},P}(v_\epsilon) \cap \{-\lambda\frac{dx}{x^2}\} = \emptyset.$$

Secondly Hörmander's theorem on the propagation of singularities is valid in this context for the 'Hamilton vector field' of the Laplacian

Proposition. *For a family v_ϵ bounded in $H^{k',l'}_{\mathrm{dz}}(X)$ for some k', l' satisfying* (6) *and any $k, l \in \mathbb{R}$, $\mathrm{WF}^{k,l}_{\mathrm{dz},P}(v_\epsilon)$ is a union of maximally extended curves in $\{\tau^2 + |\zeta|^2 = \lambda^2\}$ of the vector field $\tau\zeta \cdot \partial_\zeta - |\zeta|^2\partial_\tau$.*

Combining these two propositions it follows that $\mathrm{WF}^{k,l}_{\mathrm{dz},P}(v_\epsilon)$ is confined to the 'outgoing' surface $\lambda\frac{dx}{x^2}$. There its order can be bounded.

Proposition. *For a family v_ϵ bounded in $H^{k',l'}_{\mathrm{dz}}(X)$ for some k', l' satisfying* (6) *and any $k, l \in \mathbb{R}$ $\mathrm{WF}^{k,l}_{\mathrm{dz},P}(v_\epsilon)$*

$$\mathrm{WF}^{k,l}_{\mathrm{dz},P}(v_\epsilon) = \emptyset\ \forall\ k \text{ and } l < -\frac{1}{2}.$$

Now applying these results to $v_\epsilon = \epsilon u_\epsilon$ it follows that ϵu_ϵ is bounded in $H^{k,l}_{\text{dz}}(X)$ for $\epsilon \in P$, any k and any $l < -\frac{1}{2}$. It is therefore a precompact family in the same spaces. In fact this argument applies uniformly in λ near any given non-zero $\lambda' \in \mathbb{R}$. Thus, if $\|\epsilon u_\epsilon\|_{k',l'}$, $l' < -\frac{1}{2}$, does not tend to zero as $\epsilon \downarrow 0$ then along some subsequence it converges to a non-zero limit which is an eigenfunction and has WF_{dz} contained in the outgoing set.

Proposition. *If* $0 \neq \lambda \in \mathbb{R}$ *and* $u \in C^{-\infty}(X)$ *satisfies*

$$(\Delta - \lambda^2)u = f \in \dot{C}^\infty(X), \ \ \mathrm{WF}_{\text{dz}}(u) \in \{\lambda \frac{dx}{x^2}\}$$

then

$$u = e^{i\lambda/x} x^{n-1} u', \ \ u' \in C^\infty(X). \tag{7}$$

This explicit form of the function allows a version of Green's formula to be proved.

Lemma. *If* u *is as in the proposition above then*

$$\lim_{\delta \downarrow 0} \int_{x<\delta} \overline{u}(\Delta - \lambda^2) u dg = \int_{\partial X} |u'|^2 dh.$$

As a corollary, for an actual eigenfunction u which is outgoing it follows that $u' = 0$ on ∂X and then by a modified form of the propositions above that $u \in \dot{C}^\infty(X)$. In fact the proof shows that for $\lambda \in [\delta, 1/\delta]$, $\delta > 0$, the span of such eigenfunctions forms a subspace of $\dot{C}^\infty(X)$ with bounded sphere, it is therefore finite-dimensional.

Proposition. *For the Laplacian of a double-zero metric the embedded* L^2 *spectrum is discrete and of finite multiplicity.*

Returning now to the family u_ϵ satisfying (5) we can simply project it onto the L^2 eigenspace with eigenvalue λ^2 if this is non-zero

$$u_\epsilon = u'_\epsilon + u''_\epsilon,$$

where u''_ϵ is the part in the eigenspace. Necessarily, $\epsilon u''_\epsilon$ is bounded in $\dot{C}^\infty(X)$ for $\epsilon \in (0,1]$. Applying the operator it follows that u'_ϵ satisfies essentially the same condition in that

$$(\Delta - (\lambda - i\epsilon)^2) u'_\epsilon = g_\epsilon \text{ is bounded in } \dot{C}^\infty(X).$$

Repeating the same arguments as before but using the orthogonality to the eigenfunctions (if there are any) it follows that $u'_\epsilon \longrightarrow u$ is of the form (7). This finally proves the existence of the generalized eigenfunction as in (2) in this case.

§4: Poisson operator and scattering matrix

Construction of the Poisson operator P_λ amounts to the solution of (2) for f a delta function at a general boundary point, $\bar{y}$. This can be done for a double-zero metric in terms of Legendre distributions, much as in [10] for scattering metrics. In the present case the construction of a parametrix for P_λ is local, it takes place in a small neighbourhood of the point $\bar{y}$.

Take local coordinates near $\bar{y}$ in which $\bar{y} = 0$ and begin with the 'ansatz'

$$v = (2\pi)^{-n+1} \int \exp(i\frac{y \cdot u}{x^2} - i\frac{\alpha(u)}{x})x^l a(x, y, u)du.$$

Here $\alpha(u)$ is a smooth function of $u \in \mathbb{R}^{n-1}$ to be chosen and similarly $a(x, y, u)$ is a smooth function in $x \geq 0$ with support in $|u| < \frac{1}{2}$, $|y - \bar{y}| < \delta$. Thus the integral certainly converges for $x > 0$. Integration against a 'test function' $\phi \in C^\infty(\mathbb{R}^{n-1})$ and use of stationary phase shows that

$$\begin{aligned}\int v(x,y)\phi(y)dy &= x^l (2\pi)^{-n+1} \int \exp(i\frac{y \cdot u}{x^2} - i\frac{\alpha(u)}{x})a(x, y, u)dudy \\ &= x^{l+2(n-1)}e^{-i\alpha(0)/x}c(x),\ c \in C^\infty([0, 1)),\ c(0) = a(0, 0, 0)\phi(0).\end{aligned}$$

Thus the 'initial conditions' to give the correct incoming coefficient are

$$l = -(n-1),\ \alpha(0) = \lambda,\ a(0, 0, 0) = 1.$$

Moreover, using (3) in applying the Laplacian to v and using integration by parts shows that $(\Delta - \lambda^2)v = g$ is of the same form. If the coordinates are chosen to be Riemannian normal coordinates in the boundary and to first order off the boundary then

$$\begin{aligned} g &= (2\pi)^{-n+1} \int \exp(i\frac{y \cdot u}{x^2} - i\frac{\alpha(u)}{x})x^l b(x, y, u)du, \\ b(x, y, u) &= ((\alpha + 2u \cdot \partial_u \alpha)^2 + |u|^2 - \lambda^2)a + x^2 b',\ b'\ C^\infty. \end{aligned} \tag{8}$$

Thus the 'eikonal equation' for α is radial in the u variables and can be written in polar coordinates in the form

$$(r^{-\frac{1}{2}}\alpha(r))' = \frac{(\lambda^2 - r^2)^{\frac{1}{2}}}{2r^{\frac{3}{2}}}.$$

This has a unique smooth solution in $|u| < \lambda$ with $\alpha(0) = \lambda$.

Choosing α to be this function removes the leading term in (8). The next term, vanishing quadratically with respect to the first, is given by the transport equation and can be made to vanish by demanding that

$$u \cdot \partial_u a(0, 0, u) = 0 \text{ near } u = 0.$$

Here the constant term has been removed by the choice of $l = -(n-1)$. This also has a unique smooth solution with $a(0, 0, 0) = 1$. The presence of constant terms means that lower order transport equations have unique solutions which are smooth at $u = 0$. Thus, at an essentially formal level it can be arranged that b in (8) vanishes to all orders at $x = 0$.

There is an obstruction to continuing the solution in this form, in the singularity in the phase α which occurs at $|u| = \lambda$. Thus in practice it is necessary to insert a cut off function in a to restrict the support to $|u| \leq R < \lambda$. This leads to an error term $(\Delta - \lambda^2)v = g$ with g of the form (8) for $b = x^2 b''$ and b'' supported in an annular region $R - \delta < |u| < R + \delta$ where δ can be chosen appropriately small.

This failure amounts to the appearance of a 'caustic' for the Legendre manifold involved. Rather than deal directly with the invariance properties of oscillatory integrals of the form in the ansatz it is convenient to use the locality of the construction (so far) and to blow up the base point. Thus, instead of considering the solution on X it will be lifted to

$$[X, \{\bar{y}\}]$$

which is just X with the initial point blown up. In the local coordinates introduced above, in which $\bar{y} = 0$, this is simply the introduction of polar coordinates in the variables (x, y),

$$(x, y) = \rho(\omega_0, \omega'), \ \omega \in \mathbb{S}^{n-1}, \ \omega_0 \geq 0.$$

Thus the blown up space is a compact manifold with corner of codimension two. However, the initial parametrix is rapidly decreasing near the 'old' boundary (defined by $\omega_0 = 0$) and so attention can be limited to a neighbourhood of the interior of the new face $\{\rho = 0\}$ introduced by the blow up. In this region the simpler projective coordinates

$$x, Y = \frac{y}{x}$$

are valid.

A straightforward computation shows that in these (singular) coordinates the ansatz above gives a Legendre distribution in the sense of [10] and the Laplacian becomes an elliptic scattering differential operator in the sense of [6]. The metric of which it is the Laplacian is not quite a 'scattering metric' in the sense of [6] however, since it does not have the splitting property demanded there. It is still a smooth non-degenerate metric on the scattering tangent bundle. As a result the construction of the Poisson operator given in [10] can be mimicked to give the Poisson operator here, as a Legendre distribution. The structural difference in the metric leads to a difference in the behaviour of the geodesic flow and hence the Legendre manifold. Namely all the rescaled bicharacteristics starting from the point $Y = 0$ on the incoming radial set have an end point at $Y = 0$ on the outgoing radial set. It is not the case that the flow remains in the fibre above $Y = 0$, as it does in the original space, rather there is a natural focussing effect.

The construction of the Poisson operator as a Legendre distribution and the geometry of this Legendre manifold leads directly to the conclusion that the scattering matrix is a pseudodifferential operator, just as in [10] it is seen to be a Fourier integral operator.

References

[1] C.L. Epstein and R.B. Melrose and G. Mendoza, Resolvent of the Laplacian on strictly pseudoconvex domains, *Acta Math.* **167** (1991), 1–106.

[2] L. Hörmander, Fourier integral operators I, *Acta Math.* **127** (1971), 79–183.

[3] R. Mazzeo and R.B. Melrose, *Pseudodifferential operators on manifolds with fibred boundaries*, in preparation.

[4] R.B. Melrose, Calculus of conormal distributions on manifolds with corners, *Int. Math. Res. Notes* **21** (1992), 67–77.

[5] ——— *The Atiyah-Patodi-Singer index theorem*, A. K. Peters, 1993, Wellesley, Mass.

[6] ——— *Spectral and scattering theory for the Laplacian on asymptotically Euclidian spaces*, Spectral and scattering theory, M. Ikawa, 1994, Marcel Dekker.

[7] ——— *Geometric Scattering*, Cambridge University Press, 1995.

[8] ——— Fibrations, compactifications and algebras of pseudodifferential operators, *Partial Differential Equations and Mathematical Physics. The Danish-Swedish Analysis Seminar, 1995*, 246–261, 1996, Birkhäuser, editors Lars Hörmander and Anders Melin.

[9] ——— Differential analysis on manifolds with corners, in preparation.

[10] R.B. Melrose and M. Zworski, Scattering metrics and geodesic flow at infinity, *Invent. Math.* **124** (1996), 389–436.

Massachusetts Institute of Technology
e-mail: rbm@math.mit.edu

Hypoellipticity for a class of infinitely degenerate elliptic operators

Yoshinori Morimoto and Tatsushi Morioka

1. Introduction

In this article we shall study the hypoellipticity of second order elliptic operators with infinite degeneracy. To explain our motivation of the study we start with the classical well-known result given by Hörmander [8]: Let L be a differential operator of the form,

$$L = -\sum_{j=1}^{m} X_j^2 + X_0,$$

where X_j are real vector fields in $\mathbf{R}^n$. Then L is hypoelliptic in $\mathbf{R}^n$ if the following Lie bracket condition is satisfied;

$$(C.H) \qquad \begin{cases} X_0, X_1, \dots, X_m \text{ and their repeated} \\ \text{commutators of finite length span the tangent} \\ \text{space } T_x(\mathbf{R}^n) \text{ at any point } x \in \mathbf{R}^n \,. \end{cases}$$

The condition $(C.H)$ admits only the finite degeneracy of the operator and does not cover the infinitely degenerate case such as the following example:

$$A_0 = D_1^2 + \psi(x_1)^2 D_2^2, \quad D_j = -i\frac{\partial}{\partial x_j},$$

where $\psi(t) \in C^\infty(\mathbf{R}), \psi(t) > 0 \ (t \neq 0)$, $\psi^{(j)}(0) = 0$ for any $j = 0, 1, 2, \dots$. In 1971 Fediĭ [3] proved, as a pioneer in the study of hypoelliptic operators with infinite degeneracy, that the operator A_0 is hypoelliptic in $\mathbf{R}^2$. About this example we have the following extension:

Theorem 0. *Let* $A = D_1^2 + g(x_1)D_2^2$, $g(x_1) \in C^\infty$ *and assume*

$$g_I = \frac{1}{|I|}\int_I g(t)dt > 0 \text{ for any interval } I \subset \mathbf{R}_t,$$

where $|I|$ *denotes the length of* I. *Then* A *is hypoelliptic in* $\mathbf{R}^2$ (cf. Theorem 6.1 of [30]).

Another important result was given by Kusuoka-Stroock [17] in 1985, who studied the Malliavin calculus (the theory of stochastic differential equations) and proved as an application of their theory that the operator L_0 of the form

$$L_0 = D_1^2 + \psi(x_1)^2 D_2^2 + D_3^2 \ (= A_0 + D_3^2)$$

is hypoelliptic in $\mathbf{R}^3$ if $\psi(t)$ satisfies

$$(K.S) \qquad \lim_{t \to 0} t |\log \psi(t)| = 0$$

and moreover that the condition $(K.S)$ is necessary for the hypoellipticity if $\psi(t)$ satisfies $t\psi'(t)$, that is, $\psi(t)$ is monotone in each half axis. The $(K.S)$ condition is verified for $\psi(t) = \exp(-|t|^{-\sigma})$, $\sigma > 0$, if and only if $\sigma < 1$. We remark that other works of probabilistic methods were studied by Gaveau-Moulinier [7] and recently by Bell-Mohammed [2].

Inspired by Kusuoka-Stroock's result for L_0, one of the authors (Y.M) [19-22] studied the hypoellipticity for a class of operators containing L_0 by the method based on the logarithmic regularity up estimates, instead of subelliptic estimates in the finite degenerate case, and extended their result to the higher order case. In fact, it was shown in [22] that a second order differential operator P with real valued coefficients in $C^\infty(\mathbf{R}^n)$ is hypoelliptic in $\mathbf{R}^n$ if the principal symbol $p_2(x, \xi)$ of P is non-negative, that is, $p_2(x, \xi) \geq 0$ for any $(x, \xi) \in \mathbf{R}^{2n}$ and if

$$(*) \qquad \begin{cases} \forall \varepsilon > 0,\ \forall K \text{ compact} \subset \mathbf{R}^n,\ \exists C_{\varepsilon,K} > 0 \text{ such that} \\ \|(\log \Lambda)^2 u\| \leq \varepsilon \|Pu\| + C_{\varepsilon,K}\|u\| \\ \text{for } \forall u(x) \in C_0^\infty(K), \ \text{ where } \Lambda = (2 + |D_x|^2)^{1/2}. \end{cases}$$

After the works [19-22], methods by means of logarithmic regularity up estimates have been improved and various kinds of hypoelliptic operators with infinite degeneracy have been studied by Hoshiro [10-14], Kajitani-Wakabayashi [15], Koike [16], Suzuki [38], Wakabayashi-Suzuki [40], Taira [39] and the authors [25-28], [29-30], [31-34]. In particular, by using the Fourier integral operators with the complex phase, Wakabayashi-Suzuki [40] in 1993 gave fairly general sufficient conditions which cover the preceding results. However, their sufficient conditions seem too involved and are described as the pointwise conditions of the symbol of operators under specifying the characteristic structure of operators.

The first aim of the present article is to give simple sufficient conditions by means of energy estimates without specifying the characteristic structure of operators and to discuss how to verify such estimates for various examples of infinitely degenerate elliptic operators, in view of the average of the coefficient of operators as in Theorem 0. In fact, if we consider the uncertainty principle mathematically discussed in Fefferman-Phong [4], [6], the pointwise conditions for the symbol of operators should be replaced by some kind of stochastic average of the symbols. The second aim is to extend the criterion of the hypoellipticity for second order differential operators with the principal symbol of variable sign, studied by Zuily [41] and Lanconelli [18].

We shall state some criteria (Theorem 1-4) for the micro-hypoellipticity of the second order pseudodifferential operators in Section 2. In Section 3 we shall give some examples of hypoelliptic operators as applications of criteria. In Section 4 we shall show how to apply criteria to examples. The proofs of the theorems and the details of the applications will be given elsewhere. Some part of this research was completed during a stay at Paris VI University by one of the authors (Y.M), who would like to thank Prof. Vaillant and Prof. Chemin for their hospitality. Furthermore, the authors would like to express their hearty gratitude to Prof. Zuily for useful discussions and invaluable comments.

2. Criteria for hypoellipticity

Let $P(x,\xi)$ be a classical symbol in $S^2_{1,0} \equiv S^2_{1,0}(T^*\mathbf{R}^n)$ and let $p_j(x,\xi)$ denote the jth homogeneous part in ξ. We assume the principal symbol p_2 to be real valued. We define a pseudodifferential operator of $P^w(x,D)$ with a Weyl symbol $P(x,\xi)$ by

$$P^w(x,D)u = (2\pi)^{-n}\int e^{i(x-y)\cdot\xi}P((x+y)/2,\xi)u(y)dyd\xi, \quad u\in\mathcal{S}$$

and we assume $P^w(x,D)$ is properly supported. We refer to Chapter 18 of Hörmander [9] about the Weyl calculus of pseudodifferential operators. Let $h(x)$ be a $C_0^\infty(\mathbf{R}^n)$ function such that $0\le h\le 1$, $h(x)=1$ for $|x|\le 1/5$ and $h(x)=0$ for $|x|\ge 7/24$. For a $\delta>0$ we set $h_\delta(x)=h(x/\delta)$ and $H_\delta(x,\xi;\lambda)=h_\delta(x-x_0)h_\delta(\lambda\xi-\xi_0)$, where $0<\lambda\le 1$ is a parameter. Let δ_0 be a fixed positive real smaller than $1/100$. For a parameter $0<\lambda\le 1$, we define a microlocalized operator $P^w_\lambda(x,D)=(P_\lambda)^w$ of $P^w(x,D)$ at $\rho_0=(x_0,\xi_0)$ with $|\xi_0|=1$ by the symbol

$$P_\lambda(x,\xi)=h_{\delta_0}(x-x_0)P(x,\xi)h_{\delta_0}(\lambda\xi-\xi_0).$$

Similarly, for homogeneous symbols p_2 and $\operatorname{Re} p_1$ we define $(p_2)^w_\lambda$ and $(\operatorname{Re} p_1)^w_\lambda$, respectively. Here and in what follows we denote by $a(x,D)$

a usual pseudodifferential operator with a symbol $a(x,\xi)$. Let $(\cdot,\cdot)$ and $\|\cdot\|$ denote the inner product and the norm in $L^2(\mathbf{R}^n_x)$, respectively. Furthermore, let $\|\cdot\|_s$ denote the norm of the Sobolev space H_s.

Theorem 1. *Let $\rho_0 = (x_0,\xi_0) \in T^*(\mathbf{R}^n)$ with $|\xi_0| = 1$. Let $P^w(x,D)$ and $P^w_\lambda(x,D)$ be above operators with a fixed $\delta_0 > 0$. Assume the principal symbol of P^w is non-negative, that is,*

$$p_2(x,\xi) \geq 0 \text{ for any } (x,\xi) \in \mathbf{R}^{2n} \tag{1}$$

and that

$$\left\{\begin{array}{l} \text{there exist constants } 0 < \gamma < 1 \text{ and} \\ C_0 > 0 \text{ such that for any } 0 < \lambda \leq 1 \\ ((\gamma(p_2)^w_\lambda + (\mathrm{Re}\ p_1)^w_\lambda)u, u) \geq -C_0\|u\|^2 \quad \text{for any} \quad u \in \mathcal{S}(\mathbf{R}^n). \end{array}\right. \tag{2}$$

Furthermore, we assume that there exists a positive sequence $\{\delta_j\}_{j=1}^\infty$ tending to 0 and satisfying the following property: for each small $\delta = \delta_j$ there exists a non-negative symbol $\varphi(x,\xi;\lambda) \in S^0_{0,0}$ such that

$$\left\{\begin{array}{l} \{\lambda^{-|\alpha|}\partial_\xi^\alpha \partial_x^\beta \varphi(x,\xi;\lambda); 0 < \lambda \leq 1\} \\ \text{is a bounded set of } S^0_{0,0} \text{ for any } \alpha, \beta\ , \end{array}\right. \tag{3}$$

$$\varphi \geq 1 \text{ outside of } \mathrm{supp}\ H_{5\delta}(x,\xi;\lambda), \tag{4}$$

$$\varphi = 0 \text{ on } \mathrm{supp}\ H_\delta(x,\xi;\lambda), \tag{5}$$

and the following estimate holds: For any $\varepsilon > 0$ there exists a $C_\varepsilon > 0$ independent of $0 < \lambda \leq 1$ such that

$$\begin{aligned} \|u\|^2 &+ (\log\lambda)^2\|(H^2_\varphi p_2)(x,D;\lambda)u\|^2 \\ &+ (\log\lambda)^2|\mathrm{Re}\ ((H^2_\varphi p_2)(x,D;\lambda)u,u)| \\ &+ |\log\lambda|\ |\mathrm{Re}\ ((H_\varphi(\mathrm{Im}\ p_1))(x,D;\lambda)u,u)| \\ &\leq \varepsilon\left\{\mathrm{Re}(P^w_\lambda u,u) + \left\|\left(\frac{\lambda}{|\log\lambda|}\right)^{1/3}(\mathrm{Im}\ p_1)^w_\lambda u\right\|^2\right\} \\ &\quad +C_\varepsilon\{\lambda\|u\|^2 + \lambda^{-2}\|(1 - H_{20\delta}(x,D;\lambda))u\|^2_2\} \end{aligned} \tag{6}$$

for any $u \in \mathcal{S}(\mathbf{R}^n)$*, where* H_φ *denotes the Hamilton vector field of* φ*. Then we have*

$$\rho_0 \notin WF(P^w u) \text{ implies } \rho_0 \notin WF(u) \text{ for } \forall u \in \mathcal{D}'(\mathbf{R}^n). \tag{7}$$

Corollary 1. *Assume that there exists a constant* $C > 0$ *such that for any* $0 < \lambda \leq 1$ *and any* $u \in \mathcal{S}(\mathbf{R}^n)$

$$\begin{aligned} ||u||^2 &\leq C\{\text{Re } (P_\lambda^w u, u) \\ &+ \lambda||u||^2 + \lambda^{-2}||(1 - H_{20\delta}(x, D; \lambda))u||_2^2\}. \end{aligned} \tag{8}$$

Then the conclusion of Theorem 1 is still valid even if we remove the first term $||u||^2$ *of the left-hand-side of* (6).

Corollary 2 (Theorem 8 of [30]). *The same conclusion as Theorem 1 holds without assumptions* (1) *and* (2) *if, instead of the estimate* (6)*, we have for any* $u \in \mathcal{S}(\mathbf{R}^n)$

$$\begin{aligned} ||u||^2 &+ (\log\lambda)^2|\text{Re}((H_\varphi^2 p_2)(x, D; \lambda)u, u)| \\ &+ |\log\lambda|\ |\text{Re } ((H_\varphi(\text{Im } p_1))(x, D; \lambda)u, u)| \\ &\leq \varepsilon \text{ Re } (P_\lambda^w u, u) \\ &+ C_\varepsilon\{\lambda||u||^2 + \lambda^{-2}||(1 - H_{20\delta}(x, D; \lambda))u||_2^2\}. \end{aligned} \tag{9}$$

Remark 1. The parameter λ^{-1} corresponds to Λ. The weight $(\lambda/|\log\lambda|)^{1/3}$ of the second term of the right-hand-side of (6) was introduced by Wakabayashi-Suzuki [40] (see Lemma 4.6 of [40]), which improved the classical one, $\lambda^{1/2}$ (cf., for example, (2.6.32) of [35]).

Remark 2. It follows from condition (2) that

$$\begin{aligned} ((1-\gamma)(p_2)_\lambda^w u, u) &\leq ((p_2)_\lambda^w + (\text{Re } p_1)_\lambda^w)u, u) + C||u||^2 \\ &\leq \text{Re } (P_\lambda^w u, u) + C'||u||^2. \end{aligned}$$

If we set $p_2^{(j)} = \partial_{\xi_j} p_2$ and $p_{2,(j)} = \partial_{x_j} p_2$ then it follows from (1) that

$$|(p_2^{(j)})_\lambda(x,\xi)|^2 + |(p_{2,(j)})_\lambda(x,\xi)|^2 \leq C_0 (p_2)_\lambda(x,\xi) \text{ for } |\xi| = 1.$$

Since it follows from the expansion formula of Weyl calculus (see Theorem 18.5.4 of [9]) that $a^w(x,D)^2 - (a^2)^w(x,D)$ is L^2 bounded for $a(x,\xi) \in S^1_{1,0}$, by means of Fefferman-Phong inequality (see [5] and Theorem 18.6.8 of [9]) we see that

$$
\tag{10}
\begin{aligned}
&\sum_{j=1}^{n} \left(||(p_2^{(j)})^w_\lambda u||^2 + ||(\lambda p_{2,(j)})^w_\lambda u||^2 \right) \\
&\quad \le C_1 \left(((p_2)^w_\lambda u, u) + ||u||^2 \right) \le C_2 \left(\text{ Re } (P^w_\lambda u, u) + ||u||^2 \right).
\end{aligned}
$$

Hence the first parenthetical term of the right-hand-side of (6) in Theorem 1 can be replaced by

$$
\varepsilon \left\{ \sum_{j=1}^{n} \left(||(p_2^{(j)})^w_\lambda u||^2 + ||\lambda((p_{2,(j)})^w_\lambda u||^2 \right) + || \left(\frac{\lambda}{|\log \lambda|} \right)^{1/3} (\text{Im } p_1)^w_\lambda u||^2 \right\}.
$$

In order to explain that the above term can be replaced by that of more general form, we introduce some notation: set

$$q_0(x,\xi) = -\text{Im}p_1(x,\xi),$$

$$q_j(x,\xi) = p_2^{(j)}(x,\xi) \text{ for } j = 1, ..., n$$

$$q_j(x,\xi) = \lambda p_{2,(j-n)}(x,\xi) \text{ for } j = n+1, ..., 2n$$

and for a multi-index $J = \{j_1, j_2, ..., j_k\}$ with $j_i \in \{0, 1, ..., 2n\}$

$$q_J = H_{q_{j_1}} \cdots H_{q_{j_{k-1}}} q_{j_k},$$

$$|J| = \sum_{i=1}^{k} \max\{1, 2 - j_i\}, \quad ||J|| = \sum_{i=1}^{k} \max\{0, 1 - j_i\},$$

$$
\begin{cases}
\varepsilon_J = 2^{1-|J|}, \ \delta_J = 0 \text{ if } J \text{ satisfies } |J| = k \\
\varepsilon_J = 2^{3-|J|}/3, \ \delta_J = (2^{2-|J|} + 1 - 4^{1-||J||})/3 \text{ otherwise.}
\end{cases}
$$

Theorem 2. *Let M be a positive integer. The same conclusion as Theorem 1 holds even if the first parenthetical term of the right-hand-side of* (6) *is replaced by*

$$
\tag{11}
\varepsilon \left\{ \sum_{|J| \le M} || \frac{\lambda^{1-\varepsilon_J}}{|\log \lambda|^{\delta_J}} (q_J)^w_\lambda u||^2 + ||u||^2 \right\}.
$$

Furthermore, we can remove the first term $||u||^2$ *of the left-hand-side of the hypothetical estimate if the estimate* (8) *holds* (cf., Theorem 4.9 and its remark of [40]).

Remark. The weight in the formula (11) derived from indices ε_J and δ_J is due to Wakabayashi-Suzuki (see Definition 4.4 of [40]), which is also the improvement of the preceding results (cf., Lemma 2.6.4 of [35]). Assume the Olejnik-Radkevich type condition

$$(O.R) \qquad \sum_{|J|\le M} |q_J(x_0,\xi_0)|^2 \neq 0 \text{ for } \exists M > 0,$$

which is a generalization of $(C.H)$ condition. Then the hypothetical estimate of the theorem holds. In fact, this is a direct consequence of the Fefferman-Phong inequality and the fact that if $\delta > 0$ and $\lambda > 0$ are small then we have

$$\sum_{|J|\le M} \left| \frac{\lambda^{1-\varepsilon_J}}{|\log \lambda|^{\delta_J}} (q_J)_\lambda(x,\xi) \right|^2 \ge (\log \lambda)^5 \quad \text{on } \operatorname{supp} H_{20\delta}(x,\xi;\lambda).$$

If the finite type condition such as $(C.H)$ or $(O.R)$ is not satisfied, it is not easy to verify the hypothetical estimate of Theorem 1 or 2. Even for a simple example

$$(12) \qquad \psi(x_1)^2 D_2^2 + iD_1$$

with the same $\psi(t) \in C^\infty(\mathbf{R})$ as in the introduction, it seems not to be obvious to show

$$||u||^2 \le \varepsilon \left\{ \operatorname{Re}\,(P_\lambda^w u, u) + \sum_{|J|\le M} \left|\left| \frac{\lambda^{1-\varepsilon_J}}{|\log \lambda|^{\delta_J}} (q_J)_\lambda^w u \right|\right|^2 \right\}$$
$$+ \; C_\varepsilon\{\lambda ||u||^2 + \lambda^{-2} ||(1 - H_{20\delta}(x, D; \lambda))u||_2^2\}.$$

In view of this example, we shall give another criterion by specifying the characteristic structure of P^w. In what follows we confine ourselves to the case where P^w is a differential operator.

Theorem 3. *Let* $\rho_0 = (x_0, \xi_0) \in T^*(\mathbf{R}^n)$ *with* $|\xi_0| = 1$. *Let* $P^w(x, D)$ *be a second order differential operator satisfying* (1) *and*

$$(13) \qquad \operatorname{Re}\, p_1(x,\xi) = 0.$$

Assume that

$$(14) \qquad p_2(x_0,\xi) \equiv 0 \quad \textit{as a function of } \xi$$

and

(15) $\exists\zeta_0 \in \mathbf{R}^n \setminus 0$ *such that* $q(x_0, \zeta_0) \equiv -\mathrm{Im}\, p_1(x_0, \zeta_0) \neq 0.$

Furthermore, assume that there exists a sequence $\{\delta_j\}$ *and symbols* $\varphi(x, \xi; \lambda)$ *for each* $\delta = \delta_j$ *which are the same as in Theorem 1 except that the estimate* (6) *for any* $u \in \mathcal{S}(\mathbf{R}^n)$ *is replaced by*

$$
\begin{aligned}
&(\log\lambda)^4 ||(H_\varphi^2 p_2)(x, D; \lambda)u||^2 \\
&+(\log\lambda)^2 ||(H_\varphi(\mathrm{Im}\, p_1))(x, D; \lambda)u||^2 \\
&+(\log\lambda)^2 \sum_{j=1}^{n} ||((H_\varphi p_2)^{(j)})(x, D; \lambda)u||^2 \qquad (16)\\
&\quad \le \varepsilon \left\{ \mathrm{Re}(P_\lambda^w u, u) + \left|\left| \left(\frac{\lambda}{|\log\lambda|}\right)^{1/3} (\mathrm{Im}\, p_1)_\lambda^w u \right|\right|^2 + ||u||^2 \right\} \\
&\quad +C_\varepsilon \{\lambda ||u||^2 + \lambda^{-2} ||(1 - H_{20\delta}(x, D; \lambda))u||_2^2\}.
\end{aligned}
$$

Then we have the same conclusion (7) *as in Theorem 1. Furthermore the conclusion holds even if the first parenthetical term of the right-hand-side of* (16) *is replaced by* (11).

Remark. Set $U_\sigma = \{\, x \,;\, |x - x_0| < \sigma\}$ for $\sigma > 0$ and $W_\zeta(x) = \exp\{(x - x_0) \cdot \zeta\}$ for $\zeta \in \mathbf{R}^n$. It follows from assumptions (1) and (13-15) that for any $\varepsilon > 0$ there exist $\zeta = \zeta(\varepsilon)$ and $\sigma = \sigma(\varepsilon)$ such that $||u||^2 \le \varepsilon\, \mathrm{Re}\,(P^w u, W_\zeta(x)u)$ for any $u \in C_0^\infty(U_\sigma)$. Hence if $0 < \lambda \le 1$ we have for any $u \in C_0^\infty(U_\sigma)$

$$
\begin{aligned}
||u||^2 \;\le\;& \varepsilon\, \mathrm{Re}\,(P_\lambda^w u, W_\zeta(x)u) \qquad (17)\\
&+\; C_\varepsilon \{\lambda ||u||^2 + \lambda^{-2} ||(1 - H_{20\delta}(x, D; \lambda))u||_2^2\}.
\end{aligned}
$$

Theorem 3 is still valid even if the estimate (16) for $u \in \mathcal{S}(\mathbf{R}^n)$ is replaced by a weaker estimate for $u \in C_0^\infty(U_\sigma)$ as follows:

$$(\log\lambda)^2\Big\{|\mathrm{Re}\,((H_\varphi^2 p_2)(x,D;\lambda)u,u)|$$

$$+\,|\mathrm{Re}\,((H_\varphi^2 p_2)(x,D;\lambda)u,W_\zeta(x)u)| + ||(H_\varphi^2 p_2)(x,D;\lambda)u||^2\Big\}$$

$$+(\log\lambda)^2\sum_{j=1}^n ||(H_\varphi p_2)^{(j)})(x,D;\lambda)u||^2$$

$$+|\log\lambda|\,\{|\mathrm{Re}\,((H_\varphi(\mathrm{Im}\,p_1))(x,D;\lambda)u,u)| \tag{18}$$

$$+\,|\mathrm{Re}\,((H_\varphi(\mathrm{Im}\,p_1))(x,D;\lambda)u,W_\zeta(x)u)|\}$$

$$\leq \varepsilon\left\{\mathrm{Re}(P_\lambda^w u,u) + ||\left(\frac{\lambda}{|\log\lambda|}\right)^{1/3}(\mathrm{Im}\,p_1)_\lambda^w u||^2 + ||u||^2\right\}$$

$$+C_\varepsilon\{\lambda||u||^2 + \lambda^{-2}||(1-H_{20\delta}(x,D;\lambda))u||_2^2\}.$$

We can also replace the first parenthetical term of the right-hand-side of (18) by (11).

Our method can be applicable to second order differential operators whose principal symbols p_2 do not always satisfy (1). It is known by Theorem 2.3.1 of [35] that if P^w is hypoelliptic in an open set $\Omega \subset \mathbf{R}^n$ then

$$\begin{cases} \text{for any fixed point } x_0 \in \Omega \text{ we have} \\ \text{either } p_2(x_0,\xi) \geq 0 \text{ or } p_2(x_0,\xi) \leq 0 \text{ for all } \xi \in \mathbf{R}^n. \end{cases} \tag{19}$$

In fact, there exists a $\xi_0 \in \mathbf{R}^n \setminus 0$, otherwise, such that $p_2(x_0,\xi_0) = 0$ and $d_\xi p_2(x_0,\xi_0) \neq 0$, so that $P^w(x,D)$ is equivalent to the operator of real principal type, microlocally near (x_0,ξ_0). In view of (19), we shall consider a second order differential operator $P^w(x,D)$ whose symbol is of the form

$$P(x,\xi) = \psi(x)^m p_2^0(x,\xi) - iq(x,\xi) + p_0(x), \tag{20}$$

where $m \geq 3$ is odd and

$$\psi(x) \in C^\infty,\ \text{real valued},\ d_x\psi(x) \neq 0 \text{ on } \psi^{-1}(0), \tag{21}$$

$$p_2^0(x,\xi) \geq 0,\ \in S_{1,0}^2,\ \text{homogeneous oreder 2 in } \xi, \tag{22}$$

(23) $q(x,\xi)$ real valued, $\in S^1_{1,0}$, homogeneous of order 1 in ξ,

(24) $p_0(x)$ complex valued, $\in C^\infty$.

The operator of the form (20) was introduced by Lanconelli [18], who extended the sufficient condition for the hypoellipticity of second order differential operators with the principal symbol of variable sign, studied by Zuily [41; Part III]. It should be noted, of course, that the form (20) is particular and excludes a simple hypoelliptic operator $x_1 D_2^2 - iD_1$ in $\mathbf{R}^2$ (which is not micro-hypoelliptic on $\{x_1 = 0\}$, see Parenti-Rodino [36] for instance), though our interest in the present article is confined to the micro-hypoellipticity. Here we say $P^w(x,D)$ is micro-hypoelliptic at $\rho \in T^*(\mathbf{R}^n)\setminus 0$ if there exists a conic neighborhood V of ρ satisfying

$$\mathrm{WF}(P^w u)\cap V = \mathrm{WF}(u)\cap V \text{ for any } u \in \mathcal{D}'(\mathbf{R}^n).$$

We remark that Beals-Fefferman [1] discussed a class of hypoelliptic operators including this example. As an extension of Theorem 3.1 and 3.2 of [18], we have

Theorem 4. *Let* $\Omega^\pm = \{\, x\,; \pm\psi(x) > 0\}$ *and let* $\rho_0 = (x_0,\xi_0) \in T^*(\mathbf{R}^n)$ *with* $|\xi_0| = 1$ *and* $x_0 \in \overline{\Omega_+}\cap\overline{\Omega_-}$. *Let* $P^w(x,D)$ *be a second order differential operator whose symbol has the form of* (20) *satisfying* (21 – 24). *Assume* (13) *and*

(25) $$q(x, d_x\psi) \geq 0 \text{ on } \psi^{-1}(0).$$

Furthermore, assume that there exist a sequence $\{\delta_j\}$ and symbols $\varphi(x,\xi;\lambda)$ for each $\delta = \delta_j$ which are the same as in Theorem 1 except that the estimate (6) for any $u \in \mathcal{S}(\mathbf{R}^n)$ is replaced by

$$\begin{aligned}
&(\log\lambda)^4 ||(H_\varphi^2 p_2)(x,D;\lambda)u||^2 \\
&+(\log\lambda)^2 \sum_{j=1}^{n} ||((H_\varphi p_2)^{(j)})(x,D;\lambda)u||^2 \\
&+(\log\lambda)^2 ||(H_\varphi(\mathrm{Im}\ p_1))(x,D;\lambda)u||^2 \qquad (26)\\
&\leq \varepsilon\left\{\sum_{|J|\leq M} ||\frac{\lambda^{1-\varepsilon_J}}{|\log\lambda|^{\delta_J}}(q_J)^w_\lambda u||^2 + ||u||^2\right\} \\
&+C_\varepsilon\{\lambda||u||^2 + \lambda^{-2}||(1-H_{2\delta}(x,D;\lambda))u||_2^2\},
\end{aligned}$$

where M is a positive integer. Then we have the same conclusion (7) as in Theorem 1.

Remark 1. Set $\Omega^{\pm} = \{\, x\,;\pm\psi(x) > 0\}$ and set for $\zeta \in \mathbf{R}^n$

$$W_\zeta^{\pm}(x) = \exp\{\pm(x - x_0)\cdot\zeta\},$$

$$W_\zeta(x) = W_\zeta^{+}(x)\chi_{+}(x) - W_\zeta^{-}(x)\chi_{-}(x),$$

where $\chi_{\pm}(x)$ are the characteristic functions of sets $\Omega^{\pm}$. It follows from Lemma 3.3 of [18] that for any $\varepsilon > 0$ there exist $\zeta = \zeta_\varepsilon$ and $\sigma = \sigma_\varepsilon$ such that we have (17) for any $u \in C_0^\infty(U_\sigma)$. Similarly as stated in the remark of Theorem 3, the assumption (26) can be replaced by the estimate for $u \in C_0^\infty(U_\sigma)$ as follows:

(27)

$$\begin{aligned}
&(\log\lambda)^2\sum_{k=1}^{3}|\mathrm{Re}\;((H_\varphi^2 p_2)(x,D;\lambda)u, A_k u)| \\
&+(\log\lambda)^2\left\{||(H_\varphi^2 p_2)(x,D;\lambda)u||^2 + \sum_{j=1}^{n}||(H_\varphi p_2)^{(j)})(x,D;\lambda)u||^2\right\} \\
&+|\log\lambda|\sum_{k=1}^{3}|\mathrm{Re}\;((H_\varphi(\mathrm{Im}\;p_1))(x,D;\lambda)u, A_k u)| \\
&\le \varepsilon\left\{\sum_{k=1}^{3}\mathrm{Re}(P_\lambda^w u, A_k u) + \sum_{|J|\le M}||\frac{\lambda^{1-\varepsilon_J}}{|\log\lambda|^{\delta_J}}(q_J)_\lambda^w u||^2 + ||u||^2\right\} \\
&+C_\varepsilon\{\lambda||u||^2 + \lambda^{-2}||(1 - H_{20\delta}(x,D;\lambda)u||_2^2\},
\end{aligned}$$

where $A_1 = \chi_+(x) - \chi_-(x)$, $A_2 = \psi(x)^{m-2}$ and $A_3 = W_\zeta(x)$.

Remark 2. The estimate (26) follows from condition $(O.R)$, as stated in the remark of Theorem 2. Hence Theorem 4 is an extension of Theorem 3.1 (and moreover Theorem 3.2) of [18]. (In fact, if $(O.R)$ is satisfied with x_0 replaced by any $x \neq x_0$ in a neighborhood of x_0 then we have (26) because $H_\varphi^2 p_2, (H_\varphi p_2)^{(j)}$ and $H_\varphi(\mathrm{Im}\;p_1)$ vanish on supp $H_\delta(x,\xi;\lambda)$ by means of (5).)

3. Applications of criteria

As an application of Theorem 1 we shall first consider the parabolic version of the modified Kusuoka-Stroock operator studied in Theorem 1 of [30]. Let L_1 be a differential operator of the form

(28) $$L_1 = D_1^2 + g(x_1)D_2^2 + if(x_1)D_3 \text{ in } \mathbf{R}^3,$$

where $f, g \in C^\infty$ and

$$(A.1) \qquad f_I, g_I > 0 \text{ for any interval } I.$$

For $j = 1, 2$ and a finite interval $I_0 \in \mathbf{R}$ we say that f and g satisfy the condition $(M, f, g)_j$ in I_0 if

$$\inf_{\delta>0} \left(\sup\{(f_I^{1/2}|I|)^j |\log g_{3I}| \ ; \ I \text{ with } 3I \subset I_0 \text{ and } g_{3I} < \delta\}\right) = 0,$$

where $3I$ denotes the interval with the same center as I but with length three times that of I.

Proposition 1. *Let I_0 be an open finite interval in $\mathbf{R}_{x_1}$ such that $f, g > 0$ on ∂I_0. Let L_1 be the above operator and set $\Omega = I_0 \times \mathbf{R}^2_{x'}$, $x' = (x_2, x_3)$. Then the operator L_1 is micro-hypoelliptic at all $(\tilde{x}, \tilde{\xi}) \in T^*(\Omega)$ with $\tilde{\xi} = (0, \pm 1, 0)$ if and only if f and g satisfy $(M, f, g)_2$. Furthermore $(M, g, f)_1$ in I_0 is sufficient in order that L_1 is micro-hypoelliptic at any $(X, \Xi) \in T^*(\Omega)$ with $\Xi = (0, 0, \pm 1)$.*

Remark 1. $(M, f, g)_j$ is equivalent to $(M, f, g^r)_j$ for any $r > 1$ because it follows from the Hölder inequality that

$$g_I \le ((g^r)_I)^{1/r} \le ((\max_{I_0} |g|^{r-1}) g_I)^{1/r}.$$

Hence the condition $(M, f, g)_j$ is verified if $f = g^r$ for some $r > 0$, and even if $f^{-1}(0)$ has the positive Lebesgue measure in $\mathbf{R}$ (see Example 3.1 of [30]).

Remark 2. If f and g satisfy

$$f(t), g(t) > 0 \text{ for } t \neq 0, \quad tf'(t), tg'(t) \ge 0$$

then the condition $(M, f, g)_j$ is equivalent to

$$(K, f, g)_j \qquad \begin{cases} \lim_{t \to 0} \mu(f; t)^j |\log g(t)| = 0, \\ \mu(f; t) = \sup_{0 \le \pm z \le \pm t} f(z)^{1/2} |t - z| \text{ if } \pm t \ge 0, \end{cases}$$

which was introduced by Koike [16]. We refer to Theorem 5 of [30] about the proof of the equivalence and the further comments on $(M, f, g)_j$. A typical example of f, g given by [16] is

$$f = e^{-2/|t|}, \quad g = \exp(-|t|^{-j\sigma} e^{j/|t|}), \quad \sigma \in \mathbf{R}$$

for which $(K, f, g)_j$ is verified if and only if $\sigma < 2$.

By means of Theorem 2 we can show the following proposition which is an extension of Theorem 8 of [25] (cf., Example 5.5 of [40]).

Proposition 2. *Let L_2 be a differential operator of the form*

$$L_2 = D_1^2 + f(x_1)(D_2 + g(x_1)h(x_2)D_3)^2, \tag{29}$$

where $f, g, h \in C^\infty(\mathbf{R})$. Assume

$$f_I, |g'|_I, |h|_I > 0 \text{ for any interval } I. \tag{30}$$

If $f(t), g(t)$ and $h(t)$ satisfy $(M, f, f^2|g'|)_1$ and $(M, 1, |h|)_1$ in a finite open interval $I_0 \subset \mathbf{R}_t$ then L_2 is micro-hypoelliptic at any $\rho \in T^*(I_0^2 \times \mathbf{R}_{x_3})$.

Instead of (12) we shall consider a little more involved operator as follows:

$$L_3 = f(x_3)\left(D_1^2 + g(x_1)D_2^2\right) - iD_3, \tag{31}$$

where $f(t), g(t) \in C^\infty$ satisfy $(A.1)$. Furthermore let

$$L_4 = x_3^3 f(x_3)\left(D_1^2 + g(x_1)D_2^2\right) - iD_3. \tag{32}$$

Proposition 3. *Let L_3 and L_4 be the above differential operators with f, g satisfying $(A.1)$. If g satisfies $(M, 1, g)_1$ in a fixed open interval $I_0 \subset \mathbf{R}$ then L_3 and L_4 are micro-hypoelliptic at any $\rho \in T^*(\Omega)$, where $\Omega = I_0 \times \mathbf{R}^2_{x'}$, $x' = (x_2, x_3)$.*

The micro-hypoellipticity of L_3 (resp. L_4) can be proved by using Theorem 3 (resp. 4) and its remark.

4. Proofs of Propositions

Lemma 1. *Let I_0 be an open finite interval and let $j = 1, 2$. Assume that $f, g \in C_0^\infty(\mathbf{R})$ satisfy $(A.1)$. The condition $(M, f, g)_j$ in I_0 and the following two conditions are equivalent.*

$$(S, f, g)_j \quad \begin{cases} \text{For any } \varepsilon > 0 \text{ there exists a } \lambda_\varepsilon > 0 \text{ such that} \\ f_I(\log \lambda)^{2/j} \le \varepsilon(g_{3I}\lambda^{-2} + |I|^{-2}) \\ \text{for all } 0 < \lambda \le \lambda_\varepsilon \text{ and all } I \text{ with } 3I \subset I_0. \end{cases}$$

$$(E, f, g)_j \quad \begin{cases} \text{For any } \varepsilon > 0 \text{ there exists a } \lambda_\varepsilon > 0 \text{ such that} \\ (\log \lambda)^{2/j} \int f(t)|u(t)|^2 dt \\ \qquad \le \varepsilon \int \{|u'(t)|^2 + g(t)\lambda^{-2}|u(t)|^2\} dt \\ \text{for any } 0 < \lambda \le \lambda_\varepsilon \text{ and any } u \in C_0^1(I_0). \end{cases}$$

Proof. The equivalence between $(M,f,g)_j$ and $(S,f,g)_j$ can be proved in the almost same way as the proof of Lemma 3.1 of [30]. If we set $v(t) = f(t)(\log\lambda)^{2/j}/\varepsilon$ and $w(t) = g(t)\lambda^{-2}$, it is easy to see that $(S,f,g)_j$ and $(E,f,g)_j$ are equivalent, on account of the following lemma given by Sawyer (see Remark 5 of [37]). □

Lemma 2. *Let I_0 be an open interval in $\mathbf{R}^1_t$ and let $v(t), w(t) \geq 0$ belong to $L^1_{loc}(I_0)$. Then the estimate*

$$\int_{I_0} v(t)|u(t)|^2 dt \leq C\left\{\int_{I_0} w(t)|u(t)|^2 dt + \int_{I_0} |u'(t)|^2 dt\right\} \quad \text{for } \forall u \in C^1_0(I_0) \tag{33}$$

holds with a constant $C > 0$ if and only if

$$v_I \leq A\{w_{3I} + 2|I|^{-2}\} \text{ for any interval } I \text{ with } 3I \subset I_0 \tag{34}$$

holds with a constant $A > 0$. Moreover, if C and A are the best constants in (33) and (34), then

$$A \leq C \leq 100A. \tag{35}$$

We shall prove Proposition 1. Assume that $(M,f,g)_2$ is verified in $I_0 \subset \mathbf{R}$. We shall show the micro-hypoellipticity at $(\tilde{x},\tilde{\xi}) \in T^*(\Omega)$ with $\tilde{\xi} = (0,\pm 1, 0)$. Let $\rho_0 = (x_0,\xi_0) \in T^*(\Omega)$ with $|\xi_0| = 1$ belong to a small conic neighborhood of $(\tilde{x},\tilde{\xi})$. For the proof of the micro-hypoellipticity of L_1 at $(\tilde{x},\tilde{\xi})$ it suffices to show

$$\rho_0 \notin \mathrm{WF}(L_1 u) \text{ implies } \rho_0 \notin \mathrm{WF}(u) \text{ for } \forall u \in \mathcal{D}' \tag{36}$$

by assuming $\xi_0 = (0,\xi_{02},\xi_{03})$ with $\xi_{02} \neq 0$. In fact, $\xi_{01} \neq 0$ then (36) is obvious. For brevity we assume $x_0 = (x_{01},0,0)$ by the translation in $\mathbf{R}^2_{x'}$. Since $g^{-1}(0)$ is discrete, for any integer $j > 0$ there exist positive $\delta = \delta_j < 1/j$ and $\mu = \mu_j < 1/100$ such that

$$\{\, x_1\,; \delta(1-\mu) \leq 2|x_1 - x_{01}| \leq \delta\} \cap g^{-1}(0) = \emptyset. \tag{37}$$

For each $\delta = \delta_j$ we take $\varphi(x,\xi;\lambda) \in S^0_{0,0}$ as follows: Let $\chi(s)$ be a $C^\infty(\mathbf{R})$ function such that $0 \leq \chi \leq 1$, $\chi(s) = 1$ for $|s| \geq 1/2$ and $\chi(s) = 0$ for $|s| \leq (1-\mu)/2$, where $\mu = \mu_j$. Set

$$\varphi(x,\xi;\lambda) = \sum_{j=1}^{3}(\chi(|\lambda\xi_j - \xi_{0j}|/\delta) + \chi(|x_j - x_{0j}|/\delta)), \tag{38}$$

where $x_0 = (x_{01}, x_{02}, x_{03}) = (x_0, 0, 0)$. In view of $\mu < 1/100$, $\varphi(x, \xi; \lambda) \in S^0_{0,0}$ satisfies (3-5). Write $\chi_j(\xi_j) = \chi(|\lambda\xi_j - \xi_{0j}|/\delta)$ and $\chi_j(x_j) = \chi(|x_j - x_{0j}|/\delta)$ for brevity. Then

$$H_\varphi = \sum_{j=1}^{3}(\partial_{\xi_j}\chi_j(\xi_j)\partial_{x_j} - \partial_{x_j}\chi_j(x_j)\partial_{\xi_j}). \tag{39}$$

If we write $L_1 = P^w$ then $p_2(x, \xi) = \xi_1^2 + g(x_1)\xi_2^2$ and $\operatorname{Im} p_1 = f(x)\xi_3$. Note that

$$\begin{aligned} H_\varphi^2 p_2 &= 2(\partial_{x_1}\chi_1(x_1))^2 \\ &+ 2g(x_1)\left\{(\partial_{x_2}\chi_2(x_2))^2 - (\partial^2_{x_2}\chi_2(x_2))(\partial_{\xi_2}\chi_2(\xi_2))\xi_2\right\} \\ &+ \left\{ (\partial_{\xi_1}\chi_1(\xi_1))^2 g''(x_1)\xi_2^2 - (\partial^2_{\xi_1}\chi_1(\xi_1))(\partial_{x_1}\chi_1(x_1))g'(x_1)\xi_2^2 \right. \\ &\left. -2\partial_{\xi_1}\chi_1(\xi_1)\left(\partial_{x_2}\chi_2(x_2)g'(x_1)\xi_2 + \partial^2_{x_1}\chi_1(x_1)\xi_1\right) \right\}, \end{aligned}$$

$$H_\varphi(\operatorname{Im} p_1) = (\partial_{\xi_1}\chi_1(\xi_1))f'(x_1)\xi_3 - f(x_1)(\partial_{x_3}\chi_3(x_3)).$$

Since $\operatorname{supp} \partial_{\xi_1}\chi_1(\xi_1) \subset \{\ \delta(1-\mu) \le 2|\lambda\xi_1| \le \delta\ \}$ we have

$$\begin{aligned} |(H_\varphi^2 p_2)h_{40\delta}(\lambda\xi - \xi_0)| &\le 2(\partial_{x_1}\chi_1(x_1))^2 \\ &+ C\{g(x_1) + \Phi(\xi_1; \lambda)\}h_{40\delta}(\lambda\xi - \xi_0) \end{aligned} \tag{40}$$

and

$$|(H_\varphi(\operatorname{Im} p_1))h_{40\delta}(\lambda\xi - \xi_0)| \le C\{f(x_1) + \Phi(\xi_1, \lambda)\}h_{40\delta}(\lambda\xi - \xi_0), \tag{41}$$

where $\Phi(\xi_1; \lambda) \in C^\infty$ satisfies $|\partial^j_{\xi_1}\Phi| \le C_j\lambda^j$ for $C_j > 0$ and

$$\operatorname{supp} \Phi \subset \{\ \delta(1-\mu) \le 2|\lambda\xi_1| \le \delta\ \}. \tag{42}$$

It follows from (37) that $(\operatorname{supp} (\partial_{x_1}\chi_1)(x_1)) \cap g^{-1}(0) = \emptyset$. Consequently we have

$$\begin{cases} \forall\varepsilon > 0,\ \exists\lambda_\varepsilon > 0 \text{ such that for } 0 < \forall\lambda \le \lambda_\varepsilon \\ (\log\lambda)^2 \displaystyle\int((\partial_{x_1}\chi_1(x_1))^2 + g(x_1))|u(x_1)|^2 dx_1 \\ \le \varepsilon \displaystyle\int\{|u'(x_1)|^2 + g(x_1)\lambda^{-2}|u(x_1)|^2\}dx_1 \text{ for } \forall u \in C_0^\infty(I_0). \end{cases} \tag{43}$$

Since $(E,f,g)_2$ follows from $(M,f,g)_2$ by Lemma 1, we have in view of (2.9) of [30] together with Lemma 2

$$\text{(44)}\quad \begin{cases} \forall\varepsilon>0,\ \exists\lambda_\varepsilon>0 \text{ such that for } 0<\forall\lambda\le\lambda_\varepsilon \\ \int |u(x_1)|^2dx_1 + |\log\lambda|\int f(x_1)|u(x_1)|^2dx_1 \\ \le \varepsilon\int\{|u'(x_1)|^2 + g(x_1)\lambda^{-2}|u(x_1)|^2\}dx_1 \text{ for } \forall u\in C_0^\infty(I_0). \end{cases}$$

It follows from (42) that

$$(\log\lambda)^2\Phi(\xi_1;\lambda)h_{40\delta}(\lambda\xi-\xi_0)\le\varepsilon\xi_1^2 \tag{45}$$

if $0<\lambda\le\lambda_\varepsilon$. We have

$$\lambda^{-2}h_{20\delta}(\lambda\xi-\xi_0)\le(2/|\xi_{02}|)^2\ \xi_2^2 \tag{46}$$

if $\delta=\delta_j$ is chosen such that $0<40\delta<|\xi_{02}|/2$. In view of (45) and (46) together with Fefferman-Phong inequality, it follows from (40), (41), (43) and (44) that we have for a $C>0$ independent of ε

$$\begin{aligned} &||H_{20\delta}u||^2+(\log\lambda)^2||(H_\varphi^2\ p_2)H_{20\delta}u||^2 \\ &+(\log\lambda)^2|\mathrm{Re}\ ((H_\varphi^2\ p_2)H_{20\delta}u,\ H_{20\delta}u)| \\ &+|\log\lambda||\mathrm{Re}\ ((H_\varphi(\mathrm{Im}\ p_1))H_{20\delta}u,\ H_{20\delta}u)| \\ &\le\varepsilon\mathrm{Re}(P^wH_{20\delta}u,\ H_{20\delta}u)+C\{\varepsilon||H_{20\delta}u||^2+\lambda||u||^2\} \\ &\qquad\qquad \text{for } u\in\mathcal{S}(\mathbf{R}^3) \end{aligned} \tag{47}$$

if $0<\lambda\le\lambda_\varepsilon$ for a sufficiently small λ_ε. From this we obtain (6) because (6) for $\lambda_\varepsilon<\lambda\le1$ is trivial on account of the term $C_\varepsilon\lambda||u||^2$ on the right-hand-side and we have

$$P^wH_{20\delta}=P_\lambda^w-P_\lambda^w(1-H_{20\delta})+(P^w-P_\lambda^w)H_{20\delta}.$$

Since the sequence $\{\delta_j\}$ and $\varphi(x,\xi;\lambda)$ defined for each $\delta=\delta_j$ satisfy the assumptions of Theorem 1 we obtain (36) for $\rho_0=(x_0,\xi_0)\in T^*(\Omega)$ with $\xi_0=(0,\xi_{02},\xi_{03})$ and $\xi_{02}\ne0$. Thus we have proved the micro-hypoellipticity of L_1 at $(\tilde{x},\tilde{\xi})$ under $(M,f,g)_2$.

We can prove that if $(M,f,g)_2$ is violated then L_1 is not micro-hypoelliptic at $(x_0,\xi_0)=((x_{01},0,0),(0,1,0))\in T^*(\mathbf{R}^3)$ for some $x_{01}\in I_0$

by the quite same way as in Section 2 of [30]. Indeed, if we consider the same eigenvalue problem as (2.1) of [30] then for a certain positive increasing sequence $\{\eta_m\}_m^\infty$ we have the lowest eigenvalue $\mu(\eta_m)$ satisfying

$$0 < \mu(\eta_m) \leq C|\log \eta_m|$$

because $(E, f, g)_2$ is not satisfied by means of Lemma 1. If $v_m(x) = v(x; \eta_m)$ is the corresponding eigenfunction normalized then it suffices to consider a singular solution $u(x)$ of the form

$$u(x) = \sum_{m=1}^{\infty} \eta_m^{-4} v_m(x_1) \exp(ix_2\eta_m - x_3\mu(\eta_m)).$$

We shall prove that L_1 is micro-hypoelliptic at (X, Ξ) with $\Xi = (0, 0, \pm 1)$ if $(M, g, f)_1$ holds. If $\rho_0 = (x_0, \xi_0) \in T^*(\Omega)$ with $|\xi_0| = 1$ belongs to a conic neighborhood of (X, Ξ) then for the proof it suffices to show (36). We may assume $\xi_0 = (0, \xi_{02}, \xi_{03})$ with $\xi_{03} \neq 0$ as in the previous case. Since $(M, g, f)_1$ is equivalent to $(M, g, f^2)_1$, it follows from Lemma 1 that

$$\text{(48)} \quad \begin{cases} \forall \varepsilon > 0, \ \exists \lambda_\varepsilon > 0 \text{ such that for } 0 < \forall \lambda \leq \lambda_\varepsilon \\ (\log \lambda)^2 \int (\partial_{x_1}\chi_1(x_1))^2 + g(x_1))|u(x_1)|^2 dx_1 \\ \leq \varepsilon \int \{|u'(x_1)|^2 + \left(f(x_1)\dfrac{\lambda^{-2/3}}{|\log \lambda|^{1/3}}\right)^2 |u(x_1)|^2\} dx_1 \\ \qquad \text{for } \forall u \in C_0^\infty(I_0), \end{cases}$$

because we may assume that (supp $\partial_{x_1}\chi_1(x_1)) \cap f^{-1}(0) = \emptyset$. Note that (44) holds with $g(x_1)\lambda^{-2}$ replaced by

$$\left(f(x_1)\frac{\lambda^{-2/3}}{|\log \lambda|^{1/3}}\right)^2$$

because $(M, f, f^2)_1$ is always valid. Since we have instead of (46)

$$\left(\frac{\lambda^{-2/3}}{|\log \lambda|^{1/3}}\right)^2 h_{2\delta}(\lambda\xi - \xi_0) \leq C\left(\frac{\lambda^{1/3}}{|\log \lambda|^{1/3}}\xi_3\right)^2$$

we can see that the estimate (6) holds for $\rho_0 = (0, \xi_{02}, \xi_{03})$ with $\xi_{03} \neq 0$.

We shall prove Proposition 2. For the proof it suffices to show a formula similar to (36) for L_2 by using Theorem 2 in view of its last assertion. In fact, estimate (8) holds because of Poincaré's inequality. We may assume $\rho_0 = (x_0, \xi_0)$ with $\xi_0 = (0, \xi_{02}, \xi_{03})$. As in (37), for $x_0 = (x_{01}, x_{02}, x_{03})$ there exist positives δ and μ such that

$$\{\, x_1 \,; \delta(1-\mu) \le 2|x_1 - x_{01}| \le \delta\} \cap \left(f^{-1}(0) \cup (g')^{-1}(0)\right) = \emptyset \tag{49}$$

and

$$\{\, x_2 \,; \delta(1-\mu) \le 2|x_2 - x_{02}| \le \delta\} \cap h^{-1}(0) = \emptyset. \tag{50}$$

If φ is the same as (38) then we have

$$\begin{aligned} &|(H_\varphi^2 p_2) h_{40\delta}(\lambda\xi - \xi_0)| \\ &\le C\Big\{(\partial_{x_1}\chi_1(x_1))^2 + f(x_1)h(x_2) + f(x_1)|\partial_{x_2}\chi_2(x_2)| \\ &\qquad + (\Phi(\xi_1;\lambda) + f(x_1)\tilde{\Phi}(\xi_2;\lambda))h_{40\delta}(\lambda\xi - \xi_0)\Big\}, \end{aligned} \tag{51}$$

where $\Phi(\xi_1;\lambda)$ is the same as in (41). Here $\tilde{\Phi}(\xi_2;\lambda) \in C^\infty$ satisfies $|\partial_{\xi_2}^j \tilde{\Phi}| \le C_j \lambda^j$ for $C_j > 0$ and

$$\operatorname{supp} \tilde{\Phi} \subset \{\ \delta(1-\mu) \le 2|\lambda\xi_2 - \xi_{02}| \le \delta\ \}. \tag{52}$$

In view of (49) it follows from $(M, f, f^2|g'|)_1$ and Lemma 1 that

$$\left\{\begin{aligned} &\forall\varepsilon > 0,\ \exists\lambda_\varepsilon > 0 \text{ such that for } 0 < \forall\lambda \le \lambda_\varepsilon \\ &(\log\lambda)^2 \int ((\partial_{x_1}\chi_1(x_1))^2 + f(x_1))|u(x_1)|^2 dx_1 \\ &\le \varepsilon \int \{|u'(x_1)|^2 + (f(x_1)^2 g'(x_1)\lambda^{-1/2})^2 |u(x_1)|^2\} dx_1 \\ &\qquad\qquad \text{for } \forall u \in C_0^\infty(I_0). \end{aligned}\right. \tag{53}$$

Multiplying both sides of the above inequality by $|h(x_2)|$ we have

$$\left\{\begin{aligned} &\forall\varepsilon > 0,\ \exists\lambda_\varepsilon > 0 \text{ such that for } 0 < \forall\lambda \le \lambda_\varepsilon \\ &(\log\lambda)^2 (f(x_1)|h(x_2)|u, u) \\ &\le \varepsilon\left(\|D_1 u\|^2 + ((f(x_1)^2|g'(x_1)||h(x_2)|\lambda^{-1/2})u, u)\right) \\ &\qquad\qquad \text{for } \forall u \in C_0^\infty(I_0^2 \times \mathbf{R}_{x_3}). \end{aligned}\right. \tag{54}$$

By means of the change of $u(x_2)$ by $u(x_2)\exp(ig(x_1)h(x_2)\xi_3)$, it follows from $(M,1,|h|)_1$ that

$$(55)\quad \begin{cases} \forall\varepsilon>0,\ \exists\lambda_\varepsilon>0 \text{ such that for } 0<\forall\lambda\le\lambda_\varepsilon \\ (\log\lambda)^2\int|u(x_2)|^2dx_2 \\ \le\varepsilon\int\{|\left(\dfrac{d}{dx_2}+ig(x_1)h(x_2)\xi_3\right)u(x_2)|^2 \\ \qquad +(h(x_2)\lambda^{-1/2})^2|u(x_2)|^2\}dx_2 \quad \text{for } \forall u\in C_0^\infty(I_0). \end{cases}$$

If $q_1=\xi_1$ and $q_2=f(\xi_2+gh\xi_3)$ then

$$(56)\qquad f^2g'h\xi_3=f\{q_1,q_2\}-f'q_2.$$

If we note (51) together with (49), (50), (42) and (52), it is not difficult to check the hypothetical estimate of Theorem 2 by means of (53)-(55). By Theorem 2 we have a formula similar to (36) for L_2.

We shall sketch the proof of Proposition 3. The proof of the micro-hypoellipticity of L_3 is reduced to (18) by means of Theorem 3 and its remark. If $\rho_0=(x_0,\xi_0)$ then we may assume $x_{03}\in f^{-1}(0)$ and $\xi_{03}=0$ because the estimate is otherwise obvious. Choosing $\delta,\mu>0$ such that

$$\{\ x_3\ ;\delta(1-\mu)\le 2|x_3-x_{03}|\le\delta\}\cap f^{-1}(0)=\emptyset$$

and noting the explicit calculations for $H_\varphi^2p_2$, $(H_\varphi p_2)^{(j)}$ and $H_\varphi((\mathrm{Im})p_1)$, we can see estimate (18) holds. In fact, in the region which intersects $\{x_3=x_{03}\}$ we need only to show

$$\begin{aligned}&|\log\lambda|^2f(x_3)\int|u(x_1)|^2dx_1\\ &\qquad\le\varepsilon f(x_3)\int\{|u'(x_1)|^2+g(x_1)\lambda^{-2}|u(x_1)|^2\}dx_1,\end{aligned}$$

which follows from $(M,1,g)_1$. We can also prove the micro-hypoellipticity of L_4 by showing estimate (27) by a quite similar way.

References

[1] Beals, R. and Fefferman, C., On hypoellipticity of second order operators, *Comm. PDE* 1 (1976), 73–85.

[2] Bell, D. R. and Mohammed, S.-E. A., An extension of Hörmander theorem for infinitely degenerate second-order operators, *Duke Math. J.* **78** (1995), 453–475.

[3] Fediĭ, V. S., On a criterion for hypoellipticity, *Math. USSR Sb.* **14** (1971), 15–45.

[4] Fefferman, C., The uncertainty principle, *Bull. Amer. Math. Soc.* **9** (1983), 129–206.

[5] Fefferman, C. and Phong, D.H., On the positivity of pseudodifferential operators, *Proc. Nat. Acad. Sci.* **75** (1978), 4673–4674.

[6] Fefferman, C. and Phong, D.H., The uncertainty principle and sharp Gårding inequalities, *Comm. Pure Appl. Math.* **34** (1981), 285–331.

[7] Gaveau, B. and Moulinier, J. M., Intégrales oscillantes stochastiques et front d'onde des solutions d'équations semi-elliptiques, *C.R. Acad. Sc. Paris* **301** (1985), 825–828.

[8] Hörmander, L., Pseudo-differential operators and non-elliptic boundary problems, *Ann. of Math.* **83**, (1966), 129–209.

[9] Hörmander, L., *The Analysis of Linear Partial Differential Operators III*, Springer-Verlag, Berlin, Heiderberg, New York, Tokyo, 1985.

[10] Hoshiro, T., Hypoellipticity for infinitely degenerate elliptic and parabolic operators of second order, *J. Math. Kyoto Univ.* **28** (1988), 615–632.

[11] Hoshiro, T., Hypoellipticity for infinitely degenerate elliptic and parabolic operators II, operators of higher order, *J. Math. Kyoto Univ.* **29** (1989), 497–513.

[12] Hoshiro, T., On hypoellipticity for a certain operator with double characterisitic, *J. Math. Soc. Japan* **43** (1991), 593–603.

[13] Hoshiro, T., On Levi-type conditions for hypoellipticity of certain differential operators,*Comm. PDE* **17** (1992), 905–922.

[14] Hoshiro, T., Some examples of Hypoelliptic operators of infinitely degenerate type, *Osaka J. Math.* **30** (1993), 771–782.

[15] Kajitani, K. and Wakabayashi, S., Propagation of singularities for several classes of pseudodifferential operators, *Bull. Sc. Math.* 2e série **115** (1991), 397–449.

[16] Koike, M., A note on hypoellipticity for degenerate elliptic operators, *Publ. RIMS Kyoto Univ.* **27** (1991), 995–1000.

[17] Kusuoka, S. and Stroock, D., Applications of the Malliavin calculus, Part II, *J. Fac. Sci. Univ. Tokyo Sect. IA, Math.* **32** (1985), 1–76.

[18] Lanconelli, E., Sugli operatori ipoellittici del secondo ordine con simbolo principale di segno variable, *Boll. Un. Mat. Ital.* **16-B** (1979), 291–313.

[19] Morimoto, Y., Non-hypoellipticity for degenerate elliptic operators, *Publ. RIMS Kyoto Univ.* **22** (1986), 25–30. Erratum. *Publ. RIMS Kyoto Univ.* **30** (1994), 533–534.

[20] Morimoto, Y., Criteria for hypoellipticity of differential operators, *Publ. RIMS Kyoto Univ.* **22** (1986), 1129–1154.

[21] Morimoto, Y., Hypoellipticity for infinitely degenerate elliptic operators, *Osaka J. Math.* **24** (1987), 13–35.

[22] Morimoto, Y., A criterion for hypoellipticity of second order differential operators, *Osaka J. Math.* **24** (1987), 651–675.

[23] Morimoto, Y., The uncertainty principle and hypoelliptic operators, *Publ. RIMS Kyoto Univ.* **23** (1987), 955–964.

[24] Morimoto, Y., Propagation of wave front sets and hypoelliptic operators, *Pitman Res. Notes Math Ser.* **183** (1988), 212–224.

[25] Morimoto, Y., Estimates for degenerate Schrödinger operators and hypoellipticity for infinitely degenerate elliptic operators, *J. Math. Kyoto Univ.* **32** (1992), 333–372.

[26] Morimoto, Y., Hypoelliptic operators in $\mathbf{R}^3$ of the form $X_1^2 + X_2^2$, *J. Math. Kyoto Univ.* **32** (1992), 461–484.

[27] Morimoto, Y., Hypoelliptic operators of principal type with infinite degeneracy, *Tsukuba J. Math.* **19** (1995),187–200.

[28] Morimoto, Y., Local solvability and hypoellipticity for pseudodifferential operators of Egorov type with infinite degeneracy, *Nagoya Math. J.* **139** (1995),151–171.

[29] Morimoto, Y. and Morioka, T., Some remarks on hypoelliptic operators which are not microhypoelliptic, *Publ. RIMS Kyoto Univ.* **28** (1992), 579–586.

[30] Morimoto, Y. and Morioka, T., The positivity of Schrödinger operators and the hypoellipticity of second order degenerate elliptic operators, to appear in *Bull. Sc. Math.* 2e série.

[31] Morioka, T., Hypoellipticity for semi-elliptic operators which degenerate on hypersurface, *Osaka J. Math.* **28** (1991), 563–578.

[32] Morioka, T., Hypoellipticity for some infinitely degenerate elliptic operators of second order, *J. Math. Kyoto Univ.* **32** (1992), 373–386.

[33] Morioka, T., Some remarks on micro-hypoelliptic operators of infinitely degenerate type, *Publ. RIMS Kyoto Univ.* **28** (1992), 129–138.

[34] Morioka, T., Hypoellipticité pour un certain opérateur à caractéristiques doubles, to appear in *Tsukuba J. Math.*

[35] Olejnik, O. A. and Radkevic, E. V., Second order equations with non-negative characterisitc form, Plenum Press, New York London, 1973.

[36] Parenti, C. and Rodino, L., Examples of hypoelliptic operators which are not micro-hypoelliptic, *Boll. Um. Mat.* **17** (1980), 390–409.

[37] Sawyer, E., A weighted inequality and eigenvalue estimates for Schrödinger operators, *Indiana Univ. Math. J.* **35** (1986), 1–28.

[38] Suzuki, M., Hypoellipticity for a class of degenerate elliptic operators of secon order-, *Tsukuba J. Math.* **16** (1992), 217-234.

[39] Taira, K., On a class of hypoelliptic differential operators with double characteristics, *J. Math. Soc. Japan* **45** (1993), 391–419.

[40] Wakabayashi, S. and Suzuki, M., Microhypoellipticity for a class of pseudodifferential operators with double characteristics, *Funkciaj Ekvacioj* **36** (1993), 519–556.

[41] Zuily, C., Sur l'hypoellipticité des opérateurs différentiels du second ordre à coefficients réels, *J. Math. pures, et appl.* **55** (1976), 99–125.

Yoshinori Morimoto
Graduate School of Human and Environmental Studies
Kyoto University, Kyoto 606-01, Japan
e-mail:morimoto@math.h.kyoto-u.ac.jp, fax:81-75-753-2933

Tatsushi Morioka
Department of Mathematics, Faculty of Science
Osaka University, Osaka 560, Japan
e-mail:morioka@math.sci.osaka-u.ac.jp, fax:81-6-845-1163

Regularity of solutions to characteristic boundary value problem for symmetric systems

Tatsuo Nishitani and Masahiro Takayama

1. Introduction

Let Ω be a bounded open set in $\mathbf{R}^n$ with smooth boundary $\partial\Omega$. In Ω we study a first order symmetric system

$$Lu = \sum_{j=1}^{n} A_j(x)\partial_j u + B(x)u, \quad A_j(x), B(x) \in C^\infty(\overline{\Omega}), \quad A_j^*(x) = A_j(x)$$

where $u = (u_1, ..., u_N)$ and $\partial_j = \partial/\partial x_j$. Recall that the boundary matrix is given by

$$A_b(x) = \sum_{j=1}^{n} \nu_j(x) A_j(x), \quad x \in \partial\Omega$$

where $\nu(x) = (\nu_1(x), ..., \nu_n(x))$ is the unit outward normal to Ω. Let $H(x) \in C^\infty(\overline{\Omega})$ be symmetric positive definite. We study the following boundary value problem

$$\text{(BVP)} \qquad \begin{cases} (L + \lambda H)u = f & \text{in} \quad \Omega \\ u \in M & \text{at} \quad \partial\Omega \end{cases}$$

where $M(x)$ is a linear subspace of $\mathbf{C}^N$. We are concerned with the case in which the rank of A_b is not constant.

When $\dim\mathrm{Ker} A_b(x)$ is constant on $\partial\Omega$, one finds in Rauch [9] a detailed study of boundary value problems for positive symmetric systems, including standard classical results ([2], [4] and see also the references given there). When $\dim\mathrm{Ker} A_b(x)$ is not constant on $\partial\Omega$, in [10], the L^2 structure of (BVP) is studied (see also [5], [7], [8]). On the other hand, in a simple case that $A_b(x)$ is definite outside an embedded $n-2$ submanifold of $\partial\Omega$ on which A_b vanishes, the regularity of weak solutions is studied in [6]. In [11], the same question is studied, in a similar situation. In this paper we continue to study the same regularity question when A_b is nonsingular outside the set on which

A_b is definite, assumed to be an open set with smooth boundary. As is easily seen, solutions u to (BVP) need not be regular for smooth f (for instance, see Example 2.1 in [6]). Hence, to get regularity results, we impose further conditions. We make our assumptions more precise. Let us denote

$$(1.1) \qquad O^+(O^-) = \{x \in \partial\Omega \mid A_b(x) \text{ is positive (negative) definite}\}.$$

We assume that $O^\pm$ are open sets in $\partial\Omega$ with smooth boundaries $\gamma^\pm$ and set $\gamma = \gamma^+ \cup \gamma^-$. We assume that A_b is nonsingular on $\partial\Omega \setminus \gamma$ and $\mathrm{Ker}A_b$ is a C^∞ vector bundle over γ.

Note that the differential dA_b of A_b defines a map $A_{b,\gamma}$, up to a conormal factor of γ, going from $\mathrm{Ker}A_b$ to $\mathrm{Coker}A_b$ which can be identified with $\mathrm{Ker}A_b$ because A_b is symmetric. Let $\bar{x} \in \gamma$ and we work in a neighborhood $\bar{x} \in U \subset \partial\Omega$. Let us take $h(x) \in C^\infty(U)$, a defining function of γ, and let $v_1(x), \dots, v_p(x)$ be a smooth basis in U for $\mathrm{Ker}A_b$. Then with this basis, $A_{b,\gamma}$ is given by

$$A_{b,\gamma}(\bar{x}) = \lim_{\gamma \not\ni x \to \bar{x}} h(x)^{-1}(A_b(x)v_i(x), v_j(x))$$

which is a $p \times p$ matrix. Recall that $\sum_{j=1}^n A_j(x)\partial_j$ defines a map A_γ: $(T_\gamma^*\Omega)^N \to \mathbf{C}^N$ (just as A_b is defined with respect to $\partial\Omega$). Since $A_\gamma = A_b$ on $(T_{\partial\Omega}^*\Omega)^N$ then A_γ induces a map $A_{\gamma/b}$ so that

$$A_{\gamma/b} : (T_\gamma^*\partial\Omega)^N \cong (T_\gamma^*\Omega)^N/(T_{\partial\Omega}^*\Omega)^N \to \mathrm{Coker}A_b \cong \mathrm{Ker}A_b.$$

Since we have fixed a conormal factor of γ in $\partial\Omega$, restricting $A_{\gamma/b}$ on $\mathrm{Ker}A_b$ we have $A_{\gamma/b} : \mathrm{Ker}A_b \to \mathrm{Ker}A_b$. In local coordinates $A_{\gamma/b}$ is given by

$$A_{\gamma/b}(\bar{x}) = (A_{\tilde{h}}(\bar{x})v_i(\bar{x}), v_j(\bar{x}))$$

where $A_{\tilde{h}}(x) = \sum_{j=1}^n(\partial\tilde{h}/\partial x_j)A_j(x)$ and $\tilde{h}$ is an extension of h to a neighborhood of $\bar{x}$ in $\mathbf{R}^n$. We assume that

$$(1.2) \qquad A_{b,\gamma} \text{ and } A_{\gamma/b} \text{ have the same definiteness on } \gamma.$$

We remark that if $\gamma^+ = \gamma^-$, then (1.2) coincides with the condition assumed in [6].

As for the boundary condition we assume that the boundary space $M(x)$ is smooth in $\partial\Omega \setminus \gamma$ up to the boundary and maximal positive, that is

$$\langle A_b(x)v, v\rangle \geq 0 \quad \text{for all} \quad v \in M(x),$$

dim$M(x)$

$= \#\{$non-negative eigenvalues of $A_b(x)$ counting multiplicity$\}$.

In particular, (1.1) implies that

$$M(x) = \begin{cases} \mathbf{C}^N & \text{if} \quad x \in O^+ \\ \{0\} & \text{if} \quad x \in O^-. \end{cases}$$

Under the condition (1.2) we discuss the existence of regular solutions to (BVP) throughout the paper.

2. Main results

We denote the formal adjoint of L by L^*:

$$L^* u = -\sum_{j=1}^{n} \partial_j A_j(x) u + B^*(x) u.$$

For $u, v \in C^{0,1}(\overline{\Omega})$, Green's identity yields

$$((L + \lambda H)u, v)_{L^2(\Omega)} = (u, (L^* + \overline{\lambda} H)v)_{L^2(\Omega)} + \int_{\partial\Omega} \langle A_b u, v\rangle d\sigma.$$

The adjoint boundary space $M^*(x)$, $x \in \partial\Omega$ is defined by

$$M^*(x) = [A_b(x) M(x)]^{\perp}.$$

We recall the following definition (see [1], [2]).

Definition. *For $f \in L^2(\Omega)$, $u \in L^2(\Omega)$ is a weak solution to* (BVP) *if and only if the identity*

$$(u, (L^* + \overline{\lambda} H)\psi)_{L^2(\Omega)} = (f, \psi)_{L^2(\Omega)}$$

holds for all $\psi \in C^{0,1}(\overline{\Omega})$ with $\psi \in M^*$ at $\partial\Omega$.

Take $r(x) \in C^\infty(\overline{\Omega})$ with $dr(x) \neq 0$ on $\partial\Omega$ so that $\Omega = \{r(x) > 0\}$ and $h_\pm(x) \in C^\infty(\overline{\Omega})$ such that $O^\pm = \partial\Omega \cap \{h_\pm(x) > 0\}$ where $dh_\pm(x)$ and $\nu(x)$ are linearly independent on $\gamma^\pm$. Let us set

$$m_\pm(x) = \{r(x)^2 + h_\pm(x)^2\}^{1/2}, \quad \phi_\pm(x) = m_\pm(x) - h_\pm(x).$$

Note that $\phi_\pm(x) > 0$ if $x \in \overline{\Omega} \setminus (O^\pm \cup \gamma^\pm)$ and that $\phi_\pm(x) = 0$ if $x \in O^\pm \cup \gamma^\pm$. We now introduce the following spaces: For $q \in \mathbf{Z}_+$ and $s, t \in \mathbf{R}$ we set

$$X^q_{(s,t)}(\Omega; \partial\Omega) = \bigcap_{j=0}^{q} \phi_+^{s+q-j} \phi_-^{t+q-j} H^j(\Omega; \partial\Omega)$$

where $H^j(\Omega; \partial\Omega)$ is the conormal Sobolev space of order j (these spaces are studied in Sections 3 and 6).

Theorem 2.1. *For $q \in \mathbf{Z}_+$ there is a $\sigma(q) > 0$ such that for $s, t > \sigma(q)$ we can take a $\Lambda(q, s, t) \in \mathbf{R}$ verifying the following properties: If $f \in X^q_{(-s,t)}(\Omega; \partial\Omega) \cap \phi_- L^2(\Omega)$ and $\mathrm{Re}\lambda > \Lambda(q, s, t)$ then there exists a weak solution $u \in X^q_{(-s,t)}(\Omega; \partial\Omega) \cap \phi_- L^2(\Omega)$ to* (BVP) *which satisfies*

$$\|u\|_{X^q_{(-s,t)}(\Omega;\partial\Omega)} + \|\phi_-^{-1} u\|_{L^2(\Omega)} \le C\{\|f\|_{X^q_{(-s,t)}(\Omega;\partial\Omega)} + \|\phi_-^{-1} f\|_{L^2(\Omega)}\}$$

where $C = C(q, s, t, \lambda) > 0$ is independent of f and u.

Proposition 2.2. *There is a $\Lambda \in \mathbf{R}$ verifying the following properties: If $f \in L^2(\Omega)$ and $\mathrm{Re}\lambda > \Lambda$ then a weak solution $u \in m_- L^2(\Omega)$ to* (BVP) *is unique.*

An immediate corollary to Theorem 2.1 and Proposition 2.2 is

Corollary 2.3. *For $q \in \mathbf{Z}_+$ there is a $\sigma(q) > 0$ such that for $s, t > \sigma(q)$ one can take a $\Lambda(q, s, t) \in \mathbf{R}$ with the following properties: If $f \in X^q_{(-s,t)}(\Omega; \partial\Omega) \cap \phi_- L^2(\Omega)$ and $\mathrm{Re}\lambda > \Lambda(q, s, t)$ and if $u \in m_- L^2(\Omega)$ is a weak solution to* (BVP) *then it follows that $u \in X^q_{(-s,t)}(\Omega; \partial\Omega) \cap \phi_- L^2(\Omega)$.*

The paper is organized as follows. In Section 3 we prove that, admitting Proposition 3.3, main results (Theorem 2.1 with less precise estimates and Corollary 2.3) hold for weak solutions vanishing near O^-. In Section 4 we look for a suitable basis for $\mathbf{C}^N$ in which the system takes a simple form (Lemmas 4.3 and 4.4). We also prove a key lemma (Lemma 4.5) to a crucial weighted a priori estimate (Proposition 5.2) derived in the next section. In Section 5 we show the existence of a weak solution to (BVP) vanishing near O^-. Applying the corresponding existence results for the adjoint problem we prove Proposition 2.2 there.

Sections 6, 7, 8 and 9 are devoted to the proof of Proposition 3.3. We introduce an operator # sending $u \in L^2(\mathbf{R}^n_+)$ to $u^\# \in L^2(\mathbf{R}^n)$ and a "tangential" mollifier J_ϵ so that $(J_\epsilon u)^\# = \chi_\epsilon * u^\#$ where $\chi_\epsilon *$ is the classical mollifier in $\mathbf{R}^n$. Using J_ϵ we follow the same lines as in Tartakoff [12] to get regularity of the weak solution u. Unfortunately we do not know whether $J_\epsilon u$ is a weak solution. Instead we examine whether one can plug $\phi_+^s \phi_-^{-t} J_\epsilon u$ into the estimate if s is large. To finish the arguments in [12] we must control such terms as $m\phi_+^s \phi_-^{-t}[(L + \lambda H), J_\epsilon]u$. At this stage we take advantage of the factor $m = m_+ m_-$ to compensate for unbounded factors rm^{-2} and $h_\pm m_\pm^{-2}$ arising from $[L, J_\epsilon]$. In Section 10, to get the estimates in Theorem 2.1, we derive a priori estimates for higher order derivatives of u (Proposition 10.6) assuming that u is a smooth solution vanishing near O^-. Finally, in Section 11, using standard limiting arguments, we remove the restriction on the support of u imposed near O^- and finish the proof of Theorem 2.1.

We close this section by giving an example. We consider Tricomi's operator

$$L = \begin{bmatrix} y & 0 \\ 0 & 1 \end{bmatrix} \partial_x - \begin{bmatrix} 0 & 1 \\ 1 & 0 \end{bmatrix} \partial_y, \quad (x, y) \in \mathbf{R}^2.$$

Let us assume that the boundary of Ω situated in the upper half plane is given by $y = g(x)$. Then it is clear that

$$A_b(x, y) = \begin{bmatrix} -yg'(x) & -1 \\ -1 & -g'(x) \end{bmatrix}$$

and hence $\gamma = \{(x, g(x)) \mid g(x)g'(x)^2 = 1\}$. With $h(x, y) = 1 - yg'(x)^2$ we get

$$A_{b,\gamma} = g(x)g'(x), \quad A_{\gamma/b} = -2g(x)g'(x)(2g(x)g''(x) + g'(x)^2).$$

Lemma 2.4. *Assume that $2g''(x) + g'(x)^4 < 0$ on γ. Then $A_{b,\gamma}$ and $A_{\gamma/b}$ have the same definiteness.*

3. Regularity of weak solutions vanishing near O^-

The following proposition asserts that Corollary 2.3 holds for weak solutions u vanishing near O^-.

Proposition 3.1. *For $q \in \mathbf{Z}_+$ there is a $\sigma(q) > 0$ such that for $s, t > \sigma(q)$ we can take a $\Lambda(q, s, t) \in \mathbf{R}$ verifying the following properties: If $f \in X^q_{(-s,t)}(\Omega; \partial\Omega) \cap L^2(\Omega)$ and $\mathrm{Re}\lambda > \Lambda(q, s, t)$ and if $u \in L^2(\Omega)$ with $\mathrm{supp} u \cap (O^- \cup \gamma^-) = \emptyset$ is a weak solution to* (BVP) *then it follows that $u \in X^q_{(-s,t)}(\Omega; \partial\Omega)$.*

The next proposition, proved in Section 5, together with Proposition 3.1, shows that Theorem 2.1 holds if f vanishes near O^-.

Proposition 3.2. *There is a $\Lambda \in \mathbf{R}$ such that if $f \in L^2(\Omega)$ with $\mathrm{supp} f \cap (O^- \cup \gamma^-) = \emptyset$ and $\mathrm{Re}\lambda > \Lambda$ then there exists a weak solution $u \in L^2(\Omega)$ to* (BVP) *with $\mathrm{supp} u \cap (O^- \cup \gamma^-) = \emptyset$.*

To state Proposition 3.3 we start by studying a family of the conormal Sobolev spaces $H^q(\Omega; \partial\Omega)$, $q \in \mathbf{Z}_+$. For this purpose let us take a covering $\{U_i\}_{i=0}^k$ of Ω as follows: First we cover $\partial\Omega$ by coordinate patches U_i, $i = 1, \ldots, k$ with coordinate systems $\chi_i : U_i \cap \Omega \to \{|x| < 1,\ x_1 > 0\}$. Next we cover $\Omega \setminus \bigcup_{i=1}^k U_i$ by $U_0 \subset\subset \Omega$. Choose a partition of unity $\{\psi_i\}_{i=0}^k$ subordinate to this covering $\{U_i\}_{i=0}^k$. Then $u \in$

$H^q(\Omega;\partial\Omega)$ is characterized by $\psi_0 u \in H^q(\mathbf{R}^n)$ and $\psi_i u \in H^q(\mathbf{R}^n_+;\partial\mathbf{R}^n_+)$, $i=1,\dots,k$ in local coordinates in U_i where

$$H^q(\mathbf{R}^n_+;\partial\mathbf{R}^n_+) = \{w \in L^2(\mathbf{R}^n_+);\ Z^\alpha w \in L^2(\mathbf{R}^n_+),\ |\alpha| \le q\}$$

with $Z = (Z_1, Z_2, \dots, Z_n) = (x_1\partial_1, \partial_2, \dots, \partial_n)$. This allows us to choose the norm on $H^q(\Omega;\partial\Omega)$ as follows:

$$\|u\|^2_{H^q(\Omega;\partial\Omega)} = \|\psi_0 u\|^2_{H^q(\mathbf{R}^n)} + \sum_{i=1}^k \|\psi_i u\|^2_{H^q(\mathbf{R}^n_+;\partial\mathbf{R}^n_+)}$$

where

$$\|w\|^2_{H^q(\mathbf{R}^n_+;\partial\mathbf{R}^n_+)} = \sum_{|\alpha|\le q} \|Z^\alpha w\|^2_{L^2(\mathbf{R}^n_+)}.$$

Note that the characterization does not depend on the choice of U_i, χ_i and ψ_i and that the norms arising from different choices of U_i, χ_i and ψ_i are equivalent. Therefore it is enough to study $u \in H^q(\Omega;\partial\Omega)$ with small support near the origin.

We define the maps $\# : L^2(\mathbf{R}^n_+) \to L^2(\mathbf{R}^n)$ and $\natural : L^\infty(\mathbf{R}^n_+) \to L^\infty(\mathbf{R}^n)$ by $w^\#(x) = w(e^{x_1}, x')e^{x_1/2}$, $a^\natural = a(e^{x_1}, x')$ with $x = (x_1, x') = (x_1, x_2, \dots, x_n)$ which are norm preserving bijections. Then it is easy to see that

$$(aw)^\# = a^\natural w^\#, \quad \partial_j(a^\natural) = (Z_j a)^\natural,\ j = 1,\dots,n,$$
$$\partial_1(w^\#) = (Z_1 w)^\# + w^\#/2, \quad \partial_j(w^\#) = (Z_j w)^\#,\ j = 2,\dots,n,$$

and hence the map $\# : H^q(\mathbf{R}^n_+;\partial\mathbf{R}^n_+) \to H^q(\mathbf{R}^n)$ is an isomorphism.

We next consider the following family of norms:

$$\|w\|^2_{\mathbf{R}^n_+,q,tan,\delta} = \|w^\#\|^2_{\mathbf{R}^n,q,\delta} = \int_{\mathbf{R}^n} |(w^\#)^\wedge(\xi)|^2 \langle\xi\rangle^{2(q+1)} \langle\delta\xi\rangle^{-2} d\xi$$

for $0 < \delta \le 1$ with $\langle\xi\rangle^2 = 1 + |\xi|^2$. Here $(w^\#)^\wedge(\xi)$ denotes the Fourier transform of $w^\#(x)$ with respect to x. Noted as above, this norm is equivalent to $\|w\|_{H^q(\mathbf{R}^n_+;\partial\mathbf{R}^n_+)}$ when $0 < \delta \le 1$. For $u \in H^q(\Omega;\partial\Omega)$ let us set

$$\|u\|^2_{\Omega,q,tan,\delta} = \|\psi_0 u\|^2_{\mathbf{R}^n,q,\delta} + \sum_{i=1}^k \|\psi_i u\|^2_{\mathbf{R}^n_+,q,tan,\delta}.$$

When $\delta = 1$ we write

$$\|w\|_{\mathbf{R}^n_+,q,tan} = \|w\|_{\mathbf{R}^n_+,q,tan,1}, \quad \|u\|_{\Omega,q,tan} = \|u\|_{\Omega,q,tan,1}.$$

In order that $u \in H^q(\Omega;\partial\Omega)$ it is necessary and sufficient that $u \in H^{q-1}(\Omega;\partial\Omega)$ and that the norm $\|u\|_{\Omega,q-1,tan,\delta}$ remains bounded when $\delta \downarrow 0$. In this case we have $\|u\|_{\Omega,q-1,tan,\delta} \uparrow \|u\|_{\Omega,q,tan}$ as $\delta \downarrow 0$.

Proposition 3.1 is an immediate consequence of Proposition 3.3 below.

Proposition 3.3. *For $q \in \mathbf{Z}_+, q \geq 1$ there are $c_0 = c_0(q) > 0$ and $\sigma(q) > 0$ such that for $s, t > \sigma(q)$ we can take a $\Lambda(q,s,t) \in \mathbf{R}$ verifying the following properties: If $f \in X^{q-1}_{(-s+1,t+1)}(\Omega;\partial\Omega) \cap L^2(\Omega)$ and* $\mathrm{Re}\lambda > \Lambda(q,s,t)$ *and if $u \in X^{q-1}_{(-s+1,t+1)}(\Omega;\partial\Omega) \cap L^2(\Omega)$ with* $\mathrm{supp}\, u \cap (O^- \cup \gamma^-) = \emptyset$ *is a weak solution to* (BVP) *then it follows that*

$$\phi_+^s \phi_-^{-t} u \in H^{q-1}(\Omega;\partial\Omega), \tag{3.1}$$

and the estimate

$$\begin{aligned}(\min(s,t) - \sigma(q))&\|\phi_+^s\phi_-^{-t}u\|^2_{\Omega,q-1,tan,\delta}\\ &\leq c_0\{\|\phi_+^s\phi_-^{-t}(L+\lambda H)u\|^2_{\Omega,q-1,tan,\delta} + \|\phi_+^s\phi_-^{-t}u\|^2_{\Omega,q-1,tan,\delta}\}\\ &\quad + C_1\{\|(L+\lambda H)u\|^2_{X^{q-1}_{(-s+1,t+1)}(\Omega;\partial\Omega)} + \|(L+\lambda H)u\|^2_{L^2(\Omega)}\\ &\qquad + \|u\|^2_{X^{q-1}_{(-s+1,t+1)}(\Omega;\partial\Omega)} + \|u\|^2_{L^2(\Omega)}\}\end{aligned} \tag{3.2}$$

holds for $0 < \delta \leq 1$ where $C_1 > 0$ depends only on q, s, t, λ and $\mathrm{supp}\, u$.

Admitting Proposition 3.3 we give the proof of Proposition 3.1.

Proof of Proposition 3.1. We proceed by induction on q. The case $q = 0$ is trivial. Inductively assume that the statement is true up to $q-1$. Suppose that $f \in X^q_{(-s,t)}(\Omega;\partial\Omega) \cap L^2(\Omega)$ and that $u \in L^2(\Omega)$ with $\mathrm{supp}\, u \cap (O^- \cup \gamma^-) = \emptyset$ is a weak solution to (BVP). Set $s' = s-1$ and $t' = t+1$. Then it follows that $f \in X^{q-1}_{(-s',t')}(\Omega;\partial\Omega) \cap L^2(\Omega)$ and hence by the inductive hypothesis we have

$$\begin{aligned}u \in X^{q-1}_{(-s',t')}(\Omega;\partial\Omega) &= X^{q-1}_{(-s+1,t+1)}(\Omega;\partial\Omega)\\ &= \bigcap_{j=0}^{q-1} \phi_+^{-s+q-j}\phi_-^{t+q-j} H^j(\Omega;\partial\Omega).\end{aligned}$$

Since $\|\phi_+^s\phi_-^{-t}(L+\lambda H)u\|_{\Omega,q-1,tan,\delta} \leq c\|f\|_{X^q_{(-s,t)}(\Omega;\partial\Omega)}$ with some $c > 0$, applying (3.2) we conclude $\phi_+^s\phi_-^{-t}u \in H^q(\Omega;\partial\Omega)$. This shows the assertion for q. □

We introduce an operator which plays a crucial role in getting the estimate (3.2). Let $\chi \in C_0^\infty(\mathbf{R}^n)$ and set $\chi_\epsilon(y) = \epsilon^{-n}\chi(y/\epsilon)$ for $0 < \epsilon \leq 1$. We define $J_\epsilon : L^2(\mathbf{R}^n_+) \to L^2(\mathbf{R}^n_+)$ by

$$J_\epsilon w(x) = \int_{\mathbf{R}^n} w(x_1 e^{-y_1}, x' - y')e^{-y_1/2}\chi_\epsilon(y)dy \tag{3.3}$$

which differs from the one introduced in Rauch [9] by the factor $e^{-y_1/2}$. Then we see that

$$(J_\epsilon w)^\# = w^\# * \chi_\epsilon, \quad [Z_j, J_\epsilon] = 0.$$

Pulling back w by # and applying Theorem 2.4.1 in Hörmander [3] to $w^\#$ we get

Proposition 3.4. *For χ in (3.3) we assume that*

$$\hat{\chi}(\xi) = O(|\xi|^p), \ \xi \to 0 \quad \text{for some} \quad p \in \mathbf{Z}_+, \tag{3.4}$$

$$\hat{\chi}(t\xi) = 0 \quad \text{for all} \quad t \in \mathbf{R} \quad \text{implies} \quad \xi = 0. \tag{3.5}$$

Then for $q \in \mathbf{Z}_+, q < p$ there is a $c_0 = c_0(\chi, q) > 0$ such that

$$\begin{aligned} &c_0^{-1}\|w\|_{\mathbf{R}^n_+, q-1, tan, \delta} \\ &\le \int_0^1 \|J_\epsilon w\|_{L^2(\mathbf{R}^n_+)} \epsilon^{-2q}(1+\delta^2/\epsilon^2)^{-1} d\epsilon/\epsilon + \|w\|_{\mathbf{R}^n_+, q-1, tan} \\ &\le c_0\|w\|_{\mathbf{R}^n_+, q-1, tan, \delta} \end{aligned}$$

for all $0 < \delta \le 1$ and $w \in H^{q-1}(\mathbf{R}^n_+; \partial\mathbf{R}^n_+)$.

4. Several lemmas

Let us set

$$\begin{aligned} m_\pm(x;\kappa,\mu) &= \{\kappa r(x)^2 + (\mu r(x) - h_\pm(x))^2\}^{1/2}, \\ \phi_\pm(x;\kappa,\mu) &= m_\pm(x;\kappa,\mu) + \mu r(x) - h_\pm(x) \end{aligned}$$

for $\kappa > 0$ and $\mu \in \mathbf{R}$. The following lemma is easily checked.

Lemma 4.1. *We can choose a $c = c(\kappa, \mu) > 0$ satisfying*

$$c^{-1}m_\pm(x) \le m_\pm(x;\kappa,\mu) \le cm_\pm(x), \quad c^{-1}\phi_\pm(x) \le \phi_\pm(x;\kappa,\mu) \le c\phi_\pm(x)$$

for $x \in \Omega$.

This lemma shows that

$$\bigcap_{j=0}^{q} \phi_+(\cdot;\kappa,\mu)^{s+q-j}\phi_-(\cdot;\kappa,\mu)^{t+q-j}H^j(\Omega;\partial\Omega) = X^q_{(s,t)}(\Omega;\partial\Omega)$$

and that the norms of these two spaces are equivalent (to be precise see Lemma 6.1 below). Therefore it suffices to prove Theorem 2.1 and

Proposition 2.2 with $m_\pm(\cdot;\kappa,\mu)$, $\phi_\pm(\cdot;\kappa,\mu)$ instead of $m_\pm$, $\phi_\pm$. In what follows we simply write $m_\pm$, $\phi_\pm$ for $m_\pm(\cdot;\kappa,\mu)$, $\phi_\pm(\cdot;\kappa,\mu)$ respectively.

We now set $\phi_{\pm,\eta}(x) = \phi_\pm(x) - \eta$ for $\eta \geq 0$ and

$$A_r(x) = \sum_{j=1}^{n} (\partial_j r)(x) A_j(x), \quad A_{h_\pm}(x) = \sum_{j=1}^{n} (\partial_j h_\pm)(x) A_j(x),$$

$$G_\pm(x) = \sum_{j=1}^{n} (\partial_j \phi_\pm)(x) A_j(x).$$

Using $\partial_j \phi_\pm = m_\pm^{-1}\{(\mu\phi_\pm + \kappa r)\partial_j r - \phi_\pm \partial_j h_\pm\}$ we can prove

Lemma 4.2. *Let $G_\pm(x)$ be as above. Then*
(i) $G_\pm(x) = m_\pm(x)^{-1}\{(\mu\phi_\pm(x) + \kappa r(x))A_r(x) - \phi_\pm(x) A_{h_\pm}(x)\}$,
(ii) *$G_\pm$ is bounded on Ω,*
(iii) $\phi_{\pm,\eta}^{-s} L \phi_{\pm,\eta}^{s} = L + s\phi_{\pm,\eta}^{-1} G_\pm$, $\phi_{\pm,\eta}^{-s} L^* \phi_{\pm,\eta}^{s} = L^* - s\phi_{\pm,\eta}^{-1} G_\pm$ *for* $\eta \geq 0$ *and* $s \in \mathbf{R}$.

The following lemma will be used in Section 6.

Lemma 4.3. *For $\bar{x} \in \gamma^+$ we can choose a neighborhood $U \subset \mathbf{R}^n$ of $\bar{x}$ and a smooth invertible matrix valued function $Q(x)$ on U satisfying the following properties: Let $p = \dim \mathrm{Ker} A_b(x)$ on $\gamma^+ \cap U$. Then*

(4.1)
$$Q^*(x)A_r(x)Q(x) = \begin{pmatrix} -h_+(x)I_p & 0 \\ 0 & -I_{N-p} \end{pmatrix} + r(x)\tilde{A}(x) \quad \text{on} \quad U,$$

(4.2)
$$Q^*(x)A_{h_+}(x)Q(x) = \begin{pmatrix} B_{11}(x) & B_{12}(x) \\ B_{21}(x) & B_{22}(x) \end{pmatrix} \quad \text{on} \quad U$$

with a $p \times p$ positive definite $B_{11}(x)$ where I_p and I_{N-p} denote the identity matrix of order p and $N-p$ respectively. Furthermore

(4.3)
$$Q^{-1}(x)M(x) = \begin{cases} \mathbf{C}^N & \text{if} \quad x \in O^+ \cap U \\ \{0\} \times \mathbf{C}^{N-p} & \text{if} \quad x \in (\partial\Omega \setminus (O^+ \cup \gamma^+)) \cap U, \end{cases}$$

(4.4)
$$Q^{-1}(x)M^*(x) = \begin{cases} \{0\} & \text{if} \quad x \in O^+ \cap U \\ \mathbf{C}^p \times \{0\} & \text{if} \quad x \in (\partial\Omega \setminus (O^+ \cup \gamma^+)) \cap U. \end{cases}$$

Proof. We choose a coordinate patch U of $\bar{x}$ with coordinate system $\chi : U \to \{|x| < 1\}$, $x_1 = r \circ \chi^{-1}$, $x_2 = h_+ \circ \chi^{-1}$. Performing a change of independent variables we are led to the case that

$$U = \{|x| < 1\}, \quad \Omega = \{x_1 > 0\}, \quad \partial\Omega = \{(0, x');\ x' \in \mathbf{R}^{n-1}\},$$
$$\gamma^+ = \{(0,0,x'');\ x'' \in \mathbf{R}^{n-2}\}, \quad O^+ = \{(0,x');\ x' \in \mathbf{R}^{n-1},\ x_2 > 0\},$$
$$r = x_1, \quad h_+ = x_2$$

with $x = (x_1, x') = (x_1, x_2, x'') = (x_1, x_2, x_3, \ldots, x_n)$. Moreover we can take a smooth $\overline{M}(x')$ such that $\overline{M}(x') = M(x')$ if $x_2 < 0$. We set $V = \begin{pmatrix} I_p \\ 0 \end{pmatrix}$ which is a $N \times p$ matrix. From $\dim\mathrm{Ker}A_b(0, x'') = p$ on $\gamma^+ \cap U$ there is a smooth invertible matrix valued function $S(x')$ satisfying

$$S^{-1}(0, x'')\mathrm{Ker}A_b(0, x'') = \mathbf{C}^p \times \{0\} \quad \text{on} \quad \gamma^+ \cap U.$$

This shows that column vectors of $S(0, x'')V$ are a smooth basis for the bundle $\mathrm{Ker}A_b(0, x'')$ over $\gamma^+ \cap U$. Thus with

$$S^*(x')A_b(x')S(x') = \begin{pmatrix} x_2 A_{11}(x') & A_{12}(x') \\ A_{21}(x') & A_{22}(x') \end{pmatrix} \quad \text{on} \quad \partial\Omega \cap U$$

it follows from (1.2) that $A_{11}(x')$ is definite. Since $A_b(x')$ is positive definite on $O^+ \cap U$, $A_{11}(x')$ and $A_{22}(x')$ are also positive definite on $O^+ \cap U$, and hence $A_{11}(x')$ is positive definite on $\partial\Omega \cap U$ and $A_{22}(0, x'')$ is non-negative definite on $\gamma^+ \cap U$ by continuity.

If $v \in \mathbf{C}^p$ then noting $S(0, x'')\begin{pmatrix} v \\ 0 \end{pmatrix} \in \mathrm{Ker}A_b(0, x'')$ on $\gamma^+ \cap U$ we have

$$0 = S^*(0, x'')A_b(0, x'')S(0, x'')\begin{pmatrix} v \\ 0 \end{pmatrix} = \begin{pmatrix} 0 \\ A_{21}(0, x'')v \end{pmatrix} \quad \text{on} \quad \gamma^+ \cap U$$

which proves $A_{21}(0, x'') = 0$ and hence $A_{21}(x') = x_2 \tilde{A}_{21}(x')$. We now recall that $\dim\mathrm{Ker}A_b(0, x'') = p$ on $\gamma^+ \cap U$. This implies that $A_{22}(0, x'')$ has no zero-eigenvalues, namely, $A_{22}(0, x'')$ is positive definite on $\gamma^+ \cap U$. Taking a coordinate patch U of $\bar{x}$ small enough, if necessary, we may assume that $A_{22}(x')$ is positive definite on $\partial\Omega \cap U$. If we set

$$T(x') = \begin{pmatrix} I_p & 0 \\ -x_2 A_{22}^{-1}(x')\tilde{A}_{21}(x') & I_{N-p} \end{pmatrix}$$

then it follows that

$$T^*(x')S^*(x')A_b(x')S(x')T(x') = \begin{pmatrix} x_2 \tilde{A}_{11}(x') & 0 \\ 0 & A_{22}(x') \end{pmatrix} \quad \text{on} \quad \partial\Omega \cap U$$

with $\tilde{A}_{11}(x') = A_{11}(x') - x_2\tilde{A}_{12}(x')A_{22}^{-1}(x')\tilde{A}_{21}(x')$. Taking U small enough we may assume that $\tilde{A}_{11}(x')$ is positive definite on $\partial\Omega \cap U$. Choose smooth invertible matrix valued functions $U_{11}(x')$ and $U_{22}(x')$ on $\partial\Omega \cap U$ satisfying

$$U_{11}^*(x')\tilde{A}_{11}(x')U_{11}(x') = I_p, \quad U_{22}^*(x')A_{22}(x')U_{22}(x') = I_{N-p}$$

and set

$$U(x') = \begin{pmatrix} U_{11}(x') & 0 \\ 0 & U_{22}(x') \end{pmatrix}.$$

Then we have

$$Q^*(x')A_b(x')Q(x') = \begin{pmatrix} x_2 I_p & 0 \\ 0 & I_{N-p} \end{pmatrix} \quad \text{on} \quad \partial\Omega \cap U \tag{4.5}$$

with $Q(x') = S(x')T(x')U(x')$. Thus noting

$$A_r(x) = A_1(x) = A_1(0,x') + x_1\tilde{A}_1(x) = -A_b(x') + x_1\tilde{A}_1(x)$$

we obtain (4.1).

We now show that $\overline{M}(0,x'') \cap \mathrm{Ker} A_b(0,x'') = \{0\}$ on $\gamma^+ \cap U$. Suppose that there were a $w_0 \in \overline{M}(0,y'') \cap \mathrm{Ker} A_b(0,y'')$ with $w_0 \neq 0$ for some $(0,0,y'') \in \gamma^+ \cap U$. Then we could take a smooth $w(x_2)$ near $x_2 = 0$ such that $w(0) = w_0$ and $w(x_2) \in \overline{M}(x_2,y'')$ near $x_2 = 0$. Set $v(x_2) = Q^{-1}(x_2,y'')w(x_2)$. Since $T(0,y'') = I_N$ it follows that

$$\begin{aligned} v(0) = Q^{-1}(0,y'')w_0 \in \\ U^{-1}(0,y'')T^{-1}(0,y'')S^{-1}(0,y'')\mathrm{Ker} A_b(0,y'') = \mathbf{C}^p \times \{0\} \end{aligned}$$

and hence we can write $v(0) = \binom{v_0}{0}$ with $v_0 \in \mathbf{C}^p$, $v_0 \neq 0$. Writing $v(x_2) = v(0) + x_2\tilde{v}(x_2)$ we obtain

$$\begin{aligned} \langle A_b(x_2,y'')w(x_2), w(x_2)\rangle &= \langle Q^*(x_2,y'')A_b(x_2,y'')Q(x_2,y'')v(x_2), v(x_2)\rangle \\ &= x_2|v_0|^2 + O(x_2^2) \quad < 0 \end{aligned}$$

if $x_2 < 0$ and $|x_2|$ is small enough. On the other hand, if $x_2 < 0$ then $w(x_2) \in \overline{M}(x_2,y'') = M(x_2,y'')$. This is incompatible with the maximal positivity of M. Noticing that $\dim M(x') = N - p$ on $(\partial\Omega \setminus (O^+ \cup \gamma^+)) \cap U$ which follows from (4.5) and hence $\dim \overline{M}(x') = N - p$ on $\partial\Omega \cap U$ we have $\dim \overline{M}(0,x'') + \dim \mathrm{Ker} A_b(0,x'') = N$ on $\gamma^+ \cap U$. So we may assume that $S(x')$ satisfies

$$\begin{aligned} S^{-1}(0,x'')\mathrm{Ker} A_b(0,x'') &= \mathbf{C}^p \times \{0\} \qquad \text{on} \quad \gamma^+ \cap U, \\ S^{-1}(x')\overline{M}(x') &= \{0\} \times \mathbf{C}^{N-p} \quad \text{on} \quad \partial\Omega \cap U. \end{aligned}$$

This implies that

$$\begin{aligned}Q^{-1}(x')\overline{M}(x') &= U^{-1}(x')T^{-1}(x')S^{-1}(x')\overline{M}(x')\\ &= \{0\} \times \mathbf{C}^{N-p} \quad \text{on} \quad \partial\Omega \cap U\end{aligned}$$

which shows (4.3). Moreover if $(0, x') \in \partial\Omega \cap U$ then the identity

$$\langle Q^{-1}(x')v, Q^*(x')A_b(x')Q(x')Q^{-1}(x')w\rangle = \langle v, A_b(x')w\rangle = 0$$

holds for $v \in M^*(x')$, $w \in M(x')$. Thus (4.4) follows from (4.3) and (4.5).

It remains to show (4.2). Let $V = \begin{pmatrix} I_p \\ 0 \end{pmatrix}$. Since $T(0, x') = I_N$ it follows that

$$\begin{aligned}Q^{-1}(0, x'')\mathrm{Ker}A_b(0, x'') &= U^{-1}(0, x'')T^{-1}(0, x'')S^{-1}(0, x'')\mathrm{Ker}A_b(0, x'')\\ &= \mathbf{C}^p \times \{0\} \quad \text{on} \quad \gamma^+ \cap U.\end{aligned}$$

This gives that column vectors of $Q(0, x'')V$ are a smooth basis for $\mathrm{Ker}A_b(0, x'')$ over $\gamma^+ \cap U$. Therefore it follows from (1.2) and (4.5) that $B_{11}(0, 0, x'')$ is positive definite on $\gamma^+ \cap U$. Taking U small enough we may assume that $B_{11}(x)$ is positive definite on U, which proves (4.2). □

Similarly we obtain

Lemma 4.4. *For $\bar{x} \in \gamma^-$ we can choose a neighborhood $U \subset \mathbf{R}^n$ of $\bar{x}$ and a smooth invertible matrix valued function $Q(x)$ on U satisfying the following properties: Let $p = \mathrm{dimKer}A_b(x)$ on $\gamma^- \cap U$. Then*

$$Q^*(x)A_r(x)Q(x) = \begin{pmatrix} h_-(x)I_p & 0 \\ 0 & I_{N-p} \end{pmatrix} + r(x)\tilde{A}(x) \quad \text{on} \quad U,$$

$$Q^*(x)A_{h_-}(x)Q(x) = \begin{pmatrix} B_{11}(x) & B_{12}(x) \\ B_{21}(x) & B_{22}(x) \end{pmatrix} \quad \text{on} \quad U$$

with a $p \times p$ negative definite $B_{11}(x)$. Furthermore

$$Q^{-1}(x)M(x) = \begin{cases} \{0\} & \text{if} \quad x \in O^- \cap U \\ \mathbf{C}^p \times \{0\} & \text{if} \quad x \in (\partial\Omega \setminus (O^- \cup \gamma^-)) \cap U, \end{cases}$$

$$Q^{-1}(x)M^*(x) = \begin{cases} \mathbf{C}^N & \text{if} \quad x \in O^- \cap U \\ \{0\} \times \mathbf{C}^{N-p} & \text{if} \quad x \in (\partial\Omega \setminus (O^- \cup \gamma^-)) \cap U. \end{cases}$$

The following lemma is a key to an a priori estimate in Section 5.

Lemma 4.5. *Taking $\kappa = 1$ and $\mu > 0$ large enough we can choose neighborhoods W^+ and W^- of $O^+ \cup \gamma^+$ and $O^- \cup \gamma^-$ respectively in which we have*

$$-G_+ \gg \delta m_+^{-1} \phi_+ I_N \quad \text{on} \quad W^+ \cap \Omega, \tag{4.6}$$

$$G_- \gg \delta m_-^{-1} \phi_- I_N \quad \text{on} \quad W^- \cap \Omega \tag{4.7}$$

with some $\delta > 0$.

Proof. The proof of (4.7) is a modification of the proof of (4.6); hence only (4.7) will be proved. We first work near $\bar{x} \in \gamma^+$. Let U be a neighborhood of $\bar{x}$ as in Lemma 4.3. Take $c_0 > 0$ such that $-\tilde{A} \gg -c_0 I_N$ on U where $\tilde{A}(x)$ is given in (4.1). Since $B_{11}(x)$ in (4.2) is positive definite, choosing $\delta_1 > 0$ small enough and $c_1 > 0$ large enough, we have

$$Q^* A_{h_+} Q \gg \begin{pmatrix} 2\delta_1 I_p & 0 \\ 0 & -c_1 I_{N-p} \end{pmatrix} \quad \text{on} \quad U.$$

This implies that

$$\begin{aligned} Q^*(-G_+)Q &= m_+^{-1}\{(\mu\phi_+ + r)Q^*(-A_r)Q + \phi_+ Q^* A_{h_+} Q\} \\ &\gg m_+^{-1} \begin{pmatrix} ((\mu\phi_+ + r)h_+ + 2\delta_1\phi_+)I_p & 0 \\ 0 & (\mu\phi_+ + r - c_1\phi_+)I_{N-p} \end{pmatrix} \\ &\quad + m_+^{-1}(c_0\mu r\phi_+ + c_0 r^2)I_N \quad \text{on} \quad U. \end{aligned}$$

It is clear that $\mu\phi_+ + r - c_1\phi_+ \geq \mu\phi_+/2$ on $U \cap \Omega$ with $\mu > 0$ large enough. If $h_+ \geq 0$ then

$$(\mu\phi_+ + r)h_+ + 2\delta_1\phi_+ \geq 2\delta_1\phi_+ \quad \text{on} \quad U \cap \Omega \cap \{h_+ \geq 0\}.$$

On the other hand if $h_+ < 0$, taking a neighborhood U of $\bar{x}$ small enough, we have

$$\begin{aligned} (\mu\phi_+ + r)h_+ + 2\delta_1\phi_+ &= (\delta_1 + \mu h_+)\phi_+ + \delta_1 m_+ + \delta_1 r + (\delta_1 - r)|h_+| \\ &\geq \delta_1\phi_+/2 \quad \text{on} \quad U \cap \Omega \cap \{h_+ < 0\}. \end{aligned}$$

Moreover we obtain that $c_0\mu r\phi_+ + c_0 r^2 \leq 2\epsilon\phi_+$ on $U \cap \Omega$ with $\epsilon > 0$ small enough. Indeed since $c_0\mu r\phi_+ = (c_0\mu r)\phi_+ \leq \epsilon\phi_+$ on $U \cap \Omega$ and we may assume that $m_+ - (\mu r - h_+) \leq c_0^{-1}\epsilon$ on U taking U small enough it follows that

$$c_0 r^2 \leq \frac{\epsilon r^2}{m_+ - (\mu r - h_+)} = \epsilon\phi_+ \quad \text{on} \quad U \cap \Omega.$$

This concludes that $-G_+ \gg \delta m_+^{-1}\phi_+ I_N$ on $U \cap \Omega$ and hence we can choose a neighborhood W_1 of γ^+ such that $-G_+ \gg \delta m_+^{-1}\phi_+ I_N$ on $W_1 \cap \Omega$ with some $\delta > 0$.

We turn to $O^+ \setminus W_1$. Let us take $a > 0$ such that $h_+ \geq a$ on $O^+ \setminus W_1$ and $c_2 > 0$ so that $A_{h_+} \gg -c_2 I_N$ on Ω. Recalling that $-A_r$ is positive definite on O^+ we can choose a neighborhood W_2 of $\partial\Omega \setminus \{h_+ \geq a\}$ satisfying $-A_r \gg \delta_2 I$ on W_2 with some $\delta_2 > 0$. This yields

$$-G_+ = m_+^{-1}\{(\mu\phi_+ + r)(-A_r) + \phi_+ A_{h_+}\} \gg m_+^{-1}\{\delta_2(\mu\phi_+ + r) - c_2\phi_+\} I_N$$

on $W_2 \cap \Omega$. Since $m_+ - (\mu r - h_+) \geq \epsilon_0$ on $\partial\Omega \cap \{h_+ \geq a\}$ with some $\epsilon_0 > 0$, taking a neighborhood W_2 of $\partial\Omega \cap \{h_+ \geq a\}$ small enough, we have

$$2c_2\phi_+ = \frac{2c_2 r^2}{m_+ - (\mu r - h_+)} \leq 4c_2\epsilon_0^{-1}r^2 \leq \delta_2 r \quad \text{on} \quad W_2 \cap \Omega$$

which shows that $-G_+ \gg c_2 m_+^{-1}\phi_+ I_N$ on $W_2 \cap \Omega$. Hence $W^+ = W_1 \cup W_2$ is a desired neighborhood of $O^+ \cup \gamma^+$. □

5. Existence and uniqueness of solutions vanishing near O^-

We write $(\cdot,\cdot)_\Omega$ and $\|\cdot\|_\Omega$ for the inner product and the norm in $L^2(\Omega)$ respectively. Let us take $h_0(x) \in C^\infty(\overline{\Omega})$ so that $\gamma^+ \cap \gamma^- = \partial\Omega \cap \{h_0(x) = 0\}$ where $dh_0(x)$ and $\nu(x)$ are linearly independent on $\gamma^+ \cap \gamma^-$ and set

$$m_0(x) = \{r(x)^2 + h_0(x)^2\}^{1/2}, \quad m(x) = m_0(x)^{-1}m_+(x)m_-(x).$$

Then it is clear that $m(m_+^{-1} + m_-^{-1}) \geq \epsilon_0$ on Ω with some $\epsilon_0 > 0$. We start with

Lemma 5.1. *There are $\delta_0, \sigma_0 > 0$ and $\lambda_0 \in \mathbf{R}$ such that if $s, t \in \mathbf{R}$ and $\lambda \in \mathbf{C}$ and if $u \in C^{0,1}(\overline{\Omega})$ with $u \in M$ at $\partial\Omega$ then it follows that*

$$\begin{aligned}(\delta_0\mathrm{Re}\lambda - \lambda_0)\|m^{1/2}u\|_\Omega^2 - \sigma_0\|u\|_\Omega^2 + s(m\phi_+^{-1}(-G_+)u,u)_\Omega \\ + t(m\phi_-^{-1}G_-u,u)_\Omega \\ \leq \mathrm{Re}(m\phi_+^s\phi_-^{-t}(L+\lambda H)\phi_+^{-s}\phi_-^t u,u)_\Omega.\end{aligned}$$

Proof. Green's identity yields

$$(Lmu,u)_\Omega = (mu,L^*u)_\Omega + \int_{\partial\Omega} m\langle A_b u,u\rangle d\sigma.$$

If we write

$$R(x) = \{(B + B^*) - \sum_{j=1}^{n} (\partial_j A_j)\}/2$$

and

$$M(x) = \sum_{j=1}^{n} (\partial_j m) A_j$$

then using $L^* = -L + 2R$ we get

$$\begin{aligned}
&\mathrm{Re}(m\phi_+^s \phi_-^{-t}(L + \lambda H)\phi_+^{-s}\phi_-^t u, u)_\Omega \\
&\quad = \mathrm{Re}\lambda(mHu, u)_\Omega + (mRu, u)_\Omega - \frac{1}{2}(Mu, u)_\Omega \\
&\quad + s(m\phi_+^{-1}(-G_+)u, u)_\Omega + t(m\phi_-^{-1}G_- u, u)_\Omega + \frac{1}{2}\int_{\partial\Omega} m\langle A_b u, u\rangle d\sigma.
\end{aligned}$$

Since $H(x)$ is positive definite on Ω and $\langle A_b u, u\rangle$ is non-negative on $\partial\Omega$, the proof is complete. □

Let us choose $\psi_{\pm,1}(x), \psi_{\pm,2}(x) \in C_0^\infty(\mathbf{R}^n)$ so that $0 \le \psi_{\pm,i} \le 1$, $\psi_{\pm,1} + \psi_{\pm,2} = 1$ on Ω and $\mathrm{supp}\psi_{\pm,1} \subset W^\pm$, $\psi_{\pm,2} = 0$ on $\Omega \cap \{\phi_{\pm,\eta^*} < 0\}$ where $W^\pm$ is as in Lemma 4.5 and $\eta^* > 0$ is such that $\Omega \cap \{\phi_{\pm,\eta^*} < 0\} \subset W^\pm$.

Proposition 5.2. *There are $c_0, c_1, \sigma_0 > 0$ such that for $s, t > \sigma_0$ we can take a $\Lambda(s,t) \in \mathbf{R}$ verifying the following properties: If* $\mathrm{Re}\lambda > \Lambda(s,t)$ *and if $u \in C^{0,1}(\overline{\Omega})$ with $u \in M$ at $\partial\Omega$ then it follows that*

$$\begin{aligned}
&(\mathrm{Re}\lambda - \Lambda(s,t))\|m^{1/2}u\|_\Omega^2 + c_0(\min(s,t) - \sigma_0)\|u\|_\Omega^2 \\
&\quad \le c_1\|m\phi_+^s\phi_-^{-t}(L + \lambda H)\phi_+^{-s}\phi_-^t u\|_\Omega^2.
\end{aligned}$$

Proof. We note that

$$(m\phi_\pm^{-1}(\mp G_\pm)u, u)_\Omega = \sum_{i=1}^{2}(m\phi_\pm^{-1}(\mp G_\pm)u, \psi_{\pm,i}u)_\Omega = I_{\pm,1} + I_{\pm,2}.$$

It follows from Lemma 4.5 that

$$I_{\pm,1} \ge \delta\|\psi_{\pm,1}^{1/2} m^{1/2} m_\pm^{-1/2} u\|_\Omega^2. \tag{5.1}$$

We turn to $I_{\pm,2}$. Since $G_\pm$ is bounded on Ω we have

$$I_{\pm,2} \ge -c_2\|\psi_{\pm,2}^{1/2} m^{1/2} \phi_\pm^{-1/2} u\|_\Omega^2$$

with some $c_2 > 0$. If $x \in \text{supp}\psi_{\pm,2} \cap \Omega$ then $\phi_\pm(x) \ge \eta^*$ and $m_\pm(x) \ge \phi_\pm(x)/2 \ge \eta^*/2$, which shows that $\|\psi_{\pm,1}^{1/2} m_\pm^{-1/2} w\|_\Omega$ and $\|\psi_{\pm,1}^{1/2}\phi_\pm^{-1/2} w\|_\Omega$ are equivalent to $\|\psi_{\pm,1}^{1/2} w\|_\Omega$ for $w \in L^2(\Omega)$ and hence

$$I_{\pm,2} \ge \delta\|\psi_{\pm,2}^{1/2} m^{1/2} m_\pm^{-1/2} u\|_\Omega^2 - c_3\|\psi_{\pm,1}^{1/2} m^{1/2} u\|_\Omega^2. \tag{5.2}$$

Combining (5.1) and (5.2) we obtain that

$$\begin{aligned} &s(m\phi_+^{-1}(-G_+)u, u)_\Omega + t(m\phi_-^{-1}G_-u, u)_\Omega \\ &\ge \delta s\|m^{1/2}m_+^{-1/2}u\|_\Omega^2 + \delta t\|m^{1/2}m_-^{-1/2}u\|_\Omega^2 - c_3(s+t)\|m^{1/2}u\|_\Omega^2 \\ &\ge \epsilon_0\delta \min(s,t)\|u\|_\Omega^2 - c_3(s+t)\|m^{1/2}u\|_\Omega^2 \end{aligned}$$

and hence Lemma 5.1 proves the assertion. □

Let us set $\Omega_\eta^\pm = \Omega \cap \{\phi_{\pm,\eta} > 0\}$ for $\eta \ge 0$. Note that $\partial\Omega \cap \{\phi_{\pm,\eta} > 0\}$ is a union of smooth surfaces if $\eta > 0$ is small enough. Green's identity again yields

Lemma 5.3. *There are $\delta_0 > 0$ and $\lambda_0 \in \mathbf{R}$ such that if $\eta \ge 0$, $t \in \mathbf{R}$, $\lambda \in \mathbf{C}$ and if $u \in C^{0,1}(\overline{\Omega})$ with $u = 0$ on $\overline{\Omega} \cap \{\phi_{-,\eta} = 0\}$ and $u \in M^*$ at $\partial\Omega \cap \{\phi_{-,\eta} > 0\}$ then it follows that*

$$(\delta_0 \text{Re}\lambda - \lambda_0)\|u\|_{\Omega_\eta^-}^2 + t(\phi_{-,\eta}^{-1}G_-u, u)_{\Omega_\eta^-} \le \text{Re}(\phi_{-,\eta}^t(L^* + \overline{\lambda}H)\phi_{-,\eta}^{-t}u, u)_{\Omega_\eta^-}.$$

From this we immediately obtain

Lemma 5.4. *There is a $c > 0$ such that for $t \in \mathbf{R}$ we can take a $\Lambda(t) \in \mathbf{R}$ verifying the following properties: If $\text{Re}\lambda > \Lambda(t)$ and if $u \in C^{0,1}(\overline{\Omega})$ with $u \in M^*$ at $\partial\Omega$ then it follows that*

$$(\text{Re}\lambda - \Lambda(t))\|u\|_\Omega^2 \le c\|\phi_-^t(L^* + \overline{\lambda}H)\phi_-^{-t}u\|_\Omega^2.$$

Proof. We write $(\phi_-^{-1}G_-u, u)_\Omega = \sum_{i=1}^2(\phi_-^{-1}G_-u, \psi_{-,i}u)_\Omega = I_1 + I_2$. Arguments similar to the proof of Proposition 5.2 give

$$I_1 \ge \delta\|\psi_{-,1}^{1/2}m_-^{-1/2}u\|_\Omega^2 \ge 0, \quad I_2 \ge -c_2\|\psi_{-,2}^{1/2}\phi_-^{-1/2}u\|_\Omega^2 \ge -c_3\|\psi_{-,2}^{1/2}u\|_\Omega^2$$

and hence $(\phi_-^{-1}G_-u, u)_\Omega \ge -c_3\|u\|_\Omega^2$. Applying Lemma 5.3 with $\eta = 0$ we conclude the proof. □

The following lemma assures the existence of weak solution in $\phi_- L^2(\Omega)$.

Lemma 5.5. *There is a $c > 0$ such that for $t \geq 1$ we can take a $\Lambda(t) \in$ **R** verifying the following properties: If $f \in \phi_-^t L^2(\Omega)$ and* Re$\lambda > \Lambda(t)$ *then there exists a weak solution $u \in \phi_-^t L^2(\Omega)$ to* (BVP) *satisfying*

$$(\mathrm{Re}\lambda - \Lambda(t))\|\phi_-^{-t}u\|_\Omega^2 \leq c\|\phi_-^{-t}(L+\lambda H)u\|_\Omega^2.$$

Proof. Let us set $E = \{\phi_-^t(L^* + \overline{\lambda}H)\psi;\ \psi \in C^{0,1}(\overline{\Omega})$ with $\psi \in M^*$ at $\partial\Omega\}$ and we study the map $T : E \ni \phi_-^t(L^* + \overline{\lambda}H)\psi \mapsto (\psi, f)_\Omega \in$ **C**. From Lemma 5.4 we obtain that

$$\begin{aligned} |(\psi, f)_\Omega|^2 &\leq \|\phi_-^t\psi\|_\Omega^2\|\phi_-^{-t}f\|_\Omega^2 \\ &\leq (\mathrm{Re}\lambda - \Lambda(t))^{-1}c\|\phi_-^t(L^* + \overline{\lambda}H)\psi\|_\Omega^2\|\phi_-^{-t}f\|_\Omega^2. \end{aligned}$$

By the Hahn-Banach theorem there is a $w \in L^2(\Omega)$ such that

$$(\mathrm{Re}\lambda - \Lambda(t))\|w\|_\Omega^2 \leq \|\phi_-^{-t}f\|_\Omega^2, \quad (\phi_-^t(L^* + \overline{\lambda}H)\psi, w)_\Omega = (\psi, f)_\Omega$$

for every $\psi \in C^{0,1}(\overline{\Omega})$ with $\psi \in M^*$ at $\partial\Omega$. Then $u = \phi_-^t w$ is a desired weak solution. □

To get existence and uniqueness of solution vanishing near O^- we first show

Lemma 5.6. *There is a $\eta_0 > 0$ such that for $t \in$ **R** we can take a $\Lambda(t) \in$ **R** verifying the following properties: If $0 < \eta < \eta_0$ and* Re$\lambda > \Lambda(t)$ *and if $u \in C^{0,1}(\overline{\Omega})$ with* supp$u \cap \{\phi_{-,\eta} = 0\} = \emptyset$ *and $u \in M^*$ at $\partial\Omega \cap \{\phi_{-,\eta} > 0\}$ then it follows that*

$$\begin{aligned} (\mathrm{Re}\lambda - \Lambda(t))\|u\|_{\Omega_\eta^-}^2 &+ c_0(t - 1/4)\|\phi_{-,\eta}^{-1/2}u\|_{\Omega_\eta^-} \\ &\leq c_1\|\phi_{-,\eta}^{t+1/2}(L^* + \overline{\lambda}H)\phi_{-,\eta}^{-t}u\|_{\Omega_\eta^-} \end{aligned}$$

where $c_0, c_1 > 0$ depend only on η.

Proof. Set $\eta_0 = \eta^*/2$ and suppose $0 < \eta < \eta_0$. Let us write

$$(\phi_{-,\eta}^{-1}G_-u, u)_{\Omega_\eta^-} = \sum_{i=1}^{2}(\phi_{-,\eta}^{-1}G_-u, \psi_{-,i}u)_{\Omega_\eta^-} = I_1 + I_2.$$

Then it follows from Lemma 4.5 that $I_1 \geq \delta(\phi_{-,\eta}^{-1}m_-^{-1}\phi_-u, \psi_{-,1}u)_{\Omega_\eta^-}$. Noticing that, on Ω, m_- is bounded and $\phi_- \geq \eta$ we have

$$I_1 \geq \delta'\eta\|\psi_{-,1}^{1/2}\phi_{-,\eta}^{-1/2}u\|_{\Omega_\eta^-}^2 \tag{5.3}$$

with some $\delta' > 0$. We turn to I_2. Since $\phi_{-,\eta} \geq \eta^*/2$ on Ω, arguments similar to the proof of Proposition 5.2 show that

$$I_2 \geq \delta'\eta\|\psi_{-,2}^{1/2}\phi_{-,\eta}^{-1/2}u\|^2_{\Omega_\eta^-} - c_3\|\psi_{-,2}^{1/2}u\|^2_{\Omega_\eta^-} \tag{5.4}$$

where $c_3 = c_3(\eta_0) > 0$ is independent of η. Combining (5.3) and (5.4) we obtain

$$(\phi_{-,\eta}^{-1}G_-u, u)_{\Omega_\eta^-} \geq \delta'\eta\|\phi_{-,\eta}^{-1/2}u\|^2_{\Omega_\eta^-} - c_3\|u\|^2_{\Omega_\eta^-},$$

and hence Lemma 5.3 ends the proof. □

Proof of Proposition 3.2. By virtue of Lemma 5.6, the proof is a repetition of the proof of Proposition 4.6 in [6]. □

We now give the proof of Proposition 2.2.

Proof of Proposition 2.2. Assuming that $u \in m_-L^2(\Omega)$ is a weak solution to (BVP) with $f = 0$ we wish to show $u = 0$. Let $g \in C_0^\infty(\Omega)$. Repeating the same arguments as in the proof of Proposition 3.2 we can find $v \in L^2(\Omega)$ with $\operatorname{supp} v \cap (O^+ \cup \gamma^+) = \emptyset$ which is a weak solution to the following adjoint boundary value problem:

$$\text{(BVP*)} \qquad \begin{cases} (L^* + \overline{\lambda}H)v = g & \text{in} \quad \Omega \\ \qquad\qquad v \in M^* & \text{at} \quad \partial\Omega. \end{cases}$$

Let us choose $\chi \in C_0^\infty(\mathbf{R})$ such that $\chi = 1$ near 0 and set

$$v_k = (1 - \chi(km_-))v, \quad g_k = (L^* + \overline{\lambda}H)v_k$$

for $k > 0$. Then v_k is also a weak solution to (BVP*) with the right-hand side g_k. Since $\operatorname{supp} v_k \cap \gamma = \emptyset$ for each v_k Theorem 4 in [9] gives $\{v_{k,\epsilon}\} \subset C^1(\overline{\Omega})$ with $v_{k,\epsilon} \in M^*$ at $\partial\Omega$ such that

$$v_{k,\epsilon} \to v_k, \quad (L^* + \overline{\lambda}H)v_{k,\epsilon} \to g_k \quad \text{in} \quad L^2(\Omega) \quad \text{as} \quad \epsilon \to 0.$$

Recalling that u is a weak solution to (BVP) with $f = 0$ we obtain $(u, (L^* + \overline{\lambda}H)v_{k,\epsilon})_\Omega = 0$ and hence letting $\epsilon \to 0$ we have $(u, g_k)_\Omega = 0$. Note that $(u, g_k)_\Omega$ converges to $(u, g)_\Omega$ as $k \to \infty$. Indeed writing

$$(u, g_k)_\Omega = (u, (1 - \chi(km_-))g)_\Omega + (u, k\chi'(km_-)M_-v)_\Omega = I_1 + I_2$$

with $M_- = \sum_{j=1}^n (\partial_j m_-)A_j$, the dominated convergence theorem shows that $I_1 \to (u, g)_\Omega$. We turn to I_2. Since $u = m_-w$ with some $w \in L^2(\Omega)$ and $|\theta\chi'(\theta)| \leq c$ with some $c > 0$ the dominated convergence theorem again proves that $I_2 = (w, km_-\chi'(km_-)M_-v)_\Omega \to 0$ as $k \to \infty$. Thus we have $(u, g)_\Omega = 0$. Noticing $C_0^\infty(\Omega)$ is dense in $L^2(\Omega)$ we conclude $u = 0$. □

An immediate corollary to Proposition 2.2 and Lemma 5.5 is

Corollary 5.7. *There is a $c > 0$ such that for $t \geq 1$ we can take a $\Lambda(t) \in \mathbf{R}$ verifying the following properties: If $f \in L^2(\Omega)$ and* $\mathrm{Re}\lambda > \Lambda(t)$ *and if $u \in L^2(\Omega)$ with* $\mathrm{supp}\, u \cap (O^- \cup \gamma^-) = \emptyset$ *is a weak solution to* (BVP) *then it follows that*

$$(\mathrm{Re}\lambda - \Lambda(t))\|\phi_-^{-t}u\|_\Omega^2 \leq c\|\phi_-^{-t}(L + \lambda H)u\|_\Omega^2.$$

6. Proof of Proposition 3.3 (1)

To prove Proposition 3.3 we shall localize the problem. Let $\{U_i\}$, $\{\chi_i\}$ and $\{\psi_i\}$ be the covering of Ω, the coordinate systems and the partition of unity as in Section 3, respectively. Suppose that $f \in L^2(\Omega)$ and that $u \in L^2(\Omega)$ with $\mathrm{supp}\, u \cap (O^- \cup \gamma^-) = \emptyset$ is a weak solution to (BVP). If we set

$$u_i = \psi_i u, \quad f_i = (L + \lambda H)u_i = \psi_i f + \sum_{j=1}^n (\partial_j \psi_i) A_j u,$$

then $\mathrm{supp}\, u_i \cap (O^- \cup \gamma^-) = \emptyset$ and u_i is also a weak solution to (BVP) with right-hand side f_i. Therefore it suffices to prove Proposition 3.3 with u_i instead of u. We may assume that χ_i satisfies

$$\begin{aligned} &x_1 = r \circ \chi_i^{-1}, \quad x_2 = \pm h_\pm \circ \chi_i^{-1} && \text{if } \ U_i \cap \gamma^\pm \neq \emptyset, \\ &x_1 = r \circ \chi_i^{-1} && \text{if } \ U_i \cap \gamma = \emptyset \ \text{ but } \ U_i \cap \partial\Omega \neq \emptyset. \end{aligned}$$

If $U_i \cap \gamma = \emptyset$ then Proposition 3.3 with u_i instead of u is easily checked. Thus the interesting patches are at γ. In what follows, we simply write U, u, f for U_i, u_i, f_i. Moreover we may assume that U is as in Lemma 4.3.

Now suppose that $U_i \cap \gamma^+ \neq \emptyset$. Then performing a change of independent variables we are led to the case that

$$\begin{gathered} U = \{|x| < 1\}, \quad \Omega = \mathbf{R}_+^n, \quad \partial\Omega = \partial\mathbf{R}_+^n, \\ \gamma^+ = \{(0, 0, x'');\ x'' \in \mathbf{R}^{n-2}\}, \quad O^+ = \{(0, x');\ x' \in \mathbf{R}^{n-1},\ x_2 > 0\}, \\ r = x_1, \quad h_+ = x_2 \end{gathered}$$

$$\mathrm{supp}\, u \subset \{|x| < 1 - \epsilon_0,\ x_1 \geq 0,\ \phi_-(x) > 0\} \tag{6.1}$$

with $\epsilon_0 > 0$ small enough. Let $Q(x)$ be as in Lemma 4.3 and set

$$\begin{gathered} \tilde{L} = Q^* L Q, \quad \tilde{H} = Q^* H Q, \\ \tilde{M}(x') = Q^{-1}(0, x') M(x'), \quad \tilde{M}^*(x') = Q^{-1}(0, x') M^*(x'). \end{gathered}$$

Then $\tilde{L}$ is a first order symmetric system, $\tilde{M}$ is a maximal positive boundary space for $\tilde{L}$ and $\tilde{M}^*$ is the adjoint boundary space of $\tilde{M}$. Setting $\tilde{u} = Q^{-1}u$, $\tilde{f} = Q^* f$ it becomes that $\tilde{u}$ is a weak solution to the following boundary value problem

$$\begin{cases} (\tilde{L} + \lambda \tilde{H})\tilde{u} = \tilde{f} & \text{in } \Omega \\ \tilde{u} \in \tilde{M} & \text{at } \partial\Omega. \end{cases}$$

Therefore it suffices to prove Proposition 3.3 with $\tilde{u}$ instead of u. We now drop the tildes and work with the transformed boundary value problem. Noting Lemma 4.3 we may assume that

$$A_b(x') = \begin{pmatrix} x_2 I_p & 0 \\ 0 & I_{N-p} \end{pmatrix}, \tag{6.2}$$

$$M(x') = \begin{cases} \mathbf{C}^N & \text{if } x_2 > 0 \\ \{0\} \times \mathbf{C}^{N-p} & \text{if } x_2 < 0, \end{cases} \tag{6.3}$$

$$M^*(x') = \begin{cases} \{0\} & \text{if } x_2 > 0 \\ \mathbf{C}^p \times \{0\} & \text{if } x_2 < 0 \end{cases} \tag{6.4}$$

for $(0, x') \in \partial\mathbf{R}^n_+ = \partial\Omega$. If $U_i \cap \gamma^- \neq \emptyset$ the boundary value problem can be transformed into a similar one. We start with proving (3.1). For this purpose we make a more detailed study of the space

$$X^q_{(s,t)}(\mathbf{R}^n_+; \partial\mathbf{R}^n_+) = \bigcap_{j=0}^{q} \phi_+^{s+q-j} \phi_-^{t+q-j} H^j(\mathbf{R}^n_+; \partial\mathbf{R}^n_+).$$

If $\sigma, \tau \in \mathbf{R}$ and $\alpha \in \mathbf{Z}^n_+$ then $|\partial^\alpha(\phi_+^\sigma \phi_-^\tau)| \le c\phi_+^{\sigma-|\alpha|}\phi_-^{\tau-|\alpha|}$ on $\{|x| < 1,\ x_1 \ge 0\}$ with some $c = c(\sigma, \tau, \alpha) > 0$. This implies

Lemma 6.1. *If* $u \in X^q_{(s,t)}(\mathbf{R}^n_+; \partial\mathbf{R}^n_+)$ *with* $\operatorname{supp} u \subset \{|x| < 1,\ x_1 \ge 0\}$ *then the following two norms are equivalent:*

$$\|u\|_{X^q_{(s,t)}(\mathbf{R}^n_+;\partial\mathbf{R}^n_+)}, \quad \sum_{|\alpha| \le q} \|\phi_+^{-s-q+|\alpha|}\phi_-^{-t-q+|\alpha|} Z^\alpha u\|_{L^2(\mathbf{R}^n_+)}.$$

The following two lemmas are easily checked using Lemma 6.1.

Lemma 6.2. *If* $q \ge 1$ *and if* $u \in X^q_{(s,t)}(\mathbf{R}^n_+; \partial\mathbf{R}^n_+)$ *with* $\operatorname{supp} u \subset \{|x| < 1,\ x_1 \ge 0\}$ *then the following two norms are equivalent:*

$$\|u\|_{X^q_{(s,t)}(\mathbf{R}^n_+;\partial\mathbf{R}^n_+)}, \quad \|u\|_{X^{q-1}_{(s+1,t+1)}(\mathbf{R}^n_+;\partial\mathbf{R}^n_+)} + \sum_{|\alpha|=1} \|Z^\alpha u\|_{X^{q-1}_{(s,t)}(\mathbf{R}^n_+;\partial\mathbf{R}^n_+)}.$$

Lemma 6.3. *If* $q_1 \geq q_2$, $q_1+s_1 \geq q_2+s_2$, $q_1+t_1 \geq q_2+t_2$, *and letting* $u \in X^{q_1}_{(s_1,t_1)}(\mathbf{R}^n_+;\partial\mathbf{R}^n_+)$ *with* $\operatorname{supp} u \subset \{|x|<1,\ x_1 \geq 0\}$, *then it follows that* $u \in X^{q_2}_{(s_2,t_2)}(\mathbf{R}^n_+;\partial\mathbf{R}^n_+)$ *and*

$$\|u\|_{X^{q_2}_{(s_2,t_2)}(\mathbf{R}^n_+;\partial\mathbf{R}^n_+)} \leq c\|u\|_{X^{q_1}_{(s_1,t_1)}(\mathbf{R}^n_+;\partial\mathbf{R}^n_+)}$$

where $c>0$ *is independent of* u.

In particular, $u \in X^{q-1}_{(-s+1,t+1)}(\mathbf{R}^n_+;\partial\mathbf{R}^n_+)$ with $\operatorname{supp} u \subset \{|x|<1-\epsilon_0,\ x_1 \geq 0\}$ implies that $u \in X^{q-1}_{(-s,t)}(\mathbf{R}^n_+;\partial\mathbf{R}^n_+)$, which proves (3.1) of Proposition 3.3.

The following lemma proves that $C_0^\infty(\Omega)$ is dense in $X^q_{(s,t)}(\Omega;\partial\Omega)$.

Lemma 6.4. *If* $u \in X^q_{(s,t)}(\mathbf{R}^n_+;\partial\mathbf{R}^n_+)$ *with* $\operatorname{supp} u \subset \{|x|<1-\epsilon_0,\ x_1 \geq 0\}$ *then we can choose a* $\{u_\epsilon\} \subset C_0^\infty(\mathbf{R}^n_+)$ *with* $\operatorname{supp} u_\epsilon \subset \{|x|<1,\ x_1>0\}$ *such that*

$$u_\epsilon \to u \quad in \quad X^q_{(s,t)}(\mathbf{R}^n_+;\partial\mathbf{R}^n_+) \quad as \quad \epsilon \to 0.$$

The proof of this lemma follows from

Lemma 6.5. *If* $\chi \in C_0^\infty(\mathbf{R})$ *and if* $u \in X^q_{(s,t)}(\mathbf{R}^n_+;\partial\mathbf{R}^n_+)$ *with* $\operatorname{supp} u \subset \{|x|<1,\ x_1 \geq 0\}$ *then it follows that*

$$\chi(kx_1)u \to 0 \quad in \quad X^q_{(s,t)}(\mathbf{R}^n_+;\partial\mathbf{R}^n_+) \quad as \quad k \to \infty.$$

Proof. We proceed by induction on q. The dominated convergence theorem shows the statement for $q=0$. Inductively assume that the assertion is true up to $q-1$. Suppose that $u \in X^q_{(s,t)}(\mathbf{R}^n_+;\partial\mathbf{R}^n_+)$ with $\operatorname{supp} u \subset \{|x|<1,\ x_1 \geq 0\}$. From Lemma 6.2 it suffices to show that

$$\chi(kx_1)u \to 0 \quad \text{in} \quad X^{q-1}_{(s+1,t+1)}(\mathbf{R}^n_+;\partial\mathbf{R}^n_+), \tag{6.6}$$

$$Z^\alpha(\chi(kx_1)u) \to 0 \quad \text{in} \quad X^{q-1}_{(s,t)}(\mathbf{R}^n_+;\partial\mathbf{R}^n_+), \quad |\alpha|=1. \tag{6.7}$$

Since $u \in X^{q-1}_{(s+1,t+1)}(\mathbf{R}^n_+;\partial\mathbf{R}^n_+)$ by Lemma 6.3, we have (6.6) by inductive hypothesis. We turn to (6.7). Note that

$$Z_1(\chi(kx_1)u) = \chi(kx_1)Z_1u + kx_1\chi'(kx_1)u,$$
$$Z_j(\chi(kx_1)u) = \chi(kx_1)Z_ju, \quad j=2,\dots,n.$$

Since $Z^\alpha u \in X^{q-1}_{(s,t)}(\mathbf{R}^n_+;\partial\mathbf{R}^n_+)$, $|\alpha|=1$, by Lemma 6.2 then we have

$$\chi(kx_1)Z^\alpha u \to 0 \quad \text{in} \quad X^{q-1}_{(s,t)}(\mathbf{R}^n_+;\partial\mathbf{R}^n_+)$$

by inductive hypothesis. Furthermore writing $\tilde{\chi}(\theta)=\theta\chi'(\theta)$ we also obtain that

$$kx_1\chi'(kx_1)u = \tilde{\chi}(kx_1)u \to 0 \quad \text{in} \quad X^{q-1}_{(s,t)}(\mathbf{R}^n_+;\partial\mathbf{R}^n_+)$$

by inductive hypothesis. This concludes the proof. □

Proof of Lemma 6.4. Let $\chi \in C_0^\infty(\mathbf{R})$ and assume that $\chi = 1$ near 0. If we set $u_k = (1-\chi(kx_1))u$ for $k>0$ then it follows from Lemma 6.5 that $u_k \to u$ in $X^q_{(s,t)}(\mathbf{R}^n_+;\partial\mathbf{R}^n_+)$ as $k\to\infty$. Thus we may assume without loss of generality that $\mathrm{supp}u \subset \{|x|<1,\ x_1>0\}$. Let $\rho \in C_0^\infty(\mathbf{R}^n)$ and assume that $\rho \ge 0$ and $\int \rho(y)dy = 1$. If we write $\rho_\epsilon(y)=\epsilon^{-n}\rho(y/\epsilon)$ with $\epsilon>0$ small enough then $u_\epsilon = u*\rho_\epsilon$ is the desired sequence. □

7. Weak solutions and approximants

In this section we construct approximants for weak solutions vanishing near O^-. Let J_ϵ and χ be as in (3.3). In what follows we assume that χ satisfies

$$\mathrm{supp}\chi \subset \{|y|<\epsilon_0,\ y_1<0,\ y_2>0\} \tag{7.1}$$

where $\epsilon_0>0$ is given in (6.1). We simply write $\|\cdot\|$ for the norm in $L^2(\mathbf{R}^n_+)$ or in $L^2(\mathbf{R}^n)$ if there is no confusion. Now we introduce the following spaces:

$$\begin{aligned} D_{R,\eta}(\mathbf{R}^n_+) =& \{u\in L^2(\mathbf{R}^n_+);\ Lu\in L^2(\mathbf{R}^n_+),\\ &\quad \mathrm{supp}u \subset \{|x|<R,\ x_1\ge 0,\ \phi_-(x)>\eta\}\},\\ \|u\|_{D(\mathbf{R}^n_+)} =& \|u\| + \|(L+\lambda H)u\| \end{aligned}$$

for $0<R\le 1$, $0\le\eta\le 1$. Then we note that

$$D_{R,0}(\mathbf{R}^n_+) = \bigcup_{0<\eta\le 1} D_{R,\eta}(\mathbf{R}^n_+).$$

Moreover from Friedrichs [1], we see that $C^1(\overline{\mathbf{R}^n_+})\cap D_{R,\eta}(\mathbf{R}^n_+)$ is dense in $D_{R,\eta}(\mathbf{R}^n_+)$. We shall prove the following proposition.

Proposition 7.1. *If* $s \geq 3$, $t \in \mathbf{R}$ *and* $0 < \epsilon \leq 1$ *and if* $u \in D_{1-\epsilon_0,0}(\mathbf{R}^n_+)$ *is a weak solution to* (BVP) *then it follows that* $\phi^s_+\phi^t_- J_\epsilon u \in D_{1,0}(\mathbf{R}^n_+)$ *and that* $\phi^s_+\phi^t_- J_\epsilon u$ *is also a weak solution to* (BVP).

To prove this we use the following lemma which is easily checked.

Lemma 7.2. *For* $0 < \eta \leq 1$ *there is a* $\eta_0 = \eta_0(\eta) > 0$ *such that if* $u \in D_{1-\epsilon_0,\eta}(\mathbf{R}^n_+)$ *then*

$$\mathrm{supp} J_\epsilon u \subset \{|x| < 1,\, x_1 \geq 0,\, \phi_-(x) > \eta_0\}$$

for all $0 < \epsilon \leq 1$.

Proposition 7.3. *Suppose that* $s \geq 3$, $t \in \mathbf{R}$, $0 < \eta \leq 1$ *and that* $\eta_0 > 0$ *is as in* Lemma 7.2. *Then the map*

$$\phi^s_+\phi^t_- J_\epsilon : D_{1-\epsilon_0,\eta}(\mathbf{R}^n_+) \to D_{1,\eta_0}(\mathbf{R}^n_+)$$

is continuous.

Proof. Noticing Lemma 7.2 we may assume that $s, t \geq 3$. We wish to show that

$$\|(L + \lambda H)\phi^s_+\phi^t_- J_\epsilon u\| \leq c\|u\|_{D(\mathbf{R}^n_+)}$$

with some $c > 0$. Note that

$$\begin{aligned}\|(L + \lambda H)\phi^s_+\phi^t_- J_\epsilon u\| \leq &\|\phi^s_+\phi^t_-(L + \lambda H)J_\epsilon u\| \\ &+ s\|\phi^{s-1}_+\phi^t_- G_+ J_\epsilon u\| + t\|\phi^s_+\phi^{t-1}_- G_- J_\epsilon u\|.\end{aligned}$$

The second and the third terms on the right-hand side are bounded from above by $c\|u\|$ with some $c > 0$. We now study the first term. Since

$$\begin{aligned}\|\phi^s_+\phi^t_-(L + \lambda H)J_\epsilon u\| &\leq c\|m\phi^{s-1}_+\phi^{t-1}_-(L + \lambda H)J_\epsilon u\| \\ &\leq c'\sum_{i=1}^{2}\|x_i\phi^{s-1}_+\phi^{t-1}_-(L + \lambda H)J_\epsilon u\| \\ &\leq c'\sum_{i=1}^{2}\{\|\phi^{s-1}_+\phi^{t-1}_- J_\epsilon x_i(L + \lambda H)u\| \\ &\quad + \|\phi^{s-1}_+\phi^{t-1}_-[x_i(L + \lambda H), J_\epsilon]u\|\},\end{aligned}$$

the interesting term is $\|\phi^{s-1}_+\phi^{t-1}_-[x_i(L + \lambda H), J_\epsilon]u\|$.

Lemma 7.4. *For $s \geq 2$, $t \in \mathbf{R}$ and $0 < \eta \leq 1$ there is a $c = c(s,t,\lambda,\eta) > 0$ such that if $u \in D_{1-\epsilon_0,\eta}(\mathbf{R}^n_+)$ then it follows that*

$$\sum_{i=1}^{2} \|\phi_+^s \phi_-^t [x_i(L+\lambda H), J_\epsilon]u\|$$

$$\leq c\{\epsilon\|u\|_{D(\mathbf{R}^n_+)} + \sum_{1\leq j\leq n,\ |\alpha|=1} \|\int_{\mathbf{R}^n} a_{j,\alpha}(\cdot,y)\partial_{x_j}u^{\#}(\cdot - y)y^\alpha \chi_\epsilon(y)dy\|\}$$

for $0 < \epsilon \leq 1$ where $a_{j,\alpha}(x,y) \in \mathcal{B}^\infty(\mathbf{R}^n \times \mathbf{R}^n)$, the set of all smooth functions on $\mathbf{R}^n \times \mathbf{R}^n$ with bounded derivatives of all order.

The proof of this lemma is given in Section 8. Admitting Lemma 7.4 we end the proof of Proposition 7.3 showing

$$\|\int a(\cdot,y)\partial_{x_j}u^{\#}(\cdot - y)y^\alpha\chi_\epsilon(y)dy\| \leq c\|u\|.$$

Using $\partial_{x_j}u^{\#}(x-y) = -\partial_{y_j}u^{\#}(x-y)$ we have

$$\int a(x,y)\partial_{x_j}u^{\#}(x-y)y^\alpha\chi_\epsilon(y)dy = \int \partial_{y_j}\{a(x,y)y^\alpha\chi_\epsilon(y)\}u^{\#}(x-y)dy$$

which proves the assertion. □

Now let $\psi \in C^{0,1}(\partial\mathbf{R}^n_+)$. Then the map

$$C^1(\overline{\mathbf{R}^n_+}) \cap D_{R,0}(\mathbf{R}^n_+) \ni u \mapsto \int_{\partial\mathbf{R}^n_+} \langle A_b(0,x')u(0,x'), \psi(0,x')\rangle dx' \in \mathbf{C}$$

extends uniquely to a continuous linear map $D_{R,0}(\mathbf{R}^n_+) \to \mathbf{C}$ (see Rauch [9]). We still denote this extended continuous linear map by $\int_{\partial\mathbf{R}^n_+}\langle A_b u, \psi\rangle dx'$. Furthermore the following proposition was proved there

Proposition 7.5. *Let $u \in D_{R,0}(\mathbf{R}^n_+)$. Then the following two conditions are equivalent.*

(i) *u is a weak solution to* (BVP),

(ii) *The identity $\int_{\partial\mathbf{R}^n_+}\langle A_b u, \psi\rangle dx' = 0$ holds for all $\psi \in C^{0,1}(\partial\mathbf{R}^n_+)$ with $\psi \in M^*$ at $\partial\mathbf{R}^n_+$.*

Proof of Proposition 7.1. Let $u \in D_{1-\epsilon_0,\eta}(\mathbf{R}^n_+)$ be a weak solution to (BVP). Noticing Lemma 7.2 we may assume that $s, t \geq 3$. Let $\psi \in C^{0,1}(\partial\mathbf{R}^n_+)$ satisfy $\psi \in M^*$ at $\partial\mathbf{R}^n_+$. We wish to show that

$\int_{\partial\mathbf{R}^n_+}\langle A_b\phi^s_+\phi^t_- J_\epsilon u,\psi\rangle dx' = 0$. Choose $\{u_n\} \subset C^1(\overline{\mathbf{R}^n_+}) \cap D_{1-\epsilon_0,\eta}(\mathbf{R}^n_+)$ such that $\|u_n - u\|_{D(\mathbf{R}^n_+)} \to 0$ as $n \to \infty$. Then for each u_n we have

$$\begin{aligned}(7.2)\qquad &\int_{\partial\mathbf{R}^n_+}\langle (A_b\phi^s_+\phi^t_- J_\epsilon u_n)(0,x'),\psi(x')\rangle dx'\\ &= \int_{\partial\mathbf{R}^n_+}\langle u_n(0,x'),(J_{-\epsilon}\phi^s_+\phi^t_- A_b\psi)(0,x')\rangle dx'.\end{aligned}$$

Since Proposition 7.3 shows that $\|\phi^s_+\phi^t_- J_\epsilon u_n - \phi^s_+\phi^t_- J_\epsilon u\|_{D(\mathbf{R}^n_+)} \to 0$, the left-hand side of (7.2) goes to $\int_{\partial\mathbf{R}^n_+}\langle A_b\phi^s_+\phi^t_- J_\epsilon u,\psi\rangle dx'$ as $n \to \infty$ by definition. Thus it suffices to prove that the right-hand side of (7.2) goes to zero as $n \to \infty$.

Only the case $U \cap \gamma^+ \neq \emptyset$ will be proved. Let us set

$$\begin{aligned}\Psi(x') &= \tilde{A}(x')x_2^{-1}(J_{-\epsilon}\phi^s_+\phi^t_- A_b\psi)(0,x')\\ &= \tilde{A}(x')\int_{\mathbf{R}^n} x_2^{-1}\phi^s_+(0,x'-y')\phi^t_-(0,x'-y')\\ &\qquad\qquad \times (A_b\psi)(x'-y')\chi_{-\epsilon}(y)dy\end{aligned}$$

where $\tilde{A}(x') = \begin{pmatrix} I_p & 0\\ 0 & x_2 I_{N-p}\end{pmatrix}$. Then the right-hand side of (7.2) is

$$(7.3)\qquad \int_{\partial\mathbf{R}^n_+}\langle (A_b u_n)(0,x'),\Psi(x')\rangle dx'.$$

Since $\phi_+(0,x') = |x_2| - x_2$ and $\{y_2;\ y \in \operatorname{supp}\chi_{-\epsilon}\} = [\alpha,\beta]$ with some $\alpha < \beta < 0$, we have $a(x_2,y_2) \in C^1(\mathbf{R}\times[\alpha,\beta])$ with $a(x_2,y_2) = x_2^{-1}\{|x_2-y_2| - (x_2-y_2)\}^2$ and hence $\Psi \in C^{0,1}(\partial\mathbf{R}^n_+)$. Moreover it follows from (6.2) and (6.4) that

$$(A_b\psi)(x') \in M^*(x') = \begin{cases}\{0\} & \text{if}\quad x_2 > 0\\ \mathbf{C}^p\times\{0\} & \text{if}\quad x_2 < 0,\end{cases}$$

and hence $\Psi \in M^*$ at $\partial\mathbf{R}^n_+$. Therefore using $\|u_n - u\|_{D(\mathbf{R}^n_+)} \to 0$ we obtain that (7.3) goes to $\int_{\partial\mathbf{R}^n_+}\langle A_b u,\Psi\rangle dx'$ by definition. Since u is a weak solution to (BVP), Proposition 7.5 shows that $\int_{\partial\mathbf{R}^n_+}\langle A_b u,\Psi\rangle dx' = 0$. This concludes the proof. □

We next study approximants for weak solutions.

Proposition 7.6. *Let $u \in D_{1-\epsilon_0,0}(\mathbf{R}^n_+)$ be a weak solution to* (BVP). *Then there is a $\{u_\epsilon\} \subset H^1(\mathbf{R}^n_+; \partial\mathbf{R}^n_+)$ with* $\mathrm{supp}\, u_\epsilon \subset \{|x| < 1,\ x_1 \geq 0,\ \phi_-(x) > \eta_0\}$ *with some $\eta_0 > 0$ such that if $s \geq 3$ and $t \in \mathbf{R}$ then $\phi_+^s\phi_-^t u_\epsilon$ is also a weak solution to* (BVP) *and satisfies that*

$$u_\epsilon \to u,\ \ \phi_+^s\phi_-^t(L+\lambda H)u_\epsilon \to \phi_+^s\phi_-^t(L+\lambda H)u \ \text{ in } L^2(\mathbf{R}^n_+) \ \text{ as } \epsilon \to 0.$$

Proof. We may assume that $u \in D_{1-\epsilon_0,\eta}(\mathbf{R}^n_+)$ with some $\eta > 0$. Moreover we assume that χ in (3.3) satisfies not only (7.1) but also $\chi \geq 0$ and $\int \chi(y)dy = 1$. If we set $u_\epsilon = J_\epsilon u$ then $\{u_\epsilon\}$ is a desired sequence. Indeed we have only to show that $\phi_+^s\phi_-^t(L+\lambda H)u_\epsilon \to \phi_+^s\phi_-^t(L+\lambda H)u$ in $L^2(\mathbf{R}^n_+)$. Arguments similar to the proof of Proposition 7.3 give that

$$\begin{aligned}&\|\phi_+^s\phi_-^t(L+\lambda H)u_\epsilon - \phi_+^s\phi_-^t(L+\lambda H)u\|\\ &\quad\leq c\sum_{i=1}^{2}\{\|\phi_+^{s-1}\phi_-^{t-1}J_\epsilon x_i(L+\lambda H)u - \phi_+^{s-1}\phi_-^{t-1}x_i(L+\lambda H)u\|\\ &\qquad + \|\phi_+^{s-1}\phi_-^{t-1}[x_i(L+\lambda H), J_\epsilon]u\|\}.\end{aligned}$$

The interesting term is $\|\phi_+^{s-1}\phi_-^{t-1}[x_i(L+\lambda H), J_\epsilon]u\|$. From Lemma 7.4 it suffices to prove that

$$U_\epsilon(x) = \int a_{j,\alpha}(x,y)\partial_{x_j}u^\#(x-y)y^\alpha\chi_\epsilon(y)dy \to 0 \ \text{ in } L^2(\mathbf{R}^n) \ \text{ as } \epsilon \to 0.$$

Here we note that

$$U_\epsilon(x) = \int \partial_{y_j}\{a_{j,\alpha}(x,y)y^\alpha\chi_\epsilon(y)\}u^\#(x-y)dy.$$

Since $a_{j,\alpha}(x,y)y^\alpha\chi_\epsilon(y)$, considered as a function of y, is in $C_0^\infty(\mathbf{R}^n)$ so that

$$\int \partial_{y_j}\{a_{j,\alpha}(x,y)y^\alpha\chi_\epsilon(y)\}dy = 0$$

and hence

$$U_\epsilon(x) = \int \partial_{y_j}\{a_{j,\alpha}(x,y)y^\alpha\chi_\epsilon(y)\}\{u^\#(x-y) - u^\#(x)\}dy$$

which proves that $U_\epsilon \to 0$ in $L^2(\mathbf{R}^n)$. □

Moreover arguments similar to those in Tartakoff [12] give

Proposition 7.7. *If $w \in H^1(\mathbf{R}^n_+; \partial\mathbf{R}^n_+)$ with* $\operatorname{supp} w \subset \{|x| < 1,\ x_1 \geq 0,\ \phi_-(x) > \eta_0\}$ *and if $\phi_+^s \phi_-^t w$ with some $s > 0$ and $t \in \mathbf{R}$ is a weak solution to* (BVP) *then there is a $\{w_\epsilon\} \subset C^\infty(\mathbf{R}^n_+)$ with* $\operatorname{supp} w_\epsilon \subset \{|x| < 1,\ x_1 \geq 0,\ \phi_-(x) > \eta_0\}$ *and $w_\epsilon \in M$ at $\partial\mathbf{R}^n_+$ verifying the following properties: If $\sigma \geq 1$ and $\tau \in \mathbf{R}$ then*

$$w_\epsilon \to w, \quad \phi_+^{s+\sigma}\phi_-^{t+\tau}(L + \lambda H)w_\epsilon \to \phi_+^{s+\sigma}\phi_-^{t+\tau}(L + \lambda H)w$$

in $L^2(\mathbf{R}^n_+)$ as $\epsilon \to 0$.

Proof. Only the case $U \cap \gamma^+ \neq \emptyset$ will be proved. We now denote by I_ϵ, T_ϵ the operators defined as

$$I_\epsilon v(x) = \int v(x - \epsilon y)\rho(y)dy,$$

$$(T_\epsilon v)_j(x) = \begin{cases} v_j(x_1 - \epsilon, x') & \text{if } \ 1 \leq j \leq p \\ v_j(x_1 + \epsilon, x') & \text{if } \ p + 1 \leq j \leq N \end{cases}$$

where $\rho \in C_0^\infty(\mathbf{R}^n)$ is such that $\operatorname{supp}\rho \subset \{|y| < 1\}$, $\rho(y) = \rho(-y)$, $\rho \geq 0$, $\int \rho(y)dy = 1$ and v_j denotes the j-th component of v. We set $w_\epsilon = (T_\epsilon I_\epsilon w^0)|_{\mathbf{R}^n_+}$ where $w^0 = w$ in $\mathbf{R}^n_+$ and zero elsewhere. Then $\{w_\epsilon\}$ is a desired sequence. Indeed it suffices to show that if $\sigma \geq 1$ then

$$\phi_+^\sigma \phi_-^\tau (L + \lambda H)\phi_+^s \phi_-^t w_\epsilon \to \phi_+^\sigma \phi_-^\tau (L + \lambda H)\phi_+^s \phi_-^t w \quad \text{in} \quad L^2(\mathbf{R}^n_+).$$

Let us set $g = (L + \lambda H)\phi_+^s \phi_-^t w$. We note that

$$\begin{aligned} (7.4) \qquad & \|\phi_+^\sigma \phi_-^\tau (L + \lambda H)\phi_+^s \phi_-^t w_\epsilon - \phi_+^\sigma \phi_-^\tau g\|_{L^2(\mathbf{R}^n_+)} \\ & \leq \|[\phi_+^\sigma \phi_-^\tau (L + \lambda H)\phi_+^s \phi_-^t, T_\epsilon I_\epsilon]w^0\|_{L^2(\mathbf{R}^n_+)} \\ & \quad + \|T_\epsilon I_\epsilon \phi_+^\sigma \phi_-^\tau (L + \lambda H)\phi_+^s \phi_-^t w^0 - \phi_+^\sigma \phi_-^\tau g^0\|_{L^2(\mathbf{R}^n_+)}. \end{aligned}$$

Noticing $w \in H^1(\mathbf{R}^n_+; \partial\mathbf{R}^n_+)$ and using the same reasoning as in [1] we can obtain that the first term on the right-hand side of (7.4) goes to zero as $\epsilon \to 0$. Moreover we shall prove that as a distribution in $\mathbf{R}^n_+$

$$T_\epsilon I_\epsilon \phi_+^\sigma \phi_-^\tau (L + \lambda H)\phi_+^s \phi_-^t w^0 = T_\epsilon I_\epsilon \phi_+^\sigma \phi_-^\tau g^0$$

which shows that the second term on the right-hand side of (7.4) goes to zero as $\epsilon \to 0$. To see this, it is sufficient to prove that

$$(T_\epsilon I_\epsilon \phi_+^\sigma \phi_-^\tau (L + \lambda H)\phi_+^s \phi_-^t w^0, \psi)_{L^2(\mathbf{R}^n_+)} = (T_\epsilon I_\epsilon \phi_+^\sigma \phi_-^\tau g^0, \psi)_{L^2(\mathbf{R}^n_+)}$$

for every $\psi \in C_0^\infty(\mathbf{R}_+^n)$. Since $I_\epsilon^* = I_\epsilon$ and $T_\epsilon^* = T_{-\epsilon}$, it follows that

(7.5)
$$(T_\epsilon I_\epsilon \phi_+^\sigma \phi_-^\tau (L+\lambda H)\phi_+^s \phi_-^t w^0, \psi)_{L^2(\mathbf{R}_+^n)} - (T_\epsilon I_\epsilon \phi_+^\sigma \phi_-^\tau g^0, \psi)_{L^2(\mathbf{R}_+^n)}$$
$$= (\phi_+^s \phi_-^t w, (L^* + \overline{\lambda} H)\phi_+^\sigma \phi_-^\tau I_\epsilon T_{-\epsilon}\psi)_{L^2(\mathbf{R}_+^n)}$$
$$- (g, \phi_+^\sigma \phi_-^\tau I_\epsilon T_{-\epsilon}\psi)_{L^2(\mathbf{R}_+^n)}.$$

On the other hand we have $\phi_+^\sigma \phi_-^\tau I_\epsilon T_{-\epsilon}\psi \in C^{0,1}(\overline{\mathbf{R}_+^n})$. Furthermore it follows from $I_\epsilon T_{-\epsilon}\psi(0,x') \in \mathbf{C}^p \times \{0\}$ and $\phi_+(0,x') = |x_2| - x_2$ that $(\phi_+^\sigma \phi_-^\tau I_\epsilon T_{-\epsilon}\psi)(0,x') \in M^*(x')$ for $(0,x') \in \partial\mathbf{R}_+^n$. Therefore by the definition of weak solution the right-hand side of (7.5) is zero. □

Using Propositions 7.6 and 7.7 we obtain the following corollary.

Corollary 7.8. *Let $u \in D_{1-\epsilon_0,0}(\mathbf{R}_+^n)$ be a weak solution to* (BVP). *Then there is a $\{u_\epsilon\} \subset C^\infty(\overline{\mathbf{R}_+^n}) \cap D_{1,0}(\mathbf{R}_+^n)$ with $u_\epsilon \in M$ at $\partial\mathbf{R}_+^n$ such that if $s \geq 4$ and $t \in \mathbf{R}$ then $\phi_+^s \phi_-^t u_\epsilon$ is also a weak solution to* (BVP) *and satisfies*

$$u_\epsilon \to u, \quad \phi_+^s \phi_-^t (L+\lambda H)u_\epsilon \to \phi_+^s \phi_-^t (L+\lambda H)u \quad in\ L^2(\mathbf{R}_+^n) \quad as\ \epsilon \to 0.$$

8. Proof of Proposition 3.3 (2)

For the proof of Proposition 3.3 we must control terms such as $x_i(L + \lambda H)J_\epsilon u$, $i = 1, 2$. To this end we study $(x_i(L+\lambda H)J_\epsilon u)^\#$ in this section. The following lemma is easily checked.

Lemma 8.1. *Let $u \in D_{1-\epsilon_0,0}(\mathbf{R}_+^n)$. Then*

$$\operatorname{supp}(u^\#(x-y)\chi_\epsilon(y)) \subset \{(x,y);\ x_1 < 0,\ |x'| < 1,\ |y| < \epsilon_0\}.$$

Let us take $\psi \in C^\infty(\mathbf{R}^n \times \mathbf{R}^n)$ such that $\psi(x,y) = 1$ on $\{(x,y);\ x_1 \leq 0,\ |x'| \leq 1,\ |y| \leq \epsilon_0\}$, $\operatorname{supp}\psi \subset \{(x,y);\ x_1 < 1,\ |x'| < 2,\ |y| < 2\epsilon_0\}$. Noting Lemma 8.1, if necessary, we may cut off $u^\#(x-y)\chi_\epsilon(y)$ by ψ. In what follows we denote by $a(x,y)$, which differs from line to line, an element in $\mathcal{B}^\infty(\mathbf{R}^n \times \mathbf{R}^n)$.

Proposition 8.2. *If $u \in D_{1-\epsilon_0,0}(\mathbf{R}_+^n)$ then we can write $(x_i(L +$*

$\lambda H)J_\epsilon u)^\#$, $i = 1, 2$ as a sum of the following terms:

(8.1)
$$\int a(x,y)((L+\lambda H)u)^\#(x-y)\chi_\epsilon(y)dy,$$

(8.2)
$$\int a(x,y)u^\#(x-y)\chi_\epsilon(y)dy,$$

(8.3)
$$\lambda\int a(x,y)u^\#(x-y)y^\alpha\chi_\epsilon(y)dy,$$

(8.4)
$$\epsilon^{-1}\int a(x,y)u^\#(x-y)y^\alpha\chi_\epsilon(y)dy,$$

(8.5)
$$\int a(x,y)(m_\pm^{-2}x_i)^\natural(x-y)((L+\lambda H)u)^\#(x-y)y^\beta\chi_\epsilon(y)dy,$$

(8.6)
$$\int a(x,y)(m_\pm^{-2}x_i)^\natural(x-y)u^\#(x-y)y^\alpha\chi_\epsilon(y)dy,$$

(8.7)
$$\int a(x,y)(Z_j(m_\pm^{-2}x_i))^\natural(x-y)u^\#(x-y)y^\beta\chi_\epsilon(y)dy,$$

(8.8)
$$\lambda\int a(x,y)(m_\pm^{-2}x_i)^\natural(x-y)u^\#(x-y)y^\beta\chi_\epsilon(y)dy,$$

(8.9)
$$\epsilon^{-1}\int a(x,y)(m_\pm^{-2}x_i)^\natural(x-y)u^\#(x-y)y^\beta\chi_\epsilon(y)dy$$

where $i = 1, 2$, $j = 1, \ldots, n$, $|\alpha| = 1$, $|\beta| = 2$.

Proof. We note that

$$(x_i(L+\lambda H)J_\epsilon u)^\# = (J_\epsilon x_i(L+\lambda H)u)^\# + ([x_i(L+\lambda H), J_\epsilon]u)^\#.$$

The first term on the right-hand side can be written as (8.1). Hence we turn to the second term on the right-hand side. Using $A_1(x) = -A_b(x') + x_1\tilde{A}_1(x)$ we have

$$L = -A_b(x')\partial_1 + \tilde{A}_1(x)Z_1 + \sum_{j=2}^{n} A_j(x)Z_j + B(x). \tag{8.10}$$

We first study $([A, J_\epsilon]u)^\#$ and $([AZ_j, J_\epsilon]u)^\#$ with $A(x) \in \mathcal{B}^\infty(\mathbf{R}^n_+)$. Since

$$([A, J_\epsilon]u)^\# = A^\natural(u^\# * \chi_\epsilon) - (A^\natural u^\#) * \chi_\epsilon,$$
$$([AZ_j, J_\epsilon]u)^\# = ([A, J_\epsilon]Z_j u)^\# = A^\natural((Z_j u)^\# * \chi_\epsilon) - (A^\natural(Z_j u)^\#) * \chi_\epsilon,$$

these terms can be written as a sum of the following terms:

$$\int a(x,y)u^\#(x-y)y^\alpha \chi_\epsilon(y)dy, \tag{8.11}$$

$$\int a(x,y)\partial_{x_j} u^\#(x-y)y^\alpha \chi_\epsilon(y)dy. \tag{8.12}$$

Furthermore noticing that $\partial_{x_j} u^\#(x-y) = -\partial_{y_j} u^\#(x-y)$ we can write $([A, J_\epsilon]u)^\#$ and $([AZ_j, J_\epsilon]u)^\#$ as a sum of (8.2) and (8.4).

Thus it only remains to examine $([x_2 A_b \partial_1, J_\epsilon]u)^\#$. We note that

$$\begin{aligned}
&([x_2 A_b \partial_1, J_\epsilon]u)^\# \\
&= x_2 A_b(x') \int (\partial_1 u)^\#(x-y)e^{-y_1}\chi_\epsilon(y)dy \\
&\quad - \int (x_2 - y_2)A_b(x'-y')(\partial_1 u)^\#(x-y)\chi_\epsilon(y)dy \\
&= A_b(x') \int (x_2 - y_2)(\partial_1 u)^\#(x-y)(e^{-y_1} - 1)\chi_\epsilon(y)dy \\
&\quad + A_b(x') \int (\partial_1 u)^\#(x-y)(e^{-y_1} - 1)y_2\chi_\epsilon(y)dy \\
&\quad + \int (A_b(x') - A_b(x'-y'))(x_2 - y_2)(\partial_1 u)^\#(x-y)\chi_\epsilon(y)dy \\
&\quad + \int A_b(x'-y')(\partial_1 u)^\#(x-y)y_2\chi_\epsilon(y)dy \\
&\quad + \int (A_b(x') - A_b(x'-y'))(\partial_1 u)^\#(x-y)y_2\chi_\epsilon(y)dy
\end{aligned}$$

and hence using $x_2 - y_2 = \tilde{A}(x'-y')A_b(x'-y')$ with $\tilde{A}(x') = \begin{pmatrix} I_p & 0 \\ 0 & x_2 I_{N-p} \end{pmatrix}$ we can write $([x_2 A_b \partial_1, J_\epsilon]u)^\#$ as a sum of the following terms:

$$\int a(x,y)(A_b\partial_1 u)^\#(x-y)y^\alpha \chi_\epsilon(y)dy, \tag{8.13}$$

$$\int a(x,y)(\partial_1 u)^\#(x-y)y^\beta \chi_\epsilon(y)dy. \tag{8.14}$$

It follows from (8.10) that (8.13) can be written as a sum of (8.11), (8.12), (8.3) and the following term:

$$\int a(x,y)((L+\lambda H)u)^{\#}(x-y)y^{\alpha}\chi_{\epsilon}(y)dy, \tag{8.15}$$

which is of type (8.1). We turn to (8.14). Since $m_{\pm}(x)^2$ is a homogeneous polynomial in x_1, x_2 of order two, $\partial_1 u$ can be written as a sum of

$$m_{\pm}(x)^{-2}x_1 Z_1 u, \quad m_{\pm}(x)^{-2}x_2 Z_1 u, \quad m_{\pm}(x)^{-2}x_2\tilde{A}(x')A_b(x')\partial_1 u.$$

Therefore (8.14) can be written as a sum of (8.5), (8.8) and the following terms:

$$\int a(x,y)(m_{\pm}^{-2}x_i)^{\natural}(x-y)u^{\#}(x-y)y^{\beta}\chi_{\epsilon}(y)dy, \tag{8.16}$$

$$\int a(x,y)(m_{\pm}^{-2}x_i)^{\natural}(x-y)\partial_{x_j}u^{\#}(x-y)y^{\beta}\chi_{\epsilon}(y)dy. \tag{8.17}$$

Here we note that (8.16) is of type (8.6) and that (8.17) is written as a sum of (8.6), (8.7), (8.9). This concludes the proof. □

Next we shall give the proof of Lemma 7.4. For this purpose we show the following lemma.

Lemma 8.3. *There is a $c > 0$ such that if $u \in D_{1-\epsilon_0,0}(\mathbf{R}^n_+)$ then inequalities*

$$(\phi_+)^{\natural}(x-\theta y) \le c(\phi_+)^{\natural}(x-y), \tag{8.18}$$

$$(\phi_-^{-1})^{\natural}(x-\theta y) \le c(\phi_-^{-1})^{\natural}(x-y) \tag{8.19}$$

hold for all $(x,y) \in \operatorname{supp}(u^{\#}(x-y)\chi_{\epsilon}(y))$, $0 < \epsilon \le 1$ and $0 \le \theta \le 1$.

Proof. Only the case $U \cap \gamma^- \ne \emptyset$ will be proved. In this case (8.18) is easily checked. So we shall prove (8.19). Recalling that

$$m_-(z) = \{z_1^2 + (\mu z_1 + z_2)^2\}^{1/2}, \quad \phi_-(z) = m_-(z) + \mu z_1 + z_2$$

for $z \in \mathbf{R}^n_+$ we have

$$\partial_{z_1}\phi_-(z) = m_-(z)^{-1}\{\mu\phi_-(z) + z_1\} \ge 0,$$
$$\partial_{z_2}\phi_-(z) = m_-(z)^{-1}\phi_-(z) \ge 0$$

and hence it follows from $e^{x_1-y_1} \le e^{\epsilon_0}e^{x_1-\theta y_1}$ and $x_2 - y_2 \le x_2 - \theta y_2$ that

$$\phi_-(e^{x_1-y_1}, x' - y') \le \phi_-(e^{\epsilon_0}e^{x_1-\theta y_1}, x' - \theta y').$$

Using the notation introduced in Section 4 and noting Lemma 4.1 we get

$$\begin{aligned}\phi_-(e^{\epsilon_0}e^{x_1-\theta y_1}, x' - \theta y') &= \phi_-(e^{x_1-\theta y_1}, x' - \theta y'; e^{2\epsilon_0}, \mu e^{\epsilon_0})\\ &\le c\phi_-(e^{x_1-\theta y_1}, x' - \theta y'; 1, \mu) = c\phi_-(e^{x_1-\theta y_1}, x' - \theta y')\end{aligned}$$

which proves the assertion. □

Proof of Lemma 7.4. Noting Lemma 7.2 we may assume that $s \ge 2$ and $t \ge 0$. We note that

$$\|\phi_+^s\phi_-^t[x_i(L + \lambda H), J_\epsilon]u\| = \|(\phi_+^s\phi_-^t)^\natural([x_i(L + \lambda H), J_\epsilon]u)^\#\|.$$

The same reasoning as the proof of Proposition 8.2 shows that $([x_i(L+\lambda H), J_\epsilon]u)^\#$ can be written as a sum of (8.3), (8.5), (8.6), (8.7), (8.8), (8.9), (8.11), (8.12) and (8.15). We shall estimate (8.7) because other terms can be estimated similarly. If we write $\chi^\beta(y) = y^\beta\chi(y)$ then $y^\beta\chi_\epsilon(y) = \epsilon^{|\beta|}\chi_\epsilon^\beta(y)$. When $U\cap\gamma^- \ne \emptyset$, noticing $\mathrm{supp}\, u \subset \{\phi_-(x) > \eta\}$, we have

$$\begin{aligned}&\|(\phi_+^s\phi_-^t)^\natural(\cdot)\int a(\cdot, y)(Z_j(m_\pm^{-2}x_i))^\natural(\cdot - y)u^\#(\cdot - y)y^\beta\chi_\epsilon(y)dy\|\\ &\quad\le c\epsilon^2\|\int a(\cdot, y)(Z_j(m_\pm^{-2}x_i)u)^\#(\cdot - y)\chi_\epsilon^\beta(y)dy\|\\ &\quad\le c'\epsilon^2\|(Z_j(m_\pm^{-2}x_i)u)^\#\| = c'\epsilon^2\|Z_j(m_\pm^{-2}x_i)u\| \le c''\epsilon^2\|u\|.\end{aligned}$$

When $U \cap \gamma^+ \ne \emptyset$, using Lemma 8.3 we have

$$\begin{aligned}&\|(\phi_+^s\phi_-^t)^\natural(\cdot)\int a(\cdot, y)(Z_j(m_\pm^{-2}x_i))^\natural(\cdot - y)u^\#(\cdot - y)y^\beta\chi_\epsilon(y)dy\|\\ &\quad\le c\epsilon^2\|(\phi_+^2)^\natural(\cdot)\int a(\cdot, y)(\phi_+^{-2})^\natural(\cdot - y)\\ &\qquad\times(\phi_+^2 Z_j(m_\pm^{-2}x_i)u)^\#(\cdot - y)\chi_\epsilon^\beta(y)dy\|\\ &\quad\le c'\epsilon^2\|(\phi_+^2 Z_j(m_\pm^{-2}x_i)u)^\#\| = c'\epsilon^2\|\phi_+^2 Z_j(m_\pm^{-2}x_i)u\| \le c''\epsilon^2\|u\|.\end{aligned}$$

This concludes the proof. □

9. Proof of Proposition 3.3 (3)

We complete the proof of Proposition 3.3 in this section. The same reasoning as in Proposition 5.2 can be applied to get an an a priori estimate.

Lemma 9.1. *There are* $c_0, c_1, \sigma_0 > 0$ *such that for* $s, t > \sigma_0$ *we can take a* $\Lambda(s,t) \in \mathbf{R}$ *verifying the following properties: If* $\mathrm{Re}\lambda > \Lambda(s,t)$ *and if* $u \in C^{0,1}(\overline{\mathbf{R}^n_+})$ *with* $\mathrm{supp} u \subset \{|x| < 1,\ x_1 \geq 0\}$ *and* $u \in M$ *at* $\partial\mathbf{R}^n_+$ *then it follows that*

$$\begin{aligned}(\mathrm{Re}\lambda - \Lambda(s,t))\|m^{1/2}u\|^2 &+ c_0(\min(s,t) - \sigma_0)\|u\|^2 \\ &\leq c_1\|m\phi_+^s\phi_-^{-t}(L+\lambda H)\phi_+^{-s}\phi_-^t u\|^2.\end{aligned}$$

From this we obtain

Proposition 9.2. *There are* $c, \sigma_0 > 0$ *such that for* $s, t > \sigma_0$ *we can take a* $\Lambda(s,t) \in \mathbf{R}$ *verifying the following properties: If* $\mathrm{Re}\lambda > \Lambda(s,t)$ *and* $0 < \epsilon \leq 1$ *and if* $u \in D_{1-\epsilon_0,0}(\mathbf{R}^n_+)$ *is a weak solution to* (BVP) *then it follows that*

$$(\min(s,t) - \sigma_0)\|\phi_+^s\phi_-^{-t}J_\epsilon u\|^2 \leq c\|m\phi_+^s\phi_-^{-t}(L+\lambda H)J_\epsilon u\|^2.$$

Proof. We may assume that $u \in D_{1-\epsilon_0,\eta}(\mathbf{R}^n_+)$ with some $\eta > 0$. Then it follows from Proposition 7.1 and Lemma 7.2 that $\phi_+^{s-1}\phi_-^{-t}J_\epsilon u \in D_{1,\eta_0}(\mathbf{R}^n_+)$ and $\phi_+^{s-1}\phi_-^{-t}J_\epsilon u$ is a weak solution to (BVP). Using Proposition 7.7 and Lemma 9.1 we can prove the assertion. □

Using Proposition 9.2 we shall prove Proposition 3.3. Let $q \in \mathbf{Z}_+$, $q \geq 1$. We may assume that $s, t \geq q+1$. Suppose that $u \in X^{q-1}_{(-s+1,t+1)}(\mathbf{R}^n_+;\partial\mathbf{R}^n_+) \cap D_{1-\epsilon_0,\eta}(\mathbf{R}^n_+)$ is a weak solution to (BVP) with $f \in X^{q-1}_{(-s+1,t+1)}(\mathbf{R}^n_+;\partial\mathbf{R}^n_+)$. Moreover we assume that χ in (3.3) satisfies not only (7.1) but also (3.4) and (3.5).

The following lemma will be frequently used in the following.

Lemma 9.3. *Let* $a(x,y) \in \mathcal{B}^\infty(\mathbf{R}^n \times \mathbf{R}^n)$. *Then for* $\alpha \in \mathbf{Z}^n_+$ *there is a* $c = c(\chi, q, a, \alpha) > 0$ *verifying the following properties: If* $w \in H^{q-1}(\mathbf{R}^n_+;\partial\mathbf{R}^n_+)$ *and we set*

$$W_\epsilon(x) = \int_{\mathbf{R}^n} a(x,y)w^\#(x-y)y^\alpha\chi_\epsilon(y)dy$$

then it follows that

$$\int_0^1 \|W_\epsilon\|^2 \epsilon^{-2q}(1+\delta^2/\epsilon^2)^{-1} d\epsilon/\epsilon$$

$$\leq \begin{cases} c\|w\|^2_{\mathbf{R}^n_+, q-1, tan, \delta} & \text{if } \; |\alpha| = 0 \\ c\|w\|^2_{H^{q-|\alpha|}(\mathbf{R}^n_+; \partial \mathbf{R}^n_+)} & \text{if } \; 1 \leq |\alpha| \leq q \\ c\|w\|^2_{L^2(\mathbf{R}^n_+)} & \text{if } \; |\alpha| \geq q+1. \end{cases}$$

Proof. If we write $\chi^\alpha(y) = y^\alpha \chi(y)$ then $y^\alpha \chi_\epsilon(y) = \epsilon^{|\alpha|}\chi^\alpha_\epsilon(y)$. We first consider the case $|\alpha| \geq q+1$. Since $\|W_\epsilon\|^2 \leq c\epsilon^{2|\alpha|}\|w^\#\|^2 \leq c\epsilon^{2q+2}\|w\|^2$ it follows that

$$\int_0^1 \|W_\epsilon\|^2 \epsilon^{-2q}(1+\delta^2/\epsilon^2)^{-1} d\epsilon/\epsilon \leq c\|w\|^2 \int_0^1 \epsilon^2 (1+\delta^2/\epsilon^2)^{-1} d\epsilon/\epsilon$$
$$\leq c'\|w\|^2.$$

We turn to the case $|\alpha| \leq q$. Now we suppose that $a(x,y) = a(x) \in \mathcal{B}^\infty(\mathbf{R}^n)$. Then we can write

$$W_\epsilon(x) = a(x)\epsilon^{|\alpha|} \int w^\#(x-y)\chi^\alpha_\epsilon(y)dy = a(x)\epsilon^{|\alpha|}(w^\# * \chi^\alpha_\epsilon)(x).$$

This implies that $\|W_\epsilon\|^2 \leq c\epsilon^{2|\alpha|}\|w^\# * \chi^\alpha_\epsilon\|^2$ and hence it follows from Theorem 2.4.2 in [3] that

$$\int_0^1 \|W_\epsilon\|^2 \epsilon^{-2q}(1+\delta^2/\epsilon^2)^{-1} d\epsilon/\epsilon$$
$$\leq c \int_0^1 \|w^\# * \chi^\alpha_\epsilon\|^2 \epsilon^{-2(q-|\alpha|)}(1+\delta^2/\epsilon^2)^{-1} d\epsilon/\epsilon$$
$$\leq c'\|w^\#\|^2_{\mathbf{R}^n, q-|\alpha|-1, \delta}.$$

Here we note that

$$\|w^\#\|_{\mathbf{R}^n, q-|\alpha|-1, \delta} = \|w\|_{\mathbf{R}^n_+, q-1, tan, \delta} \qquad \text{if } \; |\alpha| = 0,$$
$$\|w^\#\|_{\mathbf{R}^n, q-|\alpha|-1, \delta} \leq c\|w\|_{H^{q-|\alpha|}(\mathbf{R}^n_+; \partial\mathbf{R}^n_+)} \qquad \text{if } \; 1 \leq |\alpha| \leq q.$$

Next let $a(x,y) \in \mathcal{B}^\infty(\mathbf{R}^n \times \mathbf{R}^n)$. If we set $k = q - |\alpha| - 1$ then using Taylor's formula we have

$$a(x,y) = \sum_{|\beta| \leq k-1} (\beta!)^{-1}(\partial^\beta_y a)(x,0)y^\beta + \sum_{|\beta|=k} R_\beta(x,y)y^\beta$$

where

$$R_\beta(x,y) = (\beta!)^{-1}k\int_0^1 (1-\theta)^{k-1}(\partial_y^\beta a)(x,\theta y)d\theta.$$

Therefore the same reasoning as above concludes the proof. □

Throughout this section we denote by c_0 constants which depend only on q and by C_1 constants which depend on q, s, t and λ.

Proof of Proposition 3.3. It follows from Proposition 9.2 that

$$(\min(s,t)-\sigma_0)\|\phi_+^s\phi_-^{-t}J_\epsilon u\|^2 \le c\|m\phi_+^s\phi_-^{-t}(L+\lambda H)J_\epsilon u\|^2.$$

Now using Taylor's formula we have

$$\begin{aligned}
&(\phi_+^s\phi_-^{-t}J_\epsilon u)^\#(x)\\
&= \int (\phi_+^s\phi_-^{-t})^\natural(x)u^\#(x-y)\chi_\epsilon(y)dy\\
&= \sum_{|\beta|\le q}(\beta!)^{-1}\int (Z^\beta(\phi_+^s\phi_-^{-t})u)^\#(x-y)y^\beta\chi_\epsilon(y)dy\\
&\quad + \sum_{|\beta|=q+1}(\beta!)^{-1}(q+1)\int \Phi_\beta(x,y)u^\#(x-y)y^\beta\chi_\epsilon(y)dy\\
&= \sum_{|\beta|\le q}U_\beta(x) + \sum_{|\beta|=q+1}U_\beta(x)
\end{aligned}$$

where

$$\Phi_\beta(x,y) = \int_0^1 (1-\theta)^q(Z^\beta(\phi_+^s\phi_-^{-t}))^\natural(x-y+\theta y)d\theta.$$

If $|\beta| = 0$ we can write

$$U_\beta(x) = \int (\phi_+^s\phi_-^{-t}u)^\#(x-y)\chi_\epsilon(y)dy = (J_\epsilon\phi_+^s\phi_-^{-t}u)^\#(x).$$

This implies that

(9.1)

$$\begin{aligned}
&(\min(s,t)-\sigma_0)\\
&\times\Big\{\int_0^1 \|J_\epsilon\phi_+^s\phi_-^{-t}u\|^2\epsilon^{-2q}(1+\delta^2/\epsilon^2)^{-1}d\epsilon/\epsilon + \|\phi_+^s\phi_-^{-t}u\|_{\mathbf{R}_+^n,q-1,tan}\Big\}\\
&\le c\int_0^1 \|m\phi_+^s\phi_-^{-t}(L+\lambda H)J_\epsilon u\|^2\epsilon^{-2q}(1+\delta^2/\epsilon^2)^{-1}d\epsilon/\epsilon\\
&\quad + C_1\Big\{\sum_{1\le|\beta|\le q+1}\int_0^1 \|U_\beta\|^2\epsilon^{-2q}(1+\delta^2/\epsilon^2)^{-1}d\epsilon/\epsilon\\
&\quad + \|\phi_+^s\phi_-^{-t}u\|^2_{\mathbf{R}_+^n,q-1,tan}\Big\}.
\end{aligned}$$

Since $\phi_+^s\phi_-^{-t}u \in H^{q-1}(\mathbf{R}_+^n; \partial\mathbf{R}_+^n)$ it follows from Proposition 3.4 that the left-hand side of (9.1) is bounded from below by

$$c_0^{-1}(\min(s,t)-\sigma_0)\|\phi_+^s\phi_-^{-t}u\|^2_{\mathbf{R}_+^n,q-1,tan,\delta}.$$

We turn to the right-hand side of (9.1). We first consider the terms which contain U_β. If $1 \le |\beta| \le q$ then using Lemma 9.3 and Lemma 6.1 we obtain that

$$\begin{aligned}\int_0^1 \|U_\beta\|^2\epsilon^{-2q}(1+\delta^2/\epsilon^2)^{-1}d\epsilon/\epsilon &\le c_0\|Z^\beta(\phi_+^s\phi_-^{-t})u\|^2_{H^{q-|\beta|}(\mathbf{R}_+^n;\partial\mathbf{R}_+^n)}\\ &\le C_1\|u\|_{X^{q-1}_{(-s+1,t+1)}(\mathbf{R}_+^n;\partial\mathbf{R}_+^n)}.\end{aligned}$$

If $|\beta| = q+1$ then it follows from Lemma 8.3 that

$$|\Phi_\beta(x,y)(\phi_+^{-s+q+1}\phi_-^{t+q+1})^\natural(x-y)|$$

is bounded from above by $C_1 > 0$ on $\mathrm{supp}(u^\#(x-y)\chi_\epsilon(y))$ and hence

$$\int_0^1 \|U_\beta\|^2\epsilon^{-2q}(1+\delta^2/\epsilon^2)^{-1}d\epsilon/\epsilon \le C_1\|\phi_+^{s-q-1}\phi_-^{-t-q-1}u\|^2 \le C_1'\|u\|.$$

Furthermore noticing $\|\phi_+^s\phi_-^{-t}u\|^2_{\mathbf{R}_+^n,q-1,tan} \le C_1\|u\|^2_{X^{q-1}_{(-s+1,t+1)}(\mathbf{R}_+^n;\partial\mathbf{R}_+^n)}$ we have

$$\begin{aligned}&(\min(s,t)-\sigma_0)\|\phi_+^s\phi_-^{-t}u\|^2_{\mathbf{R}_+^n,q-1,tan,\delta}\\ &\quad\le c_0\int_0^1 \|m\phi_+^s\phi_-^{-t}(L+\lambda H)J_\epsilon u\|^2\epsilon^{-2q}(1+\delta^2/\epsilon^2)^{-1}d\epsilon/\epsilon\\ &\quad\quad + C_1\{\|u\|^2_{X^{q-1}_{(-s+1,t+1)}(\mathbf{R}_+^n;\partial\mathbf{R}_+^n)} + \|u\|^2\}.\end{aligned}$$

We next consider the first term on the right-hand side. Note that

$$\begin{aligned}\|m\phi_+^s\phi_-^{-t}(L+\lambda H)J_\epsilon u\|^2 &\le c\sum_{i=1}^2 \|x_i\phi_+^s\phi_-^{-t}(L+\lambda H)J_\epsilon u\|^2\\ &= c\sum_{i=1}^2 \|(\phi_+^s\phi_-^{-t})^\natural(x_i(L+\lambda H)J_\epsilon u)^\#\|^2.\end{aligned}$$

Using Proposition 8.2 we can repeat the same argument as above to conclude the proof. □

10. Estimates of tangential derivatives of solutions

In this section we derive the following an a priori estimate.

Proposition 10.1. *For $q \in \mathbf{Z}_+$, $q \geq 1$ there are $c_0 = c_0(q) > 0$ and $\sigma(q) > 0$ such that for $s, t > \sigma(q)$ we can take a $\Lambda(q, s, t) \in \mathbf{R}$ verifying the following properties: If* $\mathrm{Re}\lambda > \Lambda(q, s, t)$ *and if $u \in C^{q+1}(\overline{\Omega})$ with* $\mathrm{supp}\, u \cap (O^- \cup \gamma^-) = \emptyset$ *is a weak solution to* (BVP) *then it follows that*

$$(\min(s,t) - \sigma(q))\|u\|^2_{X^q_{(-s,t)}(\Omega;\partial\Omega)}$$
$$\leq c_0\{\|(L+\lambda H)u\|^2_{X^q_{(-s,t)}(\Omega;\partial\Omega)} + |\lambda|^2\|u\|^2_{X^{q-1}_{(-s,t)}(\Omega;\partial\Omega)}\}.$$

In particular

$$\|u\|^2_{X^q_{(-s,t)}(\Omega;\partial\Omega)} \leq C_1\|(L+\lambda H)u\|^2_{X^q_{(-s,t)}(\Omega;\partial\Omega)}$$

where $C_1 = C_1(q, s, t, \lambda) > 0$.

For the proof of Proposition 10.1, with the aid of a partition of unity, we may assume that $u \in C^{q+1}(\overline{\Omega})$ is supported in a small coordinate patch. The interesting patches are at γ. Now from Lemma 9.1 we have

Lemma 10.2. *There are $c > 0$ and $\sigma_0 > 0$ such that for $s, t > \sigma_0$ we can take a $\Lambda(s,t) \in \mathbf{R}$ verifying the following properties: If* $\mathrm{Re}\lambda >$ *$\Lambda(s,t)$ and if $u \in C^{0,1}(\overline{\mathbf{R}^n_+}) \cap D_{1-\epsilon_0,0}(\mathbf{R}^n_+)$ is a weak solution to* (BVP) *then it follows that*

$$(\min(s,t) - \sigma_0)\|\phi_+^s\phi_-^{-t}u\|^2 \leq c\|m\phi_+^s\phi_-^{-t}(L+\lambda H)u\|^2.$$

Note that $[(L+\lambda H), Z_k]$, $(k = 1, \dots, n)$ is written as a sum of the following terms:

$$a(x)Z^\beta,\ |\beta| \leq 1, \quad a(x)\partial_1, \quad \lambda a(x) \tag{10.1}$$

where $a(x) \in \mathcal{B}^\infty(\mathbf{R}^n)$, which may differ from line to line.

Lemma 10.3. *$x_1[(L+\lambda H), Z^\alpha]$, $|\alpha| \leq q$ is written as a sum of the following terms:*

$$a(x)Z^\beta, \quad |\beta| \leq q, \qquad \lambda a(x)Z^{\beta'}, \quad |\beta'| \leq q-1.$$

Proof. We proceed by induction on q. By (10.1) the case $q = 1$ is trivial. Inductively assume that the statement is true up to q. We note that

$$x_1[(L+\lambda H), Z_kZ^\alpha] = x_1[(L+\lambda H), Z_k]Z^\alpha + x_1Z_k[(L+\lambda H), Z^\alpha] = I_1 + I_2.$$

By the inductive hypothesis I_1 can be written as a desired sum. We turn to I_2. Since $x_1 Z_k = Z_k x_1 + [x_1, Z_k]$, we can write I_2 as a sum of the following terms:

$$Z_k x_1[(L + \lambda H), Z^\alpha], \quad x_1[(L + \lambda H), Z^\alpha].$$

Hence by the inductive hypothesis I_2 can be written as a desired sum. This proves the assertion for $q + 1$. □

Lemma 10.4. *$x_2^q[(L + \lambda H), Z^\alpha]$, $|\alpha| \le q$ is written as a sum of the following terms:*

$$x_2^j a(x) Z^\beta, \quad |\beta| \le j + 1 \le q, \qquad \lambda x_2^{j'} a(x) Z^{\beta'}, \quad |\beta'| \le j' \le q - 1,$$
$$x_2^{j'} a(x) Z^{\beta'} (L + \lambda H), \quad |\beta'| \le j' \le q - 1.$$

Proof. We proceed by induction on q. By (10.1) and (8.10) the case $q = 1$ is trivial. Inductively assume that the statement is true up to q. We note that

$$x_2^{q+1}[(L + \lambda H), Z_k Z^\alpha] = x_2^{q+1}[(L + \lambda H), Z_k] Z^\alpha$$
$$+ x_2^{q+1} Z_k[(L + \lambda H), Z^\alpha] = I_1 + I_2.$$

By the inductive hypothesis I_1 can be written as a sum of the following terms:

$$x_2^q a(x) Z^\beta Z^\alpha, \ |\beta| \le 1, \quad \lambda x_2^q a(x) Z^\alpha, \quad x_2^q a(x)(L + \lambda H) Z^\alpha.$$

Noticing that

$$x_2^q (L + \lambda H) Z^\alpha = x_2^q Z^\alpha (L + \lambda H) + x_2^q[(L + \lambda H), Z^\alpha]$$

we can write I_1 as a desired sum. We turn to I_2. Since $x_2^{q+1} Z_k = Z_k x_2^{q+1} + [x_2^{q+1}, Z_k]$, we can write I_2 as a sum of the following terms:

$$Z_k x_2^{q+1}[(L + \lambda H), Z^\alpha], \quad x_2^q[(L + \lambda H), Z^\alpha].$$

Hence by the inductive hypothesis I_2 can be written as a desired sum. This proves the assertion for $q + 1$. □

Recalling that $m_\pm(x)^2$ is an homogeneous polynomial in x_1, x_2 of order two we obtain the following lemma.

Lemma 10.5. *$[(L+\lambda H), Z^\alpha]$, $|\alpha| \le q$ is written as a sum of the following terms:*

$$m_\pm^{-2q} x_1^i x_2^j a(x) Z^\beta, \quad |\beta| \le i+j-q-1 \le q,$$
$$\lambda m_\pm^{-2q} x_1^{i'} x_2^{j'} a(x) Z^{\beta'}, \quad |\beta'| \le i'+j'-q \le q-1,$$
$$m_\pm^{-2q} x_1^{i'} x_2^{j'} a(x) Z^{\beta'}(L+\lambda H), \quad |\beta'| \le i'+j'-q \le q-1.$$

From Lemma 6.1 it suffices to show the following proposition for the proof of Proposition 10.1.

Proposition 10.6. *For $q \in \mathbf{Z}_+$, $q \ge 1$ there are $c_0 = c_0(q) > 0$ and $\sigma(q) > 0$ such that for $s, t > \sigma(q)$ we can take a $\Lambda(q,s,t) \in \mathbf{R}$ verifying the following properties: If $\mathrm{Re}\lambda > \Lambda(q,s,t)$ and if $u \in C^{q+1}(\overline{\mathbf{R}^n_+}) \cap D_{1-\epsilon_0,0}(\mathbf{R}^n_+)$ is a weak solution to (BVP) then it follows that*

$$\begin{aligned}(\min(s,t) - \sigma(q)) &\sum_{|\alpha|\le q} \|\phi_+^{s-q+|\alpha|}\phi_-^{-t-q+|\alpha|} Z^\alpha u\|^2 \\ \le c_0\{&\sum_{|\alpha|\le q} \|\phi_+^{s-q+|\alpha|}\phi_-^{-t-q+|\alpha|} Z^\alpha (L+\lambda H) u\|^2 \\ &+ \sum_{|\alpha|\le q} \|\phi_+^{s-q+|\alpha|}\phi_-^{-t-q+|\alpha|} Z^\alpha u\|^2 \\ &+ |\lambda|^2 \sum_{|\alpha|\le q-1} \|\phi_+^{s-(q-1)+|\alpha|}\phi_-^{-t-(q-1)+|\alpha|} Z^\alpha u\|^2\}.\end{aligned}$$

Proof. Suppose that $|\alpha| = q' \le q$. Since $Z^\alpha u \in C^1(\overline{\mathbf{R}^n_+}) \cap D_{1-\epsilon_0,0}(\mathbf{R}^n_+)$ is a weak solution to (BVP) then applying Lemma 10.2 we get

$$\begin{aligned}(\min(s,t) - \sigma(q)) &\|\phi_+^{s-q+q'}\phi_-^{-t-q+q'} Z^\alpha u\|^2 \\ &\le c_0 \|m\phi_+^{s-q+q'}\phi_-^{-t-q+q'} (L+\lambda H) Z^\alpha u\|^2 \\ &\le c_0'\{\|m\phi_+^{s-q+q'}\phi_-^{-t-q+q'} Z^\alpha (L+\lambda H) u\|^2 \\ &\qquad + \|m\phi_+^{s-q+q'}\phi_-^{-t-q+q'} [(L+\lambda H), Z^\alpha] u\|^2\}.\end{aligned}$$

We consider the second term on the right-hand side. If $|\alpha| = 0$ then the second term is zero. If $|\alpha| \ge 1$ then it follows from Lemma 10.5 that the second term is bounded from above by a sum of the following terms:

$$\|m\phi_+^{s-q+q'}\phi_-^{-t-q+q'}m_\pm^{-2q'}x_1^i x_2^j Z^\beta u\|^2 = I_1,$$
$$|\beta| \le i+j-q'-1 \le q',$$
$$|\lambda|^2\|m\phi_+^{s-q+q'}\phi_-^{-t-q+q'}m_\pm^{-2q'}x_1^{i'} x_2^{j'} Z^{\beta'} u\|^2 = I_2,$$
$$|\beta'| \le i'+j'-q' \le q'-1,$$
$$\|m\phi_+^{s-q+q'}\phi_-^{-t-q+q'}m_\pm^{-2q'}x_1^{i'} x_2^{j'} Z^{\beta'}(L+\lambda H)u\|^2 = I_3,$$
$$|\beta'| \le i'+j'-q' \le q'-1.$$

Here we note that

$$\begin{aligned} I_1 &\le c_0\|\phi_+^{s-q+|\beta|}\phi_-^{-t-q+|\beta|}Z^\beta u\|^2, \\ I_2 &\le c_0|\lambda|^2\|\phi_+^{s-(q-1)+|\beta'|}\phi_-^{-t-(q-1)+|\beta'|}Z^{\beta'}u\|^2, \\ I_3 &\le c_0\|\phi_+^{s-q+|\beta'|}\phi_-^{-t-q+|\beta'|}Z^{\beta'}(L+\lambda H)u\|^2. \end{aligned}$$

This concludes the proof. □

11. Proof of main results

We start with

Lemma 11.1. *If $q \in \mathbf{Z}_+$, $q \ge 1$ and if $u \in H^q(\Omega; \partial\Omega)$ with* $\operatorname{supp} u \cap \gamma = \emptyset$ *is a weak solution to* (BVP) *with $f \in H^{q-1}(\Omega)$ then it follows that $u \in H^q(\Omega)$.*

Proof. Take a covering $\{U_i\}_{i=0}^{k+1}$ of Ω as follows: First we cover $\partial\Omega \cap \operatorname{supp} u$ by coordinate patches U_i, $i = 1, \dots, k$ with $U_i \cap \gamma = \emptyset$ and with coordinate systems $\chi_i : U_i \cap \Omega \to \{|x| < 1, x_1 > 0\}$. Next we cover $\partial\Omega \setminus \bigcup_{i=1}^k U_i$ by U_{k+1} with $\operatorname{supp} u \cap U_{k+1} = \emptyset$. Finally we cover $\Omega \setminus \bigcup_{i=1}^{k+1} U_i$ by $U_0 \subset\subset \Omega$. Choose a partition of unity $\{\psi_i\}_{i=0}^{k+1}$ subordinate to this covering $\{U_i\}_{i=0}^{k+1}$ and set

$$u_i = \psi_i u, \quad f_i = (L+\lambda H)u_i = \psi_i f + \sum_{j=1}^n (\partial_j \psi_i) A_j u.$$

Note that $u_0 \in H^q(\mathbf{R}^n)$ and $u_{k+1} = 0$. For the proof it suffices to show that

$$Z^\alpha \partial_1^p u_i \in L^2(\mathbf{R}_+^n), \quad |\alpha| + p \le q$$

for $i = 1, \dots, k$ in local coordinates in U_i. We proceed by induction on p. The case $p = 0$ is trivial. Inductively assume that the statement is true up to $p-1$. It follows from (8.10) that

$$A_b \partial_1 u_i = -f_i + \tilde{A}_1 Z_1 u_i + \sum_{j=2}^n A_j Z_j u_i + B u_i + \lambda H u_i. \tag{11.1}$$

Since $U_i \cap \gamma = \emptyset$, so A_b is non-singular on U_i and then applying $Z^\alpha \partial_1^{p-1} A_b^{-1}$ to both sides of (11.1) and using the inductive hypothesis we can prove the assertion for p. □

Corollary 11.2. *Let $q \in \mathbf{Z}_+$ and $\chi \in C^\infty(\mathbf{R}^n)$ be such that $\chi = 0$ on a neighborhood of $O^+ \cup \gamma^+$. If $u \in X^q_{(-s,t)}(\Omega; \partial\Omega)$ with* $\operatorname{supp} u \cap (O^- \cup \gamma^-) = \emptyset$ *is a weak solution to* (BVP) *with $f \in H^q(\Omega)$ then it follows that $\chi u \in H^q(\Omega)$.*

Proof. We proceed by induction on q. The case $q = 0$ is trivial. Inductively assume that the statement is true up to $q-1$. Then we note that $\chi u \in H^q(\Omega; \partial\Omega)$, $\operatorname{supp} \chi u \cap \gamma = \emptyset$ and that χu is also a weak solution to (BVP). Since the inductive hypothesis implies that

$$(L + \lambda H)\chi u = \chi f + \sum_{j=1}^{n} (\partial_j \chi) A_j u \in H^{q-1}(\Omega)$$

it follows from Lemma 11.1 that $\chi u \in H^q(\Omega)$, which proves the assertion for q. □

From Proposition 10.1 and Corollary 11.2 we have

Proposition 11.3. *For $q \in \mathbf{Z}_+$ there is a $\sigma(q) > 0$ such that for $s, t > \sigma(q)$ we can take a $\Lambda(q, s, t) \in \mathbf{R}$ verifying the following properties: If* $\operatorname{Re}\lambda > \Lambda(q, s, t)$ *and if $u \in X^{q+[n/2]+2}_{(-s,t)}(\Omega; \partial\Omega)$ with* $\operatorname{supp} u \cap (O^- \cup \gamma^-) = \emptyset$ *is a weak solution to* (BVP) *with $f \in C_0^\infty(\Omega)$ then it follows that*

$$\|u\|^2_{X^q_{(-s,t)}(\Omega;\partial\Omega)} \le C_1 \|(L + \lambda H)u\|^2_{X^q_{(-s,t)}(\Omega;\partial\Omega)}$$

where $C_1 = C_1(q, s, t, \lambda) > 0$.

Proof. Take $\chi \in C_0^\infty(\mathbf{R})$ such that $\chi = 1$ near 0 and set

$$u_k = (1 - \chi(k\phi_+))u, \quad f_k = (L + \lambda H)u_k = (1 - \chi(k\phi_+))f - k\chi'(k\phi_+)G_+ u$$

for $k > 0$. Then it follows from Corollary 11.2 that

$$u_k \in H^{q+[n/2]+2}(\Omega) \hookrightarrow C^{q+1}(\Omega).$$

Since u_k is a weak solution to (BVP), applying Proposition 10.1, we have

$$\begin{aligned}
&\|(1 - \chi(k\phi_+))u\|^2_{X^q_{(-s,t)}(\Omega;\partial\Omega)} \\
&\quad \le C_1 \{ \|(1 - \chi(k\phi_+))(L + \lambda H)u\|^2_{X^q_{(-s,t)}(\Omega;\partial\Omega)} \\
&\qquad + \|k\chi'(k\phi_+)G_+ u\|^2_{X^q_{(-s,t)}(\Omega;\partial\Omega)} \}.
\end{aligned}$$

Arguments similar to those in the proof of Lemma 6.5 show that $(1-\chi(k\phi_+))u \to u$ and

$$(1-\chi(k\phi_+))(L+\lambda H)u \to (L+\lambda H)u$$

in $X^q_{(-s,t)}(\Omega;\partial\Omega)$. Noting $u \in X^{q+[n/2]+2}_{(-s,t)}(\Omega;\partial\Omega) \hookrightarrow X^q_{(-s+1,t)}(\Omega;\partial\Omega)$ it is easy to see that

$$\|k\chi'(k\phi_+)G_+u\|_{X^q_{(-s,t)}(\Omega;\partial\Omega)} = \|k\phi_+\chi'(k\phi_+)G_+u\|_{X^q_{(-s+1,t)}(\Omega;\partial\Omega)} \to 0.$$

This concludes the proof. □

Proof of Theorem 2.1. We first suppose that $f \in C_0^\infty(\Omega)$. From Proposition 3.2 there exists a $u \in L^2(\Omega)$ with $\operatorname{supp} u \cap (O^- \cup \gamma^-) = \emptyset$ which is a weak solution to (BVP). Using Proposition 3.1 we have $u \in X^{q+[n/2]+2}_{(-s,t)}(\Omega;\partial\Omega)$ and hence it follows from Proposition 11.3 that $\|u\|^2_{X^q_{(-s,t)}(\Omega;\partial\Omega)} \le C_1\|(L+\lambda H)u\|^2_{X^q_{(-s,t)}(\Omega;\partial\Omega)}$. Moreover from Corollary 5.7 we get

$$\|\phi_-^{-1}u\|^2_{L^2(\Omega)} \le C_1\|\phi_-^{-1}(L+\lambda H)u\|^2_{L^2(\Omega)}.$$

Let $f \in X^q_{(-s,t)}(\Omega;\partial\Omega) \cap \phi_- L^2(\Omega)$. Since $C_0^\infty(\Omega)$ is dense in

$$X^q_{(-s,t)}(\Omega;\partial\Omega) \cap \phi_- L^2(\Omega)$$

the assertion can be proved by standard limiting arguments. □

References

[1] K.O.Friedrichs, The identity of weak and strong extensions of differential operators, *Trans. Amer. Math. Soc.* **55** (1944), 132–151.

[2] K.O.Friedrichs, Symmetric positive linear differential opreators, *Comm. Pure Appl. Math.* **11** (1958), 333–418.

[3] L.Hörmander, *Linear Partial Differential Operators*, Springer-Verlag, Berlin-Göttingen-Heidelberg,1963.

[4] P.D.Lax and R.S.Phillips, Local boundary conditions for dissipative symmetric linear differential operators, *Comm. Pure Appl.Math.* **13** (1960), 427–455.

[5] R.Moyer, On the nonidentity of weak and strong extensions of differential operators, *Proc. Amer. Math. Soc.* **19** (1968), 487–488.

[6] T.Nishitani and M.Takayama, A characteristic initial boundary value problem for a symmetric positive system, *Hokkaido Math. J.* **25** (1996),167–182.

[7] S.Osher, An ill-posed problem for a hyperbolic equation near a corner, *Bull. Amer. Math. Soc.* **79** (1973), 1043–1044.

[8] R.S.Phillips and L.Sarason, Singular symmetric positive first order differential operators, *J. Math. Mech.* **15** (1966), 235–272.

[9] J.Rauch, Symmetric positive systems with boundary characteristic of constant multiplicity, *Trans. Amer. Math. Soc.* **291** (1985), 167–187.

[10] J.Rauch, Boundary value problem with nonuniformly characteristic boundary, *J. Math. Pure et Appl.* **73** (1994), 347–353.

[11] P.Secchi, A symmetric positive system with non uniformly characteristic boundary, 1996, preprint.

[12] D.Tartakoff, Regularity of solutions to boundary value problems for first order systems, *Indiana Univ. Math. J.* **21** (1972),1113–1129.

Tatsuo Nishitani and Masahiro Takayama
Department of Mathematics
Graduate School of Science
Osaka University
1-16 Machikaneyama Toyonaka Osaka 560, Japan

Progress in Nonlinear Differential Equations and Their Applications

Editor
Haim Brezis
Département de Mathématiques
Université P. et M. Curie
4, Place Jussieu
75252 Paris Cedex 05
France
and
Department of Mathematics
Rutgers University
New Brunswick, NJ 08903
U.S.A.

Progress in Nonlinear Differential Equations and Their Applications is a book series that lies at the interface of pure and applied mathematics. Many differential equations are motivated by problems arising in such diversified fields as Mechanics, Physics, Differential Geometry, Engineering, Control Theory, Biology, and Economics. This series is open to both the theoretical and applied aspects, hopefully stimulating a fruitful interaction between the two sides. It will publish monographs, polished notes arising from lectures and seminars, graduate level texts, and proceedings of focused and refereed conferences.

We encourage preparation of manuscripts in some form of TeX for delivery in camera-ready copy, which leads to rapid publication, or in electronic form for interfacing with laser printers or typesetters.

Proposals should be sent directly to the editor or to: Birkhäuser Boston, 675 Massachusetts Avenue, Cambridge, MA 02139

PNLDE 1 Partial Differential Equations and the Calculus of Variations, Volume I
Essays in Honor of Ennio De Giorgi
F. Colombini, A. Marino, L. Modica, and S. Spagnolo, editors

PNLDE 2 Partial Differential Equations and the Calculus of Variations, Volume II
Essays in Honor of Ennio De Giorgi
F. Colombini, A. Marino, L. Modica, and S. Spagnolo, editors

PNLDE 3 Propagation and Interaction of Singularities in Nonlinear Hyperbolic Problems
Michael Beals

PNLDE 4 Variational Methods
Henri Berestycki, Jean-Michel Coron, and Ivar Ekeland, editors

PNLDE 5 Composite Media and Homogenization Theory
Gianni Dal Maso and Gian Fausto Dell'Antonio, editors

PNLDE 6 Infinite Dimensional Morse Theory and Multiple Solution Problems
Kung-ching Chang

PNLDE 7 Nonlinear Differential Equations and their Equilibrium States, 3
N.G. Lloyd, W.M. Ni, L.A. Peletier, J. Serrin, editors

PNLDE 8 Introduction to Γ–Convergence
Gianni Dal Maso

PNLDE 9 Differential Inclusions in Nonsmooth Mechanical Problems: Shocks and Dry Frictions
Manuel D. P. Monteiro Marques

PNLDE 10 Periodic Solutions of Singular Lagrangian Systems
Antonio Ambrosetti and Vittorio Coti Zelati

PNLDE 11 Nonlinear Waves and Weak Turbulence with Applications in Oceanography and Condensed Matter Physics
N. Fitzmaurice, D. Gurarie, F. McCaugham, and W. A. Woyczyński, editors

PNLDE 12 Semester on Dynamical Systems: Euler International Math Institute, St. Petersburg, 1991
Kuksin, Lazutkin, and Poeschel, editors

PNLDE 13 Ginzburg-Landau Vortices
F. Bethuel, H. Brezis, and F. Hélein

PNLDE 14 Variational Methods for Image Segmentation
Jean-Michel Morel and Sergio Solomini

PNLDE 15 Topological Nonlinear Analysis: Degree, Singularity, and Variations
Michele Matzeu and Alfonso Vignoli, editors

PNLDE 16 Analytic Semigroups and Optimal Regularity in Parabolic Problems
A. Lunardi

PNLDE 17 Blowup for Nonlinear Hyperbolic Equations
Serge Alinhac

PNLDE 18 The Heat Kernel Lefschetz Fixed Point Formula for the Spin-c Dirac Operator
J. Duistermaat

PNLDE 19 Chaos and Nonlinear Dynamical Systems
I. Hoveijn

PNLDE 20 Topics in Geometry: Honoring the Memory of Joseph D'Atri
Simon Gindikin, editor

PNLDE 21 Partial Differential Equations and Mathematical Physics: The Danish-Swedish Analysis Seminar, 1995
Lars Hörmander and Anders Melin, editors

PNLDE 22 Partial Differential Equations and Functional Analysis: In Memory of Pierre Grisvard
J. Cea, D. Chenais, G. Geymonat, and J.-L. Lions, editors

PNLDE 23 Geometry of Harmonic Maps
Yuanlong Xin

PNLDE 24 Minimax Theorems
Michel Willem

PNLDE 25 Variational Methods for Discontinuous Structures
Raul Serapioni and Franco Tomarelli

PNLDE 26 Algebraic Aspects of Integrable Systems: In Memory of Irene Dorfman
A. S. Fokas and I. M. Gelfand, editors

PNLDE 27 Topological Nonlinear Analysis II
Degree, Singularity, and Variations
Michele Matzeu and Alfonso Vignoli, editors

PNLDE 28 Models of Phase Transitions
Augusto Visintin

PNLDE 29 Nonlinear Partial Differential Equations in Geometry and Physics: The 1995 Barrett Lectures
Garth Baker and Alexandre Freire, editors

PNLDE 30 Parametrized Measures and Variatiional Principles
Pable Pedregal

PNLDE 31 Topics in Mathematical Modelling of Composite Materials
Andrej Cherkaev and Robert Kohn, editors

PNLDE 32 Geometrical Optics and Related Topics
Ferruccio Colombini and Nicolas Lerner, editors

The manufacturer's authorised representative in the EU is Springer Nature Customer Service Centre GmbH, Europaplatz 3, 69115 Heidelberg, Germany. If you have any concerns regarding our products, please contact ProductSafety@springernature.com

Printed and bound by CPI Group (UK) Ltd, Croydon, CR0 4YY
23/04/2026
02095607-0003